Proteomics Driven Bioremediation

Strategies for Sustainable Environment

Proteomics Driven Bioremediation

Strategies for Sustainable Environment

Edited by

Burhan Hamid

Centre of Research for Development, University of Kashmir, Srinagar, Jammu and Kashmir, India

Mohammed Kuddus

Department of Biochemistry, College of Medicine, University of Hail, Hail, Saudi Arabia

Ghazala Yunus

University of Hail, Hail, Saudi Arabia

ELSEVIER

ACADEMIC PRESS

An imprint of Elsevier

Academic Press is an imprint of Elsevier
125 London Wall, London EC2Y 5AS, United Kingdom
50 Hampshire Street, 5th Floor, Cambridge, MA 02139, United States

ISBN: 978-0-443-45294-9

For Information on all Academic Press publications
visit our website at https://www.elsevier.com/books-and-journals

Publisher: Peter Llewellyn
Acquisitions Editor: Pamela Sadhukhan
Editorial Project Manager: Avni Gupta
Production Project Manager: Sathyavani Deiva Sigamani
Cover Designer: Arumugam Kothandan

Typeset by MPS Limited, Chennai, India

Contents

List of contributors

Pritam Acharya Department of General Science, Calcutta Institute of Technology, Howrah, West Bengal, India

Saurabh Acharyya Department of General Science, Calcutta Institute of Technology, Howrah, West Bengal, India

Bilal Ahmed Agricultural and Biological Engineering, Purdue University, West Lafayette, IN, United States

Zienab Ahmed Department of Integrative Agriculture, College of Agriculture and Veterinary Medicine, United Arab Emirates University, Al Ain, United Arab Emirates

Mona Saleh Al Tami Department of Biology, College of Science, Qassim University, Buraydah, Saudi Arabia

Mohammed Aladhadh Department of Food Science and Human Nutrition, College of Agriculture and Food, Qassim University, Buraydah, Saudi Arabia

Insha Amin Microbiology Research Laboratory, Centre of Research for Development (CORD), University of Kashmir, Srinagar, Jammu and Kashmir, India

Rimsha Amin Centre of Research for Development, University of Kashmir, Srinagar, Jammu and Kashmir, India

Suhail Ashraf Department of Plant Biotechnology, Tamil Nadu Agricultural University, Coimbatore, Tamil Nadu, India

Zaffar Bashir Centre of Research for Development, University of Kashmir, Srinagar, Jammu and Kashmir, India

Prakash Bharti Department of Biological Sciences, Mahatma Gandhi Chitrakoot Gramodaya Vishwavidyalaya, Satna, Madhya Pradesh, India

Dipankar Chattopadhyay Department of Polymer Science and Technology, University of Calcutta, Kolkata, West Bengal, India

Rubiya Dar Centre of Research for Development, University of Kashmir, Srinagar, Jammu and Kashmir, India

Asha Embrandiri Department of Environmental Health, College of Medicine and Health Sciences, Wollo University, Desse, Amhara, Ethiopia

Tapas Kumar Ghosh Department of Polymer Science and Technology and Center for Research in Nanoscience and Nanotechnology, University of Calcutta, Kolkata, West Bengal, India; Applied Science and Humanities, Global Institute of Management and Technology, Krishnanagr, West Bengal, India

Khalid E. Hamed Department of Plant Protection, College of Agriculture and Food, Qassim University, Buraydah, Saudi Arabia

Burhan Hamid Centre of Research for Development, University of Kashmir, Srinagar, Jammu and Kashmir, India

Hermes Pérez Hernández Instituto Nacional de Investigaciones Forestales, Agrícolas y Pecuarias (INIFAP). Campo Experimental Edzná, Campeche, México

Akbar Hossain Soil Science Division, Bangladesh Wheat and Maize Research Institute, Dinajpur, Bangladesh

Ulfat Jahan Centre of Research for Development, University of Kashmir, Srinagar, Jammu and Kashmir, India

Md. Parvez Kabir Soil Science Division, Bangladesh Wheat and Maize Research Institute, Dinajpur, Bangladesh

Amritpal Kaur Department of Food Science and Technology, Guru Nanak Dev University, Amritsar, Punjab, India

P. Kavitha Nursing (Microbiology & Biochemistry), KMCH Institute of Health Sciences, Affliated to The TN Dr. M. G. R University- Chennai, Coimbatore, Tamil Nadu, India

Misba Khan Department of Bioengineering, Integral University, Lucknow, Uttar Pradesh, India

Mohd Sajjad Ahmad Khan Department of Basic Sciences, Deanship of Preparatory Year and Supporting Studies, Imam Abdulrahman Bin Faisal University, Dammam, Saudi Arabia

Anjani Kumar Department of Physical Sciences, Mahatma Gandhi Chitrakoot Gramodaya Vishwavidyalaya, Satna, Madhya Pradesh, India

Ashek Mohammad Entomology Division, Horticulture Research Centre, Bangladesh Agricultural Research Institute, Gazipur, Bangladesh

S. Nakkeeran Department of Plant Biotechnology, Tamil Nadu Agricultural University, Coimbatore, Tamil Nadu, India

Ruqeya Nazir Microbiology Research Laboratory, Centre of Research for Development (CORD), University of Kashmir, Srinagar, Jammu and Kashmir, India

Somna Oberoi Department of Food Science and Technology, Guru Nanak Dev University, Amritsar, Punjab, India

Gopal Palla Agricultural and Biological Engineering, College of Agriculture, Purdue University, West Lafayette, IN, United States

Suneeta Gireesh Panicker Department of Microbiology, Dr. D. Y. Patil Arts, Commerce and Science College, Pune, Maharashtra, India

M.E. Pavithra Nursing (Microbiology & Biochemistry), KMCH Institute of Health Sciences, Affliated to The TN Dr. M. G. R University- Chennai, Coimbatore, Tamil Nadu, India

Jyoti Prakash Amity Institute of Biotechnology, Amity University, Lucknow, Uttar Pradesh, India

Mariyam Rafi Department of Bioengineering, Integral University, Lucknow, Uttar Pradesh, India

Shalini Rai Department of Microbiology, DSMNRU University, Lucknow, Uttar Pradesh, India

Md. Selim Reza Soil Science Division, Bangladesh Institute of Nuclear Agriculture, Mymensingh, Bangladesh

Vinod Rishi Department of Botany, Pt. Rajaram Upadhyay Mahavidhyalaya, Chitrakoot, Uttar Pradesh, India

Asfa Rizvi Department of Botany, School of Chemical and Life Sciences, Jamia Hamdard University, New Delhi, India

Roohi Department of Bioengineering, Integral University, Lucknow, Uttar Pradesh, India

Riyaz Sayyed Department of Microbiology, PSGVP Mandal's Shri S I Patil Arts, G B Patel Science & STSKV Sangh Commerce College, Shahada, India

Alma Hortensia Serafin Muñoz Departamento de Ingeniería Civil y Ambiental, División de Ingenierías, Universidad de Guanajuato, Guanajuato, México

Thahir Basha Shaik Department of Veterinary Public Health & Epidemiology, NTR College Veterinary Science, Sri Venkateswara Veterinary University, Gannavaram, Andhra Pradesh, India

Falguni Akter Shraboni Wheat Breeding Division, Bangladesh Wheat and Maize Research Institute, Dinajpur, Bangladesh

Aisha Singh Government Medical College (Affiliated to Atal Bihari Vajpayee Medical University, Lucknow), Mirzapur, Uttar Pradesh, India

Balwinder Singh Department of Food Science and Technology, Guru Nanak Dev University, Amritsar, Punjab, India; P.G. Department of Botany, Khalsa College Amritsar, Amritsar, Punjab, India

Jaspreet Singh P.G. Department of Biotechnology, Khalsa College Amritsar, Amritsar, Punjab, India

Shalini Singh Department of Biotechnology, Jagatpur PG College (Affiliated to Kashi Vidhyapith University), Varanasi, Uttar Pradesh, India

Rabeka Sultana Smriti Soil Science Division, Bangladesh Agricultural Research Institute, Gazipur, Bangladesh

Sohail Department of Agriculture, Forestry and Bioresources, Forest Ecophysiology Lab Seoul National University, Seoul, South Korea

Edgar Vázquez-Núñez Departamento de Ingenierías Química, Electrónica y Biomédica, División de Ciencias e Ingenierías, Aplicaciones Nano y Bio Tecnológicas para la Sostenibilidad (NanoBioTS), LaSAE, Universidad de Guanajuato, León, Guanajuato, México

Ileana Vera Reyes Departamento de Biociencias y Agrotecnología, Centro de Investigación en Química Aplicada, Saltillo, México

Parvaze Ahmad Wani Centre of Research for Development, University of Kashmir, Srinagar, Jammu and Kashmir, India

Mohd. Rehan Zaheer Department of Chemistry, R.M.P.S.P. Girls Post Graduate College, Basti, Uttar Pradesh, India

About the editors

Dr. Burhan Hamid, PhD, is an assistant professor at the Centre of Research for Development (CORD) and P.G. Programme in Microbiology, University of Kashmir, J&K, India, since 2019. He has 14 years of teaching and research experience. He has worked as a researcher at the Regional Research Station (RRS) for biofertilizers at Sher-e-Kashmir University of Agricultural Sciences and Technology of Kashmir (SKUAST-K), Shalimar, India. His areas of interest in teaching and research are environmental microbiology, psychrophiles, psychrozymes, bioremediation, agricultural microbiology, and food and industrial microbiology. Currently, he is also serving as a reviewer and member of the editorial board of different journals and has published more than 70 research papers, review papers, and book chapters in peer-reviewed high-impact journals. He has published his work with internationally reputed publishers such as Springer, Elsevier, Wiley, MDPI, and Taylor & Francis. Dr. Hamid is working as an expert for the evaluation of different research projects with the National Science Centre (NCN) in Poland. He has also worked as a senior microbiologist in an ISO/IEC-17025-accredited laboratory at Baddi, Himachal Pradesh, India. He is also supervising MSc microbiology students in their dissertations and project works.

Prof. Mohammed Kuddus is a distinguished academic and researcher in the fields of biochemistry, enzymology, and biotechnology, currently serving as a professor at the Department of Biochemistry, College of Medicine, University of Hail, Hail, Saudi Arabia. With over 18 years of experience in higher education and scientific research, Prof. Kuddus has made significant contributions to the advancement of biochemistry, enzyme technology, and microbial biotechnology. Throughout his academic career, Prof. Kuddus has published extensively in peer-reviewed journals, including *The Lancet* and *Nature* journals. His research interests include medical biochemistry, food technology, global health research, and the application of biotechnology in healthcare. He has published more than 130 research articles along with 18 books and 30 book chapters. He has been awarded Young Scientist Projects from the Department of Science and Technology, India, and the International Foundation for Science, Sweden. Also, he has been ranked in Elsevier/Stanford University's World's Top 2% of Scientists since 2023.

Dr. Ghazala Yunus has completed her PhD in biophysics and is presently working as an assistant professor at the University of Hail, Hail, Saudi Arabia. Dr. Yunus's main research area includes medical physics, molecular docking, drug–DNA interaction, biosensors, and nanotechnology. She has published more than 35 research articles in reputed international journals along with two books. She has been serving as an editorial board member and reviewer of various international peer-reviewed journals.

Foreword

Proteomics-driven bioremediation stands at the forefront of modern environmental biotechnology. Today, it provides unprecedented molecular insight into how organisms respond to pollution and how they can be harnessed for ecological restoration. In the future, proteomics will enable engineered biodegradative systems, real-time environmental diagnostics, and predictive bioremediation strategies that are cleaner, faster, and more sustainable. As global pollution challenges intensify, proteomics offer not only a deeper understanding of life at the molecular level but also a powerful, innovative pathway toward a healthier and more resilient planet.

Proteomics Driven Bioremediation: Strategies for Sustainable Environment arrives at a critical moment in human history, one in which environmental degradation, pollution-driven health threats, and accelerating climate change demand solutions that are not only effective but also deeply rooted in scientific innovation. As societies worldwide confront unprecedented ecological challenges, the biological sciences have increasingly moved toward integrative, systems-level approaches. Among these, proteomics has emerged as a transformative lens through which we can understand and improve the biological mechanisms that underpin environmental resilience.

This book stands as one of the timeliest and comprehensive compilations of current knowledge at the interface of proteomics and bioremediation, offering a panoramic view of how protein-level insights can accelerate the restoration of polluted ecosystems. The chapters assembled here, authored by an international network of researchers across India, Mexico, Bangladesh, Saudi Arabia, Germany, South Korea, and the United States, demonstrate the global urgency and collaborative spirit that define modern environmental biotechnology.

Across 16 chapters, the book explores a remarkable diversity of pollutants: pesticides, petroleum hydrocarbons, heavy metals, dyes, microplastics, pharmaceutical residues, and e-waste; and showcases how proteomic technologies help decipher microbial, fungal, and plant responses to contamination. This molecular insight is not merely academic; it reveals new metabolic pathways, stress-response networks, and degradative enzymes that can be harnessed, engineered, or optimized for real-world bioremediation. The work further bridges proteomics with genomics, metabolomics, and microbial ecology, highlighting the power of integrative omics to design robust strategies for ecological restoration.

Equally important is the book's attention to emerging frontiers: proteomic biomonitoring tools for aquatic environments, proteome-based early disease diagnostics linked to environmental exposures, and the role of immobilized enzymes in tackling industrial pollutants. Such contributions underscore the dual role of proteomics, not only in solving pollution problems but also in safeguarding human and ecosystem health.

What distinguishes this volume is its deliberate blend of foundational concepts, cutting-edge research, and visionary outlook. Readers, from students and early-career researchers to seasoned scientists and policy planners, will find in these chapters both the scientific depth and the practical perspective needed to translate proteomic discoveries into sustainable environmental solutions. As the world increasingly turns toward nature-based and biotechnology-driven remediation strategies, this book provides a crucial scientific foundation and serves as a roadmap for the future. It reminds us that sustainability is not an abstract aspiration but a molecular reality, written within the proteins that govern life's response to stress and adaptation. By decoding these molecular signatures, we unlock new potential to heal ecosystems, protect communities, and shape a cleaner, more resilient planet.

This book is, without doubt, a significant and forward-looking contribution to environmental biotechnology. It illuminates pathways toward sustainability that are at once innovative, evidence-driven, and profoundly hopeful. I commend the editors and authors for their scholarly rigor and global perspective and trust that this book will inspire continued advances in proteomics-guided environmental stewardship.

Prof. (Dr.) Cristobal N. Aguilar
Distinguished Scientist by International Bioprocessing Association
Universidad Autónoma de Coahuila
Mexico

Preface

Environmental pollution has emerged as one of the most alarming global challenges of the 21st century, threatening ecosystem stability, public health, and sustainable development. As conventional remediation techniques often fall short in efficiency, specificity, and environmental compatibility, the quest for innovative and science-driven solutions has been in demand. In this context, proteomics has opened new and transformative possibilities for bioremediation.

Proteomics Driven Bioremediation: Strategies for Sustainable Environment explores how the integration of proteomic technologies with microbial and plant-based remediation systems can redefine the future of environmental cleanup. By understanding protein expression patterns, stress responses, enzymatic pathways, and metabolic networks, scientists can design targeted, adaptive, and highly efficient bioremediation strategies. These insights not only enhance our capacity to degrade pollutants but also enable the optimization of naturally occurring biological systems to address contaminants with unprecedented precision.

The chapters in this book bring together cutting-edge research and methodologies that illustrate the power of proteomics in advancing sustainable environmental practices. The book covers 16 chapters that provides updated knowledge on proteomic strategies for remediation and restoration of environmental health. This book includes several proteomic methods that could be applied to clean up contaminated environments. The chapters highlighted the potential impact of microbial proteomics in ecological safety and sustainability, along with an in-depth discussion of multidisciplinary approaches. Some chapters cover and address issues like proteomic approaches for management of e-waste, pharmaceutical waste, and petroleum and aquatic pollutions. This book also has a couple of chapters on the role of proteomics in health care. The collective expertise of contributors highlights the synergy between molecular biology, environmental engineering, and biotechnology, demonstrating how interdisciplinary approaches can generate impactful and lasting solutions.

This book intended for students, researchers, environmental professionals, and policymakers seeking a deeper understanding of how omics-based technologies can be harnessed to restore polluted ecosystems. It aims to inspire future innovation while serving as both a reference and a catalyst for further study.

We extend our sincere gratitude to all contributors whose dedication and insights have enriched this work. We also express our sincere gratitude to Elsevier for providing this opportunity. We hope that this book will contribute meaningfully to the global effort toward a cleaner, healthier, and more sustainable world.

Dr. Burhan Hamid
Dr. Mohammed Kuddus
Dr. Ghazala Yunus

Chapter 1

Proteomics based combined approach for ecological safety and sustainability

Shalini Rai[1], Suneeta Gireesh Panicker[2], and Riyaz Sayyed[3]
[1]*Department of Microbiology, DSMNRU University, Lucknow, Uttar Pradesh, India,* [2]*Department of Microbiology, Dr. D. Y. Patil Arts, Commerce and Science College, Pune, Maharashtra, India,* [3]*Department of Microbiology, PSGVP Mandal's Shri S I Patil Arts, G B Patel Science & STSKV Sangh Commerce College, Shahada, India*

1.1 Introduction

In recent decades, human pollution has become a global concern. Efforts to enhance environmental quality have increased as a result of this ongoing threat. Numerous chemicals from industrial and agricultural production processes, as well as other human activities, are discharged into the environment, posing serious threats to human health and natural ecosystems (Carrera et al., 2024). Furthermore, environmental contaminants frequently exist in mixtures that may have antagonistic or synergistic effects, and characterizing each component is a difficult process (Sharma et al., 2019). According to Force (2008) and, the "One Health" program is a transdisciplinary and holistic strategy that aims to improve conditions for humans, animals, and the environment as a single, worldwide, highly linked ecosystem. The reasoning is closely tied to ecotoxicology, the area of toxicology that evaluates the harmful impacts of contaminants on ecosystem components in a comprehensive setting (Truhaut, 1977). In fact, ecotoxicology has a direct impact on human ecological safety as well as ecosystem protection and wellness. Ecological safety is concerned with the short-term and long-term effects of pollutants on life in food chains, air, water, soil, and sediment systems. The fact that ecological safety (Fig. 1.1) incorporates disciplines from ecology to molecular biology in its attempts to forecast the effects of pollutants on the wide natural variety of biological organisms is one of its primary problems.

The capacity to elucidate the link between the genome, proteome, metabolome, and phenotypes has increased due to ongoing advancements in omics technology (Kumar et al., 2025a, 2025b). Proteomics is a branch of

Proteomics Driven Bioremediation. DOI: https://doi.org/10.1016/B978-0-443-45294-9.00011-4

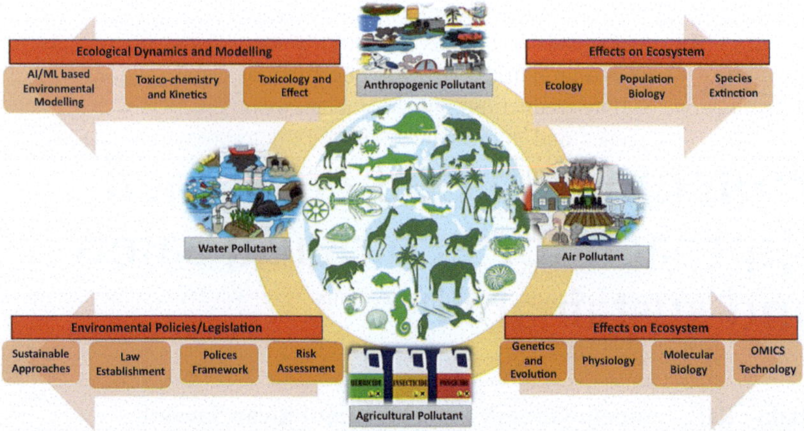

FIGURE 1.1 Ecological safety and toxic pollutants: the study of how hazardous chemicals affect populations, ecosystems, and biological species at the intersection of several disciplines. The diagram illustrates how hazardous chemicals affect our environment and its components.

biology that examines the dynamic changes in protein abundance, composition, and modification; comprehends the relationships and interactions between proteins; and discloses the function of proteins and the functioning of the law of life (Jiang et al., 2024). However, compared to genome and transcriptome investigations, the proteome's complexity and dynamic nature provide far more challenges. For instance, findings from a static snapshot of the proteome at a specific time point are not very repeatable. A major obstacle to proteome analysis is the dynamic abundance range of proteomes in a particular state, which spans six to twelve orders of magnitude (Adnane et al., 2024). Because high-abundance proteins make up a huge fraction of the proteome, it is difficult to identify low-abundance proteins that are physiologically relevant.

Despite these obstacles, the development of mass spectrometry (MS) technology and bioinformatics techniques has significantly increased our understanding of proteomics (Wilhelm et al., 2014). Protein extraction, separation, identification, and quantification, data interpretation, bioinformatics analysis, and protein function verification are some of the processes that make up a typical proteomic analysis workflow. Proteins in complicated mixtures may be separated and seen using two-dimensional polyacrylamide gel electrophoresis technology, which is the first generation of MS identification approach (Martyniuk and Alvarez, 2013). Nevertheless, in standard applications, this approach has a poor throughput. Over the past ten years, shotgun proteomics based on data-dependent acquisition has emerged as the standard method for characterizing proteins and proteomes (Dai and Shen, 2022). Despite its great capability, mass spectrometers only identify known high-abundance proteins and tend to choose the peptides with the greatest signal among randomly fragmented peptides (Price, 1991; Lopez-Casado et al., 2012). Therefore it is still difficult to quantify low-abundance peptides consistently. According to published research, the metabolic pathways controlled by high-abundance proteins are the main means by which the functional metabolic networks are linked.

The development of life has been accompanied and impacted by environmental change. Our knowledge of protein evolution is improved by the growth of genomic data, protein sequence and structure, and computational capacity (Aslam et al., 2024). Environmental succession, functional protein development, and proteome expression are intrinsically linked, which aids in understanding life's evolutionary processes and forecasts a species' role in a given environment (Gredelj et al., 2018). Additionally, the subsequent steps of protein identification, transport, and transformation are responsible for the biotransformation of contaminants. The first stage in the restricted reaction of pollutant transformation is the contact between pollutants and biological molecules (Gouveia et al., 2021). Pollutant binding, absorption, and transformation are thought to be significantly influenced by the structural features of functional proteins and the physicochemical interactions between effective proteins and pollutants (Bahamonde et al., 2016). Therefore, in order to understand the direct interaction of pollutants with these proteins, it is vital to look at the conformational changes and, consequently, the biological impacts of target proteins.

In order to evaluate the toxicity of compounds, systematic assessment procedures are needed to compare the efficacy of treatment options for reducing the toxicity of the target pollutants and to evaluate the toxicity of the pollutants and their byproducts produced via degradation. Furthermore, it is necessary to create systematic methods for screening new biomarkers that represent the effects of environmental stressors, the presence of target chemicals, and the response process of pollutant accumulation, transit, and transformation. The aforementioned scientific concerns, their obstacles, and potential proteomic answers are discussed in the sections that follow. In addition to offering a potentially insightful method for the selection of functional biomarkers and species, this chapter aims to reveal biomolecule mechanisms, protein biomarkers, and metabolism networks related to stress response, pollutant recognition, transport, and biodegradation.

1.2 Proteomics approaches for ecological safety

By examining the expression of proteins in organisms exposed to environmental stressors like pollution, proteomics is a potent technique for evaluating ecological safety. The primary approaches employed in environmental research consist of expression proteomics, ecotoxicoproteomics, and functional proteomics, and its different levels of narrow down proteomic approaches explore the ecological monitoring and safety (Fig. 1.2).

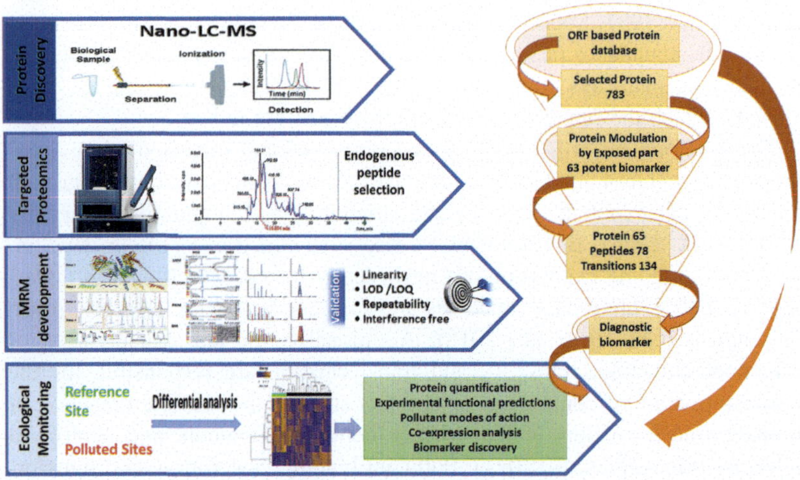

FIGURE 1.2 Different proteomic approaches to uncovering biomolecule processes, protein biomarkers, and metabolic networks associated with stress response, pollutant identification, transport, and biodegradation are illustrated. The diagram illustrates different proteomic approaches that can be applied at multiple molecular levels for biodegradation.

1.3 Expression proteomics

When expression proteomics is used in combination with environmental research, it examines how environmental elements such as pollution or temperature fluctuations affect the proteins that an organism or ecosystem produces. In order to comprehend the molecular and physiological reactions to stress, this entails analyzing patterns of protein expression, both qualitative and quantitative, using methods such as mass spectrometry to detect changes in protein levels, alterations, and interactions (Chu et al., 2009). Applications include assessing environmental risk, keeping an eye on the health of ecosystems, and comprehending microbial populations in their natural environments (Roch et al., 2022). Expression proteomics examined several facets of methods related to ecological safety.

Comparing protein expression: This is the fundamental component of the method, comparing the "proteome" (the total protein composition) under various environmental settings, such as high vs low temperatures or pollution against clean water.

- **Determining protein alterations:** To identify and measure proteins that are either upregulated or downregulated by the environmental element, researchers employ methods such as mass spectrometry.
- **Examining protein interactions and modifications:** This discipline may examine posttranslational modifications (such as phosphorylation) and protein-protein interactions, which can dramatically change how proteins operate in response to external stimuli.

- **Handling complicated settings:** Identifying proteins from unsequenced species and complex samples needs sophisticated bioinformatics, which is a significant problem when researching organisms in their natural, uncontrolled surroundings.
- **Toxicoproteomics:** This particular application focuses on using protein alterations to determine the earliest molecular processes in an organism's reaction to a toxic chemical stressor.

1.4 Ecotoxicoproteomics

A protein-centric approach called ecotoxicoproteomics is helpful for ecotoxicology and may be used in the future for environmental monitoring and chemical risk assessment (Gouveia et al., 2017). The influence that proteomics has had on managing human health, providing a wide range of diagnostic and prognostic tools, has always been the driving force behind ecotoxicoproteomics. Ecotoxicoproteomics has concentrated on examining the dynamics of proteins from environmental species using mass spectrometry (MS)-based high-throughput analytical techniques in order to ascertain the environmental impact of contaminants (Armengaud et al., 2014). The obtained protein fingerprints may be compared under various settings to determine pollutant- and species-specific biomarkers, comprehend the mechanisms of action of chemicals, and interpret the adaptation strategies used by animals (Fan et al., 2013). By examining the particular protein-level reactions (toxicity targets) of sentinel animals (such as aquatic invertebrates) to contaminants, this specialized field focuses on monitoring environmental safety (Reardon, 2024).

The goal of ecotoxicology is to protect ecosystems and the variety of biological species that sustain them by using non-model sentinel creatures with extremely different evolutionary histories (Grossmann et al., 2017). Protein identification, functional annotation, and extrapolation of toxicity across species are all part of the traditional proteomics procedure. In order to decipher molecular information and create the best databases for protein identification and pathway analysis in nonmodel species, multiomics techniques must be developed (Gouveia et al., 2017). The results of the ecotoxicoproteomics research could offer enough details about the possible pollutant's effects to be extrapolated to other species within the ecosystem. A variety of goals could be accomplished: (1) identify and record the harmful effects of the pollutant on environmental species; (2) comprehend their mechanisms of toxicity; (3) forecast the effects of exposure to anthropogenic pollutants on the environment; and (4) create techniques that could supply this integrated data. Relevant combination transcriptomic/proteomic research may be carried out for nearly any species, thanks to current equipment and software.

The past decade has seen significant advancements in ecotoxicoproteomics, particularly for the monitoring of freshwater, marine, and soil environments.

However, there is still much work to be done to determine how these tools can be incorporated into risk assessment frameworks and environmental policies. The comparison of "control versus exposed" proteomes has led to the proposal of several biomarkers; however, only a few are undergoing further validation. However, the potential of next-generation proteomics tools for the identification of novel toxic responses or adaptive pathways, the identification and validation of species-specific biomarkers, and the provision of instruments for accurate prediction and diagnosis of the health status of organisms is reinforced by the recent application of high-throughput shotgun approaches, proteogenomics, and targeted proteomics.

1.5 Functional proteomics

To comprehend the molecular processes by which environmental stressors or xenobiotics impact an organism's biology, this research examines protein functions, changes (such as phosphorylation), and protein-protein interactions (Yun et al., 2016). In order to comprehend the metabolic processes and interactions of complex environmental microbial communities (such as soil and water) with contaminants in situ, metaproteomics, also known as Microbial Community Proteomics, entails the direct recovery and analysis of the whole protein complement. Protein separation methods (such as 2D gel electrophoresis), mass spectrometry (LC-MS/MS), and bioinformatics for data processing and protein identification utilizing sequence databases are important technologies involved.

1.6 Proteomics contributes to ecological safety

Proteomics is a powerful tool for assessing ecological safety by analyzing protein expression in organisms exposed to environmental stressors like pollutants. Advanced proteomic approaches help to identify biomarkers of exposure and understand the mechanisms of toxicity, potentially leading to improved environmental monitoring and risk assessment (Table 1.1).

1.7 Identifying pollution biomarkers

In the realm of pollution biomonitoring, proteomics is a practical and efficient method for finding protein biomarkers. Proteomics and other OMICs methods may be able to find protein patterns and biomarkers, which would make them more reliable and targeted indicators of a certain pollutant (Armengaud, 2016). By finding potential biomarkers, a proteomic method may enable us to gain a more comprehensive understanding of the changes in proteins. An overview of cellular structure and function, as well as the defense mechanism against water contamination, may be obtained by proteome analysis. Following the validation procedure, these protein biomarkers can serve as substitute biomarkers for

TABLE 1.1 A summary of recent proteomic approaches performed in concern of different sectors of ecological safety against stressors. Ecological sectors, stressors, proteomic approaches, and significant achievements of different studies are described.

Sentinel species/model organisms/samples type	Stressors	Proteomic approaches/ technology	Contribution in ecological safety	Significant remark/ achievement	References
Aquatic Ecosystem (Marine/Freshwater)					
Labeorohita	As metal	2D, MALDI-TOF/TOF MS	Monitoring fish health and bioaccumulation of metals	Identified biomarkers (Apo-A1, A2ML, Wap65) indicate arsenic-induced liver toxicity	Banerjee et al. (2017)
Gasterosteus aculeatus	Anthropogenic pollutant	Data-independent acquisition	Monitoring fish health and bioaccumulation of metals	Gill proteome analysis explored molecular differences in gill proteins among ecotypes, revealing adaptation to environmental and morphological variations.	Li et al. (2018)
Scorpaenaplumieri	Anthropogenic pollutant	Nano-LC-MS/MS	Monitoring fish health and bioaccumulation of metals	Antimicrobial and proteolytic activity. Identified 391 proteins with antimicrobial and venom-related activity, including proteolytic enzymes	Borges et al. (2018)

(Continued)

TABLE 1.1 (Continued)

Sentinel species/model organisms/samples type	Stressors	Proteomic approaches/ technology	Contribution in ecological safety	Significant remark/ achievement	References
Danio rerio	Pesticide (Dieldrin)	LC-MS/MS, iTRAQ	Monitoring fish health and bioaccumulation of metals	Mitochondrial proteins showed links to diseases like Parkinson's and Huntington's.	Simmons et al. (2019)
Gammarus fossarum	Anthropogenic pollutant	Targeted scout-MRM assay	High-multiplexed freshwater monitoring and ecotoxicological models development	Protein biomarkers involved in key physiological pathways, such as molting, feeding, and general stress response	Faugere et al. (2020)
Gammarus fossarum	Anthropogenic pollutant	MRM-mass spectrometry	Active biomonitoring and Environmental molecular markers	Molecular markers in freshwater sentinel species	Leprêtre et al. (2022)
Daphnia	Chemical pollutant	High-resolution mass spectrometry, putative molecular key events (putative mKEs), sparse partial least squares discriminate analysis (sPLS-DA), multiomics factor analysis (MOFA+), and biorder canonical correlation analysis (Bi-CCA)	Biomonitoring and bioremediation of chemical pollutants. Monitoring of sentinel species for environmental health protection	Functional conservation of gene and metabolite networks, the framework enables the identification of targets of toxicity across species, guiding in vivo and in vitro validations of toxic effects in human models	Abdullahi et al. (2022)
Zebra mussel (*Dreissenapolymorpha*)	Microplastics	Fourier transform infrared spectroscopy (FTIR)	Monitoring of sentinel species	Bioindicator in the aquatic ecosystems' Microplastic pollution	Atamanalp et al. (2023)

Gammarus fossarum, *Dreissenapolymorpha*, and *Palaemon serratus*	Anthropogenic pollutant	Liquid chromatography-tandem mass spectrometry (LC-MS/MS)	Ecotoxicological investigations and environmental monitoring	Assessment of aquatic environmental risks contributes to a better understanding of molecular diversity in organisms and provides a valuable resource for future ecotoxicological studies	Leprêtre et al. (2023)
Two teleost fish (*G. aculeatus and P. flesus*), three bivalve mollusks (*D. polymorpha, M. leucophaeata, and M. edulis*), and three crustaceans (*G. fossarum, P. longirostris, and P. serratus*)	Physical stress	OMICS approach	Biomarker-based environmental monitoring	Freshwater-estuarine-coastal water continua by applying biomarker measurements in multi-species caged organisms	Slaby et al. (2024)
Anaerobic ammonium-oxidizing bacteria (anammox)	Higher copper concentrations	Mass spectrometry	Assess the treatability of industrial effluents, detect early-stage stress biomarkers for metals, organics, or even changes in physical-chemical parameters such as pH or temperature	Specific anammox proteins involved in stress response (bacterioferritin, universal stress protein, or superoxide dismutase), polysaccharide biosynthesis protein, and sulfotransferase alginate export family protein, increased expression	Guzmán-Fierro et al. (2024)

(*Continued*)

TABLE 1.1 (Continued)

Sentinel species/model organisms/samples type	Stressors	Proteomic approaches/ technology	Contribution in ecological safety	Significant remark/ achievement	References
Cyanobacterial populations	Algal bloom	Native mass spectrometry	Monitoring cyanobacterial blooms and effective strategies to mitigate this environmental challenge	Characterization of cyanobacterial populations within lakes, offering valuable insights into the dynamics of cyanobacterial species associated with harmful algae blooms	Sound et al. (2025)
Terrestrial Ecosystem (Soil/Agriculture/Forest/Food)					
White-rot fungus (*Irpex lacteus*)	Lignin/ Azo dye	Shotgun proteomics technique	Understanding stress response and pollutant toxicity	Characterization of the fungal proteome evealed important gene functions and proteome dynamics	Sun et al. (2017)
Milk and eggs	Allergens	MALDI-TOF + machine learning LCeMS/MS, MRM (LC-QQQ)	Identification of allergens and human health	Detection of known food allergens in food commodities by targeted proteomics	Monaci et al. (2020)
Food products from different sources	Food-borne pathogens and contaminants	LC-SWATH, LC-QTOF-IMS	Biomarker-based food monitoring	High-throughput diagnostic methods for the detection of parasites, fungi, and food-borne bacteria in food and feed via MALDI-MS	Kastner et al. (2021)

Food products from different sources	Pathogens and chemical pollutants	LC-MS/MS, MALDI-TOF + machine learning LCeMS/MS, MRM (LC-QQQ), MRM (LC-QTOF), LC-SWATH, LC-QTOF-IMS	Biomarker-based food monitoring	Promising biomarker in food authentication, quality, and safety	Afzaal et al. (2022)
Forest Trees (Eucalyptus, Pinus, and Quercus)	Biotic and abiotic stresses	LC-MS/MS, MALDI-TOF + machine learning LCeMS/MS	Plant responses and resilience to global climate change, contributing to climate mitigation strategies, molecular breeding programs, and forest ecology monitoring	Characterization of the variability and biodiversity, as well as wood and nonwood derived products, traceability, allergen, and bioactive peptides identification	Castillejo et al. (2023)
Food products from different sources	Mutant proteins	LC-MS/MS, MALDI-TOF + machine learning LCeMS/MS, MRM (LC-QQQ), MRM (LC-QTOF), LC-SWATH, LC-QTOF-IMS	Food-borne risk assessment	Integration of proteomics in a multiomics approach within a broader systems biology framework for augmenting current risk assessment practices of transgenics	Benevenuto et al. (2023); Varunjikar (2023)
Food products	Chemical pollutants	LC-MS/MS, MALDI-TOF + machine learning LCeMS/MS	Application of discovery untargeted proteomics to identify allergens in novel foods	Discovery food proteomics, targeted food proteomics, and proteomics-based food systems biology	Lopez-Pedrouso et al. (2023)

(*Continued*)

TABLE 1.1 (Continued)

Sentinel species/model organisms/samples type	Stressors	Proteomic approaches/technology	Contribution in ecological safety	Significant remark/achievement	References
Food products from different sources	Pathogens and chemical pollutants	LC-MS/MS, MALDI-TOF + machine learning LCeMS/MS, MRM (LC-QQQ), MRM (LC-QTOF), LC-SWATH, LC-QTOF-IMS	Food authentication and food allergy profiling	Discovery food proteomics, targeted food proteomics, and proteomics-based food systems biology	Carrera et al. (2024)
Medicinal plant (neem, *Azadirachta indica*)	Abiotic stresses	2-DE, HPLC, LC-MS/MS	Differential regulation of key metabolism	Proteins associated with oxidoreduction, energy, transcriptional, stress response, respiration, and cell division	Omar et al. (2024)
Agricultural crops	Biotic/Abiotic stressors and chemical pollutants	Gene editing and proteome profiling	Understanding stress response and pollutant toxicity	Improve crop stress tolerance (drought, salinity, disease), enhance food quality/safety, identify stress biomarkers, discover genes/proteins for breeding, and develop innovative tools like biocontrols	Kumar et al. (2025a), (2025b)
Wheat, brassica, and cotton	Abiotic stressors	2D-PAGE; MALDI-TOF/MS, iTRAQ; LC-ESI/MS, nano LC-MS/MS; IDA-DIA SWATH-MS	Stress monitoring and food security monitoring	Improve stress tolerance and yield traits	Halder et al. (2025)

regular usage (Ambrosio et al., 2018). The development of commercial kits based on protein markers would be made possible by all this knowledge.

Mass spectrometry analysis and protein separation in gel are the two primary proteomic techniques that have been developed recently (Tholey and Becker, 2017). The term "bottom-up approach" refers to electrophoretic and chromatographic methods based on previous protein digestion. The alternative approach, top-down, uses mass spectrometry and separation to analyze intact proteins. In particular, it would be simple and routine to discover and measure protein biomarkers using the second option in conjunction with tiny protein biomarkers (less than 30 kDa) (Tholey and Becker, 2017). The thorough examination of proteins would also aid in the discovery of more precise and sensitive pollution indicators. The environmental sciences now have new opportunities thanks to these innovative technologies for the understanding of whole proteomes. In order to learn more about the molecular reactions of living things to pollutants, environmental proteomics is now being researched as a crucial area of environmental studies (Li et al., 2019). For instance, research has found protein biomarkers in fish and mollusks exposed to freshwater and marine pollution, respectively. Li et al. (2019) used mussels as a worldwide bioindicator of coastal microplastic contamination in this order. Because of their extensive range, important ecological niches, vulnerability to microplastic absorption, and intimate relationship to both human health and marine predators, mussels are an appropriate bioindicator for microplastic pollution (Green et al., 2019). Mass spectrometry-based proteomics has been used in numerous studies to find biomarkers and comprehend the molecular mechanisms of toxicity from different pollutants in species such as Gammarus fossarum (Trapp et al., 2014; Gouveia et al., 2017) and Mytilus edulis (blue mussel) (Kronberg et al., 2021; Marigómez et al., 2024).

1.8 Understanding stress responses

Understanding proteome variations under diverse stress situations is crucial since proteins are directly implicated in the stress response (Ahmad et al., 2016). Signaling, energy metabolism (glycolysis, Krebs cycle, ATP biosynthesis, photosynthesis), storage proteins, protein metabolism, several other biosynthetic pathways (e.g., S-adenosylmethionine metabolism, lignin metabolism), transport proteins, proteins involved in protein folding and chaperone activities, other protective proteins (LEA, PR proteins), ROS scavenging enzymes, and proteins affect the regulation of plant growth and development are generally affected by both biotic and abiotic stressors (Sarkar and Sadhukhan, 2022). A list of proteins that have been shown to exhibit notable variations in their relative abundance or posttranslational changes across genotypes of wheat, barley, or similar species under stress conditions is provided, and their possible involvement in the differential stress response is examined. In conclusion, it is indicated that the outcomes of proteomic research may play a role in future practical applications, such as breeding for increased stress tolerance and the ability to test and employ protein markers in breeding (Mittra et al., 2024). Researchers may study how organisms adjust to and deal with environmental stresses, including pollution, temperature fluctuations, and resource constraints, by using proteomics. Researchers can pinpoint the precise processes and mechanisms behind stress responses, such as oxidative stress, immunological suppression, and metabolic alterations, by examining protein expression (Ghosh and Xu, 2014).

1.9 Evaluating pollutant toxicity

A collection of molecular methods that enable the identification of the proteins being expressed by a mixed microbial and ecological population at a particular time is collectively referred to as environmental proteomics (Zhang et al., 2013). They have been used more and more in the investigation of microbiomes involved in biotechnological processes, showing promise in identifying particular metabolic alterations in extremely complex communities (Quiton-Tapia et al., 2023) or verifying the biodegradation of particular contaminants (Kennes-Veiga et al., 2022). By analyzing how contaminants affect biological processes and protein expression, proteomics may be used to determine how harmful they are. This method can assist in identifying possible harmful impacts on ecosystems and figuring out the processes of toxicity (Poulsen et al., 2023).

1.10 Monitoring ecosystem health

Biomonitoring is becoming a crucial component of evaluating the environment's health. To better understand the effects of pollutants on various ecosystems, this monitoring necessitates the use of many species of animal and plant indicators that span the widest range of ecosystem biodiversity. As a result, proteins have been employed as monitoring instruments in a number of industries, including agriculture, aquaculture (Jitjumnong et al., 2025; Carrera et al., 2020), forensics, medicine, and environmental monitoring. The transfer of findings from proteomics investigations conducted in the lab as part of a proof-of-concept to field research is a significant problem for ecotoxicologists, despite the fact that model species are excellent laboratory models (Behmel et al., 2016). The study of ecologically significant wildlife species, which are mostly "nonmodel" creatures in the sense that they are difficult for the scientific community to share in order to compare findings on the same basis, is necessary for successful transfer. According to regulatory criteria, a minimum number of species, taxonomic groupings, and endpoints must be included when using a species-sensitivity distribution technique in risk assessment studies; hence, both standard and non-standard tests and organisms are needed (Rudén et al., 2016). Even though ecotoxicoproteomics has grown significantly over the past ten years, only a small number of species have been investigated, despite the availability of increased genetic data. Furthermore, a large number of protein sequences that are directly impacted by contaminants frequently have little to no functional information available. The molecular divergence that organisms have gained over evolution and the significance of species-specific proteins in the response to environmental stimuli are most likely the causes of this lack of knowledge.

By examining the protein profiles of sentinel species, proteomics may be used to track the condition of ecosystems. This can track the success of conservation initiatives and identify early indicators of environmental deterioration. *Mytilus edulis*, a blue mussel, and *Gammarus fossarum*, a freshwater crustacean, are used as sentinel species for environmental monitoring in proteomics investigations. To identify biomarkers and comprehend the molecular pathways of toxicity from different contaminants, researchers use mass spectrometry-based proteomics.

1.11 Applications in aquaculture

By changing proteins at the cellular level, aquatic systems—which are often complex and multifaceted—can adopt distinctive alterations and adaptations to environmental changes. Because proteins, proteoforms, and their complexity are involved in both fundamental and sophisticated biological activities, proteomics is a crucial approach for studying aquatic environments (Martinez et al., 2020). The molecular mechanisms underpinning biological activities in an aquatic environment may be analyzed more quickly using proteomics. Prior proteomic research on aquatic environments has mostly concentrated on pollution evaluations, ecotoxicology, their function in the food business, and the discovery and extraction of natural compounds. Aquatic protein indicators have been thoroughly documented and are now widely used in the medical and pharmaceutical sectors (Sidira et al., 2024).

Organisms' reactions at the cellular and molecular levels can serve as markers of environmental stressors and changes. On the other hand, environmental changes make it easier to forecast aquatic production and health, which are essential for managing and conserving ecosystems. High aquatic food production, seafood safety, and sustainable aquaculture have all benefited from recent developments in proteomics (Nissa et al., 2023). Other facets of the aquatic environment have been covered by proteomic techniques, such as species identification using protein fingerprinting. Proteomics is used to determine possible allergens, evaluate the safety and quality of aquaculture products, and determine how farming methods affect fish health.

1.12 Microbes and advancements in environmental proteomics

Proteome profiling is used in environmental proteomics, a developing discipline at the nexus of molecular biology and environmental research, to examine the varied and dynamic protein expressions in environmental materials. Researchers may learn more about the functions of microbial and ecological communities, how they interact with their surroundings, and how they play a part in different biogeochemical processes by using this approach (Wang et al., 2016). Proteome profiling has several uses in environmental microbiology and biotechnology, providing new ways to comprehend and solve environmental problems.

By offering an in-depth understanding of the roles and interactions of microbial communities, environmental proteomics has completely transformed the study of microbial e1cology (Guzmán-Fierro et al., 2024). Conventional techniques frequently depended on lab-cultured microorganisms, which only caught a small portion of the variety seen in natural settings. However, proteome profiling overcomes the drawbacks of culture-based methods by directly analyzing a community's whole protein complement from ambient samples. Studying how microbes react to environmental changes is one of the important uses of environmental proteomics. By examining variations in protein expression, for example, scientists may track how microbial populations respond to contaminants, temperature changes, or nutrition availability (Reardon, 2024). This approach has been crucial in understanding the processes behind microbial resistance to organic contaminants and heavy metals, offering important insights for bioremediation techniques.

The subject of biogeochemical cycling is another significant application. The functions of microbes in the cycles of nitrogen, carbon, sulfur, and phosphorus have been revealed by proteome profiling. Scientists can clarify the metabolic pathways and ecological functions of various microbial species by identifying important enzymes participating in these activities. Predicting and controlling how environmental changes will affect ecosystem functioning depends on this information. Proteomics is essential to the optimization and innovation of biotechnological applications in environmental biotechnology. Bioremediation, which uses microorganisms to breakdown or purify environmental contaminants, is one of the most popular uses. By identifying the precise proteins and pathways involved in the degradation processes, proteome profiling makes it possible to choose and design microbial strains with improved bioremediation capabilities.

For example, proteome investigations have revealed hydrocarbon-degrading enzymes in oil spill cleaning, resulting in the creation of microbial communities that are specialized for oil breakdown. Similar to this, proteomics can help develop more effective wastewater treatment procedures by identifying important proteins involved in sludge reduction and nutrient removal. The creation of bioindicators for environmental monitoring is another fascinating use. Biomarkers—specific proteins indicative of distinct environmental variables or stressors—can be found using proteomic analysis. A useful tool for environmental management and control, these biomarkers may be utilized to create biosensors that track pollutant levels, soil health, and water quality in real time. Technological developments in mass spectrometry, bioinformatics, and high-throughput sequencing have greatly improved the area of environmental proteomics. These developments have increased proteome profiling's sensitivity, accuracy, and speed, allowing for more thorough and in-depth analysis of environmental samples.

1.13 Challenges and future directions

Environmental proteomics has a lot of potential, but in order to fully realize its uses, a number of issues must be resolved. The complexity and variability of environmental materials, which can make protein extraction, separation, and identification more difficult, are one of the main obstacles. Improving sample preparation methods and creating standardized procedures are crucial for increasing proteomic study repeatability and comparability. The wide variety of proteins and the fluctuating range of their expression levels present another difficulty. Proteins from many different species are frequently found in environmental samples, some of which are present in incredibly small amounts. Capturing the complete proteome diversity will need better bioinformatics tools for data interpretation and increasing the sensitivity and resolution of mass spectrometry methods. In the future, there is a lot of promise for combining environmental proteomics with other strategies and cutting-edge computational techniques. By offering real-time data on community functions and reactions to environmental changes, the development of portable and field-deployable proteomic sensors might transform in situ environmental monitoring.

1.14 Conclusion

The last 10 years have seen significant advancements in environmental proteomics, particularly for the monitoring of freshwater, marine, forest, soil, and agricultural environments. However, there is still much work to be done to determine how these tools can be incorporated into risk assessment frameworks and environmental policies. The comparison of "control versus exposed" proteomes has led to the proposal of several biomarkers; however, only a ew are undergoing further validation. However, the potential of next-generation proteomics tools for the identification of novel toxic responses or adaptive pathways, the identification and validation of species-specific biomarkers, and the provision of instruments for accurate prediction and diagnosis of the health status of organisms is reinforced by the recent application of high-throughput shotgun approaches, proteogenomics, and targeted proteomics. The successful use of these instruments requires assessing their performance both inside and between laboratories using suitable quality control methods. Bioinformatics techniques and statistical analysis must be enhanced for biomarker creation in order to guarantee the removal of false positives and negatives from datasets. To be sure that the biological reaction is indeed caused by the pollutants, this must be done in conjunction with research on the impact of additional biotic and abiotic elements that are included as confounding factors when evaluating biomarker responses. Environmental proteomists need to move beyond the traditional protein expression profiles that have long dominated the discipline and include all of the current proteomics advancements in their study. Regardless of the species selected to address our research goals, it will be feasible to delve deeper into massive amounts of molecular data by creating organized "omics" studies utilizing the proper technological, methodological, and bioinformatics tools. Environmental proteomics methods should be a key component of the strategy to "Make our planet safe again" during the next ten years.

References

Abdullahi, M., Li, X., Abdallah, M.A.-E., Stubbings, W., Yan, N., Barnard, M., et al., 2022. Daphnia as a sentinel species for environmental health protection: a perspective on biomonitoring and bioremediation of chemical pollution. Environmental Science & Technology 56 (20), 14237–14248. Available at https://doi.org/10.1021/acs.est.2c01799.

Adnane, M., de Almeida, A.M., Chapwanya, A., 2024. Unveiling the power of proteomics in advancing tropical animal health and production. Tropical Animal Health and Production 56, 182.

Afzaal, M., Saeed, F., Hussain, M., Shahid, F., AzhariSiddeeg, Al-Farga, A., 2022. Proteomics as a promising biomarker in food authentication, quality and safety: a review. Food Science & Nutrition 10 (7), 2333–2346. Available at https://doi.org/10.1002/fsn3.2842.

Ahmad, P., Abdel Latef, A.A.H., Rasool, S., Akram, N.A., Ashraf, M., Gucel, S., 2016. Role of proteomics in crop stress tolerance. Frontiers in Plant Science 7, 1336.

Ambrosio, L., Russo, R., Salzano, A.M., Serpe, F.P., Ariano, A., De Tommasi, N., et al., 2018. Accumulation of polychlorinated biphenyls in mussels: a proteomic study. Journal of Food Protection 81 (2), 316–324. Available at https://doi.org/10.4315/0362-028x.jfp-17-148.

Armengaud, J., Trapp, J., Pible, O., Geffard, O., Chaumot, A., Hartmann, E.M., 2014. Non-model organisms, a species endangered by proteogenomics. Journal of Proteomics 105, 5–18. Available at https://doi.org/10.1016/j.jprot.2014.01.007, http://www.elsevier.com.

Armengaud, J., 2016. Next-generation proteomics faces new challenges in environmental biotechnology. Current Opinion in Biotechnology 38, 174–182. Available at https://doi.org/10.1016/j.copbio.2016.02.025.

Aslam, N., Li, Q., Bashir, S., Yuan, L., Qiao, L., Li, W., 2024. Integrated review of transcriptomic and proteomic studies to understand molecular mechanisms of rice's response to environmental stresses. Biology 13 (9), 659. Available at https://doi.org/10.3390/biology13090659.

Atamanalp, M., Kokturk, M., Gündüz, F., Parlak, V., Ucar, A., Alwazeer, D., et al., 2023. The use of zebra mussel (Dreissenapolymorpha) as a sentinel species for the microplastic pollution of freshwater: the case of Beyhan Dam Lake, Turkey. Sustainability 15 (2), 1422. Available at https://doi.org/10.3390/su15021422.

Bahamonde, P.A., Feswick, A., Isaacs, M.A., Munkittrick, K.R., Martyniuk, C.J., 2016. Defining the role of omics in assessing ecosystem health: perspectives from the Canadian environmental monitoring program. Environmental Toxicology and Chemistry 35 (1),

20–35. Available at https://doi.org/10.1002/etc.3218, http://www.interscience.wiley.com/jpages/0730-7268.

Banerjee, S., Mahanty, A., Mohanty, S., Mazumder, D.G., Cash, P., Mohanty, B.P., 2017. Identification of potential biomarkers of hepatotoxicity by plasma proteome analysis of arsenic-exposed carp Labeorohita. Journal of Hazardous Materials 336, 71–80. Available at https://doi.org/10.1016/j.jhazmat.2017.04.054, www.elsevier.com/locate/jhazmat.

Behmel, S., Damour, M., Ludwig, R., Rodriguez, M.J., 2016. Water quality monitoring strategies — a review and future perspectives. Science of The Total Environment 571, 1312–1329. Available at https://doi.org/10.1016/j.scitotenv.2016.06.235.

Benevenuto, R.F., Zanatta, C.B., Waßmann, F., Eckerstorfer, M.F., Agapito-Tenfen, S.Z., 2023. Integration of omicsanalyses into GMO risk assessment in Europe: a case study from soybean field trials. Environmental SciencesEurope 35, 14.

Borges, M.H., Andrich, F., Lemos, P.H., Soares, T.G., Menezes, T.N., Campos, F.V., et al., 2018. Combined proteomic and functional analysis reveals rich sources of protein diversity in skin mucus and venom from the Scorpaenaplumierifish. Journal of Proteomics 187, 200–211. Available at https://doi.org/10.1016/j.jprot.2018.08.002, http://www.elsevier.com.

Carrera, M., Piñeiro, C., Martinez, I., 2020. Proteomic strategies to evaluate the impact of farming conditions on food quality and safety in aquaculture products. Foods 9 (8), 1050.

Carrera, M., Abril, A.G., Pazos, M., Calo-Mata, P., Villa, T.G., Barros-Velázquez, J., 2024. Proteins and peptides: proteomics approaches for food authentication and allergen profiling. Current Opinion in Food Science 57. Available at https://doi.org/10.1016/j.cofs.2024.101172, https://www.sciencedirect.com/science/journal/22147993.

Castillejo, M.A., Pascual, J., Jorrín-Novo, J.V., Balbuena, T.S., 2023. Proteomics research in forest trees: a 2012-2022 update. Frontiers in Plant Science 14. Available at https://doi.org/10.3389/fpls.2023.1130665, https://www.frontiersin.org/journals/plant-science.

Chu, X., Hu, J., Ong, S.L., 2009. Application of proteomics in environmental science. Frontiers of Environmental Science & Engineering 3, 393–403.

Dai, X., Shen, L., 2022. Advances and trends in omics technology development. Frontiers in Medicine 9, 911861.

Fan, L., Wang, A., Wu, Y., 2013. Comparative proteomic identification of the hemocyte response to cold stress in white shrimp, Litopenaeusvannamei. Journal of Proteomics 80, 196–206.

Faugere, J., Gouveia, D., Ayciriex, S., Chaumot, A., Almunia, C., François, A., et al., 2020. High-multiplexed monitoring of protein biomarkers in the sentinel Gammarus fossarum by targeted scout-MRM assay, a new vision for ecotoxicoproteomics. Journal of Proteomics 226, 103901. Available at https://doi.org/10.1016/j.jprot.2020.103901.

Force, O.H.I.T., 2008. One Health: A New Professional Imperative. A.V.M. Association, Schaumburg, IL.

Ghosh, D., Xu, J., 2014. Abiotic stress responses in plant roots: a proteomics perspective. Frontiers in Plant Science 5. Available at https://doi.org/10.3389/fpls.2014.00006.

Gouveia, D., Chaumot, A., Charnot, A., Queau, H., Armengaud, J., Almunia, C., et al., 2017. Assessing the relevance of a multiplexed methodology for proteomic biomarker measurement in the invertebrate species Gammarus fossarum: a physiological and ecotoxicological study. Aquatic Toxicology (Amsterdam, Netherlands) 190, 199–209.

Gouveia, N., Kephart, J.L., Dronova, I., McClure, L., Granados, J.T., Betancourt, R.M., et al., 2021. Ambient fine particulate matter in Latin American cities: levels, population exposure, and associated urban factors. The Science of the Total Environment 772. Available at https://doi.org/10.1016/j.scitotenv.2021.145035.

Gredelj, A., AlbertoBarausse, Grechi, L., Palmeri, L., 2018. Deriving predicted no-effect concentrations (PNECs) for emerging contaminants in the river Po, Italy, using three approaches: assessment factor, species sensitivity distribution and AQUATOX ecosystem modelling. Environment International 119, 66–78. Available at https://doi.org/10.1016/j.envint.2018.06.017.

Green, D.S., Colgan, T.J., Thompson, R.C., Carolan, J.C., 2019. Exposure to microplastics reduces attachment strength and alters the haemolymph proteome of blue mussels (Mytilus edulis). Environmental Pollution 246, 423–434. Available at https://doi.org/10.1016/j.envpol.2018.12.017.

Grossmann, J., Fernández, H., Chaubey, P.M., Valdés, A.E., Gagliardini, V., Cañal, M.J., et al., 2017. Proteogenomic analysis greatly expands the identification of proteins related to reproduction in the apogamous fern Dryopteris affinis ssp. affinis. Frontiers in Plant Science 8.

Guzmán-Fierro, V., Dieguez-Seoane, A., Roeckel, M., Lema, J.M., Trueba-Santiso, A., 2024. Environmental proteomics as a useful methodology for early-stage detection of stress in anammox engineered systems. Science of The Total Environment 912, 169349. Available at https://doi.org/10.1016/j.scitotenv.2023.169349.

Jiang, Y., Rex, D.A.B., Schuster, D., Neely, B.A., et al., 2024. Comprehensive overview of bottom-up proteomics using mass spectrometry. ACS Measurement Science Au 4 (4), 338–417.

Jitjumnong, J., Taweechaipaisankul, A., Lin, J.C., Wongchanla, S., Chuwatthanakhajorn, S., Lin, C.J., et al., 2025. An overview of advancements in proteomic approaches to enhance livestock production and aquaculture. Animals 15 (13). Available at https://doi.org/10.3390/ani15131946, http://www.mdpi.com/journal/animals.

Kastner, C., Bahn, P., Sch€onfelder, R., Ozolinßa, Z., Alksne, L., Richter, M.H., et al., 2021. Development of a novel method for identification of alariaalatamesocercariae by matrix-assisted laser desorption/ionization time-of-flight mass spectrometry. Microorganisms 9, 1664.

Kennes-Veiga, D.M., Trueba-Santiso, A., Gallardo-Garay, V., Balboa, S., Carballa, M., Lema, J.M., 2022. Sulfamethoxazole enhances specific enzymatic activities under aerobic heterotrophic conditions: a metaproteomic approach. Environmental Science and Technology 56 (18), 13152–13159. Available at https://doi.org/10.1021/acs.est.2c05001, http://pubs.acs.org/journal/esthag.

Kronberg, J., Byrne, J.J., Jansen, J., Antczak, P., Hines, A., Bignell, J., et al., 2021. Modeling the metabolic profile of Mytilus edulis reveals molecular signatures linked to gonadal development, sex and environmental site. Scientific Reports 11 (1). Available at https://doi.org/10.1038/s41598-021-90494-y, www.nature.com/srep/index.html.

Kumar, R., Kumar, M., Chaudhary, V., et al., 2025a. Exploring recent advances, limitations, and future prospects of OMICS-based technologies in plant-pathogen interaction studies: a systematic review. Discover Plants 2, 284.

Kumar, M., Ahmad, A., Kumar Singh, M., Kumar Chouhan, S., Adhimoolam, P., Das, R., et al., 2025b. The power of omics: genomics and proteomics approaches for crop improvement. International Journal of Advanced Biochemistry Research 9 (5S), 27–37. Available at https://doi.org/10.33545/26174693.2025.v9.i5sa.4300.

Leprêtre, M., OlivierGeffard, Espeyte, A., Faugere, J., Ayciriex, S., Salvador, A., et al., 2022. Multiple reaction monitoring mass spectrometry for the discovery of environmentally modulated proteins in an aquatic invertebrate sentinel species, Gammarus fossarum. Environmental Pollution 315, 120393. Available at https://doi.org/10.1016/j.envpol.2022.120393.

Leprêtre, M., DegliEsposti, D., Sugier, K., Espeyte, A., Gaillard, J.C., Delorme, N., et al., 2023. Organ-oriented proteogenomics functional atlas of three aquatic invertebrate sentinel species. Nature Research, France Scientific Data 10 (1). Available at https://doi.org/10.1038/s41597-023-02545-w, https://www.nature.com/sdata/.

Li, J., Levitan, B., Gomez-Jimenez, S., Kültz, D., 2018. Development of a Gill assay library for ecological proteomics of threespine sticklebacks (Gasterosteus aculeatus). Molecular & Cellular Proteomics 17 (11), 2146–2163. Available at https://doi.org/10.1074/mcp.ra118.000973.

Li, J., Lusher, A.L., Rotchell, J.M., Deudero, S., Turra, A., Bråte, I.L.N., et al., 2019. Using mussel as a global bioindicator of coastal microplastic pollution. Environmental Pollution 244, 522–533. Available at https://doi.org/10.1016/j.envpol.2018.10.032, www.elsevier.com/inca/publications/store/4/0/5/8/5/6.

Lopez-Casado, G., Covey, P.A., Bedinger, P.A., Mueller, L.A., Thannhauser, T.W., Zhang, S., et al., 2012. Enabling proteomic studies with RNA-Seq: the proteome of tomato pollen as a test case. Proteomics 12 (6), 761–774. Available at https://doi.org/10.1002/pmic.201100164.

Lopez-Pedrouso, M., Lorenzo, J.M., De Dios, J., Alche, R.M., Franco, D., 2023. Advanced proteomic and bioinformatictools for predictive analysis of allergens in novel foods. Biology 12, 714.

Marigómez, I., Olivares, M., Zaldibar, B., Benito, D., Soto, M., Blanco-Rayón, E., et al., 2024. Biomarker responsiveness in Norwegian Sea mussels, Mytilus edulis, exposed at low temperatures to aqueous fractions of crude oil alone and combined with dispersant. Science of The Total Environment 946, 174186. Available at https://doi.org/10.1016/j.scitotenv.2024.174186.

Martinez, I., Sánchez-Alonso, I., Piñeiro, C., Careche, M., Carrera, M., 2020. Protein signatures to trace seafood contamination and processing. Foods 9 (12), 1751. Available at https://doi.org/10.3390/foods9121751.

Martyniuk, C.J., Alvarez, S., 2013. Proteome analysis of the fathead minnow (Pimephalespromelas) reproductive testes. Journal of Proteomics 79, 28–42. Available at https://doi.org/10.1016/j.jprot.2012.11.023.

Mittra, P.K., Rahman, M.A., Roy, S.K., Kwon, S.J., Yun, S.H., Kun, C., et al., 2024. Deciphering proteomic mechanisms explaining the role of glutathione as an aid in improving plant fitness and tolerance against cadmium-toxicity in Brassica napus L. Journal of Hazardous Materials 471. Available at https://doi.org/10.1016/j.jhazmat.2024.134262, https://www.sciencedirect.com/science/journal/03043894.

Monaci, L., De Angelis, E., Guagnano, R., Ganci, A.P., Garaguso, I., Fiocchi, A., et al., 2020. Validation of a MS basedproteomics method for milk and egg quantification in cookies at the lowest VITAL levels: an alternative to theuse of precautionary labeling. Foods 9, 1489.

Nissa, M.U., Banerjee, A., Pinto, N., Srivastava, S., 2023. Proteo-Metabolomic Technologies in Context of Fishery/Aquacultural Research and Applications. Elsevier BV, pp. 259–274. Available at http://doi.org/10.1016/b978-0-323-91240-2.00015-4.

Omar, S.A., Ashokhan, S., Abdul Majid, N., Karsani, S.A., Lau, B.Y.C., Yaacob, J.S., 2024. Enhanced azadirachtin production in neem (Azadirachtaindica) callus through NaCl elicitation: insights into differential protein regulation via shotgun proteomics. Pesticide Biochemistry and Physiology 199. Available at https://doi.org/10.1016/j.pestbp.2024.105778, https://www.sciencedirect.com/science/journal/00483575.

Price, P., 1991. Standard definitions of terms relating to mass spectrometry: a report from the committee on measurements and standards of the american society for mass spectrometry.

Journal of the American Society for Mass Spectrometry 2 (4), 336–348. Available at https://doi.org/10.1016/1044-0305(91)80025-3.

Quiton-Tapia, S., Trueba-Santiso, A., Garrido, J.M., Suarez, S., Omil, F., 2023. Metalloenzymes play major roles to achieve high-rate nitrogen removal in N-damo communities: lessons from metaproteomics. Bioresource Technology 385, 129476. Available at https://doi.org/10.1016/j.biortech.2023.129476.

Reardon, C., 2024. Environmental proteomics: a new frontier in microbial ecology and biotechnology. Journal of Petroleum & Environmental Biotechnology 15, 565.

Roch, S., Rebl, A., Wolski, W., Brinker, A., 2022. Combined proteomic and gene expression analysis to investigate reduced performance in rainbow trout (Oncorhynchus mykiss) caused by environmentally relevant microplastic exposure. Microplastics and Nanoplastics 2 (1). Available at https://doi.org/10.1186/s43591-022-00034-2.

Rudén, C., Adams, J., Ågerstrand, M., Brock, T.C.M., Poulsen, V., Schlekat, C.E., et al., 2016. Assessing the relevance of ecotoxicological studies for regulatory decision making. Integrated Environmental Assessment and Management 13 (4), 652–663.

Sarkar, A.K., Sadhukhan, S., 2022. Proteomics—a powerful tool for understanding saline stress response in germinating seed. In: Roy, S., Mathur, P., Chakraborty, A.P., Saha, S.P. (Eds.), Plant Stress: Challenges and Management in the New Decade. Advances in Science, Technology & Innovation. Springer, Cham. Available at https://doi.org/10.1007/978-3-030-95365-2_24.

Sharma, B.M., Bečanová, J., Scheringer, M., Sharma, A., Bharat, G.K., Whitehead, P.G., et al., 2019. Health and ecological risk assessment of emerging contaminants (pharmaceuticals, personal care products, and artificial sweeteners) in surface and groundwater (drinking water) in the Ganges River Basin, India. Republic Science of the Total Environment 646, 1459–1467. Available at https://doi.org/10.1016/j.scitotenv.2018.07.235, www.elsevier.com/locate/scitotenv.

Sidira, M., Agriopoulou, S., Smaoui, S., Varzakas, T., 2024. Omics-integrated approach (metabolomics, proteomics and lipidomics) to assess the quality control of aquatic and seafood products. Applied Sciences 14 (22), 10755. Available at https://doi.org/10.3390/app142210755.

Simmons, D.B.D., Cowie, A.M., Koh, J., Sherry, J.P., Martyniuk, C.J., 2019. Label-free and iTRAQ proteomics analysis in the liver of zebrafish (Danio rerio) following dietary exposure to the organochlorine pesticide dieldrin. Journal of Proteomics 202. Available at https://doi.org/10.1016/j.jprot.2019.04.012, http://www.elsevier.com.

Slaby, S., Geffard, A., Fisson, C., Bonnevalle-Normand, M., Allonier-Fernandes, A.-S., Amara, R., et al., 2024. Advancing environmental monitoring across the water continuum combining biomarker analysis in multiple sentinel species: a case study in the Seine-Normandie Basin (France). Journal of Environmental Management 358, 120784. Available at https://doi.org/10.1016/j.jenvman.2024.120784.

Sound, J.K., Wedgwood, H.E., Afinanisa, Q., Overton, T.W., Leney, A.C., 2025. A protein-centric mass spectrometry approach for species identification within harmful algal blooms. Journal of the American Chemical Society 147 (31), 27974–27980. Available at https://doi.org/10.1021/jacs.5c07419, http://pubs.acs.org/journal/jacsat.

Poulsen, J.S., Trueba-Santiso, A., Lema, J.M., GregersenEchers, S., Wimmer, R., Nielsen, J.L., 2023. Assessing labelled carbon assimilation from poly butylene adipate-co-terephthalate (PBAT) monomers during thermophilic anaerobic digestion. Bioresource Technology 385, 129430. Available at https://doi.org/10.1016/j.biortech.2023.129430.

Sun, S., Xie, S., Cheng, Y., Yu, H., Zhao, H., Li, M., et al., 2017. Enhancement of environmental hazard degradation in the presence of lignin: a proteomics study. Scientific Reports 7 (1). Available at https://doi.org/10.1038/s41598-017-10132-4.

Halder, T., Bhoite, R., Islam, S., Yan, G., Siddiqui, M.N., Kayess, Md.O., Siddique, K.H.M., 2025. Proteomics in allopolyploid crops: stress resilience, challenges and prospects. Proteomes 13 (4), 60. Available at https://doi.org/10.3390/proteomes13040060.

Tholey, A., Becker, A., 2017. Top-down proteomics for the analysis of proteolytic events - methods, applications and perspectives. Biochimica et Biophysica Acta (BBA) - Molecular Cell Research 1864 (11), 2191–2199. Available at https://doi.org/10.1016/j.bbamcr.2017.07.002.

Trapp, J., Geffard, O., Imbert, G., Gaillard, J.C., Davin, A.H., Chaumot, A., et al., 2014. Proteogenomics of gammarusfossarum to document the reproductive system of amphipods. Molecular and Cellular Proteomics 13 (12), 3612–3625. Available at https://doi.org/10.1074/mcp.M114.038851, http://www.mcponline.org/content/13/12/3343.full.pdf+html.

Truhaut, R., 1977. Ecotoxicology: objectives, principles and perspectives. Ecotoxicology and Environmental Safety 1 (2), 151–173.

Varunjikar, M.S., 2023. Proteomic Tools for Food and Feed Authentication. PhD Dissertation. University of Bergen, Bergen, Norway.

Wang, D.Z., Kong, L.F., Li, Y.Y., Xie, Z.X., 2016. Environmental microbial community proteomics: status, challenges and perspectives. International Journal of Molecular Sciences 17 (8). Available at https://doi.org/10.3390/ijms17081275, http://www.mdpi.com/1422-0067/17/8/1275/pdf.

Wilhelm, M., Schlegl, J., Hahne, H., Gholami, A.M., Lieberenz, M., Savitski, M.M., et al., 2014. Mass-spectrometry-based draft of the human proteome. Nature 509 (7502), 582–587. Available at https://doi.org/10.1038/nature13319, http://www.nature.com/nature/index.html.

Yun, J., Malvankar, N.S., Ueki, T., Lovley, D.R., 2016. Functional environmental proteomics: elucidating the role of a c-type cytochrome abundant during uranium bioremediation. ISME Journal 10 (2), 310–320.

Zhang, Y., Fonslow, B.R., Shan, B., Baek, M.C., Yates, J.R., 2013. Protein analysis by shotgun/bottom-up proteomics. Chemical Reviews 113 (4), 2343–2394. Available at https://doi.org/10.1021/cr3003533.

Chapter 2

Utility of proteomics in studying microbiome of polluted environments

Edgar Vázquez-Núñez[1], Ileana Vera Reyes[2], Hermes Pérez Hernández[3], and Alma Hortensia Serafin Muñoz[4]

[1]*Departamento de Ingenierías Química, Electrónica y Biomédica, División de Ciencias e Ingenierías, Aplicaciones Nano y Bio Tecnológicas para la Sostenibilidad (NanoBioTS), LaSAE, Universidad de Guanajuato, León, Guanajuato, México,* [2]*Departamento de Biociencias y Agrotecnología, Centro de Investigación en Química Aplicada, Saltillo, México,* [3]*Instituto Nacional de Investigaciones Forestales, Agrícolas y Pecuarias (INIFAP). Campo Experimental Edzná, Campeche, México,* [4]*Departamento de Ingeniería Civil y Ambiental, División de Ingenierías, Universidad de Guanajuato, Guanajuato, México*

2.1 Introduction

Microbiomes are communities of microorganisms and their collective genetic material that occupy a host or a habitat. They are present in soil, water, and air and live on and within both plants and animals, when their composition or function shifts, nutrient cycles, redox processes, and contaminant fates can change in ways that affect ecosystem health (Banerjee et al., 2025).

Conventional remediation often relies on excavation, pumping, and treatment, or strong chemical oxidants. These methods can deliver quick concentration drops; however, they usually require high energy and reagents, and may move contaminants rather than resolve the risk (Singh and Ram, 2022). Bioremediation adopts an opposite approach. It uses the metabolic breadth of microorganisms and plants to detoxify, immobilize, and mineralize pollutants. These communities perform dehalogenation, methanogenesis, denitrification, sulfate reduction, and cometabolism. They also exchange metabolites and genes and coordinate their activities through cell signaling. However, most environmental microorganisms cannot be cultured, and many of their roles and interactions remain unclear (Muhammad et al., 2024).

Omics tools allow us to interrogate these systems across the biological layers (Fig. 2.1). Genomics and metagenomics reveal genetic potential, including catabolic islands, mobile elements, and stress-response genes.

Proteomics Driven Bioremediation. DOI: https://doi.org/10.1016/B978-0-443-45294-9.00020-5

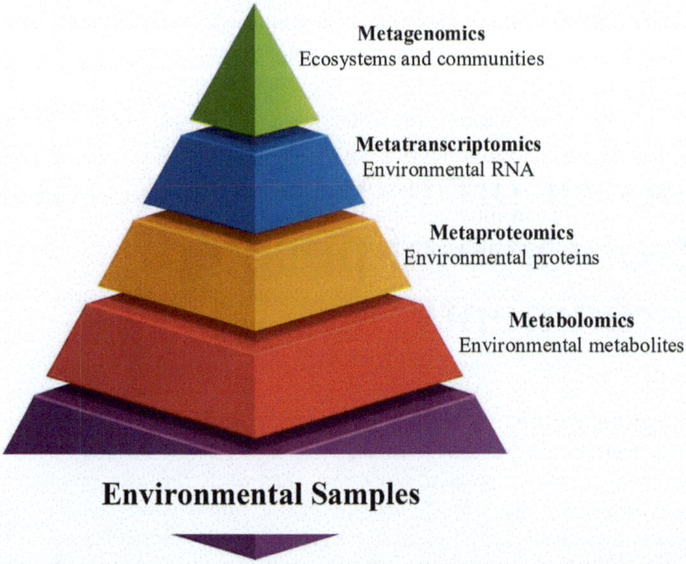

FIGURE 2.1 **Conceptual framework of omics sciences applied to environmental systems, from molecular levels to ecosystem processes.** The conceptual framework of omics sciences connects molecular level information to ecosystem-scale processes in environmental systems by integrating multiple biological layers. Environmental samples provide the foundation, capturing complex matrices such as soils, sediments, and water. Metabolomics and metaproteomics reveal active biochemical reactions and the enzymes driving metabolism and stress responses. Metatranscriptomics captures regulatory dynamics and short-term adaptations to environmental change. Metagenomics integrates these signals to explain community structure, functional potential, and ecosystem-level outcomes.

Transcriptomics and metatranscriptomics show which parts of that potential are active under particular conditions and expose the regulatory states. Proteomics and metaproteomics move closer to phenotype by quantifying the enzymes that execute metabolism and by capturing posttranscriptional regulation, protein modification, and protein interaction networks that mRNA alone cannot explain. Metabolomics closes the loop by identifying substrates, intermediates, and products, and indicating whether the transformations actually reduce hazards (Kesheri et al., 2024).

Each layer addresses a distinct question: genomes disclose capacity, transcripts indicate intention, proteins demonstrate capability, and metabolites reveal the consequences. When combined, these data link cause and effect from gene to pathway to process. In polluted habitats, this integrated view clarifies how communities assemble under stress, how functional guilds share or specialize in tasks, and how strategies such as syntrophy, cross-feeding, quorum signaling, and horizontal gene transfer sustain biodegradation in complex matrices (Kumar et al., 2021). It also exposes missing steps in canonical pathways, points to novel enzymes and electron transfer modules, and yields practical biomarkers for stress, toxicity, and remediation purposes.

Proteomics is particularly effective for mapping the physiological responses to stress. Comparative protein profiles under gradients of temperature, salinity, metals, oxidants, or organic toxicants consistently reveal coordinated shifts in chaperones, transporters, redox enzymes, and central carbon networks (Singh et al., 2021). In engineered systems, such as activated sludge, wetlands, and reactive barriers, metaproteomics extends these insights to communities. It can resolve the division of electron donor and acceptor use and the activation of modules that drive nitrification and denitrification, phosphorus removal, methanogenesis, and metal reduction (Shi et al., 2021). When paired with metabolomics, these findings verify pathway reconstructions and identify rate-limiting steps and off-target reactions that undermine the treatment goals.

Environmental proteomics and metaproteomics face several distinct challenges. Environmental matrices are chemically complex and poor in extractable proteins. Minerals and humic substances bind to proteins and co-extract inhibitors. Community complexity and the prevalence of unsequenced taxa challenge peptide identification and protein inference (Chaurasia et al., 2025). However, improvements in sample preparation, high-resolution mass spectrometry, spectral libraries, open search strategies, cross-species matching, and robust error control are expanding the coverage. Simultaneously, advances in long-read sequencing and assembly have produced better metagenome-assembled genomes and pangenomes, which strengthen the references needed to interpret environmental proteomes (Nebauer et al., 2024).

The practical benefits are already apparent. Culture-independent detection of key reductive dehalogenases, oxygenases, multiheme cytochromes, and metal transporters can inform the design and monitoring of biostimulation and bioaugmentation. Metaproteomic biomarkers distinguish functional resilience from simple taxonomic persistence after disturbances such as desiccation, freeze-thaw cycles, or starvation (Chaurasia et al., 2025). Multiomics case studies in sludges, soils, sediments, aquifers, and host-associated microbiomes, including plant roots and animal guts, now support ecosystem assessments that rely on mechanisms rather than bulk chemistry.

The next step is to ensure reliable integration and predictions. Statistical and machine learning frameworks that fuse multiomics with a geochemical context are beginning to infer metabolic fluxes and community-level emergent properties (Van Den Bossche et al., 2025). Genome-scale metabolic models and other constraint-based approaches, parameterized with proteomic abundance and metabolite availability, generate testable predictions about pathway use, nutrient and electron flow, and interaction networks under field conditions (Nebauer et al., 2024). As portable sequencing and rapid mass spectrometric phenotyping mature, timely omics will allow adaptive control of engineered bioprocesses and more nuanced risk assessment.

This chapter introduces environmental pollution as a global challenge and highlights how bioremediation can complement conventional approaches to address it. It examines how contaminants reshape microbial and plant-associated communities and how these systems adapt to stressors such as heavy metals. Proteomics and metaproteomics are powerful tools for understanding toxicity, biodegradation, community function, and ecological capacity in diverse environments. This chapter also surveys the wider omics toolkit and outlines strategies for integrating genomics, transcriptomics, proteomics, metabolomics, lipidomics, and fluxomics. Applications include pollution biomarkers, functional enzymes, and multiomics workflows for remediation and ecosystem restoration. We conclude by addressing the current technical and bioinformatic limitations and the emerging tools that are most likely to overcome these limitations. Together, these perspectives shift environmental microbiology from describing "who is there" to predicting "what they will do and under which constraints," aligning with the goals of mechanism-driven, sustainable remediation.

2.2 Environmental pollution and global challenges

Environmental quality emerges from the interaction of water, land, air, and living systems, with human activity dominating this interaction. Industrial expansion, urban growth, and intensive agriculture concentrate pollutants in air, water, and soil while depleting natural resources faster than they can recover (Rathod et al., 2024). The result is a coupled set of pressures that includes climate warming, acid deposition, ozone layer alteration, eutrophication, and the loss of biodiversity. These processes do not occur in isolation from one another. Emissions that degrade air quality also drive climate forcing and acidification, whereas nutrient and metal loads reshape food webs and weaken the capacity of ecosystems to buffer further disturbances (Mir et al., 2023).

This burden is unevenly distributed across the regions. Economies with high consumption generate large emissions and complex waste streams, whereas many rapidly growing regions confront contamination with limited infrastructure for treatment and safe disposal (Fuller et al., 2022). In both contexts, pollution alters the composition and function of microbiomes in the soil, sediment, water, and rhizosphere. Communities reorganize under stress, often with the enrichment of tolerant guilds, loss of sensitive taxa, and the reconfiguration of trophic links. Because these communities regulate redox conditions, carbon and nutrient fluxes, and the transformation or sequestration of contaminants, their responses determine whether ecosystems recover, adapt, or cross ecological thresholds (Philippot et al., 2024).

Pollution problems, such as acidified lakes, metal-impacted aquifers, and hydrocarbon-contaminated sediments, illustrate this utility. Protein inventories recover biomarkers of oxidative stress, metal resistance, oxygenase and dehalogenase activity, and multiheme cytochromes that mediate metal reduction (Bhandari et al., 2021). When these signals are interpreted together with geochemistry and metabolite profiles, they distinguish true detoxification from mere redistribution of contaminants, identify rate-limiting steps, and reveal off-target reactions that can create secondary risks (Yin et al., 2021). In engineered systems, the same measurements guide biostimulation and bioaugmentation by indicating whether the intended pathways are expressed and whether inhibitory stresses suppress performance.

Global environmental challenges are inseparable from microbial ecology, and microbial ecology in polluted settings is best understood through protein functions (Smercina et al., 2021). By placing proteomics at the center of assessment and design, we can move from describing who is present to predicting what they do under constraints and then use this knowledge to monitor, optimize, and scale sustainable bioremediation.

2.3 Microbial communities in polluted environments

2.3.1 Microbial and plant-mediated remediation as a functional continuum

Biological remediation of polluted environments emerges from the coordinated activity of microorganisms and plants operating across interconnected spatial, metabolic, and ecological scales. Microorganisms dominate the initial biochemical transformation of contaminants, whereas plants contribute through uptake, sequestration, stabilization, and metabolically assisted degradation, often in close association with rhizosphere microbiomes (Rathod et al., 2024). Rather than functioning as isolated agents, microbes and plants form integrated remediation systems in which microbial metabolism, plant physiology, and soil or sediment chemistry are closely coupled. Understanding this functional continuum requires molecular tools capable of resolving active pathways, regulatory responses, and metabolic fluxes under contaminant stress, an area where proteomics and integrated omics approaches have become central.

2.3.1.1 Microbial degradation of organic pollutants: from pathways to proteomes

Microorganisms degrade organic pollutants through highly structured enzymatic cascades that convert xenobiotics into assimilable or mineralized products. Aerobic degradation relies on oxygen-dependent enzymes, including monooxygenases, dioxygenases, and peroxidases, which initiate substrate activation and channel intermediates into β-oxidation and central carbon metabolism (Elyamine et al., 2021; Kumari and Das, 2023). In contrast, anaerobic degradation in anoxic soils and sediments proceeds via alternative activation mechanisms, such as fumarate addition, supported by nitrate-, sulfate-, or iron-reducing bacteria (Tomei et al., 2021). Although anaerobic pathways are kinetically slower, they are ecologically indispensable for pollutant turnover in stratified or compacted environments (Valsala et al., 2026).

Proteomics provides a functional lens that distinguishes potential and realized degradation. While genomics identifies catabolic gene clusters, proteomic profiling reveals which enzymes are expressed, assembled, and catalytically active under specific environmental conditions. Studies on pesticide- and hydrocarbon-degrading bacteria have demonstrated that protein-level regulation often diverges from transcript abundance, particularly under stress conditions where energy limitation, oxidative damage, or co-contaminants constrain metabolism (Aswathi et al., 2021; Gangola et al., 2023). Thus proteomics resolves the operational state of microbial degradation networks by linking enzyme abundance, posttranslational modification (PTMs), and metabolic integration.

2.3.1.2 *Microbial detoxification of xenobiotics as a multiphase metabolic strategy*

The microbial transformation of pesticides and synthetic organic pollutants follows a conserved multiphase detoxification logic. Phase I reactions introduce functional groups through oxidation, reduction, or hydrolysis, mediated by esterases, hydrolases, and cytochrome P450 monooxygenases (CYPs). Phase II conjugation reactions increase solubility and reduce toxicity through linkage to glutathione, sugars, or amino acids, whereas Phase III processes sequester or mineralize conjugates into nontoxic end products (Bose et al., 2021).

Proteomic analyses have captured the coordinated upregulation of these enzyme systems and revealed the intersection between detoxification and stress physiology. Exposure to xenobiotics consistently induces proteins involved in redox homeostasis, chaperone activity, and energy metabolism, underscoring that biodegradation is metabolically demanding and tightly coupled with cellular survival strategies (Tan et al., 2022). In this context, proteomics bridges enzymology and ecology by demonstrating how microbial cells allocate resources between detoxification, growth, and competition in polluted environments.

2.3.1.3 *Taxonomic diversity, functional redundancy, and community-level omics*

The degradation of complex pollutant mixtures is rarely the result of a single organism. Instead, they emerge from functionally redundant and metabolically complementary microbial communities. Bacterial genera such as Pseudomonas,

Burkholderia, Cupriavidus, and Rhodococcus exemplify this principle by combining plasmid-borne catabolic operons with inducible regulatory systems that respond rapidly to xenobiotic exposure (Banu et al., 2024; Guerrero Ramírez et al., 2023). Filamentous and white-rot fungi extend this capacity through extracellular oxidative enzymes capable of attacking structurally recalcitrant pollutants (Shanthi Kumari et al., 2021).

Community-level proteomics and metaproteomics are particularly powerful in this context because they identify the taxa and enzymes that dominate degradation in situ rather than under laboratory enrichment. These approaches reveal metabolic handoffs between organisms, cooperative degradation chains, and stress-sharing mechanisms that underpin the resilience of ecosystems (Philippot et al., 2024). Such insights are essential for translating strain-level capabilities into reliable, field-scale bioremediation.

2.3.1.4 Plant–microbe interfaces and omics-guided phytoremediation

Plants extend microbial remediation through the direct uptake and transformation of contaminants and by structuring rhizosphere microbiomes that enhance degradation and metal immobilization. Phytoremediation strategies, including phytoextraction, phytostabilization, rhizofiltration, phytodegradation, and phytovolatilization, depend on tightly regulated protein networks that govern transport, chelation, sequestration, and stress tolerance (Edgar et al., 2021). Proteomics has become central to resolving these networks, revealing how plants remodel photosynthesis, antioxidant defense, hormone signaling, and energy metabolism in response to metal stress (Yuan et al., 2024).

Comparative proteomic studies have consistently shown that effective phytoremediators upregulate metal transporters, chelators, heat shock proteins, and redox enzymes more rapidly and coherently than sensitive genotypes (Ningombam et al., 2024). These proteomic signatures link physiological performance to molecular mechanisms and provide actionable targets for crop selection, breeding, and genetic engineering aimed at improving phytoremediation efficiency.

2.3.1.5 Integration of proteomics with multiomics for systems-level understanding

While proteomics captures functional execution, its full explanatory power emerges when it is integrated with genomics, transcriptomics, metabolomics, and metallomics. Genomics defines the metabolic potential of microbial and plant systems, transcriptomics reveals inducible regulatory programs, metabolomics traces transformation intermediates and chelation chemistry, and metallomics resolves metal speciation and metal–protein interactions (Zhou et al., 2022). Together, these layers support mechanistic models in which contaminant exposure triggers coordinated molecular responses that reshape metabolism without the catastrophic loss of growth or ecosystem function.

In polluted environments, such integrated omics frameworks move remediation research beyond descriptive observations toward predictive and design-oriented strategies. They enable the rational assembly of microbial consortia, optimization of plant–microbe partnerships, and anticipation of ecological constraints that limit the field performance (Singh et al., 2021). In this sense, proteomics is not merely an analytical tool but a central organizing principle for understanding and engineering biological remediation systems.

2.4 Proteomics in environmental microbiology

Proteomics, a large-scale study of proteins and their structures, functions, and interactions, has emerged as a cornerstone of modern microbiology and system biology. Proteins are the primary effectors of cellular processes, acting as enzymes, structural components, signaling molecules, and regulators of gene expression (Li et al., 2025a). Unlike genomics, which provides a static blueprint of genetic potential, proteomics offers a dynamic and functional perspective, revealing how organisms respond to internal and external stimuli at the molecular level of the organism. This distinction is critical because the presence of a gene does not guarantee its expression or the activity of the corresponding protein. Therefore proteomics bridges the gap between genotype and phenotype, enabling researchers to understand the actual biological function rather than mere potential (Wang and Kong, 2016).

One of the fundamental principles of proteomics is its ability to capture the complexity of protein expression under specific physiological and environmental conditions. This capability is particularly relevant in environmental microbiology because microbial communities often inhabit ecosystems with fluctuating levels of nutrients, temperatures, pH, and pollutants. Through proteomic profiling, scientists can identify the proteins involved in stress responses, metabolic adaptation, and interspecies communication (Han et al., 2025). These insights help elucidate the mechanisms of resilience and survival, which are essential for maintaining ecosystem stability and functionality. Key principles include the following: (1) Protein complexity: Proteomes are dynamic and vary with environmental conditions, developmental stages, and stress responses; (2) PTMs: Proteins undergo chemical modifications (e.g., phosphorylation, glycosylation, and acetylation) that regulate activity and interactions; and (3) Quantitative and qualitative analysis: Proteomics aims to identify proteins and measure their relative or absolute abundance (Mandal et al., 2025; Fig. 2.2.

Proteomics also plays a pivotal role in understanding the biogeochemical cycles and ecological processes. Microorganisms are key drivers of nutrient cycling, pollutant degradation, and energy flow in natural ecosystems. By identifying enzymes and regulatory proteins associated with processes such as

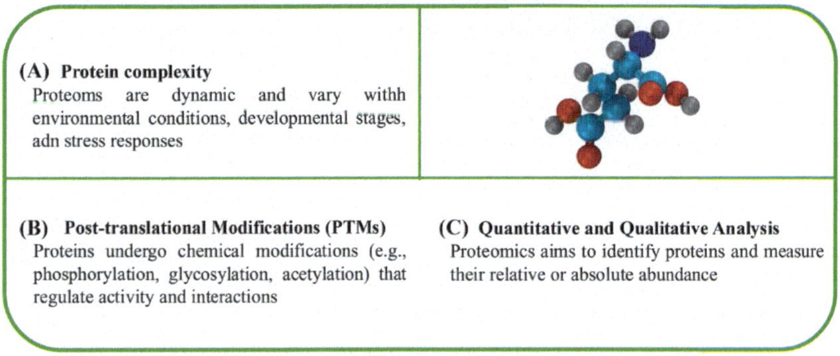

FIGURE 2.2 **Key Principles of proteomics for ecosystem resilience. (A) Protein Complexity: Proteomes are dynamic and vary according to environmental conditions, developmental stages, and stress responses. (bB) Posttranslational modifications (PTMs): Proteins undergo chemical modifications, such as phosphorylation, glycosylation, and acetylation, which regulate their activity and interactions. (C) Quantitative and Qualitative Analysis: Proteomics aims to identify proteins and measure their relative or absolute abundance, providing insights into functional responses within ecosystems.** Core principles of proteomics in environmental systems, illustrating protein complexity and dynamics, posttranslational modifications that regulate activity and interactions, and quantitative and qualitative analyses that link protein abundance to functional environmental responses.

nitrogen fixation, carbon sequestration, and bioremediation, proteomic studies provide actionable knowledge for environmental management. For example, detecting proteins linked to hydrocarbon degradation can inform strategies for cleaning oil spills, whereas profiling stress-response proteins can predict microbial behavior under climate-induced changes (Jin and Sun, 2025).

Technological advances have expanded the scope of proteomics beyond the mere identification of proteins. Modern approaches integrate quantitative analysis, PTM mapping, and interaction networks to provide a holistic view of cellular function. Techniques such as mass spectrometry, two-dimensional gel electrophoresis, and protein microarrays allow researchers to analyze thousands of proteins simultaneously, uncovering patterns that reflect the metabolic states and signaling pathways. Furthermore, coupling proteomics with other omics disciplines, such as genomics, transcriptomics, and metabolomics, enables a systems-level understanding of microbial communities and their roles in global ecological processes (Navarro et al., 2025).

In summary, the principles of proteomics highlight its dynamic, integrative, and application-oriented nature. By focusing on proteins, the true executors of biological activity, proteomics provides unparalleled insights into microbial physiology, adaptation and ecological impact. These principles advance fundamental science and support practical solutions for environmental challenges, biotechnological innovation, and sustainable resource management.

2.4.1 Principles and methodologies of proteomic analysis

2.4.1.1 Principles of proteomic analysis

Proteomics encompasses the large-scale study of the entire protein complement (proteome) of an organism or microbial community, including the identification, quantification, and characterization of proteins and their PTMs. Unlike genomic data, which reflects potential functionality, proteomics provides direct insights into cellular processes and phenotypic states. The proteome is inherently dynamic and responds to environmental stressors such as temperature, pH, pollutants, and nutrient availability. Modern proteomic workflows integrate sample preparation, protein separation, identification, and quantification, supported by advanced bioinformatics for data interpretation (Chen et al., 2022).

Sample preparation: Efficient protein extraction is critical, particularly from environmental samples containing complex mixtures from diverse taxa. Chemical lysis combined with detergents remains the standard method for solubilizing proteins from soil, water, or biofilm matrices (Kratter et al., 2025). Fractionation removes contaminants and enriches low-abundance proteins, thereby improving downstream sensitivity.

Protein separation: Two-dimensional gel electrophoresis (2D-PAGE) separates proteins based on their isoelectric point and molecular weight, while liquid chromatography (LC) provides high-resolution separation before mass spectrometry (MS) analysis (Singh et al., 2025).

Protein identification: Mass spectrometry is a cornerstone of proteomics. Bottom-up approaches digest proteins into peptides before MS analysis, whereas top-down strategies analyze the intact proteins. The coupling of LC with tandem MS (LC-MS/MS) enhances the sensitivity and resolution. Recent advances in acquisition strategies, such as data-independent acquisition (DIA), have improved reproducibility and proteome coverage (Kazakova et al., 2024).

Quantification approaches: Label-based methods (e.g., SILAC, iTRAQ, TMT) enable relative quantification, while label-free approaches rely on spectral counting or ion intensity measurements. DIA workflows now allow the simultaneous quantification of thousands of proteins from limited sample amounts (Wilkins et al., 2024).

Bioinformatics and data analysis: Proteomic datasets require robust computational pipelines for protein identification, functional annotation, and pathway mapping. Metaproteomics, which is the application of proteomics to microbial communities, leverages sequence databases and machine learning algorithms to interpret complex spectra (Heyer et al., 2025). Integration with other omics (genomics and metabolomics) facilitates system-level insights into ecosystem functioning.

In addition, there are tools such as database searches that compare mass spectra with protein databases. Functional annotation of biological functions and pathways, as well as integration with other omics, is also available. The latter combines proteomics, genomics, and metabolomics to obtain information at the system-level.

2.4.1.2 Methodologies in environmental microbiology

Environmental proteomics has emerged as a transformative approach for characterizing microbial functionality in situ, enabling the direct analysis of proteins from environmental matrices and overcoming the limitations of culture-dependent microbiology. By providing a functional perspective that complements genomic and transcriptomic data, proteomics reveals active metabolic pathways, stress responses, and ecological interactions that are critical for biogeochemical processes (Van Den Bossche et al., 2025). These methodologies are illustrated in Fig. 2.3.

Shotgun proteomics: It employs enzymatic digestion of proteins into peptides, followed by LC-MS/MS, enabling comprehensive identification in

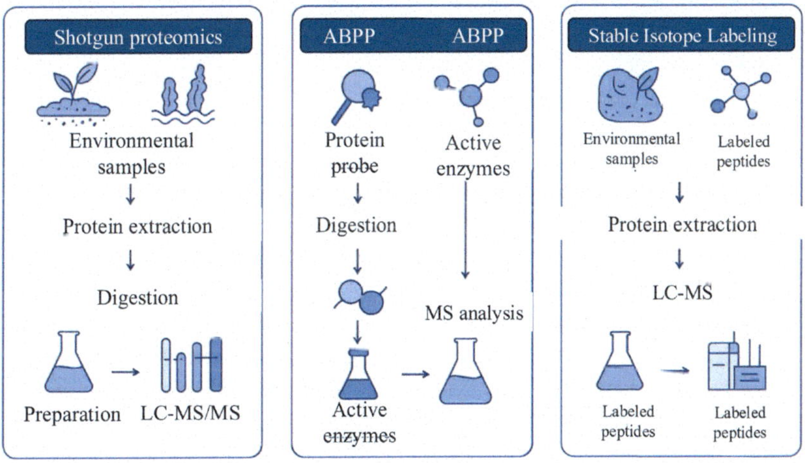

FIGURE 2.3 **Conceptual overview of environmental proteomic methodologies. This diagram illustrates three core approaches used in environmental proteomics: (1) shotgun proteomics, which involves protein extraction from environmental samples (e.g., soil, biofilms), enzymatic digestion, and subsequent LC-MS/MS analysis for high-throughput protein identification; (2) activity-based protein profiling (ABPP), where chemical probes selectively bind to active enzymes, enabling functional characterization through mass spectrometry without prior sequence annotation; and (3) Stable Isotope Labeling, which incorporates isotopic tags into peptides to monitor protein turnover and metabolic flux via LC-MS. These methodologies collectively provide insights into microbial functionality, stress responses, and biogeochemical processes in complex ecosystems.** Overview of key environmental proteomics methodologies, illustrating shotgun proteomics for broad protein profiling, ABPP for identifying catalytically active enzymes, and stable isotope labeling approaches to trace protein dynamics and metabolic activity in environmental samples.

complex samples. This approach is widely applied in metaproteomics to profile microbial communities in soils, sediments, and aquatic systems (Dashora et al., 2024).

Activity-based protein profiling (ABPP): It targets active enzymes using chemical probes that covalently bind to catalytic residues, providing functional insights beyond sequence homology.

Environmental ABPP (eABPP): Integrating metagenomics with MS-based profiling enables the direct identification of active enzymes without prior annotation, which is a breakthrough for enzyme bioprospecting in extreme habitats (Ninck et al., 2024).

Stable isotope labeling: Techniques such as SILAC and ^15 N labeling allow the tracking of protein turnover and metabolic flux, offering dynamic insights into microbial adaptation to environmental stressors (Marlow et al., 2016).

Environmental proteomics has been pivotal in elucidating the mechanisms of microbial resistance to pollutants. For example, metaproteomic analysis of acid mine drainage biofilms revealed the upregulation of efflux pumps and metallothioneins, which are key proteins conferring tolerance to heavy metals (Werner et al., 2025). These findings inform bioremediation strategies by identifying the functional biomarkers of resilience. Similarly, eABPP applied to hot spring microbiomes in Kamchatka uncovered thermostable serine hydrolases with industrial potential, demonstrating the power of function-driven enzyme discovery in extreme environments (Ninck et al., 2024).

In other studies, metaproteomic profiling of microbial consortia colonizing microplastics in the North Atlantic revealed the enrichment of oxidative stress proteins and enzymes involved in xenobiotic degradation, suggesting adaptive mechanisms for pollutant breakdown (Ginebreda et al., 2023). Proteomic analysis of soils exposed to neonicotinoids has revealed an increased abundance of detoxification enzymes, including glutathione S-transferases and ATP-binding cassette transporters, highlighting the microbial strategies for pesticide degradation (Dashora et al., 2024).

Using the urban wastewater microbiome approach, shotgun proteomics was applied to activated sludge communities. This demonstrated the upregulation of nitrifying enzymes and antibiotic-resistant proteins. This provides insight into the efficiency of wastewater treatment and the risk of emerging resistance (Van Den Bossche et al., 2025). Metaproteomic studies of microbial mats in oil-contaminated sediments have revealed elevated expression of alkane monooxygenases and dioxygenases, which are key enzymes for hydrocarbon degradation, supporting targeted bioremediation approaches (Pan et al., 2024). An interesting study on the impact of climate change on Arctic microbiomes demonstrated that proteomic surveys of thawing permafrost environments detected stress-related chaperones and enzymes associated with methane metabolism. This indicates a shift in microbial functionality under warming conditions.

In aquatic ecosystems, proteomic profiling of fish bile has emerged as a sensitive biomarker for identifying contamination. The differential expression of heat-stable proteins in bile samples from polluted and reference sites highlights the usefulness of proteomics in environmental monitoring (Azevedo-Santos et al., 2024).

The combination of eABPP and metagenomics in deep-sea hydrothermal vents has identified novel lipases and proteases with high thermostability. This has expanded the range of enzymes available for industrial biotechnology (Ninck et al., 2024). The convergence of proteomics with genomics, metabolomics, and transcriptomics will provide system-level insights into ecosystem functioning. This advancement will promote bioremediation, environmental risk assessment, and enzyme bioprospecting (Van Den Bossche et al., 2025).

These case studies underscore the versatility of environmental proteomics in addressing global challenges, such as pollution, climate change, and sustainable bioprocessing. Advances in high-resolution MS, DIA, and machine learning-driven annotation will further enhance the discovery of functional biomarkers and biocatalysts in complex ecosystems (Savalia et al., 2025).

2.4.2 Proteins as biomarkers of stress, toxicity, and pollutant degradation

Proteins are dynamic indicators of cellular responses to environmental stressors, toxins, and pollutants. Their sensitivity and specificity have positioned them as key biomarkers in ecotoxicology and environmental monitoring (Liu et al., 2025).

profoundly impacts biological systems, triggering molecular responses that can be quantified using protein-based biomarkers. Unlike genomic markers, proteins reflect real-time physiological states, making them valuable for the early detection of pollutant exposure and degradation processes (Parascanu et al., 2021).

Stress-related proteins, such as heat shock proteins (HSPs), are synthesized in response to thermal, oxidative, and chemical stresses. Members of the HSP70 and HSP90 families stabilize protein folding and prevent aggregation, acting as molecular chaperones during pollutant-induced stress (Su and Zheng, 2025). Elevated HSP expression has been correlated with exposure to heavy metals and organic pollutants, providing sensitive biomarkers for environmental monitoring (Savalia et al., 2025). In aquatic organisms, increased HSP70 levels upon Cd exposure indicate cellular stress before observable physiological damage.

Proteins involved in detoxification, such as cytochrome P450 enzymes and metallothioneins, are widely used as toxicity biomarkers. Metallothioneins bind heavy metals, including cadmium and lead, mitigating their toxic effects and serving as indicators of metal contamination (Luz et al., 2025). Similarly, cytochrome P4501A induction in fish has been validated as a biomarker for exposure to organic pollutants (Abdallah et al., 2024). Recent proteomic studies reveal that oxidative stress-related proteins, including superoxide dismutase and catalase, exhibit altered expression under pollutant exposure, linking protein biomarkers to reactive oxygen species (ROS) pathways (Mandal et al., 2025).

In addition to stress and toxicity, proteins actively participate in pollutant degradation. Enzymes such as laccases and peroxidases catalyze the breakdown of aromatic hydrocarbons and pesticides. Metaproteomic analyses of wastewater have identified microbial enzymes involved in xenobiotic degradation, highlighting their potential as biomarkers of bioremediation efficiency (Tugui et al., 2025). These analyses detected nearly 200 human and microbial proteins, including enzymes linked to pharmaceutical degradation, demonstrating the feasibility of protein-based environmental monitoring. Despite their potential, protein biomarkers face challenges, including variability across species and environmental conditions and methodological limitations in proteomic analysis. The integration of these data with transcriptomic and metabolomic data can improve their reliability. Future research should focus on standardizing biomarker panels and employing high-throughput proteomics for large-scale monitoring (Lionetto et al., 2021).

In general, proteins are powerful biomarkers for evaluating environmental stress, toxicity, and pollutant degradation. Advances in proteomics and bioinformatics are expanding their applications in ecotoxicology and environmental management fields. Continued innovation will enable more precise and real-time monitoring of ecosystem health. Table 2.1 presents several studies related to proteins as environmental monitoring biomarkers.

2.4.3 Applications of proteomics in assessing ecosystem health

Proteomics has emerged as a transformative approach for evaluating the health of ecosystems. This functional perspective is critical for understanding ecosystem resilience under pressures such as pollution, climate change, and habitat degradation (Liu et al., 2025). Proteins are the primary executors of cellular functions, and their expression patterns reflect the physiological state influenced by environmental conditions. Mass spectrometry-based proteomics, coupled with DIA and advanced bioinformatics, enables the identification and quantification of thousands of proteins in complex environmental samples. These techniques have been integrated with metagenomics and metatranscriptomics to provide a multiomics view of ecosystem functioning, bridging the gap between microbial diversity and ecological processes (Heyer et al., 2025). Proteomics applications in ecosystem health include: (1) biomarker discovery, identifying stress-related proteins as indicators of pollution or climate stress; (2) biogeochemical cycling, profiling enzymes involved in carbon, nitrogen,

TABLE 2.1 Proteins as biomarkers in environmental monitoring.

Category	Protein/Enzyme	Function	Environmental application	References
General concept	Various proteins	Dynamic indicators of cellular responses	Key biomarkers in ecotoxicology and environmental monitoring	Liu et al. (2025)
General concept	Various proteins	Reflect real-time physiological states	Early detection of pollutant exposure and degradation processes	Parascanu et al. (2021)
Stress response	Heat shock proteins (HSP70, HSP90)	Stabilize protein folding; prevent aggregation	Biomarker for thermal, oxidative, and chemical stress; elevated under heavy metal exposure	Su and Zheng (2025)
Stress response	Heat shock proteins (HSP70)	Indicate cellular stress before physiological damage	Aquatic organisms exposed to cadmium	Savalia et al. (2025)
Detoxification	Metallothioneins	Bind heavy metals (Cd, Pb) to reduce toxicity	Indicator of metal contamination in aquatic organisms	Luz et al. (2025)
Detoxification	Cytochrome P450 (CYP1A)	Catalyzes the oxidation of organic pollutants	Biomarker for exposure to hydrocarbons and pesticides	Abdallah et al. (2024)
Oxidative stress	Superoxide Dismutase (SOD), Catalase	Neutralize reactive oxygen species (ROS)	Biomarker for oxidative damage caused by pollutants	Mandal et al. (2025)
Pollutant degradation	Laccases, peroxidases	Breakdown aromatic hydrocarbons and pesticides	Used in bioremediation and wastewater treatment	Tugui et al. (2025)
Future perspectives	Various proteins	Integration with transcriptomics and metabolomics	Standardization of biomarker panels and high-throughput proteomics for large-scale monitoring	Lionetto et al. (2021)

and sulfur cycles; and (3) pollutant degradation, detecting microbial proteins linked to xenobiotic metabolism and bioremediation (Kumar et al., 2023).

Several studies have examined this topic. For example, a study demonstrated that peptide biomarkers from ocean microbes can predict harmful algal blooms, providing an early warning system for managing marine ecosystems (Mudge et al., 2025). Other studies have analyzed the proteomic plasticity of Montipora capitata gametes after thermal bleaching and identified stress-response proteins that predict coral survival in the context of climate change (Timmins-Schiffman et al., 2025). Similarly, another study emphasized the use of metaproteomics to identify microbial factors affecting ecosystems and host health, thereby reinforcing its role in integrated environmental monitoring (Heyer et al., 2025). One study used soil metaproteomics to evaluate microbial enzyme profiles in the presence of heavy metals, correlating protein signatures with pollutant degradation efficiency (Pan et al., 2024). Conversely, Tugui et al. (2025) identified microbial enzymes involved in pharmaceutical degradation through metaproteomic surveys, thereby validating proteomics as a monitoring tool for engineered ecosystems (Tugui et al., 2025). By applying data standardization andFindable, Accessible, Interoperable, Reusable (FAIR) principles (Wilkinson et al., 2019), a study reported the progress of the ProteomeXchange consortium of proteomics resources in promoting open data practices and integrating machine learning for large-scale environmental proteomics (Deutsch et al., 2025).

Environmental samples pose significant challenges owing to microbial diversity, low protein abundance, and interference from complex matrices. Additionally, the lack of complete reference genomes for many environmental microorganisms complicates protein identification. Future advancements in sample preparation, high-throughput MS technologies, and integrative multiomics approaches will enhance the resolution and ecological interpretation of the proteomic data. Environmental proteomics provides a robust framework for linking microbial functionality to ecosystem processes, offering valuable insights into sustainable environmental management and climate change mitigation.

The integration of proteomics with other omics approaches, such as metagenomics, transcriptomics, and metabolomics, provides a holistic view of microbial ecosystems. Emerging technologies, such as DIA and single-cell proteomics, promise to improve sensitivity and resolution, enabling the study of rare microbial taxa and their functional roles. Furthermore, advancements in bioinformatics and machine learning will facilitate real-time ecological monitoring and predictive modeling of microbial responses to environmental change (Emwas et al., 2022).

Ultimately, environmental proteomics offers a robust framework for linking microbial functionality to ecosystem processes, thereby supporting strategies for pollution remediation, climate change mitigation, and sustainable resource management.

2.5 Metaproteomics: community-level insights

2.5.1 Concept and scope of metaproteomics

Metaproteomics is a comprehensive large-scale analysis of the entire protein complement expressed by a microbial community in its native environment. Unlike conventional proteomics, which focuses on individual organisms or isolated cultures, metaproteomics captures the proteomes of multiple taxa simultaneously. This community-level resolution enables the direct interpretation of microbial activity, ecological roles, and interspecies interactions under real environmental conditions (Tartaglia et al., 2020). By integrating high-resolution mass spectrometry, microbial ecology, and advanced bioinformatics, metaproteomics allows taxonomic identification and quantitative assessment of protein expression. This combination enables explicit links between microbial composition, metabolic function, and ecosystem processes. In environmental matrices, such as soil, this approach has become essential for elucidating the biochemical mechanisms associated with nutrient cycling, plant–microbe symbiosis, organic matter turnover, and microbial responses to abiotic stress.

A standard metaproteomic workflow comprises: (1) optimized protein extraction from complex matrices; (2) peptide separation and identification using LC-MS/MS; and (3) integration of multiomics datasets for taxonomic and functional annotation. International efforts, such as the Metaproteomics Initiative, promote the harmonization of analytical pipelines, including standardized data acquisition, quality control procedures, and FAIR-compliant data repositories (Van Den Bossche et al., 2025). These advances represent critical milestones toward achieving reproducibility and cross-study comparability.

Recent technological advancements have broadened the analytical scope of proteomics. Innovations such as de novo peptide sequencing, database-independent identification, and machine learning-based annotation have increased the detection of proteins from uncultured or poorly characterized taxa (Van Den Bossche et al., 2025). Artificial intelligence and integrative multiomics further enable the prediction of enzymatic activity, microbial interactions, and emergent community phenotypes directly from the protein signatures. Together, these advances position metaproteomics as a predictive, model-driven framework capable of monitoring biochemical flux in real-time.

Despite its rapid progress, several challenges remain to be addressed. Protein extraction from matrices rich in minerals, polysaccharides, or humic substances can significantly reduce recovery efficiency, whereas limited genomic references continue to constrain protein annotation. Environmental heterogeneity also affects the quantitative reproducibility. Future development requires the expansion of reference genomic catalogs, the generation of benchmarking datasets, and the integration of AI-driven interpretation workflows. As these gaps are addressed, metaproteomics is anticipated to become a transformative platform in microbial ecology, environmental biotechnology, and ecosystem monitoring.

2.5.2 Analytical approaches and challenges

Metaproteomics is a key tool for deciphering the functional dynamics of microbial communities. While metagenomics provides information on genetic potential, metaproteomics identifies and quantifies the proteins that are actively expressed, enabling a real-time snapshot of metabolic activity and ecological function (Banerjee et al., 2025). This functional resolution answers not only *who is present*, but also *what microorganisms are doing* within a system (Pavelescu et al., 2024).

2.5.2.1 *Sample preparation and protein extraction*

Environmental matrices (soil, sludge, and biofilms) exhibit high chemical and physical heterogeneities. Humic substances, extracellular polymers, and inorganic particles interfere with protein solubilization and MS ionization (Herbst et al., 2016). Effective extraction protocols combine mechanical and chemical lysis, followed by contaminant removal using filter-aided sample preparation (FASP), S-Trap, or equivalent purification strategies. Monitoring the extraction yield, digestion efficiency, and LC stability is essential to ensure analytical reproducibility (Van Den Bossche et al., 2025).

2.5.2.2 *Mass spectrometry acquisition strategies*

Reversed-phase nanoLC-MS/MS remains the dominant analytical method. DIA is emerging as the preferred strategy for comparative studies because of its deeper proteome coverage, reproducibility, and lower missing value rates compared with data-dependent acquisition (DDA) (Chaurasia et al., 2025). Regardless of the strategy, strict false discovery rate (FDR) control is mandatory for confident protein identification in LC-MS/MS analysis.

2.5.2.3 *Data analysis and bioinformatics*

The data analysis stage is the most complex stage. Environmental samples frequently lack complete genomic references because most microbial taxa are uncultured (Chaurasia et al., 2025). Metaproteomics, which involves the construction of sample-specific protein databases from metagenomes, improves identification depth and reduces taxonomic ambiguity. Functional annotation using KEGG, COG, or Gene Ontology enables metabolic reconstruction and identification of functional guilds. Emerging database independent approaches, including de novo peptide sequencing, have increased the detection of previously unknown microbial lineages (Van Den Bossche et al., 2025; Liu et al., 2025). These steps are illustrated in Fig. 2.4.

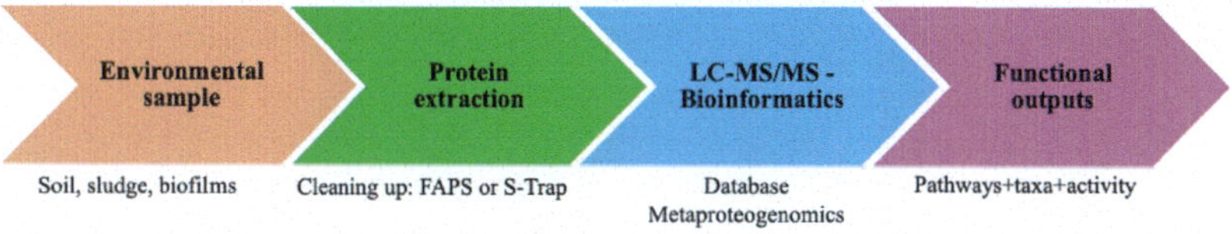

FIGURE 2.4 **Overview of the metaproteomics analytical workflow. Environmental samples (soil, sludge, biofilm, or other complex matrices) undergo optimized protein extraction and contaminant removal using lysis in combination with FASP or S-Trap purification systems. Proteins are digested into peptides and analyzed by LC-MS/MS using either data-dependent acquisition (DDA) for exploratory profiling or data-independent acquisition (DIA) for quantitative and comparative analyses. Bioinformatic processing integrates database searches, metaproteogenomics, and/or de novo sequencing to assign proteins to specific taxa and biological functions. The resulting protein profiles enable the reconstruction of metabolic pathways, identification of functionally active microbial guilds, and interpretation of community-level metabolic activity in situ in the gut.** General workflow of environmental metaproteomics, illustrating sample collection, protein extraction, mass spectrometry analysis, and bioinformatic interpretation to link community composition with functional activity in complex environmental matrices.

2.5.3 Case studies: sludge, soil, aquatic systems, and gut microbiota

To illustrate the versatility of metaproteomics, Table 2.2 summarizes representative studies from 2025 covering contrasting environments (sludge, soil, aquatic ecosystems, and gut microbiota). Each matrix requires specific analytical strategies owing to differences in biomass, matrix complexity, and genomic availability. Metaproteomics reveals metabolic pathways, pollutant degradation mechanisms, microbial stress responses, and host–microbe metabolic interactions across systems.

2.5.4 Linking community composition, function, and ecological capacity

Understanding environmental microbiomes requires an integrative perspective that connects community composition (which microorganisms are present), protein expression (which functions are performed), and ecological capacity (which ecosystem processes are supported). Metaproteomics identifies a subset of genes that are actively translated into proteins, providing direct evidence of biochemical activity in situ (Tartaglia et al., 2020).

Environmental conditions impose selective pressures that alter both microbial composition and functional protein profiles. For example, in anaerobic sludge systems, metaproteomics has revealed the differential expression of proteins associated with methanogenesis (Wintsche et al., 2018). In drought-affected soils, plants and microbes activate antioxidant and osmotic protection pathways (Tartaglia et al., 2020). In aquatic systems,

TABLE 2.2 Application of metaproteomics in contrasting environmental matrices: sludge, soil, aquatic systems, and gut microbiota

Environmental matrix	Research goal / Scientific question	Omics strategy (metaproteomics + others)	Key findings / Application	References
Industrial and municipal wastewater	Identify metabolic pathways and functional proteins enabling xenobiotic biodegradation using engineered microbial consortia.	Metaproteomics + metagenomics + AI-driven microbial design.	Identified key enzymes involved in the degradation of persistent xenobiotics; AI-assisted selection of optimal microbial consortia.	Renganathan et al. (2025)
Urban wastewater (public-health monitoring)	Track microbial and human protein biomarkers in wastewater for epidemiological surveillance.	Metaproteomics + human proteomics (low-abundance detection).	Detected microbial and human proteins with biomarker potential; supports wastewater-based epidemiology.	Tugui et al. (2025)
Anaerobic sludge / co-digestion systems	Profile shared and unique taxa during co-digestion systems differentiated by fatty-acid isomers.	Genome-centric metagenomics + metaproteomics.	Differential enrichment of taxa and proteins involved in methanogenesis and organic matter degradation.	Duan et al. (2025)
Low-biomass soils	Develop a robust protein extraction/enrichment protocol for metaproteomics in soils with extremely low biomass.	Method optimization + quantitative metaproteomics.	Enhanced recovery of proteins and detection of key functional enzymes in challenging soil matrices.	Abdullah et al. (2025)
Agricultural microbiome (rhizosphere / soil)	Understand functional interactions between soil microbiota and plant performance, stress tolerance, and nutrient cycling.	Metaproteomics + metagenomics + metabolomics.	Identified microbial metabolic pathways linked to plant productivity, nutrient availability, and environmental stress responses.	Yousuf et al. (2025)

Freshwater mollusks (aquatic ecosystem)	Investigate proteomic and phosphoproteomic profiles of freshwater mollusks as environmental sentinels and an alternative protein source.	Proteomics + phosphoproteomics.	Identified nutrient-related proteins and phosphorylation pathways linked to environmental response.	Liu et al. (2025)
Marine environments (biosensors from genomic/protein markers)	Use proteomics and genomics to develop next-generation biosensors for marine microbiome monitoring.	Proteomics + genomics (biomarker discovery).	Environmental proteins proposed as bio-reporters enabling real-time detection of biological changes in marine environments.	Singh and Ram (2022)
Gut microbiota (community coalescence in cocultures)	Determine proteomic outcomes of combining different human gut microbiomes.	Metaproteomics.	Identified dominant taxa and functional proteins determining microbial community assembly.	Sun et al. (2025)
Gut microbiota / intestinal inflammation (mouse model)	To evaluate how 2'-fucosyllactose modulates the immune and proteomic responses of the gut microbiota.	Proteomics and microbiome analysis (16 S sequencing).	2'-FL regulates immune pathways, reshapes the microbiota, reduces colitis symptoms, and identifies affected protein networks.	Dong et al. (2025)

protein signatures reveal the biosynthesis of toxins or the degradation of hydrocarbons during contamination events (Schultz et al., 2020). In healthy ecosystems, metaproteomics has been used to identify microbial enzymes involved in immunshort-chain and short-chain fatty-acid production (Li et al., 2025b).

By linking taxonomic identity to functional activity, metaproteomics enhances microbiome research beyond a simple descriptive inventory. Metaproteogenomics combines metagenomic potential with actual biological functions, allowing the quantification of microbial contributions to nutrient cycling, pollutant degradation, methane production, and plant growth promotion. Ultimately, the power of metaproteomics lies in its ability to connect microbial identity, biochemical functions, and ecological resilience.

2.6 Case studies and applications

Environmental proteomics is a fundamental tool for characterizing microbial responses in contaminated natural systems and identifying key proteins and enzymes associated with degradation processes. This section mentions some current examples that demonstrate proteomics not only as a strategy for identifying pollution biomarkers, but also for designing more precise restoration and bioremediation strategies based on the actual functionality of microbiomes.

2.6.1 Microbial biomarkers for pollution monitoring

The application of microbial biomarkers has emerged as an essential approach for assessing the environmental impact of pollutants in soils, sediments, and water bodies. Unlike the conventional physicochemical methods discussed in previous sections of this chapter, biomarkers provide information on the biological response and functional state of microbial communities to chemical stressors, such as heavy metals, hydrocarbons, and emerging contaminants. Therefore proteomic methodologies enable the identification of proteins whose abundance is linked to biogeochemical processes and microbial adaptation mechanisms, making them sensitive indicators of environmental changes.

For example, studies conducted in heavy metal-contaminated environments have demonstrated that microbial communities activate detoxification and oxidative stress pathways, which can be detected using proteomics. After isolating *Yarrowia lipolytica* from mining tailings, a significant overexpression of molecular chaperones (HSP70, HSP90), superoxide dismutase, and flux transporters associated with copper (Cu), cadmium (Cd), and chromium (Cr) was observed, identifying them as functional biomarkers of metal tolerance (Sánchez-Rojas et al., 2022). Recently, *Halomonas sp.* exposed to heavy metals were found to alter core metabolic pathways and the production of metabolites and transport proteins, demonstrating microbial adaptation mechanisms that could serve as functional biomarkers (Xu et al., 2024). In other similar studies, the proteomic response of *Penicillium chrysogenum* to lead (Pb) was investigated. The authors identified 43 differentially expressed proteins associated with metal stress, of which 13 were related to metabolism, 11 to cell signaling and processing, and 19 to information storage and processing. Researchers have highlighted the development of innovative molecular markers for heavy metal mycoremediation. They particularly emphasized the use of *P. chrysogenum* for research, as it provides valuable insights and enables the development of new molecular markers for metal bioremediation (Algahmadi et al., 2024).

Aromatic hydrocarbon pollution has been extensively studied using multiomics approaches. Following the ecological disaster in the Gulf of Mexico caused by the Deepwater Horizon oil spill, studies have revealed that seawater metaproteomes show increases in key enzymes for the degradation of alkanes and aromatics, primarily associated with genera such as Alcanivorax, Cycloclasticus, and Marinobacter. To date, the identified proteins are considered robust biomarkers for monitoring the progress of bioremediation processes in coastal and marine areas contaminated with oil (Kleindienst et al., 2015).

Several studies have indicated that the presence of hydrocarbons in the transition zone between oxic and anoxic conditions causes profound modifications in the geochemistry and microbial composition of iron mats. These changes include an increase in the oxidized forms of iron and a decrease in dissolved carbon and phosphate. Mats exposed to contamination show lower microbial diversity and are mostly composed of microorganisms associated with iron oxidation and hydrocarbon degradation. Environmental parameters such as dissolved oxygen, pH, and benzene levels significantly influenced the structure of the microbial community, with effects that varied seasonally. Taken together, the evidence suggests that these mats function as sensitive biomarkers and contribute to the natural remediation of hydrocarbons at the soil-water interface (Brooks and Field, 2024).

Therefore proteomic studies are novel for practical applications. For example, a metaproteomic study in a depleted natural gas field revealed active proteins in microbial communities, such as Methanomicrobiaceae, Methanothrix, and Smithella, which are responsible for methanogenesis and gas consumption, and have implications as functional biomarkers in industrial environments (Moore et al., 2025). In addition, by integrating proteomics, metabolomics, and morphological analysis to study the response of microalgae (*Ulothrix* sp.) to uranium (U) exposure, antioxidant and accumulation pathways were identified that could function as molecular biomarkers of radiochemical or chemical stress (Han et al., 2025).

In the field of wastewater and urban systems, environmental proteomics approaches have enabled the identification and characterization of proteins derived from microorganisms that serve as indicators of fecal contamination and markers of the activity of communities involved in the degradation of various organic and inorganic compounds. In one study, the authors reported that the identification of metabolic enzymes and structural proteins in wastewater can serve as an indicator of microbial load and the anthropogenic origin of contamination, and they proposed its use for environmental monitoring and wastewater-based epidemiology studies (Carrascal et al., 2023).

Reflecting on the examples mentioned here, we can highlight the considerable increase in studies over the last two decades that combine proteomics with other "omics" technologies (metabolomics and metagenomics) to define microbial biomarkers in contaminated environments. These studies demonstrate that the identified proteins respond to pathways of oxidative stress, metal transport, protein folding, and central metabolism, confirming that the most useful biomarkers are not always single proteins but rather functional networks. However, the validation of biomarkers in the laboratory and real-world environments provides sufficient evidence to integrate them into more functional and predictive environmental monitoring strategies.

2.6.2 Functional enzymes in biodegradation pathways

Microbiome activity in contaminated environments depends on the presence and regulation of potentially functional enzymes that catalyze key steps in the degradation of organic pollutants. In recent years, proteomic studies have identified proteins involved in the activation and transformation of hydrocarbons, aromatic compounds, and emerging contaminants, which is a key strength of using biomarkers in bioremediation assessment. Among the many studies, degradation biomarkers such as AlkB, LadA, monooxygenases, dioxygenases, and fumarate-additive enzymes have been identified (Perera et al., 2022).

Alkane monooxygenase enzymes, known as AlkB and LadA, are distinguished by their ability to introduce oxygen into aliphatic chains, producing alcohols and subsequently aldehydes and acids that enter central metabolism. Recent proteomic studies have confirmed that AlkB is specific to linear alkanes, notably its presence in marine bacteria of the genus Alcanivorax, which are effective at degrading hydrocarbons in marine spills (Guo et al., 2023). Similarly, LadA, a monooxygenase, has been identified in various bacteria, and its proteomic detection has proven to be a reliable indicator of long-chain alkane degradation in coastal and marine environments (Perera et al., 2022).

Other enzymes, such as ring dioxygenases, including catechol 1,2-dioxygenase and catechol 2,3-dioxygenase, are essential catalysts in the degradation of phenols, toluene, and other petroleum-derived aromatics. Likewise, fumarate-additive enzymes, such as benzylsuccinate synthase (BSS) and alkylsuccinate synthase (AssA), are effective for anaerobic degradation in aquifers and deep sediments contaminated with hydrocarbons (Laczi et al., 2020). Oxidoreductase enzymes are being studied from a proteomics perspective, and although research in this area is still ongoing, they show promise as efficient hydrocarbon degradation agents (Gofstein and Leigh, 2023).

A recent study demonstrated that long-chain enzymes present in the fungus *Aspergillus flavus* M11 actively participate in crude oil biodegradation. Specifically, five proteins belonging to the LadAα and LadAβ classes, which are eukaryotic homologs of bacterial long-chain alkane monooxygenase (LadA), were identified. Owing to their close structural similarity to bacterial LadA, including the elements responsible for binding to the flavin mononucleotide (FMN) cofactor, it was observed that the isoalloxazine ring of reduced FMN can establish a π-alkyl interaction with the terminal carbon atom of alkanes. Based on these findings, the authors demonstrated the ability of A. flavus LadAα monooxygenases to efficiently catalyze the degradation of long-chain alkanes (Perera et al., 2022). In addition, the researchers combined them with an organic oxidant, CaO_2. The results showed that the enzymes and CaO_2 exhibited 60%–70% degradation efficiency for naphthalene, phenanthrene, anthracene, and pyrene after a 7-day incubation period. The authors suggested that enzyme-mediated bioremediation is a viable option for PAH-contaminated soils with low biological activity (Nagy et al., 2023).

From another environmental perspective, it is important to highlight that several enzymes capable of breaking down polyethylene material have been identified. In one study, at least 21 new enzymes with PET-degrading activity were identified from soil samples, demonstrating that scientific efforts to develop alternatives for decontaminating soils contaminated with plastic waste are ongoing. Furthermore, this study underscores the importance of integrating computational methods and metagenomic approaches to generate effective solutions for real environmental problems, such as plastic pollution (Jahanshahi and Barzani, 2025). However, several researchers have previously identified and evaluated the effect of various enzymes on PET degradation; esterase, lipase, hydrolase, and cutinase are particularly noteworthy. These enzymes are derived from cultivable microorganisms (bacteria and fungi). Nevertheless, less than 1% of these organisms are cultivable, highlighting the need to implement technologies that accelerate research (Kapinusova et al., 2023). For example, the thermostable enzyme *Dm PETase* from *Deinococcus maricopensis* has demonstrated high efficiency in degrading PET within 72 h. The authors confirmed the enzyme's attributes for the efficient degradation of crystalline PET through engineering modifications (Makryniotis et al., 2023).

In another context, proteomics applied to wastewater also enables the discovery of new environmental and health biomarkers, which can ultimately lead to the development of health-monitoring devices and control of industrial landfills (Sánchez-Jiménez et al., 2023). For example, to generate early warnings about the prevalence of SARS-CoV-2 and community transmission, studies combining mass spectrometry analysis with wastewater-based epidemiological methods have demonstrated that proteomic tools are effective at detecting viral protein markers in wastewater samples. This approach goes beyond traditional PCR techniques, confirming that omics technologies are effective for complex studies and can provide accurate data (Jagadeesan et al., 2024).

2.6.3 Implications for phytoremediation design

The value of omics-guided insights becomes evident when translated into practical phytoremediation scenarios. Proteomic studies of *Brassica napus* exposed to As or Cd, for example, show that sulfur nutrition and modulation of proteins involved in glycolysis, the tricarboxylic acid cycle, and photosynthesis can alleviate As toxicity and support higher biomass production, which directly improves phytoextraction potential (Ahmed et al., 2023; Li et al., 2025c). Poplar exposed to long-term Cr pollution upregulates antioxidant and stress-response proteins, supporting its use for phytostabilization and phytoextraction along industrial corridors (Wu et al., 2022).

In Cd-polluted systems, a comparison of tolerant and susceptible sorghum or poplar genotypes highlights nitrogen supply and the upregulation of ABC transporters, 1433 proteins, heat shock proteins, and peroxidases as key determinants of successful remediation (Li et al., 2024; Wang et al., 2021). These findings suggest that combining appropriate fertilization regimes, selecting or breeding genotypes with favorable proteomic responses, and possibly inoculating with beneficial rhizosphere microbes can significantly enhance field performance.

For Pb-contaminated soils, proteomic profiling of hemp and maize indicated that genotypes with strong activation of ATP-generating pathways, aquaporins, signal transduction proteins, and structural cell wall components show higher translocation factors and better biomass maintenance (Xu et al., 2021; Alsafran et al., 2022); these traits can be used as selection criteria for phytoremediation.

Finally, understanding the proteomic and metabolomic signatures of hyperaccumulators underpins the development of biotechnological strategies. Three complementary routes can be inferred from the omics data. Overexpression of metal transporters increases uptake and translocation, enhances ligand production, and vacuolar sequestration to increase tolerance, and exploitation of metabolic pathways that convert certain elements, such as Se or Hg, into less toxic or volatile species.

2.6.4 Multiomics in bioremediation strategies

In the preceding paragraphs, we presented examples that demonstrate the importance of omics as a cutting-edge technology for identifying and understanding the diverse molecular information of microscopic organisms, including genes, proteins, metabolites, and transcripts. Therefore the implementation of multiomics strategies, including metagenomics, metatranscriptomics, proteomics, and metabolomics, is not only a crucial and powerful tool for understanding the biology of microorganisms but also for understanding their complex interactions with their environment, which are key to effective bioremediation strategies. For example, studies or monitoring of coastal and marine sediments contaminated by PAHs have reported the detection of degrading enzymes, including monooxygenases, dioxygenases, and peroxidases, where metagenomics has been the primary technological tool for accurately determining their potential to degrade hydrocarbons (Nagy et al., 2023). In addition, in studies of estuarine sediments, where metagenomics and metabolomics were integrated to evaluate the bioremediation of heavy metals, the authors demonstrated the presence of active pathways for chromium (Cr; VI) reduction and arsenic (As; III) transformation associated with genera such as *Shewanella* and *Geobacter* (Cai et al., 2024; Wen et al., 2023). As in the previous case, this study emphasizes that, in addition to identifying the specific organisms responsible for metal reduction and transformation, it allows for understanding the metabolic pathways and effectively predicting the processes involved, which can then be applied to the design and monitoring of bioremediation strategies (Boros-Lajszner et al., 2025).

In the case of wastewater treatment and quality monitoring with microalgae and bacterial consortia, studies in the last decade have used metagenomics, transcriptomics, proteomics, and metabolomics to identify changes in metabolic pathways associated with contaminant removal, carbon capture, and nutrient recovery (Kumar and Shukla, 2024).

In general, many systematic studies and recent reviews highlight that the vast amount of information in databases and the integration of multiomics platforms are being incorporated into operational control models that offer a promising future for achieving efficient and sustainable treatment and bioremediation systems (Ghai et al., 2024; Kong et al., 2025).

2.6.5 Implications for ecosystem restoration and risk assessment

The practical application of omics approaches will enable us to understand and address the problems posed by various pollutants in their interactions with microorganisms, society, and ecosystems. While traditional methods provide valuable information for evaluating physical and chemical parameters and the presence/absence of indicator species, the combination of multiomics data will allow us to understand in detail the genetic alterations at the protein and metabolite levels, thereby providing clear evidence of toxicity, detoxification pathways, and physiological changes. This tool allows us to understand remediation strategies for different scenarios involving potential pollutants and chronic exposure. The usefulness of multiomics tools can be even greater for understanding subclinical responses in aquatic and terrestrial microbial communities exposed to various pollutants, including nanomaterials, which have also been used for water and soil remediation. However, although it promises significant results in environmental decontamination, much remains to be clarified, and omics tools are crucial in this process (Shi et al., 2024).

In recent studies, the authors demonstrated that omics sciences can detect even slight changes in enzyme abundance, metabolic pathways, and membrane transporters, even when demographic variations and mortality rates are low. Therefore metabolic and protein profiles are understood to function as early biological markers, alerting us to potential ecological risks at the community or ecosystem level, and providing key information for the preventive assessment of environmental hazards (Khosrovyan et al., 2025).

In addition to all of the above, multiomics is already recognized as a solid foundation for developing legal, regulatory, and environmental management programs based on unique scientific evidence. Integrating omics data into risk assessment programs facilitates the rational use of functional thresholds or safety limits, not only based on chemical concentrations but also by considering the functional state of biota. For example, a recent international consensus proposed the inclusion of omics data in next-generation risk assessment programs to identify critical points of ecological disruption (e.g., persistent changes in key metabolic pathways, loss of essential functions, and increased stress biomarkers) as criteria for intervention. In this way, environmental decisions regarding remediation, regulation, and ecological restoration can be based on molecular evidence that is more sensitive and representative of the actual state of the ecosystem (Smith et al., 2025).

2.7 Advances and challenges in multiomics approaches

2.7.1 From genomics to metabolomics: expanding the toolkit

The expansion from genomics to metabolomics has transformed biological investigations by enabling a shift from static descriptions of genetic potential to dynamic assessments of functional activity (Singh et al., 2021). Genomics and metagenomics define the theoretical capacity of organisms and communities by cataloging genes associated with metabolism, regulation, and stress tolerance. However, genomic information alone cannot predict biological behavior in fluctuating or contaminated environments, where gene expression is tightly regulated by the context. Transcriptomics captures regulatory responses by identifying genes activated or suppressed under specific conditions; however, transcript abundance is an imperfect proxy for function due to translational control, protein turnover, and post-translational regulation (Keagy et al., 2025). Therefore proteomics represents a critical transition point, as it directly quantifies the molecular machinery responsible for catalysis, transport, and stress adaptation. Despite technical biases and incomplete coverage, proteomics is indispensable for distinguishing between potential and realized metabolic activity in situ (Singh et al., 2025; Ninck et al., 2024).

Metabolomics provides the most immediate readout of the physiological state by integrating upstream regulatory and enzymatic processes into measurable biochemical outcomes. Targeted approaches offer robust quantification of key pathways, whereas untargeted analyses reveal unexpected metabolic shifts but are limited by annotation challenges and cross-platform variability (Shah, 2022). Together, these omics layers establish a functional hierarchy that links genes to phenotype, but only when interpreted in an integrated manner.

2.7.2 Integration of multiomics frameworks

Within One Health and exposome frameworks, biological responses to contaminants and co-stressors must be resolved across molecular scales and interacting biological systems, rather than inferred from single endpoints, such as biomass reduction or pollutant concentration (Gao, 2021). The One Health idea highlights that human, animal, and environmental health are all connected. We need to look at them together to understand health problems and find solutions. Complementarily, the exposome framework looks at all the environmental exposures a living being faces throughout its life (Wild, 2005). This includes chemical, biological, and physical factors and how they affect the body. Since environmental factors that cause disease and ecosystem problems are complex and can't be fully controlled in experiments, exposome research tries to understand these influences as much as possible, similar to how we study genes. New technologies now help us better understand these factors and their interactions. This shows the importance of using systematic and combined methods to understand the non-genetic factors affecting health in living beings and ecosystems.

Multiomics approaches address this limitation by enabling the mechanistic tracking of environmental perturbations from genetic potential through regulatory activation, functional protein expression, metabolic reprogramming, and flux redistribution in microbial communities.

Genomics and metagenomics define the taxonomic and functional landscape of polluted environments by identifying catabolic operons, resistance genes, transport systems, and regulatory elements associated with biodegradation and stress tolerance (Chen et al., 2022). However, the genetic potential alone is insufficient to predict the system behavior. Transcriptomic and metatranscriptomic analyses add regulatory context by revealing which pathways are transcriptionally activated under contaminant pressure or remediation interventions; however, transcriptional signals often fail to predict functional outcomes (Adhikary et al., 2024).

Proteomics and metaproteomics are critical functional pivots that directly quantify the enzymes and protein complexes responsible for biogeochemical transformations in situ (Nebauer et al., 2024). In remediation systems, proteomic data identify active degradation pathways, reveal the reorganization of metabolic and stress-response networks during adaptation, and capture functional shifts induced by amendments such as electron donors or plant exudates (Pachauri et al., 2025). Fluxomics further extends this resolution by quantifying pathway usage and metabolic trade-offs, allowing the assessment of whether carbon and electrons are directed toward pollutant removal or diverted to cellular maintenance (Emwas et al., 2022).

Extracting mechanistic insights from these heterogeneous data requires integrative analytical frameworks that converge molecular changes at the pathway, network, and system behavior levels. Network-based analyses, constraint-based metabolic models, and enzyme-constrained formulations translate multiomics measurements into predictive representations of microbial function and interaction under environmental stress (Du et al., 2023; Jia et al., 2024). Machine learning approaches increasingly complement these models by identifying exposure signatures, forecasting contaminant fate, and proposing biomarkers of ecosystem recovery (Arend, 2024).

Despite their potential, multiomics frameworks face intrinsic challenges in their environmental applications. Environmental microbiomes are heterogeneous, exposures involve complex mixtures, and each omics layer introduces platform-specific biases, temporal mismatches, and incomplete coverage (Satya et al., 2024). Therefore meaningful integration depends on rigorous experimental design, synchronized sampling, robust normalization, and standardized analytical pipelines supported by interdisciplinary collaboration (Hayes et al., 2024).

Even with these constraints, integrated omics approaches surpass traditional methods by linking genetic capacity to functional execution and metabolic consequences. As analytical frameworks mature, multiomics is positioned to become a cornerstone of predictive environmental assessment, ecosystem monitoring, and mechanism-based remediation strategies aligned with One Health and exposome concepts (Anand and Padmanabhan, 2024).

2.7.3 Benefits and limitations of multiomics in environmental microbiology

The integration of multiomics approaches, including metagenomics, transcriptomics, proteomics, metabolomics, lipidomics, and fluxomics, has reshaped environmental microbiology by enabling a systems-level understanding of microbial communities in polluted environments. By linking genetic potential with regulatory activation, functional protein expression, and realized metabolic activity, multiomics provides mechanistic insights into processes such as xenobiotic degradation, metal transformation, nutrient cycling, and biofilm-associated functions that cannot be resolved through single-omics analyses (Anand and Padmanabhan, 2024). This integration is particularly critical in heterogeneous environments where microbial interactions and physicochemical gradients strongly modulate function.

A key strength of multiomics approaches is their ability to identify active degraders, novel catabolic pathways, and regulatory networks involved in pollutant transformation. Transcriptomic and proteomic data demonstrate pathway induction under contaminant stress, whereas metabolomics confirms functional activity through the detection of intermediates and end products (Jeevanandam and Osborne, 2022). These complementary layers are essential for interpreting microbial responses in dynamic systems shaped by redox fluctuations, substrate availability, and competition.

Multiomics is also indispensable for resolving microbial succession and functional turnover during pollution events. Time-resolved studies have revealed predictable transitions in community composition and metabolic specialization, such as shifts from primary alkane degraders to aromatic compound specialists, which can only be robustly interpreted through integrated genomic, transcriptomic, proteomic, and metabolomic analyses (Singh et al., 2021). Similar strategies have clarified the division of labor in biofilms, resilience in anaerobic digesters, and adaptive responses in soils and sediments.

Beyond mechanistic insights, multiomics supports the identification of predictive and diagnostic biomarkers for environmental monitoring and bioremediation assessment. Pathway-specific transcripts, enzymes, and metabolites enable the early detection of contamination and functional recovery, whereas systems biology models constrained by multiomics data facilitate the reconstruction of community metabolism and optimization of remediation strategies (Beale et al., 2022).

Despite these advances, there are still limitations; environmental matrices complicate sample processing, high microbial diversity obscures low-abundance but functionally important taxa, and incomplete annotation restricts the interpretation of large portions of omics data (Rodríguez del Río et al., 2024). Integration is further challenged by differences in temporal resolution, analytical noise, and data scale, which require substantial bioinformatics expertise and computational resources (Santiago-Rodriguez and Hollister, 2021). Moreover, translating multiomics insights into field-scale applications requires validation across space and time and careful consideration of ecological interactions (Planell et al., 2021).

Ongoing advances in sequencing technologies, extraction methods, databases, and integrative analytics are progressively overcoming these barriers. Emerging approaches, such as single-cell and spatial omics, isotope-informed workflows, BONCAT-based proteomics, and community-scale metabolic modeling, further enhance in situ resolution (Alcolombri et al., 2022). These developments position multiomics as a foundational framework in mechanistic, predictive, and application-oriented environmental microbiology.

2.8 Conclusions

Beyond mechanistic description, proteomics enables the direct assessment of how microbial communities function under environmental stress by identifying the enzymes and regulatory proteins that are active in situ. This functional resolution is essential for distinguishing effective contaminant transformation from transient or incomplete responses and understanding the metabolic constraints that limit ecosystem recovery. When extended to community-scale analyses, proteomics reveals cooperative and competitive interactions that shape resilience in polluted environments.

Integrating proteomics with complementary omics and environmental data supports a shift from descriptive monitoring to predictive and adaptive management. Protein-level indicators provide early signals of functional changes, allowing remediation strategies to be adjusted before critical thresholds are reached. As methodological standardization improves and reference databases expand, proteomics will increasingly inform practical decision-making in environmental monitoring, restoration, and pollution management.

Finally, proteomics is a critical link between molecular mechanisms and ecological processes, enabling the identification of functional responses that directly influence ecosystem stability. Its application in environmental science extends beyond theoretical understanding and provides practical tools for conservation strategies, pollution mitigation, and climate adaptation. By revealing active metabolic pathways and stress biomarkers, proteomics supports the early detection of ecosystem disturbances, guiding timely interventions to prevent irreversible damage to ecosystems.

Future research should emphasize long-term ecological monitoring programs that integrate proteomics with other omics technologies to develop comprehensive models of ecosystem health. Additionally, the standardization of protocols and expansion of protein databases for nonmodel organisms will enhance the reliability and comparability of data across regions and ecosystems. These efforts will position proteomics as a cornerstone of sustainable environmental management, ensuring that scientific insights translate into actionable policies for biodiversity preservation and pollution control.

Acknowledgments

E.V.-N. acknowledges SECIHTI and CIIC-UG for financial support through projects (PEE-2025-G-647 and 055/2026) and the Universidad de Guanajuato for institutional support and access to research facilities.

AI disclosure

During the preparation of this work, the author(s) used Paperpal in order to. This APP is useful for revising grammar, trimming the text, simplifying, and suggesting the structure of sentences. After using this tool/service, the author (s) reviewed and edited the content as needed and take(s) full responsibility for the content of the publication.

References

Abdallah, S.M., Muhammed, R.E., Mohamed, R.E., El Daous, H., Saleh, D.M., Ghorab, M.A., et al., 2024. Assessment of biochemical biomarkers and environmental stress indicators in some freshwater fish. Environmental Geochemistry and Health 46 (11). Available at https://doi.org/10.1007/s10653-024-02226-6, https://www.springer.com/journal/10653.

Abdullah, K., Nawrocki, A., Wilkins, D., Zhong, L., Tribbia, D.Z., Larsen, M.R., et al., 2025. Development and Application of a Meta-Proteomics Approach for Low Biomass Soils. SSRN. Available at 10.2139/ssrn.5369881, https://www.ssrn.com/index.cfm/en/.

Adhikary, S., Saha, J., Dutta, P., Pal, A., 2024. Bacterial homeostasis and tolerance to potentially toxic metals and metalloids through diverse transporters: metal-specific insights. Geomicrobiology Journal 41 (5), 496–518. Available at https://doi.org/10.1080/01490451.2024.2340517.

Ahmed, T., Masood, H.A., Noman, M., AL-Huqail, A.A., Alghanem, S.M., Khan, M.M., et al., 2023. Biogenic silicon nanoparticles mitigate cadmium (Cd) toxicity in rapeseed (*Brassica napus* L.) by modulating the cellular oxidative stress metabolism and reducing Cd translocation. Journal of Hazardous Materials 459. Available at https://doi.org/10.1016/j.jhazmat.2023.132070, www.elsevier.com/locate/jhazmat.

Alcolombri, U., Pioli, R., Stocker, R., Berry, D., 2022. Single-cell stable isotope probing in microbial ecology. ISME Communications 2 (1). Available at https://doi.org/10.1038/s43705-022-00142-3.

Algahmadi, A., Mohammed, A.E., Alfadda, A.A., Alanazi, I.O., Alwehaibi, M.A., Scaria Joy, S., et al., 2024. Proteomics of penicillium chrysogenum for a deeper understanding of lead (Pb) metal bioremediation. ACS Omega 9 (24), 26245–26256. Available at https://doi.org/10.1021/acsomega.4c02006, pubs.acs.org/journal/acsodf.

Alsafran, M., Usman, K., Ahmed, B., Rizwan, M., Saleem, M.H., Al Jabri, H., 2022. Understanding the phytoremediation mechanisms of potentially toxic elements: a proteomic overview of recent advances. Frontiers in Plant Science 13. Available at https://doi.org/10.3389/fpls.2022.881242, https://www.frontiersin.org/journals/plant-science.

Anand, S., Padmanabhan, P., 2024. Omics approaches for characterization of environmental microorganisms. Development in Waste Water Treatment Research and Processes: Role of Environmental Microbiology in Industrial Wastewater Research 1–13. Available at https://doi.org/10.1016/B978-0-443-13609-2.00001-X.

Arend, M., 2024. Comparing genome-scale models of protein-constrained metabolism in heterotrophic and photosynthetic microorganisms. Computational Biology computergestützte Biologie Metabolic Modeling metabolische Modellierung Proteomics Systembiologie Systems Biology. Available at https://publishup.uni-potsdam.de/frontdoor/index/index/docId/65147.

Aswathi, A., Pandey, A., Madhavan, A., Sukumaran, R.K., 2021. Chlorpyrifos induced proteome remodelling of Pseudomonas nitroreducens AR-3 potentially aid efficient degradation of the pesticide. Environmental Technology and Innovation 21. Available at https://doi.org/10.1016/j.eti.2020.101307, http://www.journals.elsevier.com/environmental-technology-and-innovation/.

Azevedo-Santos, V.M., Fernandes, J.A., de Souza Andrade, G., de Moraes, P.M., Magurran, A.E., Pelicice, F.M., et al., 2024. An overview of vinasse pollution in aquatic ecosystems in Brazil.

Environmental Management 74 (6), 1037–1044. Available at https://doi.org/10.1007/s00267-024-01999-x, https://www.springer.com/journal/267.

Banerjee, A., Khatun, R., Sengupta, S., Bhattacharya, M., 2025. Decoding the microbial world: an introduction to microbial omics. International Journal of Current Microbiology and Applied Sciences 14 (1), 146–154. Available at https://doi.org/10.20546/ijcmas.2025.1404.018.

Banu, M.M., Reehana, N., Imran, M.M., 2024. Microbial Degradation of pesticides in agricultural environments: a comprehensive review of mechanisms, factors and biodiversity. Molecular Sciences and ApplicationS 4, 65–101. Available at https://doi.org/10.37394/232023.2024.4.8.

Beale, D.J., Jones, O.A.H., Bose, U., Broadbent, J.A., Walsh, T.K., Van De Kamp, J., et al., 2022. Omics-based ecosurveillance for the assessment of ecosystem function, health, and resilience. Emerging Topics in Life Sciences 6 (2), 185–199. Available at https://doi.org/10.1042/ETLS20210261, https://portlandpress.com/emergtoplifesci/article/6/2/185/231127/Omics-based-ecosurveillance-for-the-assessment-of.

Bhandari, S., Poudel, D.K., Marahatha, R., Dawadi, S., Khadayat, K., Phuyal, S., et al., 2021. Microbial enzymes used in bioremediation. Journal of Chemistry 2021. Available at https://doi.org/10.1155/2021/8849512, http://www.hindawi.com/journals/chem/contents/.

Boros-Lajszner, E., Wyszkowska, J., Baćmaga, M., Kucharski, J., 2025. Analysis of changes in the microbial biodiversity of soil contaminated with Cr(III) and Cr(VI). Applied Sciences (Switzerland) 15 (20). Available at https://doi.org/10.3390/app152010951, https://www.mdpi.com/journal/applsci/.

Bose, S., Kumar, P.S., Vo, D.V.N., Rajamohan, N., Saravanan, R., 2021. Microbial degradation of recalcitrant pesticides: a review. Environmental Chemistry Letters 19 (4), 3209–3228. Available at https://doi.org/10.1007/s10311-021-01236-5, https://www.springer.com/journal/10311.

Brooks, C.N., Field, E.K., 2024. Microbial community response to hydrocarbon exposure in iron oxide mats: an environmental study. Frontiers in Microbiology 15. Available at https://doi.org/10.3389/fmicb.2024.1388973, https://www.frontiersin.org/journals/microbiology#.

Cai, Y., Chen, X., Qi, H., Bu, F., Shaaban, M., Peng, Q.A., 2024. Genome analysis of Shewanella putrefaciens 4H revealing the potential mechanisms for the chromium remediation. BMC Genomics 25 (1). Available at https://doi.org/10.1186/s12864-024-10031-9, https://bmcgenomics.biomedcentral.com/.

Carrascal, M., Sánchez-Jiménez, E., Fang, J., Pérez-López, C., Ginebreda, A., Barceló, D., et al., 2023. Sewage protein information mining: discovery of large biomolecules as biomarkers of population and industrial activities. Environmental Science & Technology 57 (30), 10929–10939. Available at https://doi.org/10.1021/acs.est.3c00535.

Chaurasia, P., Kumar, A., Sarim, K.M., 2025. Meta-Omics in Crop Improvement. Springer, Singapore. Available at http://doi.org/10.1007/978-981-95-2294-1_5.

Chen, Y., Li, E.M., Xu, L.Y., 2022. Guide to metabolomics analysis: a bioinformatics workflow. Metabolites 12 (4). Available at https://doi.org/10.3390/metabo12040357, https://www.mdpi.com/2218-1989/12/4/357/pdf.

Dashora, K., Gupta, V.K., Tripathi, G.D., Javed, Z., Gattupalli, M., 2024. Advances in metaproteomic profiling of molecular microbiology and environmental responses. Systems Microbiology and Biomanufacturing 4 (2), 463–472. Available at https://doi.org/10.1007/s43393-023-00231-x, https://www.springer.com/journal/43393.

Deutsch, E.W., Bandeira, N., Perez-Riverol, Y., Sharma, V., Carver, J.J., Mendoza, L., et al., 2025. The ProteomeXchange consortium in 2026: making proteomics data FAIR. Nucleic Acids Research. Available at https://doi.org/10.1093/nar/gkaf1146.

Dong, J., Qian, M., Zhou, D., Zhu, A., Zhang, W., 2025. 2′-Fucosyllactose modulates the intestinal immune response to gut microbiota and buffers experimental colitis in mice: an integrating investigation of colonic proteomics and gut microbiota analysis. Food Science & Nutrition 13 (6). Available at https://doi.org/10.1002/fsn3.70418.

Du, H., Li, M., Liu, Y., 2023. Towards applications of genome-scale metabolic model-based approaches in designing synthetic microbial communities. Quantitative Biology 11 (1), 15–30. Available at https://doi.org/10.15302/j-qb-022-0313.

Duan, X., Duan, H., Zhao, C., Feng, L., Zhao, G., Chen, Y., 2025. Genome-centric metagenomics and metaproteomics profiled the shared and unique taxa in isomeric fatty acid-differentiated anaerobic co-digestion of food waste and sludge. Environmental Science & Technology 59 (6), 3077–3087. Available at https://doi.org/10.1021/acs.est.4c09839.

Edgar, V.-N., Fabián, F.-L., Mario, P.-C.J., Ileana, V.-R., 2021. Coupling plant biomass derived from phytoremediation of potential toxic-metal-polluted soils to bioenergy production and high-value by-products—a review. Applied Sciences 11 (7), 2982. Available at https://doi.org/10.3390/app11072982.

Elyamine, A.M., Kan, J., Meng, S., Tao, P., Wang, H., Hu, Z., 2021. Aerobic and anaerobic bacterial and fungal degradation of pyrene: mechanism pathway including biochemical reaction and catabolic genes. International Journal of Molecular Sciences 22 (15), 8202. Available at https://doi.org/10.3390/ijms22158202.

Emwas, A.-H., Szczepski, K., Al-Younis, I., Lachowicz, J.I., Jaremko, M., 2022. Fluxomics - new metabolomics approaches to monitor metabolic pathways. Frontiers in Pharmacology 13. Available at https://doi.org/10.3389/fphar.2022.805782.

Fuller, R., Landrigan, P.J., Balakrishnan, K., Bathan, G., Bose-O'Reilly, S., Brauer, M., et al., 2022. Pollution and health: a progress update. The Lancet Planetary Health 6 (6), e535. Available at https://doi.org/10.1016/S2542-5196(22)00090-0, https://www.journals.elsevier.com/the-lancet-planetary-health.

Gangola, S., Joshi, S., Bhandari, G., Pant, G., Sharma, A., Perveen, K., et al., 2023. Exploring microbial diversity responses in agricultural fields: a comparative analysis under pesticide stress and non-stress conditions. Frontiers in Microbiology 14. Available at https://doi.org/10.3389/fmicb.2023.1271129, https://www.frontiersin.org/journals/microbiology#.

Gao, P., 2021. The exposome in the era of one health. Environmental Science & Technology 55 (5), 2790–2799. Available at https://doi.org/10.1021/acs.est.0c07033.

Ghai, M.K., Khatri, A., Kumar, K., Thakur, I.S., 2024. Multi-omics and advance technologies in biodegradation of emerging contaminants and eco-estrogens in environmental waste. Total Environment Advances 11, 200113. Available at https://doi.org/10.1016/j.teadva.2024.200113.

Ginebreda, A., Barceló, D., Picó, Y., Carrascal, M., Abián, J., 2023. Environmental proteomics: a potential tool in wastewater-based epidemiology. Water Emerging Contaminants and Nanoplastics 2 (2). Available at https://doi.org/10.20517/wecn.2023.19, https://f.oaes.cc/xmlpdf/e6419a56-bde3-4f05-b21a-028cf11ce3cd/5597_down.pdf?v=86.

Gofstein, T.R., Leigh, M.B., 2023. Metatranscriptomic shifts suggest shared biodegradation pathways for Corexit 9500 components and crude oil in Arctic seawater. Environmental Microbiology Reports 15 (1), 51–59. Available at https://doi.org/10.1111/1758-2229.13127, http://onlinelibrary.wiley.com/journal/10.1111/(ISSN)1758-2229.

Guerrero Ramírez, J.R., Ibarra Muñoz, L.A., Balagurusamy, N., Frías Ramírez, J.E., Alfaro Hernández, L., Carrillo Campos, J., 2023. Microbiology and biochemistry of pesticides biodegradation. International Journal of Molecular Sciences 24 (21). Available at https://doi.org/10.3390/ijms242115969, http://www.mdpi.com/journal/ijms.

Guo, X., Zhang, J., Han, L., Lee, J., Williams, S.C., Forsberg, A., et al., 2023. Structure and mechanism of the alkane-oxidizing enzyme AlkB. Nature Communications 14 (1). Available at https://doi.org/10.1038/s41467-023-37869-z, https://www.nature.com/ncomms/.

Han, M., Yang, X., Dong, B., Lai, J., Xi, H., Zhao, S., 2025. Integrated morphological, proteomic and metabolomic analyses reveal response mechanisms of microalgae under uranium exposure. Frontiers in Microbiology 16. Available at https://doi.org/10.3389/fmicb.2025.1679056, https://www.frontiersin.org/journals/microbiology#.

Hayes, C.N., Nakahara, H., Ono, A., Tsuge, M., Oka, S., 2024. From omics to multi-omics: a review of advantages and tradeoffs. Genes 15 (12). Available at https://doi.org/10.3390/genes15121551, http://www.mdpi.com/journal/genes/.

Herbst, F.A., Lünsmann, V., Kjeldal, H., Jehmlich, N., Tholey, A., von Bergen, M., et al., 2016. Enhancing metaproteomics-the value of models and defined environmental microbial systems. Proteomics 16 (5), 783–798. Available at https://doi.org/10.1002/pmic.201500305, http://onlinelibrary.wiley.com/journal/10.1002/(ISSN)1615-9861.

Heyer, R., Wolf, M., Benndorf, D., Uzzau, S., Seifert, J., Grenga, L., et al., 2025. Metaproteomics in the one health framework for unraveling microbial effectors in microbiomes. Microbiome 13 (1). Available at https://doi.org/10.1186/s40168-025-02119-5.

Jagadeesan, K.K., Elliss, H., Standerwick, R., Robertson, M., Barden, R., Kasprzyk-Hordern, B., 2024. Wastewater-based proteomics: a proof-of-concept for advancing early warning system for infectious diseases and immune response monitoring. Journal of Hazardous Materials Letters 5, 100108. Available at https://doi.org/10.1016/j.hazl.2024.100108.

Jahanshahi, D.A., Barzani, M.R.R., Bahram, M., Ariaeenejad, S., Kavousi, K., 2025. Metagenomic exploration and computational prediction of novel enzymes for polyethylene terephthalate degradation. Ecotoxicology and Environmental Safety 289. Available at https://doi.org/10.1016/j.ecoenv.2024.117640, https://www.sciencedirect.com/science/journal/01476513.

Jeevanandam, V., Osborne, J., 2022. Understanding the fundamentals of microbial remediation with emphasize on metabolomics. Preparative Biochemistry & Biotechnology 52 (3), 351–363. Available at https://doi.org/10.1080/10826068.2021.1946694.

Jia, C., Wang, T., Cui, D., Tian, Y., Liu, G., Xu, Z., et al., 2024. A metagene based similarity network fusion approach for multi-omics data integration identified novel subtypes in renal cell carcinoma. Briefings in Bioinformatics 25 (6). Available at https://doi.org/10.1093/bib/bbae606.

Jin, Y., Sun, Z., Gu, A.Z., Zhou, X., 2025. Proteomics technologies in toxicity screening: a review. Environmental Chemistry Letters 23 (1), 67–80. Available at https://doi.org/10.1007/s10311-024-01816-1, https://www.springer.com/journal/10311.

Kapinusova, G., Lopez Marin, M.A., Uhlik, O., 2023. Reaching unreachables: Obstacles and successes of microbial cultivation and their reasons. Frontiers in Microbiology 14. Available at http://doi.org/10.3389/fmicb.2023.1089630.

Kazakova, E., Ivanov, M., Kusainova, T., Bubis, J., Polivtseva, V., Petrikov, K., et al., 2024. Ultrafast metaproteomics for quantitative assessment of strain isolates and microbiomes. Microchemical Journal 207, 111823. Available at https://doi.org/10.1016/j.microc.2024.111823.

Keagy, J., Drummond, C.P., Gilbert, K.J., Grozinger, C.M., Hamilton, J., Hines, H.M., et al., 2025. Landscape transcriptomics as a tool for addressing global change effects across diverse species. Molecular Ecology Resources 25 (5). Available at https://doi.org/10.1111/1755-0998.13796, http://onlinelibrary.wiley.com/journal/10.1111/(ISSN)1755-0998.

Kesheri, M., Kanchan, S., Srivastava, U., Chittoori, B., Ratna-Raj, R., Sinha, R.P., et al., 2024. Ecology and environmental omics. Integrative Omics: Concept, Methodology, and

Application. Elsevier, United States, pp. 315–331. Available at https://www.sciencedirect.com/book/9780443160929.

Khosrovyan, A., Vodovnik, M., Mortimer, M., 2025. Omics approaches in environmental effect assessment of engineered nanomaterials and nanoplastics. Environmental Science: Nano 12 (5), 2551–2579. Available at https://doi.org/10.1039/d4en01169d.

Kleindienst, S., Seidel, M., Ziervogel, K., Grim, S., Loftis, K., Harrison, S., et al., 2015. Chemical dispersants can suppress the activity of natural oil-degrading microorganisms. Proceedings of the National Academy of Sciences of the United States of America 112 (48), 14900–14905. Available at https://doi.org/10.1073/pnas.1507380112, http://www.pnas.org/content/112/48/14900.full.pdf.

Kong, Y., Chen, H., Huang, X., Chang, L., Yang, B., Chen, W., 2025. Precise metabolic modeling in post-omics era: accomplishments and perspectives. Critical Reviews in Biotechnology 45 (3), 683–701. Available at https://doi.org/10.1080/07388551.2024.2390089.

Kratter, M., Beccaccioli, M., Benedetti, F., Buccini, L., Proietti, A., Zanellato, G., et al., 2025. Multidisciplinary characterization of fungi and cyanobacteria involved in biodeterioration of the Tomba delle Leonesse, Tarquinia, Italy. International Biodeterioration & Biodegradation 205, 106183. Available at https://doi.org/10.1016/j.ibiod.2025.106183.

Kumar, N., Shukla, P., 2024. Microalgal multiomics-based approaches in bioremediation of hazardous contaminants. Environmental Research 247, 118135. Available at https://doi.org/10.1016/j.envres.2024.118135.

Kumar, V., Singh, K., Shah, M.P., Singh, A.K., Kumar, A., Kumar, Y., 2021. Application of omics technologies for microbial community structure and function analysis in contaminated environment. Wastewater Treatment: Cutting-Edge Molecular Tools, Techniques and Applied Aspects. Elsevier, India, pp. 1–40. Available at https://www.sciencedirect.com/book/9780128218815.

Kumar, R., Yadav, G., Kuddus, M., Ashraf, G.M., Singh, R., 2023. Unlocking the microbial studies through computational approaches: how far have we reached? Environmental Science and Pollution Research 30 (17), 48929–48947. Available at https://doi.org/10.1007/s11356-023-26220-0, https://www.springer.com/journal/11356.

Kumari, S., Das, S., 2023. Bacterial enzymatic degradation of recalcitrant organic pollutants: catabolic pathways and genetic regulations. Environmental Science and Pollution Research 30 (33), 79676–79705. Available at https://doi.org/10.1007/s11356-023-28130-7.

Laczi, K., Erdeiné Kis, Á., Szilágyi, Á., Bounedjoum, N., Bodor, A., Vincze, G.E., et al., 2020. New frontiers of anaerobic hydrocarbon biodegradation in the multi-omics era. Frontiers in Microbiology 11. Available at https://doi.org/10.3389/fmicb.2020.590049, https://www.frontiersin.org/journals/microbiology#.

Li, M., Heng, Q., Hu, C., Wang, Z., Jiang, Y., Wang, X., et al., 2024. Phytoremediation efficiency of poplar hybrid varieties with diverse genetic backgrounds in soil contaminated by multiple toxic metals (Cd, Hg, Pb, and As). Ecotoxicology and Environmental Safety 283, 116843. Available at https://doi.org/10.1016/j.ecoenv.2024.116843.

Li, B., Zhang, Y., Du, J., Liu, C., Zhou, G., Li, M., et al., 2025a. Application of multi-omics techniques in aquatic ecotoxicology: a review. Toxics 13 (8), 653. Available at https://doi.org/10.3390/toxics13080653.

Li, L., Simopoulos, C.M.A., Mayne, J., Ning, Z., Zhang, X., Hamada, M., et al., 2025b. Systematic metaproteomics mapping reveals functional and ecological landscapes of Ex vivo human gut microbiota responses to therapeutic drugs. Nature Communications 16 (1). Available at https://doi.org/10.1038/s41467-025-64433-8, https://www.nature.com/ncomms/.

Li, L., Fan, Z., Gan, Q., Xiao, G., Luan, M., Zhu, R., et al., 2025c. Conservative mechanism through various rapeseed (Brassica napus L.) varieties respond to heavy metal (Cadmium, Lead, Arsenic) stress. Frontiers in Plant Science 15. Available at https://doi.org/10.3389/fpls.2024.1521075.

Lionetto, M.G., Caricato, R., Giordano, M.E., 2021. Pollution biomarkers in the framework of marine biodiversity conservation: state of art and perspectives. Water (Switzerland) 13 (13). Available at https://doi.org/10.3390/w13131847, https://www.mdpi.com/2073-4441/13/13/1847/pdf.

Liu, G., Wang, S., Wang, T., Zhu, C., Li, C., Zhang, X., 2025. Proteomics and phosphoproteomics of freshwater mollusk carcasses reveal novel insights as potential food source. Frontiers in Cellular and Infection Microbiology 15. Available at https://doi.org/10.3389/fcimb.2025.1568349.

Liu, H., Yao, M., Nygård, Y., 2025. Time-resolved monitoring of yeast responses to lipopolysaccharide exposure by cell-released volatile organic compounds. Applied and Environmental Microbiology 91 (9). Available at https://doi.org/10.1128/aem.00785-25.

Luz, J.Zd, Gorshkov, V., Miranda, R.R., Souza, T.Ld, Rodrigues Ribeiro, L., Duan, X., et al., 2025. Metallothionein as a biomarker of aquatic contamination in fish: an in silico and in vitro approach using zebrafish as experimental model organism. Chemosphere 376. Available at https://doi.org/10.1016/j.chemosphere.2025.144316, https://www.sciencedirect.com/science/journal/00456535.

Makryniotis, K., Nikolaivits, E., Gkountela, C., Vouyiouka, S., Topakas, E., 2023. Discovery of a polyesterase from Deinococcus maricopensis and comparison to the benchmark LCCICCG suggests high potential for semi-crystalline post-consumer PET degradation. Journal of Hazardous Materials 455, 131574. Available at https://doi.org/10.1016/j.jhazmat.2023.131574.

Mandal, B., Das, B.K., Das, B.K., Kumar, V., 2025. eDNA approaches for ecosystem health monitoring: focus on pathogens, vectors, and microbial assessment. Laboratory Techniques for Fish Disease Diagnosis. Springer Nature Singapore, Singapore, pp. 573–622. Available at https://doi.org/10.1007/978-981-96-4620-3_25.

Marlow, J.J., Skennerton, C.T., Li, Z., Chourey, K., Hettich, R.L., Pan, C., et al., 2016. Proteomic stable isotope probing reveals biosynthesis dynamics of slow growing methane based microbial communities. Frontiers in Microbiology 7. Available at https://doi.org/10.3389/fmicb.2016.00563, http://journal.frontiersin.org/article/10.3389/fmicb.2016.00563/full.

Mir, R.A., Mantoo, A.G., Sofi, Z.A., Bhat, D.A., Bashir, A., Bashir, S., 2023. Types of environmental pollution and its effects on the environment and society. Geospatial Analytics for Environmental Pollution Modeling: Analysis, Control and Management. Springer Nature, India, pp. 1–31. Available at https://link.springer.com/book/10.1007/978-3-031-45300-7.

Moore, E.K., Caldwell Eldridge, S.L., Tomaszewski, E.J., Varonka, M.S., Martini, A.M., Carley, M., et al., 2025. Metaproteomics and metagenomics reveal microbial pathways of organic matter degradation and methanogenesis in a marginally producing natural gas well. Communications Earth and Environment 6 (1). Available at https://doi.org/10.1038/s43247-025-02776-2, https://www.nature.com/commsenv/.

Mudge, M.C., Riffle, M., Chebli, G., Plubell, D.L., Rynearson, T.A., Noble, W.S., et al., 2025. Harmful algal blooms are preceded by a predictable and quantifiable shift in the oceanic microbiome. Nature Communications 16 (1). Available at https://doi.org/10.1038/s41467-025-59250-y, https://www.nature.com/ncomms/.

Muhammad, M., Batool, S., Hivare, V., Li, W.-J., Waheed, A., Sinha, D., 2024. Bioremediation Techniques—classification, Principles, Advantages, Limitations, and Prospects. Elsevier BV, pp. 1–23. Available at http://doi.org/10.1016/b978-0-443-21911-5.00003-9.

Nagy, K.K., Takács, K., Németh, I., Varga, B., Grolmusz, V., Molnár, M., et al., 2023. Novel enzymes for biodegradation of polycyclic aromatic hydrocarbons: metagenomics-linked identification followed by functional analysis. arXiv. Available at 10.48550/arXiv.2308.15914, https://arxiv.org.

Navarro, U.C., Cerón-Moreno, M.F., Fernández, F.J., 2025. Proteomic analysis: application to the study of signal transduction pathways in Penicillium chrysogenum and Acremonium chrysogenum. Whole Cell Biocatalysis: Fundamentals and Applications. Elsevier, Mexico, pp. 175–189. Available at https://www.sciencedirect.com/book/9780443239984.

Nebauer, D.J., Pearson, L.A., Neilan, B.A., 2024. Critical steps in an environmental metaproteomics workflow. Environmental Microbiology 26 (5). Available at https://doi.org/10.1111/1462-2920.16637, http://onlinelibrary.wiley.com/journal/10.1111/(ISSN)1462-2920.

Ninck, S., Klaus, T., Kochetkova, T.V., Esser, S.P., Sewald, L., Kaschani, F., et al., 2024. Environmental activity-based protein profiling for function-driven enzyme discovery from natural communities. Environmental Microbiome 19 (1). Available at https://doi.org/10.1186/s40793-024-00577-2, https://environmentalmicrobiome.biomedcentral.com.

Ningombam, L., Hazarika, B.N., Yumkhaibam, T., Heisnam, P., Singh, Y.D., 2024. Heavy metal priming plant stress tolerance deciphering through physiological, biochemical, molecular and omics mechanism. South African Journal of Botany 168, 16–25. Available at https://doi.org/10.1016/j.sajb.2024.02.032, https://www.sciencedirect.com/science/journal/02546299.

Pachauri, E., Mukeshkumar, P., Yadav, K., Nandipamu, A., Kiran, 2025. Applications of genomics and proteomics in bioremediation. Sustainable Green Technologies: Bioremediation of Environmental Pollution. Springer Nature Switzerland, Cham, pp. 189–217. Available at https://doi.org/10.1007/978-3-032-08155-1_9.

Pan, H., Wattiez, R., Gillan, D., 2024. Soil metaproteomics for microbial community profiling: methodologies and challenges. Current Microbiology 81 (8). Available at https://doi.org/10.1007/s00284-024-03781-y.

Parascanu, M.M., Sanchez, N., Sandoval-Salas, F., Carreto, C.M., Soreanu, G., Sanchez-Silva, L., 2021. Environmental and economic analysis of bioethanol production from sugarcane molasses and agave juice. Environmental Science and Pollution Research 28 (45), 64374–64393. Available at https://doi.org/10.1007/s11356-021-15471-4.

Pavelescu, L.A., Profir, M., Enache, R.M., Roşu, O.A., Creţoiu, S.M., Gaspar, B.S., 2024. A proteogenomic approach to unveiling the complex biology of the microbiome. International Journal of Molecular Sciences 25 (19). Available at https://doi.org/10.3390/ijms251910467, http://www.mdpi.com/journal/ijms.

Perera, M., Wijesundera, S., Wijayarathna, C.D., Seneviratne, G., Jayasena, S., 2022. Identification of long-chain alkane-degrading (LadA) monooxygenases in Aspergillus flavus via in silico analysis. Frontiers in Microbiology 13. Available at https://doi.org/10.3389/fmicb.2022.898456, https://www.frontiersin.org/journals/microbiology#.

Philippot, L., Chenu, C., Kappler, A., Rillig, M.C., Fierer, N., 2024. The interplay between microbial communities and soil properties. Nature Reviews Microbiology 22 (4), 226–239. Available at https://doi.org/10.1038/s41579-023-00980-5, https://www.nature.com/nrmicro/.

Planell, N., Lagani, V., Sebastian-Leon, P., van der Kloet, F., Ewing, E., Karathanasis, N., et al., 2021. STATegra: multi-omics data integration – a conceptual scheme with a bioinformatics pipeline. Frontiers in Genetics 12. Available at https://doi.org/10.3389/fgene.2021.620453.

Rathod, S.V., Saras, P., Gondaliya, S.M., 2024. Environmental Pollution: Threats and Challenges for Management Eco-Restoration of the Polluted Environment: A Biological Perspective. CRC Press, India, pp. 1–34. Available at https://www.taylorfrancis.com/books/edit/10.1201/9781003423393/eco-restoration-polluted-environment-sandip-rathod.

Renganathan, P., Gaysina, L.A., García Gutiérrez, C., Rueda Puente, E.O., Sainz-Hernández, J.C., 2025. Harnessing engineered microbial consortia for xenobiotic bioremediation: integrating multi-omics and AI for next-generation wastewater treatment. Journal of Xenobiotics 15 (4). Available at https://doi.org/10.3390/jox15040133, https://www.mdpi.com/2039-4713/11.

Rodríguez del Río, Á., Giner-Lamia, J., Cantalapiedra, C.P., Botas, J., Deng, Z., Hernández-Plaza, A., et al., 2024. Functional and evolutionary significance of unknown genes from uncultivated taxa. Nature 626 (7998), 377–384. Available at https://doi.org/10.1038/s41586-023-06955-z, https://www.nature.com/nature/.

Sánchez-Jiménez, E., Abian, J., Ginebreda, A., Barceló, D., Carrascal, M., 2023. Shotgun proteomics to characterize wastewater proteins. MethodsX 11, 102403. Available at https://doi.org/10.1016/j.mex.2023.102403.

Sánchez-Rojas, T., Espinoza-Culupú, A., Ramírez, P., Iwai, L.K., Montoni, F., Macedo-Prada, D., et al., 2022. Proteomic study of response to copper, cadmium, and chrome ion stress in yarrowia lipolytica strains isolated from andean mine tailings in Peru. Microorganisms 10 (10), 2002. Available at https://doi.org/10.3390/microorganisms10102002.

Santiago-Rodriguez, T.M., Hollister, E.B., 2021. Multi 'omic data integration: a review of concepts, considerations, and approaches. Seminars in Perinatology 45 (6). Available at https://doi.org/10.1016/j.semperi.2021.151456, http://www.elsevier.com/inca/publications/store/6/2/3/3/7/index.htt.

Satya, S., Sharma, S., Choudhary, G., Kaushik, G., 2024. Advances in environmental microbiology: a multi-omic perspective. Microbial Omics in Environment and Health. Springer Nature, India, pp. 175–204. Available at https://link.springer.com/book/10.1007/978-981-97-1769-9.

Savalia, D., Dharsandia, I., Kumar, N., Yadav, V.K., Gupta, R., Modi, S., 2025. Molecular biomarkers: their significance and application in pollution monitoring. Toxicological and Environmental Chemistry 107 (8), 1681–1722. Available at https://doi.org/10.1080/02772248.2025.2556147, http://www.tandfonline.com/loi/gtec20.

Schultz, D., Zühlke, D., Bernhardt, J., Francis, T.B., Albrecht, D., Hirschfeld, C., et al., 2020. An optimized metaproteomics protocol for a holistic taxonomic and functional characterization of microbial communities from marine particles. Environmental Microbiology Reports 12 (4), 367–376. Available at https://doi.org/10.1111/1758-2229.12842, http://onlinelibrary.wiley.com/journal/10.1111/(ISSN)1758-2229.

Shah, M.P., 2022. Microbial Community Studies in Industrial Wastewater Treatment. CRC Press. Available at http://doi.org/10.1201/9781003354147.

Shanthi Kumari, B.S., Dileep Kumar, K., Sai Geetha, K., Narasimha, G., Reddy, B.R., Srivastava, M., et al., 2021. Influence of xenobiotics on fungal ligninolytic enzymes. Bioenergy Research: Basic and Advanced Concepts. Springer Singapore, Singapore, pp. 93–117. Available at https://doi.org/10.1007/978-981-33-4611-6_4.

Shi, Z., Campanaro, S., Usman, M., Treu, L., Basile, A., Angelidaki, I., et al., 2021. Genome-centric metatranscriptomics analysis reveals the role of hydrochar in anaerobic digestion of waste activated sludge. Environmental Science & Technology 55 (12), 8351–8361. Available at https://doi.org/10.1021/acs.est.1c01995.

Shi, C., Cheng, L., Yu, Y., Chen, S., Dai, Y., Yang, J., et al., 2024. Multi-omics integration analysis: tools and applications in environmental toxicology. Environmental Pollution 360, 124675. Available at https://doi.org/10.1016/j.envpol.2024.124675.

Singh, R., Ram, K., 2022. Environmental degradation: challenges and strategies for mitigation. Environmental Remediation Technologies. Springer, Cham. Available at http://doi.org/10.1007/978-3-030-95542-7_10.

Singh, D., Geat, N., Mehriya, M., Rajawat, M.V.S., Prasanna, R., Kumar, A., et al., 2021. Omics (genomics, proteomics, metabolomics, etc.) tools to study the environmental microbiome and bioremediation. Waste to Energy: Prospects and Applications. Springer Singapore, India, pp. 235–260. Available at https://link.springer.com/book/10.1007/978-981-33-4347-4.

Singh, N., Das, S., Barman, K., Das Kurmi, B., Patel, P., 2025. Analytical mass spectrometry as a versatile tool in omics, diagnostics, and environmental monitoring. International Journal of Medical, Pharmaceutical and Health Sciences 2 (2), 275–285. Available at https://doi.org/10.62946/ijmphs/2.2.275-285.

Smercina, D.N., Bailey, V.L., Hofmockel, K.S., 2021. Micro on a macroscale: relating microbial-scale soil processes to global ecosystem function. FEMS Microbiology Ecology 97 (7). Available at https://doi.org/10.1093/femsec/fiab091, http://femsec.oxfordjournals.org/.

Smith, K., Pochon, X., Melvin, S., Wheeler, T., Tremblay, L., 2025. Integration of "omics"-based approaches in environmental risk assessment to establish cause and effect relationships: a review. Toxics 13 (9), 714. Available at https://doi.org/10.3390/toxics13090714.

Su, Y., Zheng, X., 2025. HSP70-mediated autophagy-apoptosis-inflammation network and neuroprotection induced by heat acclimatization. Biology 14 (7), 774. Available at https://doi.org/10.3390/biology14070774.

Sun, B., Yuan, J., Zhang, X., Ma, X., Hao, Z., Wang, L., et al., 2025. Metaproteomics reveals community coalescence outcomes in co-cultured human gut microbiota. Proteomics 25 (17-18), 6–18. Available at https://doi.org/10.1002/pmic.70009.

Tan, Y.S., Zhang, R.K., Liu, Z.H., Li, B.Z., Yuan, Y.J., 2022. Microbial adaptation to enhance stress tolerance. Frontiers in Microbiology 13. Available at https://doi.org/10.3389/fmicb.2022.888746, https://www.frontiersin.org/journals/microbiology#.

Tartaglia, M., Bastida, F., Sciarrillo, R., Guarino, C., 2020. Soil metaproteomics for the study of the relationships between microorganisms and plants: a review of extraction protocols and ecological insights. International Journal of Molecular Sciences 21 (22), 8455. Available at https://doi.org/10.3390/ijms21228455.

Timmins-Schiffman, E.B., Khanna, R., Brown, T., Dilworth, J., MacLean, B.X., Mudge, M.C., et al., 2025. Proteomic plasticity in the coral montipora capitata gamete bundles after parent thermal bleaching. Journal of Proteome Research 24 (3), 1317–1328. Available at https://doi.org/10.1021/acs.jproteome.4c00946, http://pubs.acs.org/journal/jprobs.

Tomei, M.C., Mosca Angelucci, D., Clagnan, E., Brusetti, L., 2021. Anaerobic biodegradation of phenol in wastewater treatment: achievements and limits. Applied Microbiology and Biotechnology 105 (6), 2195–2224. Available at https://doi.org/10.1007/s00253-021-11182-5, link.springer.de/link/service/journals/00253/index.htm.

Tugui, C.G., Cordesius, F., van Holthe, W., van Loosdrecht, M.C.M., Pabst, M., 2025. Wastewater metaproteomics: tracking microbial and human protein biomarkers. bioRxiv. Available at 10.1101/2025.02.08.637285, https://www.biorxiv.org.

Valsala, A., Vincent, S.G.T., Karthikeyan, S., 2026. Anaerobic microbial degradation of persistent organic pollutants in aquatic sediments: implications of climate change. Archives of Microbiology 208 (1). Available at https://doi.org/10.1007/s00203-025-04565-z, https://www.springer.com/journal/203.

Van Den Bossche, T., Beslic, D., van Puyenbroeck, S., Suomi, T., Holstein, T., Martens, L., et al., 2025. Metaproteomics beyond databases: addressing the challenges and potentials of de novo sequencing. Proteomics 25 (17-18), 51–61. Available at https://doi.org/10.1002/pmic.202400321, https://onlinelibrary.wiley.com/journal/10.1002/(ISSN)1615-9861.
Van Den Bossche, T., Armengaud, J., Benndorf, D., Blakeley-Ruiz, J.A., Brauer, M., Cheng, K., et al., 2025. The microbiologist's guide to metaproteomics. iMeta 4 (3). Available at https://doi.org/10.1002/imt2.70031, https://onlinelibrary.wiley.com/journal/2770596X.
Wang, D.Z., Kong, L.F., Li, Y.Y., Xie, Z.X., 2016. Environmental microbial community proteomics: status, challenges and perspectives. International Journal of Molecular Sciences 17 (8). Available at https://doi.org/10.3390/ijms17081275, https://www.mdpi.com/1422-0067/17/8/1275/pdf.
Wang, S., Huang, Dy, Zhu, Qh, Li, Bz, Xu, C., Zhu, Hh, et al., 2021. Agronomic traits and ionomics influence on Cd accumulation in various sorghum (Sorghum bicolor (L.) Moench) genotypes. Ecotoxicology and Environmental Safety 214. Available at https://doi.org/10.1016/j.ecoenv.2021.112019, https://www.elsevier.com/inca/publications/store/6/2/2/8/1/9/index.htt.
Wen, S., Yin, F., Liu, C., Dang, Y., Sun, D., Li, P., 2023. Integrated analysis of transcriptomic and protein-protein interaction data reveals cadmium stress response in Geobacter sulfurreducens. Environmental Research 218, 115063. Available at https://doi.org/10.1016/j.envres.2022.115063.
Werner, A.Z., Giannone, R.J., Keller, M.J., Plavchak, C., Carper, D.L., Abraham, P.E., et al., 2025. A distinct subpopulation of membrane vesicles in Pseudomonas putida is enriched in enzymes for lignin catabolism. Applied and Environmental Microbiology 91 (10). Available at https://doi.org/10.1128/aem.01617-25, https://journals.asm.org/doi/10.1128/aem.01617-25.
Wild, C.P., 2005. Complementing the genome with an "exposome": the outstanding challenge of environmental exposure measurement in molecular epidemiology. Cancer Epidemiology Biomarkers & Prevention 14 (8), 1847–1850. Available at https://doi.org/10.1158/1055-9965.EPI-05-0456.
Wilkins, J.M., Mangalaparthi, K.K., Netzel, B.C., Sherman, W.A., Guo, Y., Kalinowska-Lyszczarz, A., et al., 2024. Proteomics analysis of periplaque and chronic inactive multiple sclerosis lesions. Frontiers in Molecular Neuroscience 17. Available at https://doi.org/10.3389/fnmol.2024.1448215, https://www.frontiersin.org/journals/molecular-neuroscience.
Wilkinson, M.D., Dumontier, M., Jan Aalbersberg, I., Appleton, G., Axton, M., Baak, A., et al., 2019. Addendum: the FAIR guiding principles for scientific data management and stewardship. Scientific Data 6 (1), 6. Available at https://doi.org/10.1038/s41597-019-0009-6.
Wintsche, B., Jehmlich, N., Popp, D., Harms, H., Kleinsteuber, S., 2018. Metabolic adaptation of methanogens in anaerobic digesters upon trace element limitation. Frontiers in Microbiology 9. Available at https://doi.org/10.3389/fmicb.2018.00405.
Wu, X., Chen, Q., Chen, L., Tian, F., Chen, X., Han, C., et al., 2022. A WRKY transcription factor, PyWRKY75, enhanced cadmium accumulation and tolerance in poplar. Ecotoxicology and Environmental Safety 239, 113630. Available at https://doi.org/10.1016/j.ecoenv.2022.113630.
Xu, Y., Deng, G., Guo, H., Yang, M., Yang, Q., 2021. Accumulation and sub cellular distribution of lead (Pb) in industrial hemp grown in Pb contaminated soil. Industrial Crops and Products 161, 113220. Available at https://doi.org/10.1016/j.indcrop.2020.113220.
Xu, X., Song, F., Zhang, G., Ma, L., Yang, N., 2024. Proteomic insights into the response of Halomonas sp. MNB13 to excess Mn(II) and the role of H2S in Mn(II) resistance.

Environmental Research 246, 118157. Available at https://doi.org/10.1016/j.envres.2024.118157.

Yin, Q., Feng, Z., Hu, Y., Zhan, X., Wu, G., 2021. Microbial interactions in pollution control ecosystems. Current Pollution Reports 7 (2), 104–114. Available at https://doi.org/10.1007/s40726-021-00181-9.

Yousuf, S., Wang, Y., Yu, P., Liu, Y.-X., 2025. Metagenomic analysis for unveiling agricultural microbiome—2nd edition. Agronomy 15 (6), 1419. Available at https://doi.org/10.3390/agronomy15061419.

Yuan, Z., Cai, S., Yan, C., Rao, S., Cheng, S., Xu, F., et al., 2024. Research progress on the physiological mechanism by which selenium alleviates heavy metal stress in plants: a review. Agronomy 14 (8), 1787. Available at https://doi.org/10.3390/agronomy14081787.

Zhou, C., Song, X., Wang, Y., Wang, H., Ge, S., 2022. The sorption and short-term immobilization of lead and cadmium by nano-hydroxyapatite/biochar in aqueous solution and soil. Chemosphere 286 (Pt 3), 131810. Available at https://doi.org/10.1016/j.chemosphere.2021.131810.

Chapter 3

Application of proteomics in remediation of petroleum products

Ulfat Jahan[1], Sohail[2], Burhan Hamid[1], Zaffar Bashir[1], and Rubiya Dar[1]
[1]*Centre of Research for Development, University of Kashmir, Srinagar, Jammu and Kashmir, India,* [2]*Department of Agriculture, Forestry and Bioresources, Forest Ecophysiology Lab Seoul National University, Seoul, South Korea*

3.1 Introduction

Petroleum is a naturally occurring, yellow-to-black liquid mixture of hydrocarbons found in geological formations. The term covers both unrefined crude oil and the refined products made from it (Al-Rubaye et al., 2023). Petroleum is generated when large quantities of marine organisms, primarily algae and zooplankton, become buried beneath sedimentary layers and, over geological time, are transformed by elevated temperatures and pressures (AL-Rubaye et al., 2022). Petroleum hydrocarbons are used globally as a fuel source, and owing to their high demand, numerous contaminations arise during exploration, production, maintenance, transportation, and storage, as well as from accidental releases leading to serious ecological consequences (Ahmed and Fakhruddin, 2018). Environmental contamination by petroleum-derived hydrocarbons is a major global concern, with potentially catastrophic effects on human health and ecosystem integrity (Spierings and Mulder, 2017). Much of this pollution stems from intensified activities in petroleum exploration, transport, and processing, while inadequate recycling of waste oil and the disposal of hazardous oil wastes into landfills without proper management further expand the number of contaminated sites (Koshlaf and Ball, 2017). Crude oil is dominated by hydrocarbons, including hazardous compounds such as benzene and polycyclic aromatic hydrocarbons (PAHs); inhalation of these substances can result in serious adverse health effects (Al-Rubaye et al., 2023). There is a clear and urgent need to remediate petroleum hydrocarbon-contaminated areas worldwide. Several traditional physicochemical methods, such as soil washing, soil vapor extraction, incineration, the use of oil booms,

Proteomics Driven Bioremediation. DOI: https://doi.org/10.1016/B978-0-443-45294-9.00008-4

and solidification, have been employed for the remediation of oil spills and contaminated soils (Koshlaf and Ball, 2017). However, these approaches are often costly, energy-intensive, and may lead to secondary environmental impacts, thereby highlighting the importance of exploring more sustainable and eco-friendly alternatives such as bioremediation (Tarfeen et al., 2022; Hamid et al., 2023). Through bioremediation, toxic compounds are transformed into non-toxic end products, including carbon dioxide (CO_2), water (H_2O), and fatty acids (Ahmed and Fakhruddin, 2018). However, several contaminants have been found to be highly recalcitrant, meaning that microorganisms either fail to metabolize them or convert them into intermediate metabolites that subsequently accumulate in the environment (Esteve-Núñez et al., 2001). Proteomics, the large-scale study of proteins, provides molecular-level insights that enhance our understanding of the interactions between organisms and their environment (Sriprapat et al., 2020). Proteomics focuses on identifying and quantifying the full complement of proteins expressed by a cell, tissue, or organism under specific conditions (Nesvizhskii, 2014). Unlike the relatively stable genome, protein expression is highly dynamic, varying across time, space, and environmental context. Because proteins possess complex three-dimensional structures and lower stability than nucleic acids, their extraction and analysis present greater technical challenges compared with genomic or transcriptomic studies. Despite these difficulties, proteomics remains indispensable for elucidating the protein expression patterns and functional pathways that underpin microbial biodegradation processes.

3.2 Petroleum and its byproducts

Petroleum is a naturally occurring hydrocarbon resource formed through geological processes and commonly found in association with specific subsurface formations. It is frequently co-generated with natural gas, leading to their combined consideration in industrial and scientific discourse, often collectively referred to as the "oil and gas" sector (Robinson and Hsu, 2017).

3.2.1 Crude oil

Crude oil is a complex mixture composed of several thousand distinct molecules, with hydrocarbons constituting the predominant components. Approximately 10% of crude oil consists of heteroatom-containing compounds, primarily incorporating sulfur, oxygen, and nitrogen (Hedgpeth et al., 2021). These hydrocarbons are commonly classified into four primary fractions—Saturates, Aromatics, Resins, and Asphaltenes (SARA)—which collectively define the light and heavy constituents of crude oil (Al-Rubaye et al., 2023).

3.2.2 Diesel

Diesel fuels are middle distillates of crude petroleum separated by fractional distillation. Other middle distillates include kerosene and aviation fuel. The carbon number of diesel oil hydrocarbons is between 11 and 25, and the distillation range is between 180°C and 380°C. Diesel oil contains 2000–4000 hydrocarbons, which cannot be totally separated by gas chromatography (Kuppusamy et al., 2020).

3.2.3 Gasoline

Gasoline, encompassing automotive gasoline (petrol) and traditional aviation gasoline (avgas), is a refined petroleum product composed of a blend of hydrocarbons and performance-enhancing additives, including various blending agents. It is predominantly utilized as a fuel in spark-ignition engines, particularly in motor vehicles and select types of aircraft (Hsu and Robinson, 2006).

3.2.4 Kerosene

Kerosene, also referred to as fuel oil No. 1, paraffin oil, lamp oil, or coal oil, is a combustible liquid hydrocarbon widely utilized as a fuel source (Speight, 2020). Kerosene is a petroleum-derived fuel widely employed in domestic heating systems, lamps, and furnaces. In addition to household applications, it serves as a fuel component in diesel and tractor engines, jet propulsion systems, and certain rocket formulations. Kerosene is also utilized industrially as a solvent for greases and in insecticide formulations, and it remains a commonly used energy source for cooking in many regions (Dioha et al., 2012).

3.3 Effects of petroleum on the environment

Petroleum and its derivatives introduce toxic hydrocarbons, heavy metals, and persistent organic pollutants into the environment, leading to long-term soil and water contamination (Varjani, 2017; Tarfeen et al., 2022; Hamid et al., 2023). The accumulation of petroleum hydrocarbons alters soil physicochemical properties, reducing nutrient availability and microbial activity essential for soil fertility (Hentati et al., 2013). Exposure to petroleum pollutants inhibits seed germination and plant growth by restricting oxygen diffusion and inducing oxidative stress in plant tissues (Merkl et al., 2005). In aquatic environments, oil contamination disrupts oxygen transfer, impairs fish respiration, and causes bioaccumulation of toxic compounds across food chains (Beyer et al., 2016). Chronic exposure to petroleum products also poses significant health risks to humans, including carcinogenic, mutagenic, and neurotoxic effects (IARC, 2012). The effect of petroleum on plants, animals, and human beings and its impact outcome are represented in Tables 3.1–3.3.

3.4 Remedial approaches

Soil contamination by petroleum-derived hydrocarbons, resulting from intensified exploration and processing activities alongside improper waste-oil management and landfill disposal of hazardous residues, remains a pervasive environmental concern with profound implications for ecosystem stability and public health (Koshlaf and Ball, 2017).

According to Cole (2018), technologies available for the remediation of petroleum hydrocarbons can be grouped in a few ways as follows:

(a) **Based on the general category:**
Treatment (technologies that destroy or treat petroleum contaminants in the environment) and containment (technologies designed solely to restrict the movement or release of petroleum hydrocarbons).

(b) **Based on the location of the application:**
In situ (treatment technologies applied on-site without removal of the contaminated medium) and ex situ (treatment technologies that require excavation of the contaminated medium and off-site remediation).

TABLE 3.1 Effect of petroleum on plants.

Observed effects	Mechanism / cause	Impact outcome	References
Inhibited seed germination and reduced growth	Hydrocarbon coating restricts oxygen and water uptake	Poor crop establishment and biomass reduction	Merkl et al. (2005)
Chlorosis and reduced photosynthetic activity	Toxic hydrocarbons disrupt chlorophyll synthesis and stomatal function	Decline in plant productivity	Mishra et al. (2021)
Altered root morphology and reduced root penetration	Soil structure changes and toxicity stress root cells	Limited nutrient absorption	Merkl et al. (2005)
Soil fertility decline	Reduction of beneficial soil	Long-term soil degradation	Hentati et al. (2013)

TABLE 3.2 Effect of petroleum on animals.

Observed effects	Mechanism / cause	Impact outcome	References
Respiratory distress in aquatic organisms	Oil films block oxygen diffusion in water bodies	Reduced oxygen availability leading to suffocation	Beyer et al. (2016)
Bioaccumulation of toxic compounds	Ingestion or dermal absorption of hydrocarbons	Toxicity transfer through the food chain	Beyer et al. (2016)
Organ damage (liver, gills, immune system)	Metabolic breakdown of petroleum releases toxic metabolites	Increased mortality and reduced immunity	Almeida et al. (2023)
Reproductive and behavioral disruption	Neurotoxic and endocrine-disrupting properties	Reduced fertility and altered migration/behavior	Almeida et al. (2023)

TABLE 3.3 Effect of petroleum on humans.

Observed effects	Mechanism / cause	Impact outcome	References
Increased cancer risk	Presence of carcinogenic compounds (e.g., benzene, PAHs)	Long-term carcinogenic effects	IARC (2012)
Neurological and developmental toxicity	Neurotoxic chemical exposure through inhalation or skin contact	Cognitive and nervous system impairment	Kim et al. (2013)
Respiratory illness	Inhalation of petroleum vapors and combustion pollutants	Asthma, bronchitis, lung irritation	Kim et al. (2013)
Skin and eye disorders	Direct dermal contact with petroleum products	Irritation, dermatitis, inflammation	IARC (2012)

(c) Based on the type of processes:

Physical (treatment processes that change the physical form of petroleum contaminants to remove them from the medium), chemical (processes involving chemical reactions that alter or destroy petroleum hydrocarbons present in the environment), thermal (processes using energy/heat to destroy or detoxify petroleum hydrocarbons), and biological (processes that use biological agents to mineralize petroleum hydrocarbons at contaminated sites).

Conventional physicochemical approaches applied to sites contaminated with total petroleum hydrocarbons include excavation and landfill disposal (dig-and-dump), soil washing and soil vapor extraction, while regulatory frameworks typically favor thermal methods such as thermal desorption and incineration. Nevertheless, these technologies are often costly, invasive, and environmentally disruptive, and may result in incomplete degradation of petroleum hydrocarbons. In contrast, bioremediation, in which the use of microorganisms to detoxify or remove pollutants presents a promising alternative, thanks to its environmentally friendly nature, relatively low cost, and minimal operational equipment (April et al., 1999). Bioremediation also represents a green-technology solution for petroleum hydrocarbon contamination. The most commonly employed bioremediation strategies for petroleum hydrocarbon-impacted sites include natural attenuation, landfarming, slurry bioreactors, composting, bioaugmentation, biostimulation, biopiling, bioventing, and phytoremediation (Nur Zaida and Piakong, 2019) Fig. 3.1 demonstrates most commonly employed bioremediation strategies for petroleum contamination.

3.4.1 Bioremediation

Hydrocarbon-degrading (hydrocarbonoclastic) microorganisms play a central role in the biodegradation of petroleum hydrocarbons due to their specialized metabolic capacities. Previous investigations on petrogenic hydrocarbon degradation have demonstrated that a wide range of microbial taxa, predominantly bacteria and fungi, are capable of utilizing hydrocarbons as their primary carbon and energy source. Among these, bacteria are widely recognized as the most metabolically versatile and active degraders, functioning as key agents in the initial breakdown of diverse hydrocarbon constituents in contaminated matrices, including soil, aquatic environments, and sludge (Brooijmans, Pastink, and Siezen, 2009; Hamid et al., 2023). Given the wide diversity of indigenous soil microorganisms capable of degrading various hydrocarbon fractions under differing environmental conditions, bioremediation strategies must often be tailored to the specific characteristics of each contaminated site (Koshlaf and Ball, 2017).

3.4.2 Natural attenuation

Natural attenuation is the most fundamental form of bioremediation, relying on indigenous microbial communities, including bacteria and fungi, to degrade, transform, or detoxify petroleum and other hydrocarbon contaminants. Through inherent metabolic processes, these microorganisms convert harmful pollutants into less toxic or non-toxic by-products, thereby contributing to the

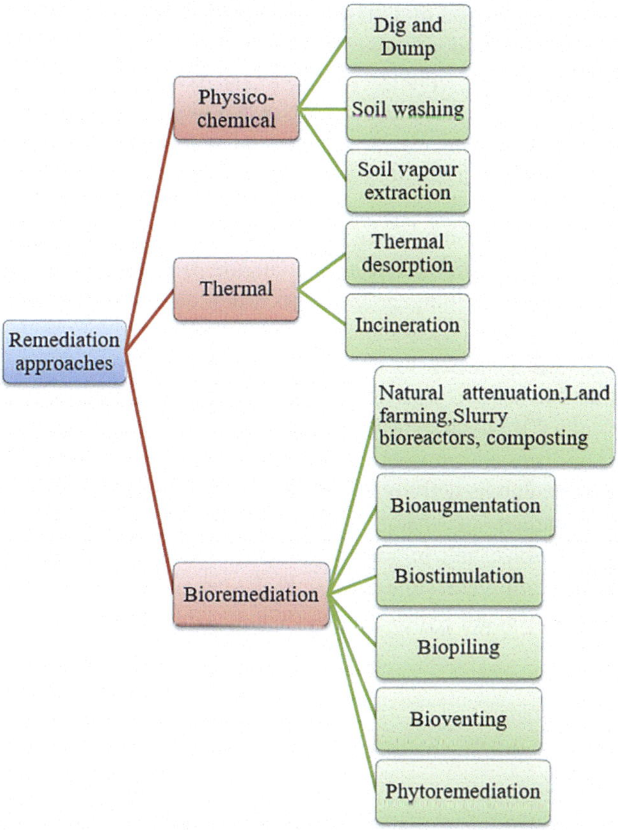

FIGURE 3.1 **Illustration of the major remediation approaches of petroleum contamination.** This figure demonstrates major remediation approaches used for the decontamination of a petroleum-polluted environment.

gradual remediation and risk reduction of contaminated environments. During this process, the indigenous microbes utilize hydrocarbons as the sole carbon source based on their natural metabolic pathways (Koshlaf and Ball, 2017). This remediation process occurs spontaneously at many contaminated locations, particularly at sites with relatively low levels of petroleum hydrocarbon pollution, and is often employed where more intensive remediation technologies are impractical or unsuitable. Natural attenuation offers notable advantages for the management of TPH-impacted sites, primarily due to its minimal operational requirements, as it largely relies on routine monitoring rather than active engineered intervention (Pilon-Smits, 2005). Biodegradation studies have demonstrated the potential of natural attenuation as a viable strategy for the remediation of TPH-contaminated environments. For example, Lv et al. (2018) reported an approximately 60% reduction in a petroleum hydrocarbon contaminant plume over several years, attributed to natural attenuation processes in both soil and groundwater. On a broader scale, it has been estimated that nearly one-quarter of petroleum-contaminated soils worldwide have been remediated primarily through natural attenuation mechanisms (Holden et al., 2002). A comparative investigation has indicated that natural attenuation can perform at a level comparable to, or in some cases surpassing, that of biostimulation and bioaugmentation strategies. The study further highlighted that indigenous hydrocarbon-degrading microorganisms, including those inherently associated with the petroleum source, possess the intrinsic metabolic capacity to degrade hydrocarbons without the need for external amendments or microbial enhancement (Makadia et al., 2011; Sheppard et al., 2012). Despite its benefits, natural attenuation is frequently constrained by limited nutrient availability, which can restrict microbial growth and degradative activity. Furthermore, contaminated sites may lack sufficiently abundant or metabolically competent hydrocarbon-degrading microbial populations, and in some cases, the resident microorganisms may not possess the complete suite of catabolic genes needed to achieve full degradation of petroleum hydrocarbons (Kuppusamy et al., 2020).

3.4.2.1 Landfarming

Landfarming is a bioremediation technique in which excavated Petroleum hydrocarbon-contaminated soil is evenly spread in thin layers over a treatment area. To enhance microbial activity, amendments such as nutrients, minerals, moisture, and oxygen are routinely supplied to stimulate indigenous hydrocarbon-degrading microorganisms. These native microbial communities metabolize petroleum hydrocarbons as their primary carbon and energy source, ultimately transforming them into benign end products such as carbon dioxide and water. During landfarming, petroleum hydrocarbons are primarily reduced in soil through a combination of volatilization, biodegradation, and adsorption processes (Hejazi et al., 2003). Landfarming has been reported to be highly effective in reducing hydrocarbon contamination. (Maila and Cloete, 2004) observed a 95% decrease in total petroleum hydrocarbons following landfarming treatment. Similarly, (Mmom and Deekor, 2010) documented reductions of up to 82% in petroleum hydrocarbons and 50% in PAHs after applying the same remediation technique. A major limitation of the landfarming technique is its requirement for extensive treatment areas, which can elevate the potential for human exposure to total petroleum hydrocarbons (TPHs). Nonetheless, this exposure is typically transient, as TPHs undergo biodegradation when favorable environmental conditions are maintained.

3.4.3 Bioreactors

A bioreactor is an engineered system designed to enhance biochemical transformations by optimizing the activity of microorganisms or isolated cellular components such as enzymes. In soil remediation, slurry-phase bioreactors are commonly employed to treat oil-contaminated soils. This approach is considered one of the most rapid methods for degrading total petroleum hydrocarbons, as the slurry conditions improve mass transfer and facilitate efficient contact between pollutants and microbial cells (Onwurah et al., 2007). However, they are generally more expensive than in situ or land-based treatment methods due to the specialized equipment and addational operational support (Ward et al., 2003; Chikere et al., 2012) reported more than 95% degradation of total petroleum hydrocarbons in crude oil-contaminated marine sediments treated in a bioreactor enriched with both inorganic and organic fertilizers. Similarly, Kuyukina et al. (2003) observed an 88% reduction in oil concentration within two months in an aerobic slurry bioreactor amended with an oleophilic biofertilizer derived from *Rhodococcus* surfactant complexes.

3.4.4 Composting

Composting is another valuable bioremediation technique for managing oil-contaminated environments. The process entails mechanically blending polluted soil or sediment with compost rich in hydrocarbon-degrading microorganisms, under controlled aerobic and thermophilic conditions that promote microbial activity. Onwurah et al. (2007); Ramavandi et al. (2018) demonstrated that a cocomposting system composed of oily sludge, wheat-straw bulking material, and mature compost achieved significant remediation performance, removing approximately 65% of total petroleum hydrocarbons within 63 days. Likewise, Sinha et al. (2008) reported that vermicomposting using a mixed population of *Eisenia fetida*, *Eisenia euginae,* and *Perionyx excavatus* of varying ages and sizes substantially enhanced the degradation of PAHs in heavily contaminated gas-work site soil, which initially contained 11,820 mg kg^{-1} of total PAHs, over a 12-week period.

3.4.5 Bioaugmentation

Bioaugmentation represents an alternative to in situ bioremediation strategies. It entails introducing selected microbial strains or consortia, primarily bacteria, and occasionally fungi, with demonstrated hydrocarbon-degrading capabilities to contaminated sites. When supported by appropriate environmental conditions, these microorganisms enhance and accelerate the biodegradation of total petroleum hydrocarbons (Alvarez and Illman, 2006). The primary objective of bioaugmentation is to enrich the contaminated environment with an efficient microbial community, thereby enhancing the overall biodegradation performance and functional capacity of the system (El Fantroussi and Agathos, 2005). It is important to recognize that microorganisms vary widely in their ability to degrade different petroleum hydrocarbons. Consequently, selecting appropriate microbial strains is critical for achieving effective TPH remediation. This requires thorough screening and evaluation of their biodegradation potential before conducting bioremediation experiments. Historically, the introduction of cultured or nonindigenous microorganisms capable of degrading petroleum hydrocarbons has been considered a viable strategy for managing petroleum-contaminated environments (Kuppusamy et al., 2020).

Throne-Holst et al. (2007) demonstrated that a single *Acinetobacter* culture was capable of degrading petroleum hydrocarbons spanning carbon chain lengths from C10 to C40. (Gentry et al., 2004) further reported that microbial strains capable of producing biosurfactants exhibited enhanced petroleum hydrocarbons degradation efficiency. Numerous bioaugmentation studies have identified both Gram-negative bacteria, particularly species within the genera *Pseudomonas*, *Flavobacterium*, *Sphingomonas*, and *Achromobacter* (El Fantroussi and Agathos, 2005), and Gram-positive bacteria such as *Mycobacterium*, *Bacillus*, and *Rhodococcus* as effective degraders of total petroleum hydrocarbons. Fungal species, including *Aspergillus*, *Penicillium*, and *Verticillium,* have also been applied successfully in petroleum hydrocarbons bioaugmentation. Although yeasts and algae possess the ability to degrade crude oil, available studies remain limited (Ezeji et al., 2007). Among these microbial groups, bacteria are generally favored for petroleum hydrocarbons bioremediation due to their high abundance, metabolic versatility, and robust degradative capabilities. Soil pH, temperature, moisture levels, and nutrient availability all play critical roles in determining the efficiency of petroleum hydrocarbons degradation. According to Bossert (1984), temperatures between 30°C and 40°C are optimal for crude oil biodegradation. At lower temperatures, degradation slows because increased oil viscosity limits dispersion and reduces the available surface area for microbial attack. Nutrient amendments, such as slow-release fertilizers, have been shown to enhance biodegradation rates in contaminated soils (Xu et al., 2005). Conversely, extreme pH conditions, insufficient oxygen, and inadequate moisture suppress microbial growth and activity, ultimately resulting in reduced petroleum hydrocarbons degradation (Ezeji et al., 2007).

3.4.6 Biostimulation

Another widely applied bioremediation approach that leverages the natural ability of microorganisms to degrade or detoxify TPHs in contaminated environments is known as biostimulation. This approach enhances the growth and metabolic activity of indigenous petroleum hydrocarbon-degrading microorganisms by supplying essential nutrients, thereby accelerating the natural biodegradation process (Nikolopoulou and Kalogerakis, 2019). Numerous studies have shown that sites contaminated with high concentrations of petroleum hydrocarbons, often composed of up to 80% carbon, quickly become depleted of essential nutrients such as nitrogen and phosphorus. This nutrient imbalance suppresses the growth and metabolic activity of TPH-degrading microorganisms, ultimately slowing the degradation process (Walworth et al., 2007). In such nutrient-limited environments, biostimulation, typically involving the addition of nutrients and electron acceptors such as nitrogen, phosphorus, carbon sources, and oxygen, is recommended as an effective strategy to enhance microbial activity and improve bioremediation outcomes (Li et al., 2007). The primary advantage of biostimulation is that it relies on native microbial communities that are already well adapted to the local environmental conditions, enabling more efficient and sustained biodegradation. A variety of organic and inorganic amendments such as nutrient supplements, surfactants, fresh or composted sewage sludge, and manure have been identified as effective biostimulating agents for enhancing petroleum degradation (Blyth et al., 2015).

3.5 Proteomics and its application in remediation of petroleum hydrocarbons

During bioremediation, contaminants at polluted sites can be removed through their breakdown by microorganisms. A wide range of bacteria, fungi, and yeasts have been employed to remediate polluted environments (Goutam et al., 2018; Saxena and Bharagava, 2016). Nevertheless, microbial species differ considerably in their metabolic abilities to degrade, transform, or detoxify pollutants. The efficiency of these processes is further influenced by the bioavailability and accessibility of contaminants to the degrading organisms (Antizar-Ladislao, 2010; Singh et al., 2018). To obtain microbial strains with reliable bioremediation potential, pure cultures with demonstrable degradation capabilities are often isolated and characterized (Fulekar, 2005). Advances in genomics have significantly facilitated the isolation and development of pure microbial cultures for the bioremediation of both organic and inorganic pollutants (Merman and Nelson, 2002). Genome sequences for many microorganisms known to participate in bioremediation are now publicly available (Lovley, 2003). Access to complete genome information has greatly enhanced our ability to examine gene expression patterns under diverse environmental conditions through DNA microarray analyses (Merman and Nelson, 2002). Furthermore, next-generation sequencing (NGS) technologies provide high-throughput sequencing capabilities, enabling the characterization of microorganisms that cannot be cultured but remain viable in environmental samples. For microorganisms that participate in both in situ and ex situ bioremediation, examining their global proteomic profiles is essential. Proteomic analyses, which assess the complete complement of proteins expressed within a cell, offer valuable information about microbial functional capabilities. Contemporary research increasingly relies on proteomics, as it provides deep insights into the physiological and metabolic states of microorganisms (Singh and Nagaraj, 2006).

3.5.1 Proteomics

The concepts of proteomics and the proteome were first introduced in 1995 (Wasinger et al., 1995), arising as major post-genomic developments following the rapid expansion of large and complex genome-sequencing datasets. Mass spectrometry has become a central tool for defining the protein composition of cells and for examining how it is shaped by post-translational modifications such as phosphorylation. These modifications often function as regulatory switches that modulate protein activity, stability, and interactions, thereby exerting substantial influence on cellular function and physiological outcomes (Yunusa and Umar, 2022). Encompassed within this broader field are proteogenomics, which applies proteomic data to improve gene annotation, and structural genomics, which seeks to determine protein three-dimensional structures through experimental and computational approaches (Zhao and Poh, 2008). Proteomics provides a more complete depiction of the expressed phenotype of a microbial species than genomic information alone, offering direct insight into the proteins produced under specific conditions (Singh and Nagaraj, 2006). Traditionally, proteomic analyses have relied on multidimensional polyacrylamide gel electrophoresis and mass spectrometry. However, advances in analytical technology have led to the development of multidimensional protein identification technology (MudPIT), which has emerged as a powerful approach for large-scale proteome characterization (Paoletti et al., 2004).

3.5.2 Role of proteomics in bioremediation

Proteomic analyses have emerged as a powerful tool for identifying biomarkers indicative of environmental pollutants like PAHs (Aardema and MacGregor, 2002). During the microbial bioremediation of PAHs, profound modifications occur in cell surface receptors and other proteins of the degrading organism. By applying proteomic analyses, such molecular-level alterations can be systematically monitored and quantified, enabling the detection of significant functional changes (Yunusa and Umar, 2022). Proteomics plays a direct role in understanding microbial biodegradation, because the expression of many proteins is regulated in response to external environmental stimuli—such as the presence of pollutants (Wilkins et al., 2001; Kim et al., 2002). Several bacterial genera, including *Pseudomonas*, *Sphingomonas*, *Mycobacterium*, and *Rhodococcus*, are well known for their capacity to degrade PAHs. These organisms frequently possess catabolic genes encoding enzymes responsible for the initial oxidation and subsequent ring-cleavage reactions necessary for PAHs breakdown (Davletgildeeva and Kuznetsov, 2024). Seong-Jae Kim and colleagues conducted simulations of the Deepwater Horizon (DWH) oil spill and examined the bioremediation potential of *Mycobacterium vanbaalenii* PYR-1. After 30 days of treatment, a substantial portion of the petroleum constituents had been fully degraded. Macroproteomic profiling throughout the process revealed pronounced upregulation of multiple protein groups, including catalytic enzymes, transcriptional regulators, and stress-response proteins. These findings indicate that PYR-1 undergoes extensive physiological adjustments when exposed to petroleum contaminants and actively contributes to hydrocarbon degradation (Kim et al., 2015). In a separate investigation focused on benzene-responsive proteins in *Bacillus* sp. M4, Kureel et al. (2018) identified seven such proteins. Five of these had previously been characterized and included a 2Fe–2 S ferredoxin-type iron–sulfur cluster–binding domain, an ABC transporter, a methionine ABC transporter ATP-binding subunit, a peptidase S24-like protein, and a polyketide synthase dehydratase. Phylogenetic analysis of the two hypothetical proteins suggested that they are likely associated with pore formation and membrane localization. Collectively, all seven identified proteins appear to contribute significantly to the benzene biodegradation process (Kureel et al., 2018). Many constituents of petroleum hydrocarbons, including chlorinated and aromatic compounds, exert toxic effects on microorganisms. When exposed to such stressors, microbial cells not only produce catalytic enzymes to breakdown these harmful compounds but also upregulate stress-related proteins to safeguard cellular integrity. As a membrane-associated molecule, glycerol can further enhance a microbe's ability to cope with environmental stress, supporting cellular resilience under toxic conditions (Yang et al., 2019). Proteomic analysis by Prashanth Bhaganna et al. demonstrated this phenomenon: *Pseudomonas putida* grown in benzene showed a 15-fold increase in chaotrope-stress-response proteins when supplemented with glycerol, enhancing protection of cellular macromolecules (Bhaganna et al., 2016). Similarly, Marlén Pöritz and colleagues examined the proteome of *Dehalococcoides mccartyi*strain DCMB5 during dechlorination of multiple chlorinated substrates. In addition to identifying several dehalogenases, they observed strong upregulation of proteins involved in fimbrial formation, suggesting that fimbriae assembly is a characteristic feature of the dehalogenation process (Pöritz et al., 2015). Advances in multiomics technologies have markedly accelerated research on petroleum-pollution bioremediation, greatly expanding databases of microorganisms, genes, and proteins involved in hydrocarbon degradation. These tools have also deepened our understanding of how microbial communities evolve and how genetic and protein-level adaptations underpin biodegradation processes (Gillan et al., 2015; Khomarbaghi et al., 2019). However, it is important to recognize that omics approaches primarily enhance our ability to observe and interpret microbial functions; they do not enable the creation of "superbugs" capable of degrading all petroleum hydrocarbons. Our knowledge of microbial physiology and particularly of interactions within and between microbial populations remains limited. Greater emphasis on microbial communication and interspecies interactions will be crucial for developing a systems-level understanding and for guiding the rational design of effective artificial microbial consortia for hydrocarbon degradation (Yang et al., 2019).

3.6 Conclusion

Petroleum contamination continues to pose significant ecological and public health challenges, necessitating remediation strategies that are both effective and environmentally sustainable. Bioremediation stands out as a promising approach due to its reliance on naturally occurring microorganisms capable of transforming hazardous hydrocarbons into benign end products. However, the complexity of petroleum mixtures and the variable metabolic capacities of microbial communities require deeper insight into the biochemical processes governing hydrocarbon degradation. Proteomics provides this critical understanding by revealing the proteins, pathways, and regulatory networks activated in microorganisms during pollutant exposure. Studies employing proteomic analyses have identified key enzymes involved in hydrocarbon catabolism, stress-response systems that enhance microbial resilience, and structural proteins linked to pollutant interaction and uptake. These molecular insights not only improve our ability to assess biodegradation potential but also lay the groundwork for designing optimized remediation strategies, including tailored microbial consortia and targeted biostimulation/bioaugmentation approaches. As omics technologies continue to advance, the integration of proteomics with genomics, metabolomics, and metagenomics will further refine our understanding of microbial ecology in contaminated environments. Ultimately, such multidisciplinary approaches will strengthen the scientific foundation required to develop efficient, scalable, and sustainable solutions for petroleum hydrocarbon remediation.

References

AL-Rubaye, A.H., Suwaid, M.A., Al-Muntaser, A.A., Varfolomeev, M.A., Rakhmatullin, I.Z., Hakimi, M.H., et al., 2022. Intensification of the steam stimulation process using bimetallic oxide catalysts of MFe2O4 (M=Cu, Co, Ni) for in-situ upgrading and recovery of heavy oil. Federation Journal of Petroleum Exploration and Production Technology 12 (3), 577–587. Available at https://doi.org/10.1007/s13202-021-01311-1, http://www.springerlink.com/content/2190-0566/.

Aardema, M.J., MacGregor, J.T., 2002. Toxicology and genetic toxicology in the new era of "toxicogenomics": impact of "-omics" technologies. Mutation Research/Fundamental and Molecular Mechanisms of Mutagenesis 499 (1), 13–25.

Ahmed, F., Fakhruddin, A.N.M., 2018. A review on environmental contamination of petroleum hydrocarbons and its biodegradation. International Journal of Environmental Sciences & Natural Resources 11 (3), 1–7.

Al-Rubaye, A.H., Jasim, D.J., Ameen, H.F.M., Al-Robai, H.A., Al-Assal, J.R., 2023. The impacts of petroleum on environment. IOP Conference Series: Earth and Environmental Science 3, 17551315. Available at 10.1088/1755-1315/1158/3/032014, https://iopscience.iop.org/journal/1755-1315.

Almeida, J.R., Oliveira, M., Gravato, C., Guilhermino, L., 2023. Toxicological effects of petroleum exposure on aquatic organisms: physiological and ecological perspectives. Environmental Toxicology and Chemistry 42 (3), 512–529.

Alvarez, P.J., Illman, W.A., 2006. Geochemical attenuation mechanisms. Bioremediation and Natural Attenuation: Process Fundamentals and Mathematical Models. John Wiley & Sons, Ltd, Hoboken, pp. 25–48.

Antizar-Ladislao, B., 2010. Bioremediation: working with bacteria. Elements 6 (6), 389–394. Available at https://doi.org/10.2113/gselements.6.6.389.

April, T.M., Foght, J.M., Currah, R.S., 1999. Hydrocarbon-degrading filamentous fungi isolated from flare pit soils in northern and western Canada. Canadian Journal of Microbiology 46 (1), 38–49.

Beyer, J., Trannum, H.C., Bakke, T., Hodson, P.V., Collier, T.K., 2016. Environmental effects of the Deepwater Horizon oil spill: a review. Marine Pollution Bulletin 110 (1), 28–51. Available at https://doi.org/10.1016/j.marpolbul.2016.06.027, www.elsevier.com.

Bhaganna, P., Bielecka, A., Molinari, G., Hallsworth, J.E., 2016. Protective role of glycerol against benzene stress: insights from the Pseudomonas putida proteome. Current Genetics 62 (2), 419–429.

Blyth, W., Shahsavari, E., Morrison, P.D., Ball, A.S., 2015. Biosurfactant from red ash trees enhances the bioremediation of PAH contaminated soil at a former gasworks site. Journal of Environmental Management 162, 30–36. Available at https://doi.org/10.1016/j.jenvman.2015.07.041, https://www.sciencedirect.com/journal/journal-of-environmental management

Bossert, I., 1984. The fate of petroleum in soil ecosystem. Petroleum Microbiology. 355–398.

Brooijmans, R.J.W., Pastink, M.I., Siezen, R.J., 2009. Hydrocarbon-degrading bacteria: the oil-spill clean-up crew. Microbial Biotechnology 2 (6), 587 594. Available at https://doi.org/10.1111/j.1751-7915.2009.00151.x.

Chikere, C.B., Chikere, B.O., Okpokwasili, G.C., 2012. Bioreactor-based bioremediation of hydrocarbon-polluted Niger Delta marine sediment, Nigeria. 3 Biotech 2 (1), 53–66. Available at https://doi.org/10.1007/s13205-011-0030-8, https://www.springer.com/journal/13205.

Cole, G.M., 2018. Assessment and Remediation of Petroleum Contaminated Sites. CRC Press.

Davletgildeeva, A.T., Kuznetsov, N.A., 2024. Bioremediation of polycyclic aromatic hydrocarbons by means of bacteria and bacterial enzymes. Federation Microorganisms 12 (9). Available at https://doi.org/10.3390/microorganisms12091814, www.mdpi.com/journal/microorganisms.

Dioha, I.J., Ikeme, C.H., Tijjam, N., Dioha, E.C., 2012. Comparative studies of ethanol and kerosene fuels and cook stove performance. Journal of Natural Sciences Research 2 (6), 34–39.

El Fantroussi, S., Agathos, S.N., 2005. Is bioaugmentation a feasible strategy for pollutant removal and site remediation? Current Opinion in Microbiology 8 (3), 268–275. Available at https://doi.org/10.1016/j.mib.2005.04.011.

Esteve-Núñez, A., Caballero, A., Ramos, J.L., 2001. Biological degradation of 2, 4, 6-trinitrotoluene. Microbiology and Molecular Biology Reviews 65 (3), 335–352.

Ezeji, U., Anyadoh, S.O., Ibekwe, V.I., 2007. Clean up of crude oil-contaminated soil. Terrestrial and Aquatic Environmental Toxicology 1 (2), 54–59.

Fulekar, M.H., 2005. Bioremediation technologies for environment. Indian Journal of Environmental Protection 25 (4), 358–364. Available at http://ijep.co.in/archives.

Gentry, T., Rensing, C., Pepper, I., 2004. New approaches for bioaugmentation as a remediation technology. Critical Reviews in Environmental Science and Technology 34 (5), 447–494. Available at https://doi.org/10.1080/10643380490452362.

Gillan, D.C., Roosa, S., Kunath, B., Billon, G., Wattiez, R., 2015. The long-term adaptation of bacterial communities in metal-contaminated sediments: a metaproteogenomic study. Environmental Microbiology 17 (6), 1991–2005. Available at https://doi.org/10.1111/1462-2920.12627Belgium, http://onlinelibrary.wiley.com/journal/10.1111/(ISSN)1462-2920.

Goutam, S.P., Saxena, G., Singh, V., Yadav, A.K., Bharagava, R.N., Thapa, K.B., 2018. Green synthesis of TiO2 nanoparticles using leaf extract of Jatropha curcas L. for photocatalytic degradation of tannery wastewater. Chemical Engineering Journal 336, 386–396. Available at https://doi.org/10.1016/j.cej.2017.12.029, www.elsevier.com/inca/publications/store/6/0/1/2/7/3/index.htt.

Hamid, B., Majeed, N., Ganai, B.A., Hassan, S., Bashir, Z., Wani, P.A., et al., 2023. Heavy-metal tolerant bacterial strains isolated from industrial sites and scrap yards in Kashmir, India. Journal of Basic Microbiology 63 (12), 1361–1372. Available at https://doi.org/10.1002/jobm.202300349, http://onlinelibrary.wiley.com/journal/10.1002/(ISSN)1521-4028.

Hedgpeth, B.M., McFarlin, K.M., Prince, R.C., 2021. Crude oils and their fate in the environment. Petrodiesel Fuels: Science, Technology, Health, and Environment. CRC Press, United States, pp. 891–910. Available at http://doi.org/10.1201/9780367456252-5, https://www.taylorfrancis.com/books/edit/10.1201/9780367456252/petrodiesel-fuels-ozcan-konur.

Hejazi, R.F., Husain, T., Khan, F.I., 2003. Landfarming operation of oily sludge in arid region - Human health risk assessment. Journal of Hazardous Materials 99 (3), 287–302. Available at https://doi.org/10.1016/S0304-3894(03)00062-1, www.elsevier.com/locate/jhazmat.

Hentati, O., Lachhab, R., Ayadi, M., Ksibi, M., 2013. Toxicity assessment for petroleum-contaminated soil using terrestrial invertebrates and plant bioassays. Environmental Monitoring and Assessment 185 (4), 2989–2998. Available at https://doi.org/10.1007/s10661-012-2766-y.

Holden, P.A., LaMontagne, M.G., Bruce, A.K., Miller, W.G., Lindow, S.E., 2002. Assessing the role of Pseudomonas aeruginosa surface-active gene expression in hexadecane biodegradation in sand. Applied and Environmental Microbiology 68 (5), 2509–2518. Available at https://doi.org/10.1128/aem.68.5.2509-2518.2002.

Hsu, C.S., Robinson, P.R. (Eds.), 2006. Practical Advances in Petroleum Processing 1. Springer, New York, pp. 23–34.

IARC, A., 2012. Review of human carcinogens: chemical agents and related occupations. IARC Monographs on the Evaluation of Carcinogenic Risks to Humans 100, 225–248.

Khomarbaghi, Z., Shavandi, M., Amoozegar, M.A., Dastgheib, S.M.M., 2019. Bacterial community dynamics during bioremediation of alkane- and PAHs-contaminated soil of Siri island, Persian Gulf: a microcosm study. International Journal of Environmental Science and Technology 16 (12), 7849–7860. Available at https://doi.org/10.1007/s13762-018-02198-y.

Kim, K.H., Jahan, S.A., Kabir, E., Brown, R.J.C., 2013. A review of airborne polycyclic aromatic hydrocarbons (PAHs) and their human health effects. Environment International 60, 71–80. Available at https://doi.org/10.1016/j.envint.2013.07.019, www.elsevier.com/locate/envint.

Kim, S.I., Kim, S.J., Nam, M.H., Kim, S., Ha, K.S., Oh, K.H., et al., 2002. Proteome analysis of aniline-induced proteins in Acinetobacter lwoffii K24. Current Microbiology 44 (1), 61–66. Available at https://doi.org/10.1007/s00284-001-0075-8.

Kim, S.J., Kweon, O., Sutherland, J.B., Kim, H.L., Jones, R.C., Burback, B.L., et al., 2015. Dynamic response of Mycobacterium vanbaalenii PYR-1 to BP deepwater horizon crude oil. Applied and Environmental Microbiology 81 (13), 4263–4276. Available at https://doi.org/10.1128/AEM.00730-15, http://aem.asm.org/content/81/13/4263.full.pdf.

Koshlaf, E., Ball, A.S., 2017. Soil bioremediation approaches for petroleum hydrocarbon polluted environments. AIMS Microbiology 3 (1), 25–49. Available at https://doi.org/10.3934/MICROBIOL.2017.1.25, http://www.aimspress.com/article/doi/10.3934/microbiol.2017.1.25.

Kuppusamy, S., Maddela, N.R., Megharaj, M., Venkateswarlu, K., 2020. Total Petroleum Hydrocarbons: Environmental Fate, Toxicology, and Remediation. Springer.

Kureel, M.K., Geed, S.R., Rai, B.N., Singh, R.S., 2018. Novel investigation of the performance of continuous packed bed bioreactor (CPBBR) by isolated Bacillus sp. M4 and proteomic study. Bioresource Technology 266, 335–342.

Kuyukina, M.S., Ivshina, I.B., Ritchkova, M.I., Philp, J.C., Cunningham, C.J., Christofi, N., 2003. Bioremediation of crude oil-contaminated soil using slurry-phase biological treatment and land farming techniques. Federation Soil and Sediment Contamination 12 (1), 85–99. Available at https://doi.org/10.1080/713610962, http://www.tandfonline.com/toc/bssc20/current.

Li, H., Zhao, Q., Boufadel, M.C., Venosa, A.D., 2007. A universal nutrient application strategy for the bioremediation of oil-polluted beaches. Marine Pollution Bulletin 54 (8), 1146–1161. Available at https://doi.org/10.1016/j.marpolbul.2007.04.015.

Lovley, D.R., 2003. Cleaning up with genomics: applying molecular biology to bioremediation. Nature Reviews. Microbiology 1 (1), 35–44. Available at https://doi.org/10.1038/nrmicro731.

Lv, H., Su, X., Wang, Y., Dai, Z., Liu, M., 2018. Effectiveness and mechanism of natural attenuation at a petroleum-hydrocarbon contaminated site. Chemosphere 206, 293–301. Available at https://doi.org/10.1016/j.chemosphere.2018.04.171.

Malla, M.P., Cloete, T.E., 2004. Bioremediation of petroleum hydrocarbons through landfarming: are simplicity and cost-effectiveness the only advantages? Reviews in Environmental Science and Biotechnology 3 (4), 349–360. Available at https://doi.org/10.1007/s11157-004-6653-z.

Makadia, T.H., Adetutu, E.M., Simons, K.L., Jardine, D., Sheppard, P.J., Ball, A.S., 2011. Re-use of remediated soils for the bioremediation of waste oil sludge. Journal of Environmental Management 92 (3), 866–871. Available at https://doi.org/10.1016/j.jenvman.2010.10.059, https://www.sciencedirect.com/journal/journal-of-environmental-management.

Merkl, N., Schultze-Kraft, R., Infante, C., 2005. Phytoremediation in the tropics – influence of heavy crude oil on root morphological characteristics of graminoids. Environmental Pollution 138 (1), 86–91. Available at https://doi.org/10.1016/j.envpol.2005.02.023.

Merman, W.C., Nelson, K.E., 2002. Genomics for applied microbiology. Advances in Applied Microbiology 51, 201–245. Available at https://doi.org/10.1016/S0065-2164(02)51007-8, http://www.sciencedirect.com/science/journal/00652164.

Mishra, S., Lin, Z., Pang, S., 2021. Plant responses to petroleum hydrocarbon contamination: a review. Chemosphere 277, 130–135.

Mmom, P.C., Deekor, T., 2010. Assessing the effectiveness of land farming in the remediation of hydrocarbon polluted soils in the Niger Delta, Nigeria. Research Journal of Applied Sciences, Engineering and Technology 2 (7), 654–660. Available at http://maxwellsci.com/print/rjaset/v2-654-660.pdf.

Nesvizhskii, A.I., 2014. Proteogenomics: concepts, applications and computational strategies. Nature Methods 11 (11), 1114–1125. Available at https://doi.org/10.1038/NMETH.3144, http://www.nature.com/nmeth/.

Nikolopoulou, M., Kalogerakis, N., 2019. Biostimulation strategies for enhanced bioremediation of marine oil spills including chronic pollution. In: Consequences of Microbial Interactions with Hydrocarbons, Oils, and Lipids: Biodegradation and Bioremediation. Springer, Cham, pp. 89–98.

Nur Zaida, Z., Piakong, M.T., 2019. Bioaugmentation of petroleum hydrocarbon in contaminated soil: a review. Microbial Action on Hydrocarbons 415–439.

Onwurah, I.N. E., Ogugua, V.N., Onyike, N.B., Ochonogor, A.E., & Otitoju, O.F., 2007. Crude Oil Spills in the Environment, Effects and Some Innovative Clean-up Biotechnologies.

Paoletti, A.C., Zybailov, B., Washburn, M.P., 2004. Principles and applications of multidimensional protein identification technology. Expert Review of Proteomics 1 (3), 275–282. Available at https://doi.org/10.1586/14789450.1.3.275.

Pilon-Smits, E., 2005. Phytoremediation. Annual Review of Plant Biology 56 (1), 15–39. Available at https://doi.org/10.1146/annurev.arplant.56.032604.144214.

Pöritz, M., Schiffmann, C.L., Hause, G., Heinemann, U., Seifert, J., Jehmlich, N., et al., 2015. Dehalococcoides mccartyi strain DCMB5 respires a broad spectrum of chlorinated aromatic compounds. Applied and Environmental Microbiology 81 (2), 587–596. Available at https://doi.org/10.1128/AEM.02597-14, http://aem.asm.org/content/81/2/587.full.pdf.

Ramavandi, B., Ghafarizadeh, F., Alavi, N., Babaei, A.A., Ahmadi, M., 2018. Biotreatment of total petroleum hydrocarbons from an oily sludge using co-composting approach. Soil and Sediment Contamination: An International Journal 27 (6), 524–537. Available at https://doi.org/10.1080/15320383.2018.1489371.

Robinson, P.R., Hsu, C.S., 2017. Petroleum and its products. Handbook of Industrial Chemistry and Biotechnology. Springer International Publishing, Cham, pp. 13–106.

Saxena, G., Bharagava, R.N., 2016. Ram Chandra: advances in biodegradation and bioremediation of industrial waste. Clean Technologies and Environmental Policy 18 (3), 979–980. Available at https://doi.org/10.1007/s10098-015-1084-9.

Sheppard, P.J., Simons, K.L., Kadali, K.K., Patil, S.S., Ball, A.S., 2012. The importance of weathered crude oil as a source of hydrocarbonoclastic microorganisms in contaminated seawater. Journal of Microbiology and Biotechnology 22 (9), 1185–1192. Available at https://doi.org/10.4014/jmb.1201.01049Australia, http://www.jmb.or.kr/journal/download.php?Filedir=./submission/Journal/022/&num=4983.

Singh, O.V., Nagaraj, N.S., 2006. Transcriptomics, proteomics and interactomics: Unique approaches to track the insights of bioremediation. Briefings in Functional Genomics and Proteomics 4 (4), 355–362. Available at https://doi.org/10.1093/bfgp/eli006.

Singh, V.K., Singh, A.L., Singh, R., Kumar, A., 2018. Iron oxidizing bacteria: insights on diversity, mechanism of iron oxidation and role in management of metal pollution. Environmental Sustainability 1 (3), 221–231.

Sinha, R.K., Bharambe, G., Ryan, D., 2008. Converting wasteland into wonderland by earthworms - a low-cost nature's technology for soil remediation: a case study of vermiremediation of PAHs contaminated soil. The Environmentalist 28 (4), 466–475. Available at https://doi.org/10.1007/s10669-008-9171-7.

Speight, J.G., 2020. The Refinery of the Future. Gulf Professional Publishing.

Spierings, E.L., Mulder, M.J., 2017. Persistent orofacial muscle pain: its synonymous terminology and presentation. CRANIO® 35 (5), 304–307.

Sriprapat, W., Roytrakul, S., Thiravetyan, P., 2020. Proteomic studies of plant and bacteria interactions during benzene remediation. Journal of Environmental Sciences 94, 161–170. Available at https://doi.org/10.1016/j.jes.2020.03.052.

Tarfeen, N., Nisa, K.U., Hamid, B., Bashir, Z., Yatoo, A.M., Dar, et al., 2022. Microbial remediation: a promising tool for reclamation of contaminated sites with special emphasis on heavy metal and pesticide pollution: a review. Processes 10 (7), 1358.

Throne-Holst, M., Wentzel, A., Ellingsen, T.E., Kotlar, H.K., Zotchev, S.B., 2007. Identification of novel genes involved in long-chain n-alkane degradation by Acinetobacter sp. strain DSM 17874. Applied and Environmental Microbiology 73 (10), 3327–3332. Available at https://doi.org/10.1128/AEM.00064-07.

Varjani, S.J., 2017. Microbial degradation of petroleum hydrocarbons. Bioresource Technology 223, 277–286. Available at https://doi.org/10.1016/j.biortech.2016.10.037.

Walworth, J., Pond, A., Snape, I., Rayner, J., Ferguson, S., Harvey, P., 2007. Nitrogen requirements for maximizing petroleum bioremediation in a sub-Antarctic soil. Cold Regions Science and Technology 48 (2), 84–91. Available at https://doi.org/10.1016/j.coldregions.2006.07.001.

Ward, O., Singh, A., Van Hamme, J., 2003. Accelerated biodegradation of petroleum hydrocarbon waste. Journal of Industrial Microbiology and Biotechnology 30 (5), 260–270. Available at https://doi.org/10.1007/s10295-003-0042-4.

Wasinger, V.C., Cordwell, S.J., Poljak, A., Yan, J.X., Gooley, A.A., Wilkins, M.R., et al., 1995. Progress with gene-product mapping of the Mollicutes: mycoplasma genitalium. Electrophoresis 16 (1), 1090–1094. Available at https://doi.org/10.1002/elps.11501601185.

Wilkins, J.C., Homer, K.A., Beighton, D., 2001. Altered protein expression of Streptococcus oralis cultured at low pH revealed by two-dimensional gel electrophoresis. Applied and Environmental Microbiology 67 (8), 3396–3405. Available at https://doi.org/10.1128/AEM.67.8.3396-3405.2001.

Xu, R., Yong, L.C., Lim, Y.G., Obbard, J.P., 2005. Use of slow-release fertilizer and biopolymers for stimulating hydrocarbon biodegradation in oil contaminated beach sediments. Marine Pollution Bulletin 51 (8-12), 1101–1110. Available at https://doi.org/10.1016/j.marpolbul.2005.02.037.

Yang, R., Liu, G., Chen, T., Li, S., An, L., Zhang, G., et al., 2019. Characterization of the genome of a Nocardia strain isolated from soils in the Qinghai-Tibetan Plateau that specifically degrades crude oil and of this biodegradation. Genomics 111 (3), 356–366. Available at https://doi.org/10.1016/j.ygeno.2018.02.010.

Yunusa, Y.R., Umar, Z.D., 2022. Effective microbial bioremediation via the multi-omics approach: an overview of trends, problems and prospects. UMYU Journal of Microbiology Research 6 (1), 127–145.

Zhao, B., Poh, C.L., 2008. Insights into environmental bioremediation by microorganisms through functional genomics and proteomics. Proteomics 8 (4), 874–881. Available at https://doi.org/10.1002/pmic.200701005.

Chapter 4

Mycoremediation of heavy metals by proteomic techniques: mechanism and applications

Mohd Sajjad Ahmad Khan

Department of Basic Sciences, Deanship of Preparatory Year and Supporting Studies, Imam Abdulrahman Bin Faisal University, Dammam, Saudi Arabia

4.1 Introduction

Heavy metal contamination in soil and water bodies has become one of the most pressing environmental concerns globally. Heavy metals enter the environment by natural and anthropogenic means, which includes: natural weathering of the earth's crust, mining, soil destruction, industrial expulsion, city overflow, sewage, and pesticide applications to plants and air pollution fallout (Ming-Ho, 2005). Unlike organic pollutants, heavy metals are non-degradable and tend to bioaccumulate (Ningombam et al., 2025). Accelerated industrialization and urban expansion have significantly contributed to the accumulation of these pollutants in soil and aquatic systems, posing serious threats to ecosystems and human health. Industrial activities, agricultural practices, and improper waste disposal contribute significantly to the release of heavy metals like lead (Pb), mercury (Hg), cadmium (Cd), arsenic (As), and chromium (Cr) into the environment. These toxic metals persist for long periods in the ecosystem, leading to bioaccumulation, affecting soil quality, water resources, and overall biodiversity (Bala et al., 2022). At low concentrations, essential heavy metals are either non-toxic or considerably less harmful, such as Zinc (Zn), Copper (Cu), Cobalt (Co), and Iron (Fe), since necessary for the metabolism of the human body. On the other hand, even at low concentrations, nonessential metals are extremely hazardous, such as Lead (Pb), Cadmium (Cd), Mercury (Hg), Chromium (Cr), Thallium (Tl), and Arsenic (As), as they are not involved in any metabolic pathway of the human body (Vaishali Kumar et al., 2025). Heavy metal contamination has been identified as a growing global concern, with estimates indicating that approximately 14% to 17% of the world's arable land is contaminated by at

Proteomics Driven Bioremediation. DOI: https://doi.org/10.1016/B978-0-443-45294-9.00010-2

least one heavy metal, translating to nearly 242 million hectares of affected land (Ondrasek et al., 2025). This widespread contamination poses substantial risks to agricultural productivity and public health. Heavy metals exert profound and diverse effects on biota, impacting organisms across various ecosystems. These effects range from sub-lethal to lethal, affecting individual organisms, populations, and entire communities. The contamination of food with heavy metals is a significant public health concern. Crops grown in contaminated soil or irrigated with polluted water can accumulate heavy metals, posing risks to human health when consumed (Khan et al., 2008). It has been estimated that between 900 million and 1.4 billion people reside in regions classified as high-risk zones for heavy metal exposure, with numerous studies indicating their potential to cause severe health issues in humans and wildlife (Das et al., 2025).

Thus, it is crucial to develop strategies to combat this ever-growing problem. The conventional physico-chemical methods of remediating heavy metals, including chemical precipitation, ion exchange, adsorption, membrane filtration, coagulation-flocculation, and flotation, often have limitations related to cost, efficiency, and environmental impact (Dey et al., 2022; Li et al., 2019). Due to the observed negative impact, alternatives to chemical and physical methods for heavy metal removal need to be addressed. An environmental-friendly remediation method, bioremediation, has recently gained great concern due to no possible expected drawbacks. Since it relies on microorganisms to remove contaminants from wastewater and soil, bioremediation is regarded as a safe, sustainable technology and a more affordable process (Bwapwa, 2022). Microorganisms like bacteria, archaebacteria, yeasts, algae, and fungi showed high ability in the bioremediation process (Bala et al., 2022). Among the bioremediation techniques, mycoremediation is a promising technology that has the potential to ameliorate these hazardous chemicals from various environments. The superiority of fungi over single-celled microbes, such as bacteria, to remediate these recalcitrant heavy metals is well-documented (Deshmukh et al., 2016). In addition, the fungi are omnipresent, multifarious, and have a wider arsenal to acclimatize to environmental limitations, such as immoderations of temperature, extremes of pH, higher metal concentrations, and low nutrient accessibility, due to their morphological diversity (Anand et al., 2006). Besides, fungal mycelia have enhanced enzymatic and mechanical contact with the pollutant due to a greater cell-to-surface ratio (Sagar and Singh, 2011).

Mycoremediation, the use of living or dead fungal biomass to degrade, immobilize, or detoxify heavy metals—offering low cost, sustainability, and effectiveness across heavy metal pollution remediation of soil, mine tailings, and industrial wastewater (Dey et al., 2022). Their enzymatic machinery includes extracellular oxidoreductases such as laccases, manganese peroxidases, lignin peroxidases, and versatile peroxidases, which degrade complex aromatic compounds and xenobiotics (Pointing, 2001). Fungal systems possess unique biochemical pathways, cell wall architectures, and metabolic flexibility that enable them to tolerate and remediate metals such as Cd, Pb, Cr, Ni, As, and Hg. This bioremediation process leverages the unique biochemical capabilities of fungi to absorb, transform, or accumulate harmful metals, thereby reducing their concentration in contaminated environments (Vaksmaa et al., 2023). The use of fungi for the remediation of heavy metals is not limited to laboratory studies. Numerous applications have been developed for in situ remediation of contaminated environments. This process has been used successfully in the remediation of several heavy metal-polluted soils, such as lead, cadmium, and arsenic-contaminated soils (Gul et al., 2024; Singh et al., 2021a,b). In a field study, *Phanerochaete chrysosporium*, a white-rot fungus, was shown to reduce the concentration of lead in contaminated soils by over 50% within a few months, primarily through biosorption and bioaccumulation (Liu et al., 2025). Similarly, *Lentinula edodes* (shiitake mushrooms) has been used to remediate cadmium-contaminated soils, where it effectively sequesters the metal in its mycelial network (Chen et al., 2025; Wang et al., 2016).

In recent days, genetic engineering coupled with multi-OMICS approaches (i.e., genomics, transcriptomics, proteomics, and metabolomics) has been found to be more effective than the conventional microbial process (Sreedevi et al., 2022). Modern molecular tools, especially proteomics, now enable mechanistic understanding of fungal metal-handling at the protein level, illuminating transporters, metal-binding proteins (metallothioneins), secreted detoxification enzymes, and stress proteins that govern performance and can be leveraged to design effective remediation systems, and it provides biomarkers and engineering targets to improve efficacy (Dey et al., 2022). Recent comparative proteomic studies have clarified how fungi modulate cellular pathways under single- and multimetal stress, and how environmental co-contaminants affect remediation capacity. However, understanding the molecular mechanisms that underpin fungal tolerance requires deeper insight, where proteomics, the large-scale study of proteins and their modifications, has become indispensable. Proteomic analyses reveal dynamic protein expression patterns and biochemical pathways activated during fungal–metal interactions, thereby elucidating the mechanistic basis of mycoremediation (Dey et al., 2022; Algahmadi et al., 2024).

This chapter explores the potential of mycoremediation for heavy metal contamination, focusing on the mechanisms by which fungi handle toxic metals, and the role of proteomic techniques in deciphering these processes, including gel-based proteomics, LC-MS/MS, quantitative proteomics, PTM profiling, secretome analysis, and integrative proteogenomic strategies. We have also discussed the applications of fungi in the remediation of heavy metals, also limitations and future directions to improve the rate and efficiency of mycoremediation are suggested. The manuscript describes the strategies that can be used as a future framework to address the global problem of heavy metal pollution.

4.2 Mycoremediation: overview and mechanistic basis

The Conventional physicochemical remediation (chemical precipitation, ion exchange, reverse osmosis) can be costly and produce secondary wastes. The process of mycoremediation involves several distinct biological strategies that fungi employ to reduce or neutralize the toxicity of heavy metals (Dey et al., 2022). Fungi remediate heavy metals through enzymatic processes such as biotransformation and biodegradation by extracellular or intracellular enzymatic reactions, and nonenzymatic processes such as biosorption and bioaccumulation (Dinakarkumar et al., 2024; Chakraborty et al., 2025; Das et al., 2018). Fungal cell walls contain various proteins, including enzymes, that can interact with heavy metals. These proteins can act as binding sites for metal ions, facilitating their biosorption and bioaccumulation. The composition and abundance of these proteins can vary depending on the fungal species and the environmental conditions (Khan and Alam, 2018). Some fungi have been shown to increase the production of certain proteins, such as oxidoreductases, in response to heavy metal stress. These proteins can be involved in various cellular processes, including detoxification and stress response, which are crucial for fungal survival in contaminated environments (Gunjal, 2021; Chakraborty et al., 2025; Khan and Alam, 2018). The effectiveness of mycoremediation is dependent on the selection and use of an appropriate fungus species for the target heavy metal or other pollutant (Dinakarkumar et al., 2024). Here, we have briefly summarized the various fungi used in remediation technology and their enzymatic and nonenzymatic capability to remediate heavy metals. In some cases, fungi can also degrade organic pollutants that are associated with heavy metal contamination, further enhancing their efficacy in remediation (Chakraborty et al., 2025; Colmo et al., 2025).

4.2.1 Examples of fungi used in mycoremediation

Fungal groups mainly filamentous Ascomycetes (*Aspergillus*, *Penicillium*), Basidiomycetes (white-rot fungi: *Trametes*, *Pleurotus*, *Phanerochaete*), Zygomycetes, and yeasts (*Saccharomyces*, *Candida*) are mostly found in association with bioremediation research studies or commercial exploitation (Troiano et al., 2020). Microbial metabolism in bioremediation can involve primary metabolism, where the pollutant serves as a nutrient and energy source, or cometabolism, a process in which microbes degrade non-nutritive pollutants incidentally while metabolizing another primary substrate (Tegli et al., 2014; Zengguang et al., 2015). The degradation efficiency of pollutants by fungal enzymes is significantly influenced by various environmental and biochemical parameters, which determine their catalytic performance in situ (Betancor et al., 2013). Classical reviews and surveys provide lists of high-performing isolates for different metals. Fungi (white-rot, brown-rot, filamentous, and yeasts) are promising agents for the remediation of heavy metal-contaminated environments due to robust cell wall chemistry, secretion systems, and metabolic plasticity (Singh and Singh, 2024). Choice of species depends on target metal, matrix, and whether living or dead biomass is used (Dinakarkumar et al., 2024). Yeasts, in particular, employ multiple mechanisms for heavy metal detoxification, including metal binding to functional groups on their cell walls, sequestration within intracellular organelles, and interactions with extracellular polymeric substances (EPSs). These pathways enable the reduction of metal bioavailability through adsorption, active transport, and complexation mechanisms (Shao et al., 2025). *Aspergillus niger*isknown for producing organic acids like citric acid, which can solubilize and chelate heavy metals, making them easier to remove (Yang et al., 2019). *TrichodermaHarzianum* is effective in bioremediation due to its ability to secrete organic acids and its ability to form extensive mycelial networks (Dinakarkumar et al., 2024). *Pleurotus ostreatus* can remove heavy metals like lead, zinc, chromium, cobalt, copper, and nickel from effluents, and has shown increased antioxidant enzyme activity in response to heavy metal accumulation (Akhtar and Amin-ul Mannan, 2020). White-rot fungi, like those that produce lignin-degrading enzymes, can also degrade certain pollutants and immobilize heavy metals (Dinakarkumar et al., 2024; Zengguang et al., 2015). Table 4.1 lists various filamentous fungi reported in the removal of heavy metals.

TABLE 4.1 List of various filamentous fungi reported in the removal of heavy metals.

Heavy metals	Fungi
Cd	*Aspergillus niger, A. clavatus DESM, Penicillium simplicissimum, P. chrysogenum, P. canescens*
Cu, Pb, As	*A. niger*
Cr	*A. niger, A. foetidus*
Ni, Pb bioleaching	*A. niger*
Zn	*P. digitatum, P. simplicissimum, P. chrysogenum, P. spinulosum*
Pb	*P. simplicissimum, P. chrysogenum*
Cu	*P. chrysogenum, P. cyclopium, P. spinulosum, P. italicum*
As, Hg, Cr	*P. canescens*
Mn, Fe, Ni, Co.	*P. italicum*
Pb, Zn	*V. insectorum J3*, Basidiomycota

Source: Adapted from Ghosh, S., Rusyn, I., Dmytruk, O.V., Dmytruk, K.V., Onyeaka, H., Gryzenhout, M., et al., 2023. Filamentous fungi for sustainable remediation of pharmaceutical compounds, heavy metal and oil hydrocarbons. Frontiers in Bioengineering and Biotechnology 11, Available from: http://journal.frontiersin.org/journal/bioengineering-and-biotechnology#archive. https://doi.org/10.3389/fbioe.2023.1106973.

4.3 Nonenzymatic strategies adapted by fungi in mycoremediation

Fungi also utilize biosorption, bioaccumulation, and other strategies to remove heavy metals and other inorganic pollutants. In biosorption, fungal cell wall components like chitin, glucans, and proteins passively adsorb contaminants. In contrast, bioaccumulation involves the active uptake and internal sequestration or transformation of pollutants within fungal cells (Ningombam et al., 2025). Redox transformations are also vital in fungal remediation. These reactions entail electron transfer between the fungal cells and the pollutant, resulting in either the oxidation or reduction of the contaminant (Dinakarkumar et al., 2024). Overall, the mycoremediation shown by various fungi can be broadly classified into three categories, including biosorption, bioaccumulation, and biotransformation. Fungi employ these mechanisms, often simultaneously to handle heavy metals.

4.3.1 Biosorption

Biosorption refers to the passive adsorption of heavy metals by fungal cell walls. This process does not require energy and is often dependent on the physical and chemical properties of the fungal cell surface (Ghosh et al., 2023). Fungi can secrete extracellular compounds such as proteins, polysaccharides, and phenolic acids that interact with metal ions, facilitating their binding to the fungal cells (Colmo et al., 2025). Passive binding of metal ions to functional groups (carboxyl, hydroxyl, phosphate, amino, sulfate) on the cell wall (chitin, glucans, mannoproteins) or EPS of living or dead biomass. This is often rapid and reversible (Gunjal, 2021). Absorption is the predominant mechanism of bioremediation of the heavy metals Copper (Cu) and Cobalt (Co.), described for *Trichoderma*, *Penicillium*, and *Aspergillus* (Dusengemungu et al., 2020). Key factors influencing biosorption include pH, temperature, fungal species, and metal concentration. The binding of metals to the fungal surface can result in a decrease in their bioavailability, preventing their uptake by other organisms in the ecosystem (Aslam et al., 2025).

4.3.2 Bioaccumulation

Bioaccumulationor Intracellular sequestration refers to the active uptake of heavy metals by fungal cells, often through transporters present in the fungal membrane. This process allows fungi to sequester metals inside their cells, where they can be stored in vacuoles or bound to intracellular ligands like metallothioneins, phytochelatins, or other proteins (Dinakarkumar et al., 2024). Fungi can concentrate metals to a much higher level than the surrounding environment, thus reducing their toxicity in the soil or water. Bioaccumulation involves energy-dependent mechanisms, such as ion pumps and symporters, which transport metals into the fungal cells (Diep et al., 2018). Active intracellular sequestration is driven by intracellular ligands like Metal transporters (e.g., ZIP, NRAMP-like proteins), Vacuolar import systems, Metallothioneins (MTs), and Phytochelatin-like peptides. Notably, some fungi species exhibit high selectivity for certain metals, providing an avenue for the development of metal-specific mycoremediation strategies (Chojnacka, 2010; Sácký et al., 2014).

4.3.2.1 *Biomineralization/precipitation*

Fungi can induce precipitation reactions by converting soluble metal ions into insoluble forms. It is a conversion of heavy metals into less soluble mineral forms (carbonates, phosphates, sulfides) often mediated by secreted metabolites (Ghosh et al., 2023). For instance, fungi can secrete organic acids that react with metal ions to form metal oxalates or phosphates, which precipitate out of solution. Oxalate is a key metabolite that plays a significant role in many metal and mineral transformations mediated by fungi. Oxalic acid is produced by a variety of filamentous fungi, and it is generally stated that metal oxalate formation may confer protection from the potentially toxic effects of such metals (Gadd et al., 2014). *Aspergillus* spp have been documented to facilitate the precipitation of lead, copper, and zinc as oxalate, effectively removing them from contaminated media (Sutjaritvorakul et al., 2016; Niehaus et al., 2018; Frank-Kamenetskaya et al., 2021).

4.3.2.2 *Chelation and siderophore-mediated interactions*

Siderophores and secreted organic acids from fungi influence metal mobility, complexation, and sometimes immobilization (Zelaya-Molina et al., 2025). For example, phosphate-solubilizing fungi, such as *Aspergillus niger*, secrete organic acids like citric acid and oxalic acid, which play a crucial role in chelation and precipitation of heavy metals. Citric acid can form stable complexes with metals such as lead, cadmium, and aluminum, reducing their bioavailability and toxicity. For instance, *A. niger* can produce up to 15 g/L of citric acid, which effectively chelates metals like cadmium, facilitating their immobilization and subsequent removal from the environment (Yang et al., 2019).

4.3.2.3 *Mycorrhizal associations*

Fungi that form symbiotic associations with plants, such as mycorrhizal fungi, enhance the bioremediation process by improving the uptake and translocation of contaminants within the plant-fungal system. Mycorrhizal fungi extend the root system of plants, increasing the surface area for nutrient and pollutant absorption. These fungi can secrete organic acids and chelating agents that facilitate the uptake of metals and their subsequent stabilization within the plant tissues. For instance, arbuscular mycorrhizal fungi have been shown to improve the phytoremediation of heavy metals like arsenic and lead by enhancing their uptake and immobilization in plant roots (Niehaus et al., 2018).

4.3.2.4 Cometabolism and synergistic interactions

Fungi can also engage in co-metabolism, where the degradation of one compound facilitates the breakdown of another. This process is particularly useful in environments contaminated with complex mixtures of pollutants. For example, fungi can degrade hydrocarbons while simultaneously enhancing the breakdown of heavy metals by altering the pH and redox conditions in the soil (Burton et al., 2019). Moreover, fungi can interact synergistically with other microorganisms, such as bacteria, to enhance the overall bioremediation process. These interactions can lead to the sequential degradation of pollutants, where fungi breakdown complex molecules into simpler intermediates that are further degraded by bacteria.

4.3.2.5 Fungal enzymes for bioremediation

Fungi, recognized as nature's primary decomposers, secrete a wide range of extracellular enzymes such as laccases, peroxidases, esterases, cutinases, lipases, amylases, nucleases, proteases, and others that degrade cell walls, DNA, and other components. This approach is used to remediate the environment by transforming organic waste into less harmful or inert compounds (Balabanova et al., 2018). These enzymes facilitate the degradation of heavy metals, organic pollutants, aromatic hydrocarbons, and polychlorinated compounds without disrupting the natural soil environment.

Below are some of the commonly used fungal enzymes for the bioremediation of various pollutants.

4.3.3 Remediation through extracellular enzymes

Enzymatic fungal-mediated biotransformation, in turn, can be carried out by means of extracellular and intracellular enzymes (Ghosh et al., 2023). Extracellular enzymes catalyze various reactions (e.g., oxidative coupling products, hydroxylation, CO_2 emission, ether cleavage, aromatic ring fission, quinoid products formation, and others), resulting in biotransformation of toxic compounds, facilitating bioremediation processes (Singh et al., 2021a,b). Secondary utilization or gratuitous metabolism refers to the transformation of contaminants without any direct metabolic benefit to the organism (Sista et al., 2019). These processes are particularly committed to filamentous fungi and often involve the secretion of nonspecific extracellular enzymes like lignin-modifying enzymes, laccases, manganese peroxidases, lignin peroxidase, dye decolorizing peroxidases, and peroxygenases that breakdown a wide range of substrates, including those not directly useful to the microbe (El-Gendi et al., 2021).

Biotransformation: Biotransformation refers to the ability of fungi to chemically alter heavy metals, thereby reducing their toxicity, such as the redox transformation of metals, for example, enzymes (reductases, oxidases, laccases, peroxidases) can alter metal valence (e.g., Cr (VI) → Cr (III)), affecting solubility/toxicity (Yang et al., 2019). This can occur through a variety of processes, including oxidation, reduction, methylation, and complexation (Harish et al., 2024; Naziębło and Dobrzyński, 2025). For instance, fungi like *Aspergillus* species can convert highly toxic forms of mercury (Hg^{2+}) into less toxic methylated forms (CH_3Hg^+) through the action of specific enzymes. This biotransformation not only detoxifies the metals but may also facilitate their further sequestration or immobilization (Karunasagar et al., 2003; Nirmal Kumar et al., 2010).

Enzymatic degradation: Fungi produce a wide array of extracellular enzymes that facilitate the degradation of complex organic pollutants. These enzymes include laccases, manganese peroxidases, and lignin peroxidases, which catalyze the oxidation of various pollutants. For instance, white-rot fungi like *Phanerochaete chrysosporium* secrete ligninolytic enzymes that degrade lignin and other complex organic molecules, including polycyclic aromatic hydrocarbons (PAHs) and synthetic dyes (Das et al., 2018). Laccases, for example, catalyze the oxidation of phenolic compounds, leading to the breakdown of complex organic molecules into smaller, less harmful compounds. This process is particularly effective in the degradation of persistent organic pollutants such as PAHs and PCBs, which are resistant to conventional degradation processes (Khan and Alam, 2018).

Below, we have briefly summarized the significance of various extracellular enzymes produced by fungi that are being exploited in the remediation of heavy metals.

4.3.3.1 Peroxidases

In fungi, the primary heme-containing peroxidases are essential enzymes involved in the breakdown of lignocellulose, enabling the degradation of both organic materials, such as plant biomass, and certain inorganic-contaminated substrates (Balabanova et al., 2018). Specific enzymes such as lignin peroxidase (LiP) and manganese peroxidase (MnP) utilize hydrogen peroxide and manganese as cofactors in oxidizing recalcitrant pollutants. These enzymes have been identified as being used by the basidiomycetes and white-rot fungus to breakdown hazardous chemicals that are resistant to breakdown (Zengguang et al., 2015).

4.3.3.2 Laccases

Laccases are those enzymes that are based on copper and are produced by fungi. Laccases also carry out reactions such as depolymerization, decarboxylation, and demethylation of aromatic pollutants, making them ideal for the breakdown of xenobiotics (Varjani and Patel, 2017). Laccases oxidize phenolic and nonphenolic compounds, with fungi such as *T. versicolor* showing an approximately 20-fold increase in activity. Various fungal genera, which catalyze the oxidation of ortho and polyphenols, produce both internal and external laccases (Redman et al., 2002).

4.3.3.3 Oxidoreductases

The oxidoreductase enzymes are mainly oxygenases and help various fungi in detoxifying organic contaminants through an oxidative coupling mechanism (Echeveria et al., 2020). Fungal enzymes facilitate the transfer of electrons from a reduced organic substrate by breaking the chemical bonds of substrates through biochemical reactions and releasing energy (Zhang et al., 2019). The pollutants are then oxidized to nontoxic molecules through both oxidation and reduction reactions. Oxidoreductases convert harmful xenobiotics into harmless substances through the detoxification process (Singh et al., 2020).

4.3.3.4 Catalase

In biological systems, the accumulation of reactive oxygen species (ROS) leads to cellular damage, adversely affecting macromolecules. To mitigate the toxicity of ROS, fungi produce monofunctional catalases and bifunctional peroxidase/catalase enzymes as an initial defense mechanism (Ningombam et al., 2025). The generation of reactive oxygen species in microorganisms has also been associated with heavy metals such as lead, copper, cadmium, and zinc. Numerous studies have shown a distinct correlation between the presence of heavy metals and the generation of reactive oxygen species, which triggers the production of antioxidative enzymes (Iram et al., 2015; Ningombam et al., 2025).

4.4 Remediation through intracellular enzymes

Intracellular enzymes participating in the bioremediation of toxic compounds can be divided in two groups. First–cytochromes and second–glutathione transferases. The cytochrome P450 system is a large family of enzymes, mainly monooxygenases, catalyzing reactions of hydroxylation, oxygenation, dealkylation, epoxidation of C=C bonds, reduction, and dehalogenation (Díaz-Cruz et al., 2014). It was shown that P450 plays an indirect but crucial role in the process of clearance of cadmium and lead in filamentous fungi *Phanerochaete chrysosporium* belonging to the Basidiomycota phylum (Zhang et al., 2015). It was demonstrated that several *Aspergillus* species are capable of degrading a carcinogenic contaminant, Benzo [*a*]pyrene, by the action of the cytochrome P450 monooxygenase (Loss et al., 2019). Fungal glutathione S-transferases (GST) are detoxification enzymes that can catalyze the conjugation of glutathione to non-polar compounds containing an electrophilic carbon, nitrogen, or sulfur atom. By this mechanism, glutathione S-transferases are able to metabolize drugs, pesticides, xenobiotics, heavy metals, and other toxic compounds (Pocsi et al., 2004). GST has been reported to contribute to the defense against heavy metal stress in different filamentous fungi, including *A. nidulans* (Fraser et al., 2002). Glutathione transferases play an important role in the intracellular reduction of Cr (VI) in fungi (Garcia-Hernandez et al., 2017).

4.5 Recent advancements in mycoremediation

Strategies like synthetic ecology and directed evolution are gaining momentum in designing optimized fungal consortia, which may offer synergistic pollutant degradation and broader metabolic capacities (Ningombam et al., 2025). Cocultivation of white-rot fungi (e.g., *Trametes versicolor*) with *Aspergillus* or *Penicillium* species has shown up to 90% degradation of textile dyes within 5–7 days in bench-scale studies, underscoring the value of microbial synergy (Singh et al., 2025). Genetic and metabolic engineering are recent advanced technologies used to enhance the bioremediation process using living organisms to cleanup environmental contaminants (Liang et al., 2020). Metagenomics offers a way to analyze genetic material directly from environmental samples, revealing the microbial communities, including uncultivable species. It aids in identifying microbes, exploring gene clusters for novel enzymes, and discovering new microbial products (Zhang et al., 2021). Metabolic engineering modifies existing pathways to improve bioremediation by targeting microbial enzymes such as esterases, phenoloxidases, oxidoreductases, and monooxygenases that are involved in various degradation processes (Singh et al., 2020). Enzymatic bioremediation offers several advantages as an environmentally friendly method. The genetic approach enhances the possibility of obtaining robust intracellular and extracellular enzymes. Long-term monitoring of mycoremediation sites is crucial to evaluate ecological impacts, enzyme persistence, and shifts in soil microbiomes (Ningombam et al., 2025; Fig. 4.1. Role of various fungi in mycoremediation strategies and significance of proteomics in enhancing its efficiency. Current pilot studies have reported a reduction in PAH levels by 60%–85% over 30 days in soil amended with lignocellulosic substrates and fungal biomass, but field trials at scale remain limited (Bianco et al., 2023). Improving bioremediation techniques through bioaugmentation, biostimulation, and formulation of fungal inoculants with protective carriers can help ensure fungal survival and activity under field conditions (Bala et al., 2022). Molecular and metagenomic tools should also be employed to track fungal colonization and pollutant degradation pathways in situ (Zhang et al., 2021). From identifying metal-transport proteins to characterizing oxidative stress pathways and secretome dynamics, proteomic approaches provide unparalleled insights that enhance remediation strategies, strain

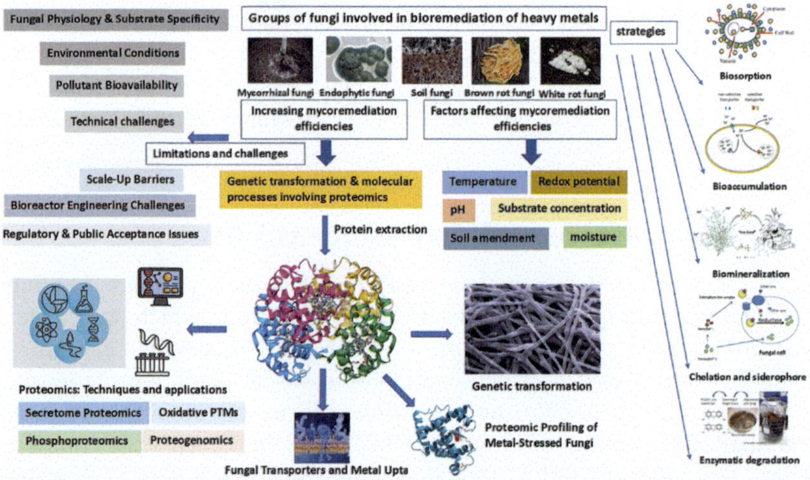

FIGURE 4.1 **Role of various fungi in mycoremediation strategies and significance of proteomics in enhancing its efficiency.** The diagram illustrates the application of fungi in remediation and the importance of proteomics in enhancing remediation strategies.

development, and field applications (Bianco and Perrotta, 2015; Dey et al., 2022; Khatiwada et al., 2020). As proteomic technologies continue to advance, they will play an increasingly central role in optimizing fungal systems for environmental cleanup.

4.6 Role of proteomics in understanding mycoremediation

Proteomics enables the identification of the proteins that regulate each of these processes. Proteomics, the large-scale study of proteins, has proven to be an invaluable tool for understanding the molecular basis of mycoremediation (Chandramouli and Qian, 2009). Proteomics allows researchers to identify and quantify the proteins involved in various cellular processes that mediate heavy metal uptake, transformation, and detoxification in fungi. By analyzing the proteome, researchers can uncover the key enzymes, transporters, and regulatory networks that fungi employ to mitigate metal toxicity (Bianco and Perrotta, 2015). Proteomics provides molecular resolution of fungal responses to heavy metal stress. It deciphers:

- Differentially expressed proteins under specific metal stresses
- Signaling pathways activated (MAPK, oxidative stress, apoptosis-like responses)
- Metal-binding proteins
- Detoxification enzymes
- Stress-related posttranslational modifications (PTMs)
- Secretome dynamics influencing extracellular metal sequestration

4.6.1 Proteomic techniques applied to fungal heavy-metal remediation

Recent advances in proteomics have provided insights into the underlying mechanisms of mycoremediation, enabling a deeper understanding of fungal interactions with heavy metals at the molecular level (Bianco and Perrotta, 2015; Dey et al., 2022).

4.6.1.1 Gel-based proteomics (2D-PAGE, 2D-DIGE)

- Useful for profiling global protein changes.
- Differentially expressed protein spots are excised and identified by MALDI-TOF or LC-MS/MS.
- Strength: good visualization of abundant proteins.
- Limitation: low sensitivity for membrane proteins and extreme pI proteins.

4.6.1.1.1 Applications

- Cd-induced proteins in *Aspergillus nidulans*: Cadmium is an exceptionally toxic industrial and environmental pollutant classified as a human carcinogen. As AtfA is the main transcription factor governing stress responses in *A. nidulans*, Emri et al. (2021) examined genome-wide expression responses of wild-type and the *atfA* null mutant exposed to $CdCl_2$. They concluded that the regulatory network of *A. nidulans* has a high flexibility, allowing the fungus to adapt efficiently to stress both in the presence and absence of this important transcription factor. The P-type ATPase CrpA is an important Cu^{2+} /Cd^{2+} pump in the Aspergilli, significantly contributing to the heavy metal stress tolerance of these ascomycetous fungi. A study from Boczonádi et al. (2020) showed that deletion of crpA resulted in Cu^{2+} /Cd^{2+} -sensitive phenotypes in *Aspergillus nidulans* on stress agar plates inoculated with conidia.
- Cu-responsive stress proteins in *Trametes versicolor*: Heavy metals, such as copper (Cu), are prevalent in the environment and pose a substantial threat to human health. White-rot fungi, especially Trametes spp., display prominent Cu tolerance and removal capacity. Vasina et al. (2016) carried out secretome analysis for white-rot basidiomycete *Trametes hirsuta* cultivated on a synthetic medium and a lignocellulose substrate. They demonstrated that *T. hirsuta* st. 072 produces multiple extracellular ligninolytic, cellulolytic, hemicellulolytic, peroxide-generating, and proteolytic enzymes, as well as cerato-platanins. An accessory enzyme, glyoxal oxidase, was found among the proteins secreted in the media during submerged cultivation of *T. hirsuta* both in the presence and in the absence of copper. Recently, a comparative transcriptome analysis conducted by Wang et al. (2025) identified 812, 1898, and 2110 differentially expressed genes (DEGs) at the Cu concentrations of 1.25,

12.5, and 25 mg/L, respectively. qRT-PCR confirmed the alterations in gene expression and demonstrated that glutathione S-transferases, catalases, cytochrome P450s, and laccases were involved in counteracting Cu-induced stress. Many transcription factors were enriched under Cu stress, including the Zn2Cys6 family transcription factor GME8421_g (TH8421), which was significantly upregulated at the Cu concentration of 12.5 mg/L. ChIP-seq identified five antioxidant enzyme-encoding genes as direct targets of TH8421, forming a regulatory network that protects against Cu stress. These findings offer insights into the molecular mechanisms driving Cu toxicity tolerance in *Trametes* fungi.

- Cr-responsive redox enzymes in *Phanerochaete chrysosporium*: *Phanerochaete chrysosporium* is considered the model fungus for white-rot fungi. It is the first basidiomycete whose genome has been completely sequenced. Its importance lies in the fact that its enzymatic system comprises the major enzymes involved in lignin degradation (Konan et al., 2024). A study from Murugavelh and Mohanty (2013) investigated the bioreduction capacity of the live and immobilized *P. chrysosporium.* A maximum of 98.3% of bioreduction of Cr (VI) was obtained with an initial Cr (VI) concentration of 10 mg L^{-1} at pH 5 and a temperature of 25°C.

4.6.1.2 *Shotgun proteomics (LC-MS/MS)*

The most widely used modern method:

- Digested peptides are separated by nano-LC and analyzed by tandem MS.
- Enables high-throughput identification of thousands of proteins.
- Suitable for membrane proteins, transporters, and secreted proteins.

4.6.1.2.1 Applications

- Discovery of Zn- and Cd-binding proteins in *Saccharomyces cerevisiae*: Zinc is an essential cofactor for many proteins. The first eukaryotic zinc uptake transporter was discovered in the yeast *Saccharomyces cerevisiae.* Since then, this organism has been an invaluable tool for the discovery of genes involved in zinc homeostasis. Genomic and proteomic studies have revealed an abundance of Zn^{2+}-regulated genes and Zn^{2+}-binding proteins (Regalla and Lyons, 2005). A key mechanism of zinc homeostasis during deficiency is "zinc sparing," in which specific zinc-binding proteins are repressed to reduce the cellular requirement. Wang et al. (2018) evaluated zinc sparing across the zinc proteome of *Saccharomyces cerevisiae* and discovered widespread zinc sparing mechanisms and obtained evidence of a high accumulation of zinc proteins that lack their cofactor during deficiency. Inouhe et al. (1989) induced a Cd-binding protein in the Cd^{2+}-resistant strain 301 N of *Saccharomyces cerevisiae* by administration of 0.5 mM $CdSO_4$. and results suggested that metallothionein generally found in animals, also occurs in Cd-adapted yeast cells and thus has a role in its Cd-resistance.

4.6.1.3 *Quantitative protcomics*

4.6.1.3.1 Techniques

- Label-free quantification (LFQ)
- Isobaric tags (iTRAQ, TMT)
- SILAC (less common in fungi but powerful for model species)

4.6.1.3.2 Advantages

- Precise quantification of protein abundance changes
- Detection of subtle stress-specific signatures

4.6.1.3.3 Key discoveries via quantitative proteomics

- Induction of antioxidant enzymes (SOD, CAT, GST), upregulation of metallothioneins and metal-transport ATPases, suppression of ribosomal proteins under acute metal stress, and activation of autophagy and protein degradation machinery (Bianco and Perrotta, 2015; Jiang et al., 2024; Chandramouli and Qian, 2009; Yang et al., 2024).

4.6.1.3.4 PTM proteomics (phosphoproteomics, oxidative PTMs)

Heavy metal stress often induces changes in phosphorylation, nitrosylation, and carbonylation. Phosphoproteomics reveals MAPK phosphorylation cascades under Cd, Cr, and Pb stress, Activation of protein kinases and phosphatases, regulation of metal transporter trafficking (Jiang et al., 2024; Arribas et al., 2025). Whereas oxidative PTM studies show protein carbonylation of mitochondrial enzymes, enhanced nitrosylation affecting cellular respiration, redox-regulated transcription factors, and PTM profiling are crucial for understanding functional protein regulation rather than just abundance (Dey et al., 2022; Jiang et al., 2024).

4.6.1.3.5 Secretome proteomics

Since fungi secrete enzymes and organic acids that mobilize or immobilize metals, secretome studies are essential. Secretome analysis reveals Oxalate biosynthesis enzymes involved in metal precipitation, Extracellular laccases and peroxidases regulating redox transformations, EPS-associated proteins that increase metal binding (Gadd et al., 1999; Lebrun et al., 2011).

4.6.1.3.6 Proteogenomics

Integration of genomic data with proteomics enables the discovery of novel metal-responsive proteins, accurate annotation of detoxification pathways, and identification of strain-specific proteins in wild fungal isolates used for bioremediation (Jiang et al., 2024).

4.6.1.3.7 Proteomic profiling of metal-stressed fungi

When fungi are exposed to heavy metals, they activate various stress response pathways, leading to changes in the proteome. These changes include the upregulation of proteins involved in metal binding, transport, antioxidant defense, and cellular repair. Using techniques like two-dimensional gel electrophoresis (2-DE), liquid chromatography-mass spectrometry (LC-MS), and protein microarrays, researchers can identify differentially expressed proteins in metal-stressed fungal species (Bianco and Perrotta, 2015; Chandramouli and Qian, 2009; Jiang et al., 2024). For example, Trichoderma and Penicillium species have been shown to upregulate proteins involved in the binding of metals like copper (Cu), zinc (Zn), and lead (Pb). These proteins include metallothioneins, which bind to metal ions and prevent their toxic effects, and antioxidant enzymes like superoxide dismutase (SOD), which mitigate oxidative stress caused by metal exposure (Khan et al., 2022; Konieczna et al., 2024; Yang et al., 2024; Zapata-Sarmiento et al., 2025).

4.7 Identification of key enzymes in biotransformation

Proteomic analysis has also been used to identify enzymes involved in the biotransformation of heavy metals (Dey et al., 2022). For example, fungal species capable of methylating mercury, such as *Fusarium* and *Aspergillus*, express a unique class of enzymes called mercury reductases, which play a crucial role in reducing mercury ions to less toxic forms. Proteomic studies have highlighted the upregulation of mercury reductases under mercury exposure, providing insights into the mechanism of mercury detoxification (Khan et al., 2022; Jiang et al., 2024). Similarly, fungi that are capable of arsenic biotransformation often express arsenic reductases, which catalyze the reduction of arsenate (As (V)) to arsenite (As (III)), a less toxic form. Understanding the regulation and expression of these enzymes can lead to the development of fungal strains with enhanced metal-remediation capabilities (Su et al., 2011; Konieczna et al., 2024).

4.8 Fungal transporters and metal uptake

Fungal transporters play a crucial role in the uptake of heavy metals from the environment. By using proteomic tools, researchers have identified key metal transporters, such as cation-efflux pumps, ABC transporters, and metal ion channels, which mediate the movement of metal ions into and out of fungal cells. These transporters are often regulated by metal-responsive transcription factors, and their activity is critical for the efficiency of mycoremediation (Pocsi et al., 2004) (Jiang et al., 2024; Yang et al., 2024). For instance, *Saccharomyces cerevisiae*, a well-studied yeast model, has been used to identify a variety of transport proteins involved in the uptake of cadmium and copper. Proteomic analysis has revealed how these transporters are regulated under different metal stress conditions, providing insights that can be applied to engineering fungi for enhanced metal absorption and detoxification (Revel et al., 2022; Dežulović et al., 2025).

4.9 Limitations and challenges

Despite its promising potential, mycoremediation faces several scientific, technical, and regulatory challenges that must be addressed to enable effective implementation at scale. These include the following.

4.9.1 Technical challenges

- Robust protein extraction from recalcitrant fungal biomass and contaminated matrices.
- Incomplete fungal proteome databases for environmental isolates complicate identifications, and integrating transcriptomics often becomes necessary.
- PTMs and metalloproteo forms are complex and under-characterized.

4.9.2 Environmental conditions

- Factors such as pH, temperature, moisture content, and nutrient availability significantly influence fungal colonization, enzymatic activity, and pollutant degradation efficiency.
- Maintaining optimal environmental conditions in the field can be difficult, particularly in heterogeneous or extreme environments.
- Multicontaminant environments, such as co-metals, organics, variable pH, and salinity, alter fungal responses, and lab findings need validation in field settings.
- Metal speciation and bioavailability in soils strongly affect remediation but are often not fully addressed by proteomic studies.

4.9.3 Pollutant bioavailability

- The chemical form, concentration, and physical distribution of pollutants in soil or water matrices directly affect the accessibility of these compounds to fungal hyphae and secreted enzymes.
- Low solubility and pollutant adsorption to soil particles limit fungal interactions, reducing degradation efficiency.

4.9.4 Fungal physiology and substrate specificity

- Different fungal strains vary in their growth kinetics, enzyme secretion patterns, and pollutant tolerance. Many fungi exhibit substrate specificity,

which may not be present in all contaminated environments. This limits their practical utility unless co-substrates are supplemented, which can increase cost and complexity.
- Genetic engineering, metabolic pathway modification, and coculture strategies with other fungi or bacteria are currently being explored to broaden substrate range and improve pollutant degradation versatility.

4.9.5 Scale-up barriers

- Transitioning mycoremediation from laboratory to field scale presents major obstacles. These include inconsistent performance under variable field conditions, susceptibility to competition with indigenous microbial communities, and challenges in maintaining fungal viability and activity over time.
- Additionally, high production costs for fungal inocula, enzyme extraction, and formulation for large-scale application pose economic constraints.

4.9.6 Bioreactor engineering challenges

- In engineered systems such as bioreactors, fungi can be deployed more efficiently, but their use introduces unique design considerations. Maintaining optimal oxygen transfer, preventing clogging by fungal biomass, and controlling shear stress on hyphae are critical.
- Fungi often require solid or semisolid substrates for growth, complicating liquid-phase bioreactor operations.

4.9.7 Regulatory and public acceptance issues

- Regulatory approval for the use of genetically modified fungi or nonnative species in environmental applications remains limited.
- Public perception and safety concerns further restrict field trials and commercialization efforts.
- Clear regulatory frameworks, environmental risk assessments, and pilot-scale demonstrations are needed to build confidence and enable wider deployment.

4.10 Future directions

To overcome these challenges, future research in mycoremediation should emphasize the discovery and characterization of novel fungal strains, particularly extremophiles with high tolerance to pollutants. However, potential ecological risks, such as horizontal gene transfer and unintended interactions with native microbiota, must be carefully assessed through contained field trials and biosafety regulations. Future progress requires multi-omics integration, field validation, and development of proteomics-driven monitoring tools to realize the full potential of fungi for sustainable heavy metal remediation. However, the following are certain future directives for researchers and policymakers aiming to implement mycoremediation for an effective and sustainable environment.

- Integration of proteomics with metabolomics and genomics (multiomics) for systems-level predictive models.
- Engineering of fungi (synthetic biology) targeted by proteomics to over-express remediation effectors.
- Development of rapid, field-deployable proteomic assays for real-time monitoring.
- Single-cell proteomics for hyphal-level resolution.
- Spatial proteomics to map metal distribution across hyphae.
- Top-down proteomics for intact protein analysis.
- Integration with metabolomics and transcriptomics.
- Machine-learning–based proteomic classification for strain selection.

4.11 Conclusions

Heavy metal contamination poses a persistent global threat due to industrial discharge, mining, agricultural runoff, and urbanization. Heavy metal contamination (Cd, Pb, Cr, Hg, As, Cu, Zn, Ni) is a human-health concern because these elements are persistent, bioaccumulative, and often toxic even at low concentrations. Mycoremediation, employing fungi or their derivatives for the remediation of environmental pollutants, is a comparatively cost-effective, eco-friendly, and effective method. It has advantages over other conventional and bioremediation methods. Fungi, with their high surface-area hyphae, robust cell walls, and enzymatic versatility, represent powerful biological tools for metal detoxification. Mycoremediation of heavy metals encompasses several biological processes. Mycoremediation represents a promising, sustainable, and environmental-friendly approach for mitigating heavy metal pollution. The integration of proteomic techniques has revolutionized our understanding of the molecular mechanisms involved in fungal metal uptake, transformation, and detoxification. By identifying key proteins, enzymes, and transporters. Proteomics has transformed our understanding of mycoremediation by revealing the intricate protein networks and mechanisms that enable fungi to detoxify heavy metals. In conclusion, while mycoremediation offers a low-cost and sustainable solution to environmental pollution, advances in genetic engineering, systems biology, enzyme immobilization, and long-term ecological assessments are essential for unlocking its full potential and facilitating its widespread application in real-world scenarios.

Acknowledgments

Author M.S.A. Khan is thankful to the Department of Scientific Research, Imam Abdulrahman Bin Faisal University, for the cooperation received during this work.

References

Akhtar, N., Amin-ul Mannan, M., 2020. Mycoremediation: expunging environmental pollutants. Biotechnology Reports 26, e00452.

Algahmadi, A., Mohammed, A.E., Alfadda, A.A., Alanazi, I.O., Alwehaibi, M.A., Scaria Joy, S., et al., 2024. Proteomics of Penicillium chrysogenum for a deeper understanding of lead (Pb) metal bioremediation. ACS Omega 9 (24), 26245–26256. Available at https://doi.org/10.1021/acsomega.4c02006, http://pubs.acs.org/journal/acsodf.

Anand, P., Isar, J., Saran, S., Saxena, R.K., 2006. Bioaccumulation of copper by Trichoderma viride. Bioresource Technology 97 (8), 1018–1025. Available at https://doi.org/10.1016/j.biortech.2005.04.046.

Arribas, V., Borrajo, A., Hernáez, M.L., Martínez, R., Monteoliva, L., Gil, C., et al., 2025. Integrative phosphoproteomic and proteomic analysis of Candida albicans exposed to oxidative stress. Journal of Proteome Research 24 (7), 3484–3497. Available at https://doi.org/10.1021/acs.jproteome.5c00137. http://pubs.acs.org/journal/jprobs.

Aslam, A., Kanwal, F., Javied, S., Nisar, N., Torriero, A.A.J., 2025. Microbial biosorption: a sustainable approach for metal removal and environmental remediation. International Journal of Environmental Science and Technology 22 (13), 13245–13276. Available at https://doi.org/10.1007/s13762-025-06611-1.

Bala, S., Garg, D., Thirumalesh, B.V., Sharma, M., Sridhar, K., Inbaraj, B.S., et al., 2022. Recent strategies for bioremediation of emerging pollutants: a review for a green and sustainable environment. Toxics 10 (8). Available at https://doi.org/10.3390/toxics10080484. http://www.mdpi.com/journal/toxics.

Balabanova, L., Slepchenko, L., Son, O., Tekutyeva, L., 2018. Biotechnology potential of marine fungi degrading plant and algae polymeric substrates. Frontiers in Microbiology 9. Available at https://doi.org/10.3389/fmicb.2018.01527.

Betancor, L., Johnson, G.R., Luckarift, H.R., 2013. Stabilized laccases as heterogeneous bioelectrocatalysts. ChemCatChem 5 (1), 46–60.

Bianco, L., Perrotta, G., 2015. Methodologies and perspectives of proteomics applied to filamentous fungi: from sample preparation to secretome analysis. International Journal of Molecular Sciences 16 (3), 5803–5829. Available at https://doi.org/10.3390/ijms16035803. http://www.mdpi.com/1422-0067/16/3/5803/pdf.

Bianco, F., Race, M., Papirio, S., Esposito, G., 2023. A critical review of the remediation of PAH-polluted marine sediments: current knowledge and future perspectives. Resources, Environment and Sustainability 11, 100101. Available at https://doi.org/10.1016/j.resenv.2022.100101.

Boczonádi, I., Török, Z., Jakab, Á., Kónya, G., Gyurcsó, K., Baranyai, E., et al., 2020. Increased Cd2+ biosorption capability of Aspergillus nidulans elicited by crpA deletion. Journal of Basic Microbiology 60 (7), 574–584. Available at https://doi.org/10.1002/jobm.202000112. http://onlinelibrary.wiley.com/journal/10.1002/(ISSN)1521-4028.

Burton, J.L., Smith, E.A., Peterson, S.M., Brown, M.T., 2019. Degradation of polycyclic aromatic hydrocarbons by Phanerochaete chrysosporium. Journal of Environmental Chemical Engineering 7 (5).

Bwapwa, J., 2022. Factors affecting the bioremediation of industrial and domestic wastewaters. Microbial Biodegradation and Bioremediation. Elsevier, pp. 461–472.

Chakraborty, S., Talukdar, A., Dey, S., Bhattacharya, S., 2025. Role of fungi, bacteria and microalgae in bioremediation of emerging pollutants with special reference to pesticides, heavy metals and pharmaceuticals. Discover Environment 3 (1). Available at https://doi.org/10.1007/s44274-025-00217-7.

Chandramouli, K., Qian, P.Y., 2009. Proteomics: challenges, techniques and possibilities to overcome biological sample complexity. Human Genomics and Proteomics 239204.

Chen, G., Han, B., Nan, W., Dong, X., 2025. Cadmium tolerance and detoxification mechanisms of lentinula edodes: physiology, subcellular distribution, and chemical forms. Microorganisms 13 (1), 62. Available at https://doi.org/10.3390/microorganisms13010062.

Chojnacka, K., 2010. Biosorption and bioaccumulation – the prospects for practical applications. Environment International 36 (3), 299–307. Available at https://doi.org/10.1016/j.envint.2009.12.001.

Colmo, C., Tegelaar, M., Ayres, P., 2025. Bioremediation of polycyclic aromatic hydrocarbons (PAHs) in aqueous environments: a review of biofiltration, biosorption, and biodegradation strategies using living fungal mycelium. Fermentation 11, 573.

Das, P., Pal, S., Ghosh, S.K., 2018. Ligninolytic enzyme production by *Phanerochaete chrysosporium* and its application in hydrocarbon degradation. International Journal of Environmental Science and Technology 15 (5), 927–935.

Das, K., Masud, M.A.A., Sarker, A., Arafa, R.A., Patel, M., 2025. Unveiling the sustainable and biological remediation of heavy metals contaminations in soils and water ecosystems through potential microbes—a review. Sustainability 17 (16), 7357. Available at https://doi.org/10.3390/su17167357.

Deshmukh, R., Khardenavis, A.A., Purohit, H.J., 2016. Diverse metabolic capacities of fungi for bioremediation. Indian Journal of Microbiology 56, 247–264. Available at https://doi.org/10.1007/s12088-016-0584-6.

Dey, P., Malik, A., Singh, D.K., Haange, S.B., von Bergen, M., Jehmlich, N., 2022. Insight into the molecular mechanisms underpinning the mycoremediation of multiple metals by proteomic technique. Frontiers in Microbiology 13. Available at https://doi.org/10.3389/fmicb.2022.872576. https://www.frontiersin.org/journals/microbiology#.

Dežulović, L., Ćurko-Cofek, B., Čanadi Jurešić, G., 2025. Proteomic research of the stress response of *Saccharomyces cerevisiae W303* yeast to metal ions eluted from orthodontic appliances. Microorganisms 13 (9), 2200.

Diep, P., Mahadevan, R., Yakunin, A.F., 2018. Heavy metal removal by bioaccumulation using genetically engineered microorganisms. Frontiers in Bioengineering and Biotechnology 6, 157.

Dinakarkumar, Y., Ramakrishnan, G., Gujjula, K.R., Vasu, V., Balamurugan, P., Murali, G., 2024. Fungal bioremediation: an overview of the mechanisms, applications and future perspectives. Environmental Chemistry and Ecotoxicology 6, 293–302. Available at https://doi.org/10.1016/j.enceco.2024.07.002.

Dusengemungu, L., Kasali, G., Gwanama, C., Ouma, K.O., 2020. Recent advances in biosorption of copper and cobalt by filamentous fungi. Frontiers in Microbiology 11. Available at https://doi.org/10.3389/fmicb.2020.582016. https://www.frontiersin.org/journals/microbiology#.

Díaz-Cruz, M.S., Gago-Ferrero, P., Badia-Fabregat, M., Caminal, G., Vicent, T., Barceló, D., 2014. Fungal-mediated biodegradation of ingredients in personal care products. Personal Care Products in the Aquatic Environment 295–317.

Echeveria, L., Gilmore, S., Harrison, S., Heinz, K., Chang, A., Nunz-Conti, G., et al., 2020. Versatile bio-organism detection using microspheres for future biodegradation and bioremediation studies. Laser Resonators Microresonators Beam Control XXII 11266, 205–212.

El-Gendi, H., Saleh, A.K., Badierah, R., Redwan, E.M., El-Maradny, Y.A., El-Fakharany, E.M., 2021. A comprehensive insight into fungal enzymes: structure, classification, and their role in mankind's challenges. Journal of Fungi (Basel) 8 (1), 23.

Emri, T., Gila, B., Antal, K., Fekete, F., Moon, H., Yu, J.-H., et al., 2021. AtfA-independent adaptation to the toxic heavy metal cadmium in Aspergillus nidulans. Microorganisms 9 (7), 1433. Available at https://doi.org/10.3390/microorganisms9071433.

Frank-Kamenetskaya, O.V., Zelenskaya, M.S., Izatulina, A.R., Vereshchagin, O.S., Vlasov, D.Y., Himelbrant, D.E., et al., 2021. Copper oxalate formation by lichens and fungi. Scientific Reports 11, 24239.

Fraser, J.A., Davis, M.A., Hynes, M.J., 2002. A gene from *Aspergillus nidulans* with similarity to URE2 of *Saccharomyces cerevisiae* encodes a glutathione S-transferase which contributes to heavy metal and xenobiotic resistance. Applied and Environmental Microbiology 68, 2802–2808.

Gadd, G.M., Bahri-Esfahani, J., Li, Q., Rhee, Y.J., Wei, Z., Fomina, M., et al., 2014. Oxalate production by fungi: significance in geomycology, biodeterioration and bioremediation. Fungal Biology Reviews 28 (2-3), 36–55. Available at https://doi.org/10.1016/j.fbr.2014.05.001. http://www.elsevier.com.

Garcia-Hernandez, M.A., Villarreal-Chiu, J.F., Garza-Gonzalez, M.T., 2017. Metallophilic fungi research: an alternative for its use in the bioremediation of hexavalent chromium. International Journal of Environmental Science and Technology 14, 2023–2038.

Gadd, G.M., 1999. Fungal production of citric and oxalic acid: importance in metal speciation, physiology and biogeochemical processes. In: In: Poole, R.K. (Ed.), Advances in Microbial Physiology Volume 41. Academic Press, pp. 47–92 ISSN 0065-2911.

Ghosh, S., Rusyn, I., Dmytruk, O.V., Dmytruk, K.V., Onyeaka, H., Gryzenhout, M., et al., 2023. Filamentous fungi for sustainable remediation of pharmaceutical compounds, heavy metal and oil hydrocarbons. Frontiers in Bioengineering and Biotechnology 11. Available at https://doi.org/10.3389/fbioe.2023.1106973. http://journal.frontiersin.org/journal/bioengineering-and-biotechnology#archive.

Gul, I., Adil, M., Lv, F., Li, T., Chen, Y., Lu, H., et al., 2024. Microbial strategies for lead remediation in agricultural soils and wastewater: mechanisms, applications, and future directions. Frontiers in Microbiology 15. Available at https://doi.org/10.3389/fmicb.2024.1434921. https://www.frontiersin.org/journals/microbiology#.

Gunjal, A.B., 2021. Mechanism and pretreatment effect of fungal biomass on the removal of heavy metals. Biotechnology for Sustainable Environment. Springer, Singapore, India, pp. 155–172.. https://link.springer.com/book/10.1007/978-981-16-1955-7. Available at https://doi.org/10.1007/978-981-16-1955-7_6.

Harish, B.S., Lekshmi, K.M., Jayapal, M., 2024. Simultaneous removal of heavy metals and production of fungal enzymes. In: Uppuluri, K.B., Selvasembian, R. (Eds.), Bioprospecting of Multi-Tasking Fungi for a Sustainable Environment. Springer, Singapore.

Inouhe, M., Hiyama, M., Tohoyama, H., Joho, M., Murayama, T., 1989. Cadmium-binding protein in a cadmium-resistant strain of Saccharomyces cerevisiae. Biochimica et Biophysica Acta (BBA) - General Subjects 993 (1), 51–55. Available at https://doi.org/10.1016/0304-4165(89)90142-6.

Iram, S., Shabbir, R., Zafar, H., Javaid, M., 2015. Biosorption and bioaccumulation of copper and lead by heavy metal-resistant fungal isolates. Arabian Journal for Science and Engineering 40 (7), 1867–1873. Available at https://doi.org/10.1007/s13369-015-1702-1.

Jiang, Y., Rex, D.A.B., Schuster, D., Neely, B.A., Rosano, G.L., Volkmar, N., et al., 2024. Comprehensive overview of bottom-up proteomics using mass spectrometry. ACS Measurement Science Au 4 (4), 338–417. Available at https://doi.org/10.1021/acsmeasuresciau.3c00068.

Karunasagar, D., Arunachalam, J., Rashmi, K., Naveena Lavanya Latha, J., Maruthi Mohan, P., 2003. Biosorption of inorganic and methyl mercury by a biosorbent from Aspergillus niger. World Journal of Microbiology and Biotechnology 19 (3), 291–295. Available at https://doi.org/10.1023/a:1023610425758.

Khan, S.K., Alam, M.S., 2018. Role of ligninolytic enzymes in fungal bioremediation. Environmental Technology Reviews 6 (1), 1–14.

Khan, S., Cao, Q., Zheng, Y.M., Huang, Y.Z., Zhu, Y.G., 2008. Health risks of heavy metals in contaminated soils and food crops irrigated with wastewater in Beijing, China. Environmental Pollution (Barking, Essex: 1987) 152, 686–692.

Khan, Z., Elahi, A., Bukhari, D.A., Rehman, A., 2022. Cadmium sources, toxicity, resistance and removal by microorganisms-a potential strategy for cadmium eradication. Journal of Saudi Chemical Society 26 (6), 101569 ISSN 1319-6103.

Khatiwada, B., Hasan, M.T., Sun, A., Kamath, K.S., Mirzaei, M., Sunna, A., et al., 2020. Proteomic response of Euglena gracilis to heavy metal exposure – identification of key proteins involved in heavy metal tolerance and accumulation. Algal Research 45, 101764.

Konan, D., Ndao, A., Koffi, E., Elkoun, S., Robert, M., Rodrigue, D., et al., 2024. Biodecomposition with *Phanerochaete chrysosporium*: a review. AIMS Microbiology 10 (4), 1068–1101.

Konieczna, W., Turkan, S., Warchoł, M., Skrzypek, E., Dąbrowska, G.B., Mierek-Adamska, A., 2024. The contribution of *Trichoderma viride* and metallothioneins in enhancing the seed quality of *Avena sativa* L. in cd-contaminated soil. Foods 13, 2469.

Kumar, V., Singh, V., Pandit, S., 2025. Advanced omics approach and sustainable strategies for heavy metal microbial remediation in contaminated environments. Bioresource Technology Reports 29, 102040.

Lebrun, J.D., Demont-Caulet, N., Cheviron, N., Laval, K., Trinsoutrot-Gattin, I., Mougin, C., 2011. Secretion profiles of fungi as potential tools for metal ecotoxicity assessment: a study of enzymatic system in Trametes versicolor. Chemosphere 82 (3), 340–345.

Li, C., Zhou, K., Qin, W., Tian, C., Qi, M., Yan, X., et al., 2019. A review on heavy metals contamination in soil: effects, sources, and remediation techniques. Soil and Sediment Contamination: An International Journal 28 (4), 380–394.

Liang, Y., Jiao, S., Wang, M., Yu, H., Shen, Z., 2020. A CRISPR/Cas9-based genome editing system for Rhodococcusruber TH. Metabolic Engineering 57, 13–22.

Liu, Y., Hu, L., He, N., Jiang, Z., Gong, J., Jiang, C., et al., 2025. Lead and cadmium adsorption by Phanerochaete chrysosporium under the protection of phosphorus-containing biochar: effects and mechanisms. Journal of Environmental Chemical Engineering 13 (4), 117330.

Loss, E.M.O., Lee, M.K., Wu, M.Y., Martien, J., Chen, W.P., Amador-Noguez, D., et al., 2019. Cytochrome P450 monooxygenase-mediated metabolic utilization of benzo[a]Pyrene by *Aspergillus* species. mBio 10 (3), e00558–19.

Ming-Ho, Y., 2005. Biological and health effects of pollutants, chap.12. In: Environmental Toxicology. CRC Press LLC, Boca Raton, USA, ISBN 1-56670-670-2, 2nd ed. pp. 1–3.

Murugavelh, S., Mohanty, K., 2013. Bioreduction of Cr (VI) using live and immobilized Phanerochaete chrysosporium. Desalination and Water Treatment 51 (16–18), 3482–3488 ISSN 1944-3986.

Naziębło, A., Dobrzyński, J., 2025. Biotransformation of As, Cr, Hg, and Mn by Pseudomonadota: chances and risks. Biodegradation 36 (4), 60.

Niehaus, J.W., Lim, J.M., Kim, H.S., 2018. Fungal mediated precipitation of Lead from contaminated water: a case study with *Aspergillus Niger*. Chemosphere 211, 1–9.

Ningombam, L., Mana, T., Pradhan, S., et al., 2025. Fungal bioremediation in environmental pollution and recent strategies. Discover Environment 3, 100.

Nirmal Kumar, J.I., George, B., Kumar, R.N., Sajish, P.R., Viyol, S., 2010. Biosorption of mercury and lead by dried *Aspergillus niger* Tiegh. isolated from estuarine sediments. International Journal of Environmental Studies 67 (5), 735–746.

Ondrasek, G., Rengel, Z., Hengl, T., Mikac, N., Gaji´c, B.A., Kistler, H.C., et al., 2025. Metal contamination—a global environmental issue: sources, implications and advances in mitigation. RSC Advances 15, 7894–7915.

Pocsi, I., Prade, R.A., Penninckx, M.J., 2004. Glutathione, altruistic metabolite in fungi. Advanced Microbial Physiology 49, 1–76.

Pointing, S.B., 2001. Feasibility of bioremediation by white-rot fungi. Applied Microbiology and Biotechnology 57 (1–2), 20–33.

Redman, R.S., Sheehan, K.B., Stout, R.G., Rodriguez, R.J., Henson, J.M., 2002. Thermotolerance generated by plant/fungal symbiosis. Science (New York, N.Y.) 298 (5598), 1581.

Regalla, L.M., Lyons, T.J., 2005. Zinc in yeast: mechanisms involved in homeostasis. In: In: Tamas, M.J., Martinoia, E. (Eds.), Molecular Biology of Metal Homeostasis and Detoxification. Topics in Current Genetics, vol 14 Springer, Berlin, Heidelberg.

Revel, B., Catty, P., Ravanel, S., Bourguignon, J., Alban, C., 2022. High-affinity iron and calcium transport pathways are involved in U(VI) uptake in the budding yeast Saccharomyces cerevisiae. Journal of Hazardous Materials 422, 126894.

Sagar, V., Singh, D.P., 2011. Biodegradation of lindane pesticide by non white- rots soil fungus Fusarium sp. World Journal of Microbiology and Biotechnology 27, 1747–1754.

Shao, Q., Yan, S., Sun, X., Chen, H., Lu, Y., Li, S., et al., 2025. Applications of yeasts in heavy metal remediation. Fermentation 11, 236.

Singh, V.K., Singh, R., 2024. Role of white rot fungi in sustainable remediation of heavy metals from the contaminated environment. Mycology 15 (4), 585–601.

Singh, R.K., Tripathi, R., Ranjan, A., Srivastava, A.K., 2020. Fungi as potential candidates for bioremediation. In: Singh, P., Kumar, A., Borthakur, A. (Eds.), Abatement of Environmental Pollutants. Elsevier, Amsterdam, pp. 177–191.

Singh, S., Jha, P., Jobby, R., 2021a. Chapter 6 - Fungi: a promising tool for bioremediation of toxic heavy metals. In: Saxena, G., Kumar, V., Shah, M.P. (Eds.), Bioremediation for Environmental Sustainability. Elsevier, pp. 123–144.

Singh, A., Kumari, R., Yadav, A.N., 2021b. Fungal secondary metabolites for bioremediation of hazardous heavy metals. In: Yadav, A.N. (Ed.), Recent Trends in Mycological Research: Volume 2: Environmental and Industrial Perspective. Springer International Publishing, Cham, pp. 65–98.

Singh, J., Mandal, A., Nandabalan, Y.K., 2025. Synergistic effect on lignolytic enzyme production through co-culturing of white rot fungi during solid state fermentation of agricultural residues. Bulletin of the National Research Centre 49 (1), 3.

Sista, Kameshwar, A.K., Qin, W., 2019. Systematic review of publicly available non-Dikarya fungal proteomes for understanding their plant biomass-degrading and bioremediation potentials. Bioresources and Bioprocessing 6 (1), 1–19.

Sreedevi, P.R., Suresh, K., Jiang, G., 2022. Bacterial bioremediation of heavy metals in wastewater: a review of processes and applications. Journal of Water Process Engineering 48, 102884.

Su, S., Zeng, X., Bai, L., Li, L., Duan, R., 2011. Arsenic biotransformation by arsenic-resistant fungi Trichoderma asperellum SM-12F1, Penicillium janthinellum SM-12F4, and Fusarium oxysporum CZ-8F1. The Science of the Total Environment 409 (23), 5057–5062.

Sutjaritvorakul, T., Gadd, G.M., Whalley, A.J.S., Suntornvongsagul, K., Sihanonth, P., 2016. Zinc oxalate crystal formation by *Aspergillus nomius*. Geomicrobiology Journal 33, 289–293.

Sácký, J., Leonhardt, T., Borovička, J., Gryndler, M., Briksí, A., Kotrba, P., 2014. Intracellular sequestration of zinc, cadmium and silver in Hebeloma mesophaeum and characterization of its metallothionein genes. Fungal Genetics and Biology: FG & B 67, 3–14.

Tegli, S., Cerboneschi, M., Corsi, M., Bonnanni, M., Bianchini, R., 2014. Water recycle as a must: decolorization of textile wastewaters by plant associated fungi. Journal of Basic Microbiology 54 (2), 120–132.

Troiano, D., Orsat, V., Dumont, M.J., 2020. Status of filamentous fungi in integrated biorefineries. Renewable and Sustainable Energy Reviews 117, 109472.

Vaksmaa, A., Guerrero-Cruz, S., Ghosh, P., Zeghal, E., Hernando-Morales, V., Niemann, H., 2023. Role of fungi in bioremediation of emerging pollutants. Frontiers in Marine Science 10, 1070905. Available at https://doi.org/10.3389/fmars.2023.1070905.

Varjani, S.J., Patel, R.K., 2017. Fungi: a remedy to eliminate environmental pollutants. In: Prasad, R. (Ed.), Mycoremediation and Environmental Sustainability. Springer, Cham, pp. 53–67.

Vasina, D.V., Pavlov, A.R., Koroleva, O.V., 2016. Extracellular proteins of Trametes hirsuta st. 072 induced by copper ions and a lignocellulose substrate. BMC Microbiology 16 (1), 106.

Wang, Y., Wang, C., Cheng, W., Bian, Y., Guo, P., 2016. Removal of cadmium from contaminated *Lentinula edodes* by optimized complexation and coagulation. Food and Science & Nutrition 5 (2), 215–222.

Wang, Y., Weisenhorn, E., MacDiarmid, C.W., Andreini, C., Bucci, M., Taggart, J., et al., 2018. The cellular economy of the Saccharomyces cerevisiae zinc proteome. Metallomics 10 (12), 1755–1776.

Wang, C., Wu, K., Pang, N., Zhao, H., Liu, S., Zhang, X., et al., 2025. Transcriptome analysis reveals the mechanism of tolerance to copper toxicity in the white rot fungus Trametes hirsuta AH28-2. Ecotoxicology and Environmental Safety 296, 118194. Available at https://doi.org/10.1016/j.ecoenv.2025.118194.

Yang, L., Li, T., Li, J., Chen, J., 2019. Chelation of heavy metals by phosphate-solubilizing fungi for bioremediation. Ecotoxicology and Environmental Safety 175, 1–8.

Yang, R., Roshani, D., Gao, B., Li, P., Shang, N., 2024. Metallothionein: a comprehensive review of its classification, structure, biological functions, and applications. Antioxidants (Basel) 13 (7), 825.

Zapata-Sarmiento, D.H., Rodríguez-Hernández, A.A., Sepúlveda-Jiménez, G., Rodríguez-Monroy, M., 2025. Tolerance and antioxidant response to heavy metals are differentially activated in *Trichoderma asperellum* and *Trichoderma longibrachiatum*. PeerJ 13, e19016.

Zelaya-Molina, L.X., Chávez-Díaz, I.F., Urrieta-Velázquez, J.A., Aragón-Magadan, M.A., Puente-Valenzuela, C.O., Blanco-Camarillo, M., et al., 2025. Microbial metallophores in the productivity of agroecosystems. Microbiological Research 16, 67.

Zengguang, X., Yanlong, L., Junrui, C., Ronggao, Q., Rong, Y., 2015. Bioremediation modeling of an aquifer contaminated by benzene using the slow-release oxygen source technique. Arabian Journal for Science and Engineering 40, 2457–2463.

Zhang, Q.H., Zeng, G.M., Chen, G.Q., Yan, M., Chen, A.W., Du, J.J., et al., 2015. The effect of heavy metal-induced oxidative stress on the enzymes in white rot fungus *Phanerochaete chrysosporium*. Applied Biochemistry and Biotechnology 175, 1281–1293.

Zhang, B., Wang, S., Diao, M., Fu, J., Xie, M., Shi, J., et al., 2019. Microbial community responses to vanadium distributions in mining geological environments and bioremediation assessment. Journal of Geophysical Research: Biogeosciences 124 (3), 601–615.

Zhang, L., Chen, F., Zeng, Z., Xu, M., Sun, F., Yang, L., et al., 2021. Advances in metagenomics and its application in environmental microorganisms. Frontiers in Microbiology 12, 766364.

Chapter 5

Integrative omics for bioremediation: proteomics with metabolomics and genomics

Md. Parvez Kabir[1], Falguni Akter Shraboni[2], Md. Selim Reza[3], Rabeka Sultana Smriti[4], Ashek Mohammad[5], and Akbar Hossain[1]

[1]Soil Science Division, Bangladesh Wheat and Maize Research Institute, Dinajpur, Bangladesh, [2]Wheat Breeding Division, Bangladesh Wheat and Maize Research Institute, Dinajpur, Bangladesh, [3]Soil Science Division, Bangladesh Institute of Nuclear Agriculture, Mymensingh, Bangladesh, [4]Soil Science Division, Bangladesh Agricultural Research Institute, Gazipur, Bangladesh, [5]Entomology Division, Hortic ulture Research Centre, Bangladesh Agricultural Research Institute, Gazipur, Bangladesh

5.1 Introduction

The 21st century is already overwhelmed by an ecological disaster that has been driven by industrialization, urban growth, and unsustainable agriculture (Mishra, 2025). The discharge of heavy metals, persistent organic pollutants, and synthetic chemicals threatens the ecosystem balance, resulting in disruptions of microbial diversity and a challenge to both human and animal health (Thakur et al., 2025). The traditional methods of treatment (e.g., incineration, chemical neutralization, and physical containment) are frequently expensive and energy intensive; in addition, the possibility that secondary pollution will be generated during processing the contaminated soil or water, drives the imperative for environment-friendly alternatives (Olanbiwoninu et al., 2025; Varshney et al., 2022).

The bioremediation techniques employing microorganisms, plants, or their enzymatic systems for detoxification and remediation of pollutants have been considered as a green approach (Arora, 2018; Bala et al., 2022). Microorganisms in particular have large metabolic versatility so that they are able to degrade hydrocarbons, pesticides, dyes, and heavy metals under diverse environmental conditions (Aragaw, 2021; Kour et al., 2021). However, the culture-based techniques of conventional microbiology depict nothing more than a small portion of the microbial world since most environmental microorganisms are unculturable in the lab (Czaplicki and Gunsch, 2016).

Proteomics Driven Bioremediation. DOI: https://doi.org/10.1016/B978-0-443-45294-9.00009-6

With the appearance of omics (genomic, transcriptomic, proteomic, and metabolomic) tools, bioremediation research was transformed (Singh et al., 2021). These approaches enable a detailed study of microbial communities and their metabolism without the need for isolation. Genomics provides information on genetic potential and metabolic blueprints, proteomics studies functioning enzymes and proteins, and finally, metabolomics tracks the dynamic chemical output of microbial metabolism (Yunusa and Umar, 2022). The integration of such datasets with multiomics strategies will support the more close-combat perception in microbial processes and assist in connecting the well-known gap between genetic potential, protein functioning, and metabolic activity by a complementary systems perspective on microorganisms' approaches (Hayes et al., 2024; Malla et al., 2018).

Integration of multiple omics analysis facilitates finding regulatory genes and enzymes, mapping of metabolic fluxes, and the development of predictive computational models in order to optimize biodegradation performance (Mariam et al., 2025). In this complex network, proteomics plays a central role as proteins are indeed the actors of these biochemical changes (Larance and Lamond, 2015). Together with metabolomics and genomics, proteomic data reflect enzyme activities, metabolite-abundance changes, and gene regulations for a holistic view on microbial functions (Wanichthanarak et al., 2015).

The shift of paradigm in environmental biotechnology was substantiated from empirical observations toward computational modeling and machine learning (ML) for more prediction rather than system biology-based approaches (Choubey et al., 2025). Simulating the cross-talk between genes, proteins, and metabolites allows researchers to trigger responses of microorganisms, to predict degradation kinetics, or to design tailored microbial consortia. These integrating approaches are compatible with the circular economy and UN Sustainable Development Goals (SDGs) and rationalize bioremediation toward a data driven and sustainable framework for environmental restoration.

5.2 Overview of omics technologies

Omics applications have revolutionized environmental microbiology by providing a holistic understanding of complex biological systems at different molecular scales (Kumar et al., 2021). Each omics discipline (genomic, proteomic, or metabolomic) provides a complementary window on how microorganisms sense (to detect), respond to, and detoxify contaminants in the polluted environments (Malla et al., 2018). The cellular levels at which these omics platforms operate within a bacterial cell are schematically illustrated in Fig. 5.1. In this section, we outline the concepts, methodology, and functions of these omics tools in bioremediation studies.

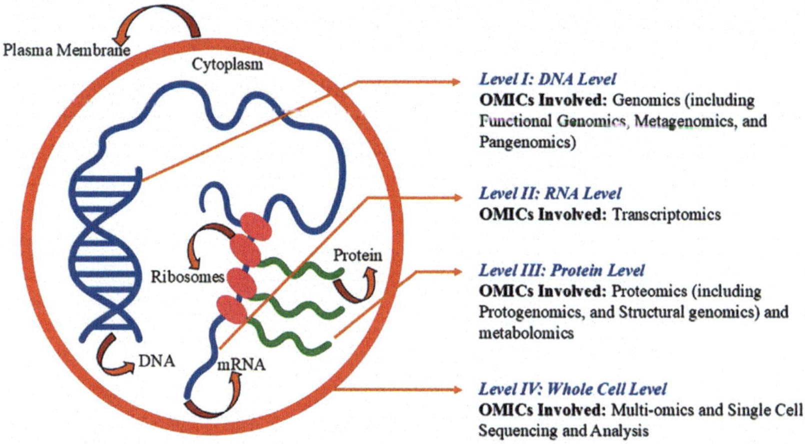

FIGURE 5.1 **Cellular levels at which omics platforms operate within a bacterial cell.** Simplified schematic diagram of a typical bacterial cell highlighting the cellular levels at which different omics technologies are applied.

5.2.1 Genomics—the blueprint of bioremediation potential

Genomics is the characterization of an organism's entire genetic material, a blueprint for its biological potential (Chaitanya, 2019). In bioremediation, the genes for pollutant degradation, stress tolerance, and metal resistance are uncovered in a genomics approach (Das et al., 2016).

5.2.1.1 Genome sequencing and annotation

The rapid advancement of next-generation sequencing (NGS) technology, such as Illumina, PacBio, and Oxford Nanopore, has had a profound influence on microbial genomic research, where costs of DNA sequencing have been massively reduced, throughput increased, and accuracy improved (Satam et al., 2023). These emerging methods of sequencing facilitate deep sequencing at the single-organism level from environmental isolates and bacterial communities, leading to a new history for their genetic potential in the degradation of pollutants (Chakraborty and Das, 2016; Shekhar et al., 2019). Whole-genome sequences (WGS) can be used to identify genes encoding essential catabolic enzymes, such as oxygenases, hydrolases, and dehalogenases in detoxification pathways (Oyewusi et al., 2021).

Furthermore, metagenomic studies extend this potential to microbial communities uncultured so far, revealing metabolic activities unable to be inferred before and novel degradative gene clusters (Liu et al., 2022). For decoding these big datasets, functional annotation tools, for example, Prokka, RAST, and KEGG Mapper are used to find association of sequences with metabolic pathways, catering pollutant transformations (Wani et al., 2022; Kumar et al., 2020). For instance, novel alkB genes that encode the alkane monooxygenase system involved in the initial oxidation and subsequent metabolism of petroleum hydrocarbons have been identified by metagenomic analysis of hydrocarbon-contaminated soils, highlighting the potential of genome-based discovery of new biocatalysts for environmental remediation (Somee et al., 2022; Di Canito, 2019).

5.2.1.2 *Comparative and functional genomics*

The observations contribute to comparative genomics by cataloging conserved and novel genes for pollution degradation, providing insights into evolutionary adaptation (Chen et al., 2021). Functional genomics also analyzes gene expression in stress conditions by RNA sequencing (RNA-seq) or genomes, including assessments, discovering regulatory mechanisms that trigger degradation pathways (Razzaq et al., 2021). These genome-based insights have provided a foundation for the development of microbial strains characterized with improved bioremediation capacity and stress tolerance.

5.2.2 Proteomics: the functional landscape of bioremediation

While genomics indicates what an organism can do, proteomics reveals what it actually does under specific environmental conditions. Proteomics focuses on the identification, quantification, and characterization of proteins expressed by a cell, tissue, or microbial community (Karlsson et al., 2015).

5.2.2.1 *Analytical approaches in proteomics*

The proteomic studies up to now in bioremediation have been based mainly on mass spectrometry (MS) after the extraction and enzymatic digestion of proteins (Singh et al., 2021). 2D gel electrophoresis (2D-GE), one of the predominant analysis methodologies, separates proteins according to their isoelectric point (IP) and molecular weight (MW), thus enabling expression changes in response to pollutant exposure to be visualized (Mortimer et al., 2022). The latest generation of liquid chromatography–tandem mass spectrometry (LC-MS/MS) allows a large-scale identification and quantification of complex protein mixtures with high sensitivity, accuracy, and resolution (Chaudhary et al., 2025).

Additionally, the isobaric tagging strategies such as iTRAQ and tandem mass tag (TMT) are globally applied to analyses by multiplexed detection, thereby realizing comparative proteomics profiling across multiple environmental or experimental conditions (Sivanich et al., 2022). Taken together, these approaches identify the protein or enzyme systems responsible for pollutant metabolism and those stress-response proteins that protect microbial cells from toxic pollutants, as well as revealing potential global determinants of successful bioremediation.

5.2.2.2 *Functional proteomics in bioremediation*

Functional proteomics is capable of providing valuable insight into the biochemical systems that microorganisms degrade and tolerate environmental pollutants (López et al., 2023). Various enzymes, including some that play leading roles in degradative processes (i.e., dioxygenases, monooxygenases responsible for oxidizing hydrocarbons and aromatic compounds, reductase for metal ion detoxification, laccase and peroxidase for dye and phenol substances) have been identified by protein profiling (Bhardwaj et al., 2022). However, the occurrence of oxygenases and chaperones more abundantly presented in the extracellular proteome of toluene-exposed *Pseudomonas putida* indicates that tolerance to solvent and effective pollutant degradation is an example of adaptive response (Segura et al., 2005).

Besides enzyme identification, posttranslational modifications (PTMs) such as phosphorylation or acetylation can also provide information on the regulation of enzyme targets, like signaling networks and enzymatic activities, during pollutant stress (Sharma et al., 2023). However, these results demonstrate the important role of functional proteomics in monitoring and identifying dynamic microbial responses and designing successful bioremediation strategies. The following figure (Fig. 5.2) illustrates the various omics approaches used to analyze polycyclic aromatic hydrocarbon (PAH)-contaminated soil samples for bioremediation.

5.2.2.3 *Environmental and community proteomics*

Metaproteomics has emerged as a powerful tool to investigate community-level functions in contaminated ecosystems (Maron et al., 2007). It helps guide to active degraders, protein interaction networks, and functional redundancy in the complex microbiome that all things single-species studies don't take account of (Lohmann et al., 2020).

5.2.3 Metabolomics: the chemical fingerprints of microbial function

Metabolomics represents a snapshot of the small-molecule metabolites (<1500 Da) produced by cellular processes that reflect the physiological state of an organism (Muthubharathi et al., 2021). Metabolites are created from the interaction of genes and proteins,which means that, by using metabolomics, knowledge on how microorganisms respond to pollutants (Singh, 2006).

5.2.3.1 *Metabolomic profiling techniques*

Metabolomics research integrates various advanced analytical platforms for profiling small molecular metabolites of microbes in response to pollutants (Kumar et al., 2021). Volatile and thermally stable molecules, such as fatty acids and organic acids, can be profiled using gas chromatography–mass spectrometry

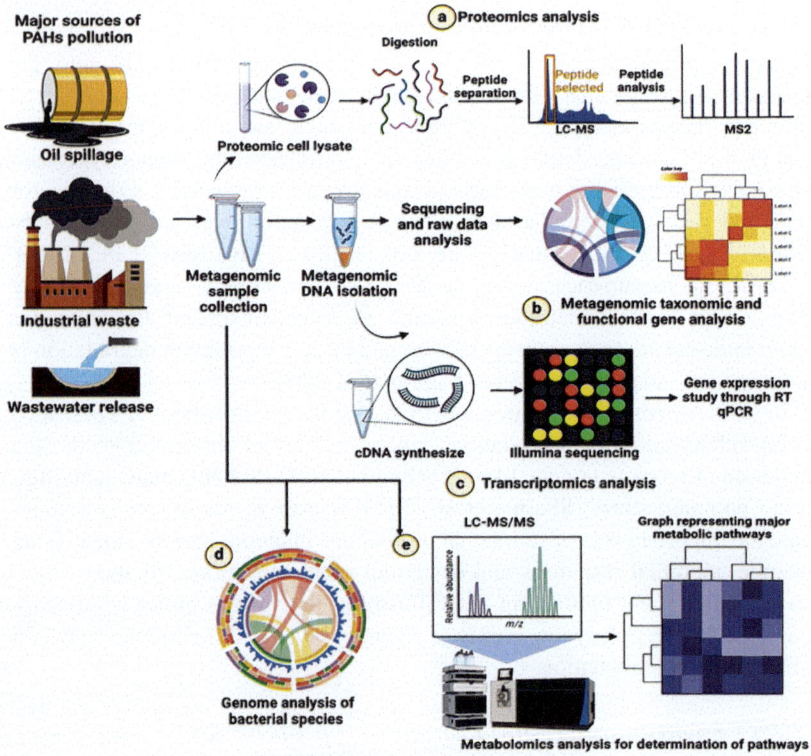

FIGURE 5.2 Omics approaches used to analyze PAH contaminated soil samples. Various omics approaches were employed to analyze contaminated soil samples from PAH contaminated environments. *PAH*, Polycyclic aromatic hydrocarbon.
Adapted from Das et al., (2021).

(GC-MS) (Lindeque, 2024). Liquid chromatography–mass spectrometry (LC-MS) is suitable for a wider range of polar and non-polar metabolites, with the latter providing not only high sensitivity but also high versatility (Patidar and Kamble, 2025). On the other hand, nuclear magnetic resonance (NMR) spectroscopy offers a noninvasive and highly reproducible method for the identification and quantification of metabolites from complex biological mixtures (Crook and Powers, 2020). Both methods have their specific advantages in resolution, quantification, and compound specificity; thus, a hybrid approach is particularly well suited to metabolite profiling, which provides an understanding of the system-wide metabolism during bioremediation.

5.2.3.2 *Metabolomics in bioremediation research*

Metabolomics is key in understanding the biochemical processes of pollutant degradation pathways by profiling intermediate metabolites and pathway kinetics (Jeevanandam and Osborne, 2022). Analysis of metabolic intermediates allows researchers to rebuild degradation pathways and discover toxic byproducts (Dangi et al., 2019). In hydrocarbon degradation, the occurrence of catechol, succinate, and maleate provides clear evidence of aromatic ring cleavage in confirming catabolic activity (Rabus et al., 2016; Kumari and Das, 2023). These concepts are exemplified in Fig. 5.3, which illustrates the three main pathways by which fungi and bacteria degrade PAHs, highlighting the key metabolic intermediates involved in these biotransformation processes.

Similarly, during heavy-metal detoxication by the microorganisms, increased levels of glutathione, phytochelatins, and organic acids hint at the microbial detoxication responses and metal-binding mechanisms (Devi et al., 2022). In addition, the metabolomic profile can be tracked over time to monitor microbial adaptation to stress tolerance and degradable capability of microbes (Kaya, 2024). In this regard, the knowledge obtained in the current

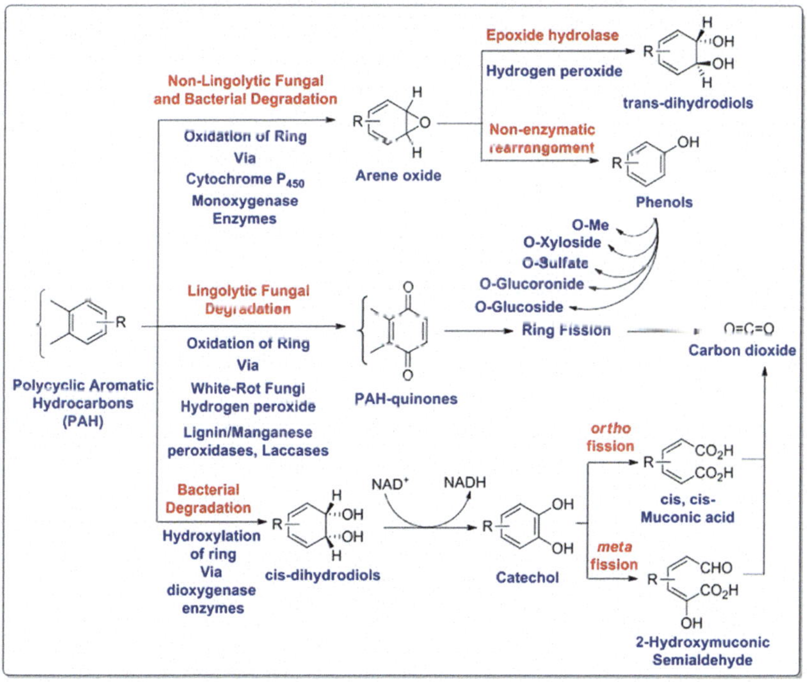

FIGURE 5.3 Three main pathways for polycyclic aromatic hydrocarbon degradation by fungi and bacteria.
Adapted from Kour, D., Kaur, T., Devi, R., Yadav, A., Singh, M., Joshi, D., et al., 2021. Beneficial microbiomes for bioremediation of diverse contaminated environments for environmental sustainability: present status and future challenges. Environmental Science and Pollution Research 28 (20), 24917–24939, doi: 10.1007/s11356-021-13252-7..

metabolomic study substantially strengthens system-level comprehension of microbe-converted pollutants and information easily applicable to bioremediation optimization or environmental recovery monitoring.

5.2.4 Synergistic discoveries—multiomics strata

The integration of genomics, proteomics, and metabolomics provides a comprehensive view of microbial bioremediation, linking genetic potential to observable functional outcomes (Singh et al., 2021). Genomic analyses reveal the existence of genes for several important degradative enzymes, and proteomic studies support their expression patterns and regulation in response to pollutant stress (Zhao and Poh, 2008). In addition to these, metabolomic profiling validates the intermediates and terminus products existing in response to genomic and proteomic evidence (Prosser et al., 2014). This vertical integration across the molecular strata supports causal relationships between genotype, protein function, and metabolic function that enable system-level insights into how microorganisms conduct complex detoxification transformations (Islam, 2021). In the following sections, methodologies, computational tools, and a few cases of these integrated omics analyses have been discussed for improving bioremediation research and environmental restoration studies.

5.3 Integrative omics approach in bioremediation

5.3.1 The concept and rationale of multiomics integration

Bioremediation studies have moved from parameter to system level and can now provide all tools enabling the mimicry of subtle relationships between microbial communities in polluted environments (Desai et al., 2010). Although genomics, proteomics, and metabolomics can provide information about individual layers of the molecular system, their combination (multi-omics or integrated omics) allows for a comprehensive picture of biological systems in action (Misra et al., 2019). This approach also links genotype with phenotype by showing how that genetic potential is expressed in the form of functional and nonfunctional proteins, as well as a metabolic response.

In bioremediation, the multiomics integration can also identify molecular players and regulatory sites responsible for pollutant degradation, infer gene, protein, and metabolites, and predict microbial performance in varied environmental niches (Ghai et al., 2024). This holistic understanding is particularly crucial for the study of microbial communities, as complex interspecies interactions are likely to govern the overall degradation performance (Nawaz et al., 2022). By connecting these scales of molecular information, multi-omics integration renders bioremediation operationalized from a descriptive towards a predictive science that might allow to tailor targeted environmental interventions.

5.3.2 Integration of frameworks and strategies

The (multi-) omics data integration is the association among heterogeneous data of gene expression and/or protein synthesis levels measured using different experimental platforms (Santiago-Rodriguez and Hollister, 2021). Together with these findings, Fig. 5.4 illustrates how integrative omics combines genomic, transcriptomic, proteomic, and metabolomic data to provide a comprehensive framework for understanding and optimizing bioremediation. Integration can be broadly divided into two strategies, horizontal integration and vertical integration, considering how the datasets are aligned and analyzed.

5.3.2.1 Horizontal integration

Horizontal integration, blending data from datasets of the same omics layer (proteomics, transcriptomic and metabolomics) across various microbial

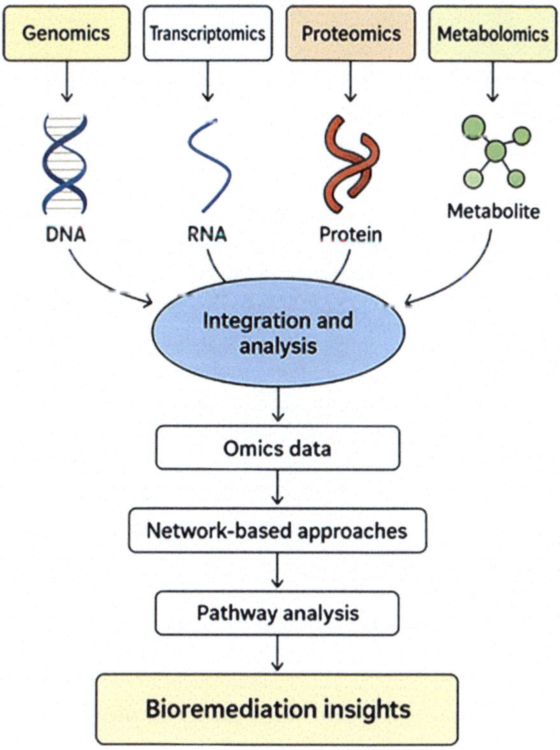

FIGURE 5.4 Integrative omics for bioremediation. Integrative omics combines genomic, transcriptomic, proteomic, and metabolomic data to provide a comprehensive framework for understanding and optimizing bioremediation.

strains, species, or environmental conditions to explore conserved and divergent responses (Satya et al., 2024). In bioremediation research, such a comparative approach may allow for screening of stress-response proteins being common among the ones involved in combating across various types of pollutants (potential universal adaptation), as well as unique to certain strain-soil- or pollutant-specific adaptations (Samuel et al., 2013). It may also be useful to obtain data about the evolutionary dynamics of degradation pathways, revealing how different microbes take advantage of their diverse enzymatic systems to degrade pollutants. In particular, comparative proteomic strategies on *Pseudomonas, Bacillus*, or *Rhodococus* strains upon hydrocarbon exposure have allowed us to draw the profile of common enzyme families (dioxygenases and hydrolases) and unique detoxification mechanisms, concerning the ecological and functional diversity (Zampolli, 2017). Chaurasia et al. (2022). Through profiling such horizontal connections, researchers may set the groundwork to decipher microbial flexibility and to identify optimal strains or consortia for targeted detoxification purposes.

5.3.2.2 *Vertical integration*

Vertical integration connects multiple omics layers (presumably with genomics, proteomics, and metabolomics) to form a continuous cascade linking genetic information to functional metabolic outcomes (Savitha et al., 2025). This approach is essentially based on mapping genes to proteins (by using big databases such as KEGG, Uniprot, and STRING) that will account for the genetic basis of pollutant degradation (Mahallea and Dafale, 2021; Huang, 2023; Menaa et al., 2025). Next, it needs to incorporate proteins and metabolites with pathway analysis and flux balance analysis (FBA) models so that the metabolic flows and active biochemical routes during pollutant degradation could be predicted (Qiu et al., 2023). Later interaction networks are imputed via visualization and integration tools such as Cytoscape, MetaboAnalyst, OmicsNet, and Pathway Tools to depict the dynamics of cross-layer connections and regulations (Eicher et al., 2020; Zhou et al., 2020; Joshi et al., 2024). By combining genetic potential with protein function and metabolic expression, vertical integration presents the most complete representation of microbial function in bioremediation, capturing the flow of biological information across hierarchical molecular scales.

5.3.3 Computational tools and data analysis pipelines

Efficient multiomics integration necessitates innovative computational approaches and bioinformatics tools for addressing complex, high-dimensional datasets (Vahabi and Michailidis, 2022). Preprocessing of data, including normalization, filtering, and noise suppression, is performed by dedicated software such as MS-DIAL for metabolomic data, and MaxQuant in proteomic analysis (Wang et al., 2020; Prudhomme, 2023). The most prominent algorithms making use of these technologies for dimensionality reduction and cluster analysis are the Principal Component Analysis (PCA), Partial Least Squares (PLS), or t-distributed Stochastic Neighbor Embedding (t-SNE) (Debik et al., 2022; Medina-Ortiz et al., 2020). Network analysis software is then applied to construct maps of gene–protein–metabolite connections that represent functional and regulatory relationships among the elements of microbial systems (Bassel et al., 2012). Some of the key software and platforms used for multi-omics data analysis are summarized in Table 5.1

ML algorithms, such as random forests, neural networks, and Bayesian inference, are also increasingly applied to predict microorganism behavior, pollutant degradation rates, and system outputs from integration of multiomics data (Trivedi et al., 2025; McElhinney et al., 2022). Furthermore, the genomic reconstruction of metabolic pathways using bioinformatics tools such as COBRA Tools, BioCyc, and ModelSEED opens up the possibility to predict pollutant degradation mechanisms in response to various environmental conditions (Hasan et al., 2025; Navid, 2021; Desai et al., 2021). Such pipelines of data analysis can be conducted within programming languages such as R, Python, and MATLAB® through integration of software tools to achieve a flexible, reproducible, and scalable analysis in multi-omics applications for bioremediation (Anand and Padmanabhan, 2024; Guo et al., 2025). A range of specialized databases supporting multiomics analyses in bioremediation, including those for sequences, pathways, metabolic models, and biodegradation-specific information, is summarized in Table 5.2

5.4 Application of integrative omics in bioremediation

The era of integrated omics has transformed bioremediation from a descriptive to a mechanistic science with the use of genetics, protein expression, and metabolite data as an integrated systems-level framework (Islam, 2021). For example, multiomics characterization of hydrocarbon-contaminated soils has identified novel uncharacterized dioxygenases and hydrolases against PAHs, which indicates new enzymatic pathways that are amenable to discoveries through multiomics approach (Das et al., 2025). Furthermore, proteomic and metabolomic studies on microbial communities under exposure to the heavy metals were indicative of the upregulation of sulfur and glutathione-dependent detoxification enzymes as well as for an accumulation of thiol-metabolites in situ serving as response patterns, which are feasible indicators of remediation progress within contaminated sites (Cummins et al., 2011; Talukdar, 2015).

The integrative strategy also allows the rational design of microbial consortia: through the mapping of synergistic metabolic functions among strains from combined omics data, scientists can develop mixed-culture systems in which one organism serves to initiate pollutant degradation and another completes mineralization, increasing degradation performance

TABLE 5.1 Some key Software and platforms used for the analysis of multi-omics data.

Name	Function	URL	References
KEGG	Used for genome analysis, enzyme annotation, and metabolic pathway reconstruction	http://www.kegg.jp/	Kanehisa et al. (2017)
PLINK	Used for whole-genome association studies (WGAS)	http://pngu.mgh.harvard.edu/purcell/plink/	Purcell et al. (2007)
Bioconductor packages for R	Contains comprehensive tools for omics data analyses	http://www.bioconductor.org/	Gentleman et al. (2004)
MEGA 4.0	Molecular evolutionary genetics analysis	http://www.megasoftware.net/	Tamura et al. (2007)
Velvet	Algorithms for de novo short read assembly using de Bruijn graphs	https://www.ebi.ac.uk/~zerbino/velvet/	Zerbino, Birney (2008)
R	All types of statistical analyses	http://r-project.org	Neupane et al. (2014)
BlastKOALA & GhostKOALA	Used for the functional characterization of genome and metagenome sequences	http://www.kegg.jp/blastkoala/	Kanehisa et al. (2016)
ProteoWizard	Software for rapid proteomics	http://proteowizard.sourceforge.net/	Kessner et al. (2008)
SB Workbench & BioSPICE	Integrated software tool for modeling, analysis, visualization, and general data manipulation	http://SystemsBiologyw.sourceforge.net	Finney and Hucka (2003)
HISAT	Transcript-level expression analysis of RNA-sequences	http://ccb.jhu.edu/software.shtml	Pertea et al. (2016)

(Saraiva et al., 2021; Timofeeva et al., 2023). Collectively, integrative omics also open the vistas of (1) knowledge on microbial metabolism in a holistic perspective; (2) discovery of novel biocatalyst and pathway routes; (3) determination of experimentally trustworthy molecular biomarkers; and hence (4) design optimization of successful microbial consortia, adding great advantage to prediction capability and practical deployment for bioremediation

TABLE 5.2 Databases used in multiomics-based bioremediation strategies.

Name	Database	URL	References
NCBI databases	DNA	http://www.ncbi.nlm.nih.gov/gquery/	Brown et al. (2015)
Ensambl	RNA	http://www.ensembl.org/index.html	Aken et al. (2016)
IntAct	Proteins	http://www.ebi.ac.uk/intact/	Orchard et al. (2014)
BioRadBase	Bioremediation of radioactive waste generated from nuclear plants	http://biorad.igib.res.in/	Reena et al. (2012)
BioFNet	Database for analysis and synthesis of biosystems	http://kurata22.bio.kyutech.ac.jp/db/pub/pub_main.php?Ver¼3.8	Kurata et al. (2014)
AromaDeg	Phylogenomics for bioremediation of aromatic compounds	http://aromadeg.siona.helmholtz-hzi.de/	Duarte et al. (2014)
KBase	Search and organize microbial communities, plants, genomics, and SB data	https://kbase.us/	Arkin et al. (2016)
BioCyc	Contains full information about organism-specific pathway/genome databases	http://biocyc.org/	Caspi et al. (2016)
MEMOSys 2.0	Genome-scale models (GEMs) database	https://memosys.i-med.ac.at/MEMOSys/home.seam	Pabinger et al. (2014)
BSD	Biodegradative Strain Database	http://bsd.cme.msu.edu/	Urbance et al. (2003)
Bionemo	Provides molecular information about biodegradation metabolism	https://omictools.com/bionemo-tool	Carbajosa et al. (2009)

strategies. To illustrate these advantages in practice, Table 5.3 presents a few examples of integrative omics techniques used in the bioremediation of different toxic pollutants and the microorganisms involved.

TABLE 5.3 Integrative omics-based strategies for the bioremediation of toxic pollutants.

Microorganisms	Techniques	Toxic compounds	References
Anaeromyxobacter sp. Fw109-5	Genomics	Nitrate and uranium	Hwang et al. (2015)
Bacillus subtilis HUK15	Genomics	Hexachlorocyclohexane (HCH)	Gasc et al. (2016)
Brevibacterium epidermidis EZ-K02	Genomics	Benzoate, p-hydroxybenzoate, aceto-phenone, catechol, gentisate, arsenic, cobalt, and cadmium	Ziganshina et al. (2018)
Rhodococcus aetherivorans BCP1 and *Rhodococcus opacus* R7	Phenomics	Antimicrobial compounds	Cappelletti et al. (2016)
Miscanthus sinensis	Proteomics	Antimony (SB)	Xue et al. (2015)
Sphingomonas sp. GY2B	Proteomics	Phenanthrene	Liu et al. (2017)
Pseudomonas putida	Metabolomics	Antibiotics and aromatic hydrocarbons	Sayqal et al. (2016)
Acinetobacter venetianus RAG-1	Transcriptomics	Alkanol (dodecanol)	Nikel and Chavarría (2015)
Thiobacillus sp.	Metagenomics	Thiocyanate (SCN^-)	Rahman et al. (2017)
Pseudomonas putida KT2440	Fluxomics	Hydrocarbons	Kothari et al. (2016)

5.5 Case studies of integrative omics in bioremediation

Integrative omics (genomic, proteomic, and metabolomic) is a powerful approach that has been employed with much success for elucidation of the mechanisms of degradation of pollutants (e.g., hydrocarbons, heavy metals, pesticides, and dyes) by microorganisms (Singh et al., 2021; Kumar and Thakur, 2022). In hydrocarbon remediation, metagenomic analyses of oil-polluted soils, for instance, identified new gene families coding for mono- and dioxygenases, alkB and nah gene families of microbes, like *Alcanivorax borkumensis* and *Pseudomonas putida* genera (Liu et al., 2024; Karthikeyan et al., 2021; Baburam, 2020). In support, the comparison of the proteomic profiles from *P. putida* during toluene exposure observed increases in abundance of both toluene monooxygenase and heat-shock proteins, which indicate coordinated stress-response and biodegradation facilitation (Segura et al., 2005; Nogales et al., 2019).

In heavy-metal detoxification systems, genomics demonstrated the presence of efflux pump genes (e.g., czcA) and metal-binding peptides such as the metallothioneins, and metabolomic profiling led to the identification of increments in glutathione, phytochelatins, and organic acids associated with detoxification pathways (Biswas et al., 2022; Raza et al., 2023; Anand et al., 2023). At the level of pesticide degradation, omics analysis has been seen uncovering organophosphorus hydrolase (opd) genes and their translation into proteins performed by organisms like *Sphingomonas paucimobilis* with early intermediates like 3,5,6- trichloro-2-pyridinol (TCP) detection at metabolomic level along with downstream products (Sharma and Pandit, 2023; Guerrero Ramírez et al., 2023; Ravintheran et al., 2019).

Genomic mapping of laccases and azoreductases in bacteria such as *Bacillus subtilis* was complemented in textile dye removal by ontogeny with proteomic and metabolomics evidence for aromatic amine intermediates and benign acid end-products (Ikram et al., 2023; Tyagi et al., 2025). In the end, within consortium-based systems, metagenomic and metaproteomic profiling of petroleum-sludge communities identified cooperating hydrocarbon-degrading taxa and biosurfactants production with metabolomics confirming transformation of hydrocarbons to organic acids along with CO_2 (Chen et al., 2022; Mohapatra and Phale, 2021). Altogether, these case studies illustrate that proteomics-guided integration of multi-omics intermediates genetic potential and metabolic output to yield the design-based prediction of remediation strategies rather than a purely descriptive approach. The diverse roles of microorganisms in the bioremediation of different environmental pollutants are summarized in Table 5.4

5.6 Challenges in multiomics integration

While integrative omics has transformed bioremediation studies, complexities that range from technical to computational and conceptual prevent it from widespread implementation (Singh et al., 2021). Common issues include the heterogeneity of data as a result of each omics: genomics, transcriptomics, proteomics, and metabolomics returning datasets with different scales, formats, and sensitivities (Cambiaghi et al., 2017). Fusing these layers necessitates sophisticated normalization and harmonization methods that can accommodate differences in the level of coverage and timespan (Misra et al., 2019). Recent studies emphasize that missing or incomplete data across omics layers remain a critical constraint, prompting the need for additional work in artificial intelligence (AI)-based imputation and fusion models, which can enhance continuity of data and accuracy (Vincenzo, 2024; Bunne et al., 2024).

Computational complexity presents another challenge (Hériché et al., 2019). Multiomics studies produce data at terabyte scales, which require powerful computing facilities, robust storage systems, and an efficient pipeline for data integration (Serrano et al., 2025). Scalability has been advanced by open-source platforms such as OmicsNet and Multi Assay Experiment, but interpretability and reproducibility are stifled by the lack of universal analytical standards (Hayes et al., 2024). As emphasized by RSC mechanisms and modeling, synchronized experimental design and balanced biological replicates are critical for reducing noise and increasing cross-platform comparability (Tarazona et al., 2020; Lu and Tan, 2023; Godwin, 2019).

Biological interpretation is equally important because molecular insight only makes sense within the context of microbial ecology and physiology (Prosser, 2015). To translate statistical correlations to causal mechanisms of pollutant degradation requires domain knowledge in microbiology, systems biology, and environmental chemistry (Handy, 2008). Concurrently, high cost and access limitations (expensive instrumentation, costly maintenance, and limited bioinformatic infrastructure) are still limiting the general use of such technology, particularly in developing countries (Helmy et al., 2016).

The obstacles may be tackled only through interdisciplinary contributions of microbiologists, bioinformaticians, chemists, and environmental engineers (Ejaz et al., 2025; National Academies of Sciences, Engineering, and Medicine, 2019). Novel methods, like AI-assisted data fusion, cloud databanks, and the establishment of metadata standards, are leading to the development of reproducible, scalable, and globally accessible multiomics integration frameworks for predictive and sustainable bioremediation.

5.7 Emerging trends and future directions

The next generation of bioremediation is in the nexus of AI, systems biology, and multi-omics integration, where predictability is normal rather than chance when it comes to pollutant degradation (Ameen et al., 2024). In recent studies, ML techniques such as random forest, convolutional neural network, and Bayesian inference have been successfully used to integrate genomics, proteomics, and metabolomics data for the discovery of nonlinear regulatory patterns while predicting pollutant degradation under a variety of

TABLE 5.4 Functions of microbes in the bioremediation of environmental pollutants.

Microorganisms	Function	Environmental pollutant	References
Achromobacter spp.	Bioaugmentation	Crude oil, PAHs, and pesticides	Dehnavi and Ebrahimipour (2022)
Alcanivorax borkumensis	Biodegradation	Petrol and oil	Lee et al. (2017)
Pseudomonas putida	Biodegradation and biotransformation	Industrial cyanide-containing waste, or the 2,4,6-trinitrotoluene	Wibberg et al. (2016)
Bacillus, Pseudomonas, Vibrio, Micrococcus, and *Alcaligenes*	Biodegradation	Crude oil	Ijah and Antai (2003)
Halorientalis hydrocarbonoclasticus	Biodegradation	Hexadecane, PAHs (naphthalene, anthracene, phenanthrene, pyrene, and benzo[a] anthracene)	Zhao et al. (2017)
Irpex lacteus	Bioremediation	Aromatic dye	Sun et al. (2016)
Pseudomonas sp. (strain CPSB21)	Redox reaction	Chromium (Cr^{6+})	Gupta et al. (2018)
Rhodococcus spp.	Biodegradation	Organic pollutants, including nitriles, halogenated hydrocarbons	DeLorenzo et al. (2018)
Penicillium oxalicum	Bioremediation	Polycyclic aromatic hydrocarbons (anthracene)	Camacho-Morales et al. (2018)
Pseudomonas sp.	Biodegradation	BTEX (benzene, toluene, ethylbenzene, and xylene)	Hassan and Aly (2018)

(Continued)

TABLE 5.4 (Continued)

Microorganisms	Function	Environmental pollutant	References
Naegleria, Vorticella, Arabidopsis, Asarum, and *Populus*	Bioremediation	Hydrocarbon	Kachienga et al. (2018)
Bacillus licheniformis M2–7	Biotransformation	Benzo-pyrene	Guevara-Luna et al. (2018)
Bacillus sorensis strain.	Biodegradation	Hydrocarbon	Oualha et al. (2019)
Bacillus, Coprothermobacter, Rhodobacter, Pseudomonas, Achromobacter, Desulfitobacter, Desulfosporosinus, T78, Methanobacterium, Methanosaeta	Bioremediation	Sulfate-reducing, CO_2-assimilating, hydrocarbon-rich petroleum refinery waste	Roy et al. (2018)
Planococcus sp. S5 strain	Immobilization	Naproxen	Dzionek et al. (2018)
Rhodococcus erythropolis strains	Bioaugmentation	Petroleum product	Pacwa-Płociniczak et al. (2019)
Pseudomonas aeruginosa san ai	Proteomics	Cadmium	Izrael-Živković et al. (2018)
Acinetobacter radioresistens APH1	Biodegradation	Phenol	Liu et al. (2020)
Stenotrophomonas maltophilia strain AJH1	Biodegradation	Anthracene, phenanthrene, naphthalene, fluorine, pyrene	Arulazhagan et al. (2017)

environmental conditions (Trivedi et al., 2025; Ogunjobi et al., 2024). Advances in single-cell omics have enabled this functional division within microbial populations to be uncoupled, furthering our ability to understand the role of each individual cell in maintaining community-level resilience and pollutant degradation (Mauger et al., 2022; Sahoo et al., 2021).

Simultaneously, spatial omics and metabolic fluxomics create dynamic maps of metabolite distribution and turnover right on the locations in contaminated matrices where it is happening amidst molecular activity and environmental context (Volkova et al., 2020; Law et al., 2024). The development of cloud-based resources (e.g., MGnify, PRIDE, and MetaboLights) in which to share such data globally enables reproducible and collaborative means of analyzing, resulting in highly efficient discovery at speed with a level of standardization across research consortia (Khoonsari et al., 2019; Finn et al., 2025; Vince et al., 2025).

Finally, combining omics with synthetic biology and computational metabolic modeling allows the construction of customized engineered microbial consortia for targeted pollutant removal. Taken as a whole, these new technologies are turning bioremediation into a predictive and design-based engineering operation that brings towards the possibility of generating self-optimizing and adaptive microbial systems and strategies with which can re-establish environmental balance in an efficient and sustainable manner.

5.8 Conclusion

Traditional bioremediation has evolved into a predictive, systems-level science due to advancements in genetics, proteomics, and metabolomics. Multilayered omics data integrates genetic potential, protein concentrations, and metabolic flux to elucidate how bacteria detect, respond to, and neutralize pollutants. This methodology offers a molecular and regulatory framework of metabolic biodegradation networks to pinpoint essential enzymes, adaptive pathways, and predictive biomarkers for hydrocarbons, heavy metals, pesticides, and dyes. Proteomics is crucial for correlating genetic information with metabolic functions and translating predictions into biological actions. Integrative omics facilitates the systematic design of modified strains and synthetic microbial communities for in situ remediation. Notwithstanding advantages, data harmonization, computational advancements, and interdisciplinary collaborations continue to pose challenges. Advanced bioinformatics, AI, standardized analytical pipelines, and ethical data sharing will be requisite. In the forthcoming decades, omics tools, synthetic biology, ML techniques, and nanobiotechnologies will yield highly dependable self-regulating systems capable of detecting and eliminating toxins in real time. Integrated omics will facilitate predictive and sustainable bioremediation, wherein microbial ecology propels global environmental restoration and planetary well-being.

References

Aken, B.L., Ayling, S., Barrell, D., Clarke, L., Curwen, V., Fairley, S., et al., 2016. The ensembl gene annotation system. Database 2016, baw093. Available at https://doi.org/10.1093/database/baw093.

Ameen, F., Bhat, S.A., Kumar, V. (Eds.), 2024. Microbial Bioremediation and Multiomics Technologies for Sustainable Development: Recent Trends. RSC.

Anand, S., Padmanabhan, P., 2024. Omics approaches for characterization of environmental microorganisms. Development in Waste Water Treatment Research and Processes. Elsevier, pp. 1–13.

Anand, S., Singh, A., Kumar, V., 2023. Recent advancements in cadmium-microbe interactive relations and their application for environmental remediation: a mechanistic overview. Environmental Science and Pollution Research 30 (7), 17009–17038.

Aragaw, T.A., 2021. Functions of various bacteria for specific pollutants degradation and their application in wastewater treatment: a review. International Journal of Environmental Science and Technology 18 (7), 2063–2076. Available at https://doi.org/10.1007/s13762-020-03022-2.

Arkin, A.P., Stevens, R.L., Cottingham, R.W., Maslov, S., Henry, C.S., Dehal, P. et al. 2016. The DOE Systems Biology Knowledgebase (KBase). BioRxiv, 096354.

Arora, N.K., 2018. Bioremediation: a green approach for restoration of polluted ecosystems. Environmental Sustainability 1 (4), 305–307. Available at https://doi.org/10.1007/s42398-018-00036-y.

Arulazhagan, P., Al-Shekri, K., Huda, Q., Godon, J.J., Basahi, J.M., Jeyakumar, D., 2017. Biodegradation of polycyclic aromatic hydrocarbons by an acidophilic Stenotrophomonas maltophilia strain AJH1 isolated from a mineral mining site in Saudi Arabia. Extremophiles: Life Under Extreme Conditions 21 (1), 163–174. Available at https://doi.org/10.1007/s00792-016-0892-0.

Baburam, C. 2020. Engineering of an Enzyme Cocktail for Biodegradation of Petroleum Hydrocarbons Based on Known Enzymatic Pathways and Metagenomic Techniques (Doctoral Dissertation, Vaal University of Technology (South Africa)).

Bala, S., Garg, D., Thirumalesh, B.V., Sharma, M., Sridhar, K., Inbaraj, B.S., et al., 2022. Recent strategies for bioremediation of emerging pollutants: a review for a green and sustainable environment. Toxics 10 (8). Available at https://doi.org/10.3390/toxics10080484, http://www.mdpi.com/journal/toxics.

Bassel, G.W., Gaudinier, A., Brady, S.M., Hennig, L., Rhee, S.Y., De Smet, I., 2012. Systems analysis of plant functional, transcriptional, physical interaction, and metabolic networks. The Plant Cell 24 (10), 3859–3875. Available at https://doi.org/10.1105/tpc.112.100776, http://www.plantcell.org/content/24/10/3859.full.pdf+html.

Bhardwaj, I., Verma, R., Kumar, V., Bhardwaj, N., 2022. The science of microbial enzymes as detoxification tool for inorganic and organic pollutants. Microbial Bioremediation: Sustainable Management of Environmental Contamination. International Publishing, India, pp. 283–303. Available at https://doi.org/10.1007/978-3-031-18017-0, https://doi.org/10.1007/978-3-031-18017-0_16.

Biswas, J.K., Mondal, M., Kumar, V., Vithanage, M., Selvasembian, R., Ambade, B., et al., 2022. Omics reflection on thebacterial escape from the toxic trap of metal(loid)s: cracking the code of contaminants stress, resistance repertoire, and remediation. Omics for Environmental Engineering and Microbiology Systems. CRC Press, India, pp. 333–354. Available at https://doi.org/10.1201/9781003247883-16, https://www.routledge.com/Omics-for-Environmental-Engineering-and-Microbiology-Systems/Kumar-Garg-Kumar-Biswas/p/book/9781032162836.

Brown, G.R., Hem, V., Katz, K.S., Ovetsky, M., Wallin, C., Ermolaeva, O., et al., 2015. Gene: a gene-centered information resource at NCBI. Nucleic Acids Research 43 (1), D36. Available at https://doi.org/10.1093/nar/gku1055, http://nar.oxfordjournals.org/.

Bunne, C., Schiebinger, G., Krause, A., Regev, A., Cuturi, M., 2024. Optimal transport for single-cell and spatial omics. Nature Reviews Methods Primers 4 (1). Available at https://doi.org/10.1038/s43586-024-00334-2.

Camacho-Morales, R.L., García-Fontana, C., Fernández-Irigoyen, J., Santamaría, E., González-López, J., Manzanera, M., et al., 2018. Anthracene drives sub-cellular proteome-wide alterations in the degradative system of Penicillium oxalicum. Ecotoxicology and Environmental Safety 159, 127–135.

Cambiaghi, A., Ferrario, M., Masseroli, M., 2017. Analysis of metabolomic data: tools, current strategies and future challenges for omics data integration. Briefings in Bioinformatics 18 (3), 498–510.

Cappelletti, M., Fedi, S., Zampolli, J., Di Canito, A., D'Ursi, P., Orro, A., et al., 2016. Phenotype microarray analysis may unravel genetic determinants of the stress response by Rhodococcus aetherivorans BCP1 and Rhodococcus opacus R7. Research in Microbiology 167 (9-10), 766–773. Available at https://doi.org/10.1016/j.resmic.2016.06.008.

Carbajosa, G., Trigo, A., Valencia, A., Cases, I., 2009. Bionemo: molecular information on biodegradation metabolism. Nucleic Acids Research 37 (Database), D598. Available at https://doi.org/10.1093/nar/gkn864.

Caspi, R., Billington, R., Ferrer, L., Foerster, H., Fulcher, C.A., Keseler, I.M., et al., 2016. The MetaCyc database of metabolic pathways and enzymes and the BioCyc collection of pathway/genome databases. Nucleic Acids Research 44 (1), D471. Available at https://doi.org/10.1093/nar/gkv1164, http://nar.oxfordjournals.org/.

Chaitanya, K.V., 2019. Genome and Genomics. Springer Singapore, Singapore, pp. 1–30.

Chakraborty, J., Das, S., 2016. Molecular perspectives and recent advances in microbial remediation of persistent organic pollutants. Environmental Science and Pollution Research 23 (17), 16883–16903. Available at https://doi.org/10.1007/s11356-016-6887-7.

Chaudhary, S., Passi, A., Jindal, S., Goyal, K., 2025. LC-MS/MS: a powerful tool for modern analytical science: fundamentals, techniques, applications and innovations. Journal of Liquid Chromatography & Related Technologies 48 (11-15), 191–201. Available at https://doi.org/10.1080/10826076.2025.2491475.

Chaurasia, P., Jasuja, N.D., Kumar, S., 2022. Bioremediation assessment in industrial wastewater treatment: the omics approach. Omics Insights in Environmental Bioremediation. Springer Nature Singapore, Singapore, pp. 455–485.

Chen, M.Y., Teng, W.K., Zhao, L., Hu, C.X., Zhou, Y.K., Han, B.P., et al., 2021. Comparative genomics reveals insights into cyanobacterial evolution and habitat adaptation. ISME Journal 15 (1), 211–227. Available at https://doi.org/10.1038/s41396-020-00775-z, http://www.nature.com/ismej/marketing/aims_scope.html.

Chen, K., He, R., Wang, L.'ao, Liu, L., Huang, X., Ping, J., et al., 2022. The dominant microbial metabolic pathway of the petroleum hydrocarbons in the soil of shale gas field: carbon fixation instead of CO2 emissions. Science of The Total Environment 807 (Pt 2), 151074. Available at https://doi.org/10.1016/j.scitotenv.2021.151074.

Choubey, J., Choudhari, J.K., Sahariah, B.P., 2025. Future directions on systems biology. *Systems Biology and In-Depth Applications for Unlocking Diseases*. Academic Press, pp. 319–328.

Crook, A.A., Powers, R., 2020. Quantitative NMR-based biomedical metabolomics: current status and applications. Molecules (Basel, Switzerland) 25 (21). Available at https://doi.org/10.3390/molecules25215128, https://www.mdpi.com/1420-3049/25/21/5128/pdf.

Cummins, I., Dixon, D.P., Freitag-Pohl, S., Skipsey, M., Edwards, R., 2011. Multiple roles for plant glutathione transferases in xenobiotic detoxification. Drug Metabolism Reviews 43 (2), 266–280. Available at https://doi.org/10.3109/03602532.2011.552910.

Czaplicki, L.M., Gunsch, C.K., 2016. Reflection on molecular approaches influencing state-of-the-art bioremediation design: culturing to microbial community fingerprinting to omics. Journal

of Environmental Engineering 142 (10). Available at https://doi.org/10.1061/(asce)ee.1943-7870.0001141.
Dangi, A.K., Sharma, B., Hill, R.T., Shukla, P., 2019. Bioremediation through microbes: systems biology and metabolic engineering approach. Critical Reviews in Biotechnology 39 (1), 79–98. Available at https://doi.org/10.1080/07388551.2018.1500997, http://www.tandfonline.com/loi/ibty20.
Das, S., Dash, H.R., Chakraborty, J., 2016. Genetic basis and importance of metal resistant genes in bacteria for bioremediation of contaminated environments with toxic metal pollutants. Applied Microbiology and Biotechnology 100 (7), 2967–2984.
Das, N., Kumar, V., Chaure, K., Pandey, P., 2025. Environmental restoration of polyaromatic hydrocarbon-contaminated soil through sustainable rhizoremediation: insights into bioeconomy and high-throughput systematic analysis. Environmental Science: Advances 4 (6), 842–883. Available at https://doi.org/10.1039/d4va00203b.
Debik, J., Sangermani, M., Wang, F., Madssen, T.S., Giskeødegård, G.F., 2022. Multivariate analysis of NMR-based metabolomic data. NMR in Biomedicine 35 (2). Available at https://doi.org/10.1002/nbm.4638.
Dehnavi, S.M., Ebrahimipour, G., 2022. Comparative remediation rate of biostimulation, bioaugmentation, and phytoremediation in hydrocarbon contaminants. International Journal of Environmental Science and Technology 19 (11), 11561–11586. Available at https://doi.org/10.1007/s13762-022-04343-0.
DeLorenzo, D.M., Rottinghaus, A.G., Henson, W.R., Moon, T.S., 2018. Molecular toolkit for gene expression control and genome modification in Rhodococcus opacus PD630. ACS Synthetic Biology 7 (2), 727–738. Available at https://doi.org/10.1021/acssynbio.7b00416, http://pubs.acs.org/journal/asbcd6.
Desai, C., Pathak, H., Madamwar, D., 2010. Advances in molecular and "-omics" technologies to gauge microbial communities and bioremediation at xenobiotic/anthropogen contaminated sites. Bioresource Technology 101 (6), 1558–1569.
Desai, T., Ahmad, A., Srivastava, S., 2021. Role of systems biology in enhancing efficiency of biocatalysts. Biomass for Bioenergy and Biomaterials. CRC Press, pp. 127–151.
Devi, R., Behera, B., Raza, M.B., Mangal, V., Altaf, M.A., Kumar, R., et al., 2022. An insight into microbes mediated heavy metal detoxification in plants: a review. Journal of Soil Science and Plant Nutrition 22 (1), 914–936. Available at https://doi.org/10.1007/s42729-021-00702-x, https://www.springer.com/life+sciences/plant+sciences/journal/42729.
Di Canito, A. 2019. Genomic and Functional Analysis of Rhodococcus Strains to Identify Genes and Degradative Functions for Soil Quality Evaluation.
Duarte, M., Jauregui, R., Vilchez-Vargas, R., Junca, H., Pieper, D.H., 2014. AromaDeg, a novel database for phylogenomics of aerobic bacterial degradation of aromatics. Database 2014. Available at https://doi.org/10.1093/database/bau118, http://database.oxfordjournals.org/.
Dzionek, A., Wojcieszyńska, D., Hupert-Kocurek, K., Adamczyk-Habrajska, M., Guzik, U., 2018. Immobilization of Planococcus sp. S5 strain on the loofah sponge and its application in naproxen removal. Catalysts 8 (5), 176.
Eicher, T., Kinnebrew, G., Patt, A., Spencer, K., Ying, K., Ma, Q., et al., 2020. Metabolomics and multi-omics integration: a survey of computational methods and resources. Metabolites 10 (5). Available at https://doi.org/10.3390/metabo10050202, https://www.mdpi.com/2218-1989/10/5/202/pdf.
Ejaz, M., Hassan, L., Imtiaz, U., Rukan, M., Ali, S., 2025. Future microbiology education. Microbiology in the Era of Artificial Intelligence 291–313.

Finney, A., Hucka, M., 2003. Systems biology markup language: level 2 and beyond. Biochemical Society Transactions 31 (6), 1472–1473. Available at https://doi.org/10.1042/bst0311472.

Finn, R.D., Balech, B., Burgin, J., Chua, P., Corre, E., Cox, C.J., et al., 2025. Establishing the ELIXIR microbiome community. F1000Research. 13. Available at https://doi.org/10.12688/f1000research.144515.2, https://f1000research.com/.

Gasc, C., Richard, J.Y., Peyret, P., 2016. Genome sequence of Bacillus subtilis strain HUK15, isolated from hexachlorocyclohexane-contaminated soil. Genome Announcements 4 (2), 10–1128.

Gentleman, R.C., Carey, V.J., Bates, D.M., Bolstad, B., Dettling, M., Dudoit, S., et al., 2004. Bioconductor: open software development for computational biology and bioinformatics. Genome Biology 5 (10), R80.

Ghai, M.K., Khatri, A., Kumar, K., Thakur, I.S., 2024. Multi-omics and advance technologies in biodegradation of emerging contaminants and eco-estrogens in environmental waste. Total Environment Advances 11. Available at https://doi.org/10.1016/j.teadva.2024.200113, https://www.sciencedirect.com/journal/total-environment-advances.

Godwin, D.R. I. 2019. Experimental Reproducibility in High-throughput Multi-omic Analysis Systems.

Guerrero Ramírez, J.R., Ibarra Muñoz, L.A., Balagurusamy, N., Frías Ramírez, J.E., Alfaro Hernández, L., Carrillo Campos, J., 2023. Microbiology and biochemistry of pesticides biodegradation. International Journal of Molecular Sciences 24 (21). Available at https://doi.org/10.3390/ijms242115969, http://www.mdpi.com/journal/ijms.

Guevara-Luna, J., Alvarez-Fitz, P., Ríos-Leal, E., Acevedo-Quiroz, M., Encarnación-Guevara, S., Moreno-Godinez, M.E., et al., 2018. Biotransformation of benzo [a] pyrene by the thermophilic bacterium Bacillus licheniformis M2-7. World Journal of Microbiology and Biotechnology 34 (7), 88.

Guo, D.,. Chen, Y., Wu, Y., Cheng, J., Lai, W., Ma, W. et al., 2025. Multi-omics Characterization of the Skin Microbiota Reveals the Anti-aging Roles of Stenotrophomonas maltophilia. bioRxiv, China bioRxiv. Available from. https://www.biorxiv.org. doi: 10.1101/2025.07.09.663908.

Handy, R.D., 2008. Systems toxicology: using the systems biology approach to assess chemical pollutants in the environment. Advances in Experimental Biology 2, 249–281. Available at https://doi.org/10.1016/S1872-2423(08)00007-0.

Gupta, P., Rani, R., Chandra, A., Kumar, V., 2018. Potential applications of Pseudomonas sp. (strain CPSB21) to ameliorate Cr6+ stress and phytoremediation of tannery effluent contaminated agricultural soils. Scientific Reports 8, 4860. Available at 10.1038/s41598-018-23322-5.

Hasan, M.E., Osman, M.A., Hemeida, A.A., Elhefnawi, M., 2025. Bioinformatics and artificial intelligence approaches in metabolic pathways. Metabolic Dynamics in Host-Microbe Interaction. Springer Nature, Egypt, pp. 17–32. Available at https://doi.org/10.1007/978-981-96-1305-2_2, https://link.springer.com/book/9789819613052.

Hassan, H.A., Aly, A.A., 2018. Isolation and characterization of three novel catechol 2,3-dioxygenase from three novel haloalkaliphilic BTEX-degrading Pseudomonas strains. International Journal of Biological Macromolecules 106, 1107–1114. Available at https://doi.org/10.1016/j.ijbiomac.2017.08.113, https://www.sciencedirect.com/journal/international-journal-of-biological-macromolecules.

Hayes, C.N., Nakahara, H., Ono, A., Tsuge, M., Oka, S., 2024. From omics to multi-omics: a review of advantages and tradeoffs. Genes 15 (12). Available at https://doi.org/10.3390/genes15121551, http://www.mdpi.com/journal/genes/.

Helmy, M., Awad, M., Mosa, K.A., 2016. Limited resources of genome sequencing in developing countries: challenges and solutions. Applied & Translational Genomics 9, 15–19.

Huang, S., 2023. Efficient analysis of toxicity and mechanisms of environmental pollutants with network toxicology and molecular docking strategy: acetyl tributyl citrate as an example. Science of The Total Environment 905, 167904. Available at https://doi.org/10.1016/j.scitotenv.2023.167904.

Hwang, C., Copeland, A., Lucas, S., Lapidus, A., Barry, K., Glavina del Rio, T., et al., 2015. Complete genome sequence of Anaeromyxobacter sp. Fw109-5, an anaerobic, metal-reducing bacterium isolated from a contaminated subsurface environment. Genome Announcements 3 (1). Available at https://doi.org/10.1128/genomea.01449-14.

Hériché, J.K., Alexander, S., Ellenberg, J., 2019. Integrating imaging and omics: computational methods and challenges. Annual Review of Biomedical Data Science 2 (1), 175–197.

Ijah, U.J.J., Antai, S.P., 2003. Removal of Nigerian light crude oil in soil over a 12-month period. International Biodeterioration & Biodegradation 51 (2), 93–99. Available at https://doi.org/10.1016/s0964-8305(01)00131-7.

Ikram, M., Zahoor, M., Naeem, M., Islam, N.U., Shah, A.B., Shahzad, B., 2023. Bacterial oxidoreductive enzymes as molecular weapons for the degradation and metabolism of the toxic azo dyes in wastewater: a review. Zeitschrift fur Physikalische Chemie 237 (1-2), 187–209. Available at https://doi.org/10.1515/zpch-2022-0150, http://www.degruyter.com/view/j/zpch.

Islam, M.M. 2021. *Metabolic Modeling and Omics-Integrative Analysis of Single and Multi-Organism Systems: Discovery and Redesign* (Doctoral Dissertation, The University of Nebraska-Lincoln).

Izrael-Živković, L., Rikalović, M., Gojgić-Cvijović, G., Kazazić, S., Vrvić, M., Brčeski, I., et al., 2018. Cadmium specific proteomic responses of a highly resistant Pseudomonas aeruginosa san ai. RSC Advances 8 (19), 10549–10560. Available at https://doi.org/10.1039/c8ra00371h.

Jeevanandam, V., Osborne, J., 2022. Understanding the fundamentals of microbial remediation with emphasize on metabolomics. Preparative Biochemistry & Biotechnology 52 (3), 351–363. Available at https://doi.org/10.1080/10826068.2021.1946694.

Joshi, A., Kumar, A., Kaushik, V., 2024. Functional genomics and network biology. Advances in Bioinformatics. Springer Nature Singapore, Singapore, pp. 71–96.

Kachienga, L., Jitendra, K., Momba, M., 2018. Metagenomic profiling for assessing microbial diversity and microbial adaptation to degradation of hydrocarbons in two South African petroleum-contaminated water aquifers. Scientific Reports 8 (1), 7564.

Kanehisa, M., Sato, Y., Morishima, K., 2016. BlastKOALA and GhostKOALA: KEGG tools for functional characterization of genome and metagenome sequences. Journal of Molecular Biology 428 (4), 726–731.

Kanehisa, M., Furumichi, M., Tanabe, M., Sato, Y., Morishima, K., 2017. KEGG: new perspectives on genomes, pathways, diseases and drugs. Nucleic Acids Research 45 (D1), D353. Available at https://doi.org/10.1093/nar/gkw1092.

Karlsson, R., Gonzales-Siles, L., Boulund, F., Svensson-Stadler, L., Skovbjerg, S., Karlsson, A., et al., 2015. Proteotyping: proteomic characterization, classification and identification of microorganisms - a prospectus. Systematic and Applied Microbiology 38 (4), 246–257. Available at https://doi.org/10.1016/j.syapm.2015.03.006, www.urbanfischer.de/journals/sam/mic_biol.htm.

Karthikeyan, S., Hatt, J.K., Kim, M., Spain, J.C., Huettel, M., Kostka, J.E., et al., 2021. A novel, divergent alkane monooxygenase (alkB) clade involved in crude oil biodegradation.

Environmental Microbiology Reports 13 (6), 830–840. Available at https://doi.org/10.1111/1758-2229.13018, http://onlinelibrary.wiley.com/journal/10.1111/(ISSN)1758-2229.

Kaya, C., 2024. Microbial modulation of hormone signaling, proteomic dynamics, and metabolomics in plant drought adaptation. Food and Energy Security 13 (1). Available at https://doi.org/10.1002/fes3.513.

Kessner, D., Chambers, M., Burke, R., Agus, D., Mallick, P., 2008. ProteoWizard: open source software for rapid proteomics tools development. Bioinformatics (Oxford, England) 24 (21), 2534–2536. Available at https://doi.org/10.1093/bioinformatics/btn323.

Khoonsari, P.E., Moreno, P., Bergmann, S., Burman, J., Capuccini, M., Carone, M., et al., 2019. Interoperable and scalable data analysis with microservices: applications in metabolomics. Bioinformatics (Oxford, England) 35 (19), 3752–3760. Available at https://doi.org/10.1093/bioinformatics/btz160.

Kothari, A., Charrier, M., Wu, Y.W., Malfatti, S., Zhou, C.E., Singer, S.W., et al., 2016. Transcriptomic analysis of the highly efficient oil-degrading bacterium Acinetobacter venetianus RAG-1 reveals genes important in dodecane uptake and utilization. FEMS Microbiology Letters 363 (20). Available at https://doi.org/10.1093/femsle/fnw224, http://femsle.oxfordjournals.org/.

Kour, D., Kaur, T., Devi, R., Yadav, A., Singh, M., Joshi, D., et al., 2021. Beneficial microbiomes for bioremediation of diverse contaminated environments for environmental sustainability: present status and future challenges. Environmental Science and Pollution Research 28 (20), 24917–24939. Available at https://doi.org/10.1007/s11356-021-13252-7.

Kumari, S., Das, S., 2023. Bacterial enzymatic degradation of recalcitrant organic pollutants: catabolic pathways and genetic regulations. Environmental Science and Pollution Research 30 (33), 79676–79705. Available at https://doi.org/10.1007/s11356-023-28130-7.

Kumar, V., Thakur, I.S., 2022. Omics Insights in Environmental Bioremediation Omics Insights in Environmental Bioremediation. Springer Nature, India, pp. 1–702. Available at https://doi.org/10.1007/978-981-19-4320-1, https://link.springer.com/book/10.1007/978-981-19-4320-1.

Kumar, A., Pillay, B., Olaniran, A.O., 2020. Genome sequence and metabolic analysis revealed the catabolic pathways for the degradation of 1, 2-dichloroethane and other related Xenobiotics in Ancylobacter aquaticus strain UV5. Gene Reports 21, 100969.

Kumar, V., Singh, K., Shah, M.P., Singh, A.K., Kumar, A., Kumar, Y., 2021. Application of omics technologies for microbial community structure and function analysis in contaminated environment. Wastewater Treatment: Cutting-Edge Molecular Tools, Techniques and Applied Aspects. Elsevier, India, pp. 1–40. Available at https://doi.org/10.1016/B978-0-12-821881-5.00001-5, https://www.sciencedirect.com/book/9780128218815.

Kurata, H., Maeda, K., Onaka, T., Takata, T., 2014. BioFNet: biological functional network database for analysis and synthesis of biological systems. Briefings in Bioinformatics 15 (5), 699–709. Available at https://doi.org/10.1093/bib/bbt048.

Larance, M., Lamond, A.I., 2015. Multidimensional proteomics for cell biology. Nature Reviews Molecular Cell Biology 16 (5), 269–280. Available at https://doi.org/10.1038/nrm3970, http://www.nature.com/molcellbio.

Law, S.R., Mathes, F., Paten, A.M., Alexandre, P.A., Regmi, R., Reid, C., et al., 2024. Life at the borderlands: microbiomes of interfaces critical to one health. FEMS Microbiology Reviews 48 (2). Available at https://doi.org/10.1093/femsre/fuae008, https://academic.oup.com/femsre/pages/About.

Lee, J., Han, I., Kang, B.R., Kim, S.H., Sul, W.J., Lee, T.K., 2017. Degradation of crude oil in a contaminated tidal flat area and the resilience of bacterial community. Marine Pollution

Bulletin 114 (1), 296–301. Available at https://doi.org/10.1016/j.marpolbul.2016.09.043, www.elsevier.com/locate/marpolbul.

Lindeque, J.Z., 2024. Targeted analysis of organic acids with GC-MS/MS: Challenges and prospects. Analytical Biochemistry 694, 115620.

Liu, S., Guo, C., Dang, Z., Liang, X., 2017. Comparative proteomics reveal the mechanism of Tween80 enhanced phenanthrene biodegradation by Sphingomonas sp. GY2B. Ecotoxicology and Environmental Safety 137, 256–264. Available at https://doi.org/10.1016/j.ecoenv.2016.12.015.

Liu, Y., Wang, W., Shah, S.B., Zanaroli, G., Xu, P., Tang, H., 2020. Phenol biodegradation by Acinetobacter radioresistens APH1 and its application in soil bioremediation. Applied Microbiology and Biotechnology 104 (1), 427–437. Available at https://doi.org/10.1007/s00253-019-10271-w, link.springer.de/link/service/journals/00253/index.htm.

Liu, S., Moon, C.D., Zheng, N., Huws, S., Zhao, S., Wang, J., 2022. Opportunities and challenges of using metagenomic data to bring uncultured microbes into cultivation. Microbiome 10 (1). Available at https://doi.org/10.1186/s40168-022-01272-5, http://www.microbiomejournal.com/.

Liu, Q., He, W., Zhang, W., Wang, L., Tang, J., 2024. Metagenomic analysis reveals the microbial response to petroleum contamination in oilfield soils. Science of The Total Environment 912, 168972. Available at https://doi.org/10.1016/j.scitotenv.2023.168972.

Lohmann, P., Schäpe, S.S., Haange, S.B., Oliphant, K., Allen-Vercoe, E., Jehmlich, N., et al., 2020. Function is what counts: how microbial community complexity affects species, proteome and pathway coverage in metaproteomics. Expert Review of Proteomics 17 (2), 163–173. Available at https://doi.org/10.1080/14789450.2020.1738931, https://www.tandfonline.com/loi/ieru20?open=9&year=2012&repitition=0#vol_9_2012.

Lu, Y., Tan, Y., 2023. Coordination mechanism of revenue sharing contracts in port supply chains: a case study of China's Nantong Port. Sustainability 15 (14), 11248. Available at https://doi.org/10.3390/su151411248.

López, A.M.Q., dos Santos Silva, A.L., 2023. Proteomics and bioremediation using prokaryotes. Genomics Approach to Bioremediation: Principles, Tools, and Emerging Technologies 485–502.

Mahallea, S., Dafale, N.A., 2021. Understanding the role of genetic and protein networking involved. Bioremediation of Environmental Pollutants: Emerging Trends and Strategies 187.

Malla, M.A., Dubey, A., Yadav, S., Kumar, A., Hashem, A., Abd-Allah, E.F., 2018. Understanding and designing the strategies for the microbe-mediated remediation of environmental contaminants using omics approaches. Frontiers in Microbiology 9. Available at https://doi.org/10.3389/fmicb.2018.01132, https://www.frontiersin.org/articles/10.3389/fmicb.2018.01132/full.

Mariam, I., Rova, U., Christakopoulos, P., Matsakas, L., Patel, A., 2025. Data-driven synthetic microbes for sustainable future. npj Systems Biology and Applications 11 (1). Available at https://doi.org/10.1038/s41540-025-00556-4.

Maron, P.A., Ranjard, L., Mougel, C., Lemanceau, P., 2007. Metaproteomics: a new approach for studying functional microbial ecology. Microbial Ecology 53 (3), 486–493. Available at https://doi.org/10.1007/s00248-006-9196-8.

Mauger, S., Monard, C., Thion, C., Vandenkoornhuyse, P., 2022. Contribution of single-cell omics to microbial ecology. Trends in Ecology & Evolution 37 (1), 67–78.

McElhinney, J.M.W.R., Catacutan, M.K., Mawart, A., Hasan, A., Dias, J., 2022. Interfacing machine learning and microbial omics: a promising means to address environmental challenges. Frontiers in Microbiology 13. Available at https://doi.org/10.3389/fmicb.2022.851450, https://www.frontiersin.org/journals/microbiology#.

Medina-Ortiz, D., Contreras, S., Quiroz, C., Olivera-Nappa, Á., 2020. Development of supervised learning predictive models for highly non-linear biological, biomedical, and general datasets. Frontiers in Molecular Biosciences 7. Available at https://doi.org/10.3389/fmolb.2020.00013.

Menaa, B., Hachicha, R., Dubessay, P., Abdelkafi, S., Fendri, I., Michaud, P., 2025. Evolution of Chlorella sp. RCC288 proteome and transcriptome during its adaptation to oil mill wastewater. Biochimica et Biophysica Acta (BBA) - Proteins and Proteomics 1873 (4), 141077. Available at https://doi.org/10.1016/j.bbapap.2025.141077.

Mishra, H., 2025. Environmental degradation and impacts on agricultural production: a challenge to urban sustainability. In: Parray, J.A., Shameem, N., Haghi, A.K. (Eds.), Sustainable Urban Environment and Waste Management. Urban Sustainability. Springer, Singapore. Available at https://doi.org/10.1007/978-981-96-1140-9_3.

Misra, B.B., Langefeld, C., Olivier, M., Cox, L.A., 2019. Integrated omics: tools, advances and future approaches. Journal of Molecular Endocrinology 62 (1), R21. Available at https://doi.org/10.1530/JME-18-0055, https://jme.bioscientifica.com/downloadpdf/journals/jme/62/1/JME-18-0055.xml.

Mohapatra, B., Phale, P.S., 2021. Microbial degradation of naphthalene and substituted naphthalenes: metabolic diversity and genomic insight for bioremediation. Frontiers in Bioengineering and Biotechnology 9. Available at https://doi.org/10.3389/fbioe.2021.602445, http://journal.frontiersin.org/journal/bioengineering-and-biotechnology#archive.

Mortimer, M., Fang, W., Zhou, X., Vodovnik, M., Guo, L.H., 2022. Omics approaches in toxicological studies. Advances in Toxicology and Risk Assessment of Nanomaterials and Emerging Contaminants. Springer Nature, China, pp. 61–94. Available at https://doi.org/10.1007/978-981-16-9116-4, https://doi.org/10.1007/978-981-16-9116-4_4.

Muthubharathi, B.C., Gowripriya, T., Balamurugan, K., 2021. Metabolomics: small molecules that matter more. Molecular Omics 17 (2), 210–229.

National Academies of Sciences, Engineering, and Medicine, 2019. Environmental Engineering for the 21st Century: Addressing Grand Challenges. National Academies Press.

Navid, A., 2021. Curating COBRA models of microbial metabolism. Microbial Systems Biology: Methods and Protocols. Springer US, New York, NY, pp. 321–338.

Nawaz, M.Z., Sasidharan, R.S., Alghamdi, H.A., Dang, H., 2022. Understanding interaction patterns within deep-sea microbial communities and their potential applications. Marine Drugs 20 (2). Available at https://doi.org/10.3390/md20020108, https://www.mdpi.com/1660-3397/20/2/108/pdf.

Neupane, B., Richer, D., Bonner, A.J., Kibret, T., Beyene, J., Gibas, C., 2014). Network meta-analysis using r: a review of currently available automated packages. PLoS One 9 (12), e115065. Available at https://doi.org/10.1371/journal.pone.0115065.

Nikel, P.I., Chavarría, M., 2015. Quantitative physiology approaches to understand and optimize reducing power availability in environmental bacteria. Hydrocarbon and Lipid Microbiology Protocols: Synthetic and Systems Biology-Tools. Springer, Berlin, Heidelberg, pp. 39–70.

Nogales, J., García, J.L., Díaz, E., 2019. Degradation of aromatic compounds in Pseudomonas: a systems biology view. Aerobic Utilization of Hydrocarbons, Oils, and Lipids. Springer, Cham, pp. 639–687.

Ogunjobi, T.T., Ohaeri, P.N., Akintola, O.T., Atanda, D.O., Orji, F.P., Adebayo, J.O., et al., 2024. Bioinformatics applications in chronic diseases: a comprehensive review of genomic, transcriptomics, proteomic, metabolomics, and machine learning approaches. Medinformatics.

Olanbiwoninu, A.A., Adebayo, E.A., Popoola, B.M., Adeyemi, O.A., 2025. Different eco-friendly biological means for soil remediation. Environmental Science and Engineering 256, 109–146. Available at https://doi.org/10.1007/978-3-031-76886-6_5, https://www.springer.com/series/7487.

Orchard, S., Ammari, M., Aranda, B., Breuza, L., Briganti, L., Broackes-Carter, F., et al., 2014. The MIntAct project - IntAct as a common curation platform for 11 molecular interaction databases. Nucleic Acids Research 42 (1), D358. Available at https://doi.org/10.1093/nar/gkt1115.

Oualha, M., Al-Kaabi, N., Al-Ghouti, M., Zouari, N., 2019. Identification and overcome of limitations of weathered oil hydrocarbons bioremediation by an adapted Bacillus sorensis strain. Journal of Environmental Management 250, 109455. Available at https://doi.org/10.1016/j.jenvman.2019.109455.

Oyewusi, H.A., Wahab, R.A., Huyop, F., 2021. Whole genome strategies and bioremediation insight into dehalogenase-producing bacteria. Molecular Biology Reports 48 (3), 2687–2701.

Pabinger, S., Snajder, R., Hardiman, T., Willi, M., Dander, A., Trajanoski, Z., 2014. MEMOSys 2.0: an update of the bioinformatics database for genome-scale models and genomic data. Database 2014. Available at https://doi.org/10.1093/database/bau004.

Pacwa-Płociniczak, M., Czapla, J., Płociniczak, T., Piotrowska-Seget, Z., 2019. The effect of bioaugmentation of petroleum-contaminated soil with Rhodococcus erythropolis strains on removal of petroleum from soil. Ecotoxicology and Environmental Safety 169, 615–622.

Patidar, A., Kamble, P., 2025. A comprehensive review on liquid chromatography-mass spectrometry (LC-MS): a hyphenated technique. Asian Journal of Pharmaceutical Research and Development 13 (1), 95–103.

Pertea, M., Kim, D., Pertea, G.M., Leek, J.T., Salzberg, S.L., 2016. Transcript-level expression analysis of RNA-seq experiments with HISAT, StringTie and Ballgown. Nature Protocols 11 (9), 1650–1667. Available at https://doi.org/10.1038/nprot.2016.095, http://www.natureprotocols.com/.

Prosser, G.A., Larrouy-Maumus, G., de Carvalho, L.P.S., 2014. Metabolomic strategies for the identification of new enzyme functions and metabolic pathways. EMBO Reports 15 (6), 657–669.

Prosser, J.I., 2015. Dispersing misconceptions and identifying opportunities for the use of 'omics' in soil microbial ecology. Nature Reviews Microbiology 13 (7), 439–446. Available at https://doi.org/10.1038/nrmicro3468, http://www.nature.com/nrmicro/index.html.

Prudhomme, N. 2023. Multi-Omics Identification of Yield Limiting Factors in Molecular Pharming (Doctoral Dissertation, University of Guelph).

Purcell, S., Neale, B., Todd-Brown, K., Thomas, L., Ferreira, M.A.R., Bender, D., et al., 2007. PLINK: a tool set for whole-genome association and population-based linkage analyses. American Journal of Human Genetics 81 (3), 559–575. Available at https://doi.org/10.1086/519795, http://www.elsevier.com/wps/find/journaldescription.cws_home/713561/description#description.

Qiu, S., Yang, A., Zeng, H., Ouzounis, C.A., 2023. Flux balance analysis-based metabolic modeling of microbial secondary metabolism: current status and outlook. PLoS Computational Biology 19 (8), e1011391. Available at https://doi.org/10.1371/journal.pcbi.1011391.

Rabus, R., Boll, M., Heider, J., Meckenstock, R.U., Buckel, W., Einsle, O., et al., 2016. Anaerobic microbial degradation of hydrocarbons: from enzymatic reactions to the environment. Journal of Molecular Microbiology and Biotechnology 26 (1-3), 5–28. Available at https://doi.org/10.1159/000443997, www.karger.com/journals/mmb/mmb_jh.htm.

Rahman, S.F., Kantor, R.S., Huddy, R., Thomas, B.C., van Zyl, A.W., Harrison, S.T.L., et al., 2017. Genome-resolved metagenomics of a bioremediation system for degradation of thiocyanate in mine water containing suspended solid tailings. MicrobiologyOpen 6 (3). Available at https://doi.org/10.1002/mbo3.446.

Ravintheran, S.K., Sivaprakasam, S., Loke, S., Lee, S.Y., Manickam, R., Yahya, A., et al., 2019. Complete genome sequence of Sphingomonas paucimobilis AIMST S2, a xenobiotic-degrading bacterium. Scientific Data 6 (1). Available at https://doi.org/10.1038/s41597-019-0289-x, www.nature.com/sdata/.

Raza, A., Bashir, S., Salehi, H., Jamla, M., Charagh, S., Rad, A.C., et al., 2023. Advanced techniques in omics research in relation to heavy metal/metalloid toxicity and tolerance in plants. Heavy Metal Toxicity and Tolerance in Plants: A Biological, Omics, and Genetic Engineering Approach. wiley, China, pp. 35–57. Available at https://doi.org/10.1002/9781119906506.ch2, https://onlinelibrary.wiley.com/doi/book/10.1002/9781119906506.

Razzaq, M.K., Aleem, M., Mansoor, S., Khan, M.A., Rauf, S., Iqbal, S., et al., 2021. Omics and crispr-cas9 approaches for molecular insight, functional gene analysis, and stress tolerance development in crops. International Journal of Molecular Sciences 22 (3), 1–13. Available at https://doi.org/10.3390/ijms22031292, https://www.mdpi.com/1422-0067/22/3/1292/pdf.

Reena, R., Majhi, M.C., Arya, A.K., Kumar, R., Kumar, A., 2012. BioRadBase: a database for bioremediation of radioactive waste. African Journal of Biotechnology 11 (35), 8718–8721.

Roy, A., Dutta, A., Pal, S., Gupta, A., Sarkar, J., Chatterjee, A., et al., 2018. Biostimulation and bioaugmentation of native microbial community accelerated bioremediation of oil refinery sludge. Bioresource Technology 253, 22–32.

Sahoo, H., Kisku, K., Naik, U.C., 2021. Application of omics tools for microbial community structure and function analysis. Wastewater Treatment. Elsevier, pp. 433–456.

Samuel, J., Paul, M.L., Ravishankar, H., Mathur, A., Saha, D.P., Natarajan, C., et al., 2013. The differential stress response of adapted chromite mine isolates Bacillus subtilis and Escherichia coli and its impact on bioremediation potential. Biodegradation 24 (6), 829–842. Available at https://doi.org/10.1007/s10532-013-9631-8.

Santiago-Rodriguez, T.M., Hollister, E.B., 2021. Multi 'omic data integration: a review of concepts, considerations, and approaches. Seminars in Perinatology, Vol. 45. WB Saunders, pp. 151456.

Saraiva, J.P., Worrich, A., Karakoç, C., Kallies, R., Chatzinotas, A., Centler, F., et al., 2021. Mining synergistic microbial interactions: a roadmap on how to integrate multi-omics data. Microorganisms 9 (4), 840. Available at https://doi.org/10.3390/microorganisms9040840.

Satam, H., Joshi, K., Mangrolia, U., Waghoo, S., Zaidi, G., Rawool, S., et al., 2023. Next-generation sequencing technology: current trends and advancements. Biology 12 (7). Available at https://doi.org/10.3390/biology12070997, www.mdpi.com/journal/biology.

Satya, S., Sharma, S., Choudhary, G., Kaushik, G., 2024. Advances in environmental microbiology: a multi-omic perspective. Microbial Omics in Environment and Health. Springer Nature, India, pp. 175–204. Available at https://doi.org/10.1007/978-981-97-1769-9_7, https://link.springer.com/book/10.1007/978-981-97-1769-9.

Savitha, S., Keerthana, R., Logeswaran, K., Keerthika, P., Sharmila, V., Sangeetha, M., 2025. Integration of multi-omics data: genomics, proteomics, metabolomics. Harnessing AI and Machine Learning for Precision Wellness. IGI Global Scientific Publishing, pp. 149–184.

Sayqal, A., Xu, Y., Trivedi, D.K., AlMasoud, N., Ellis, D.I., Muhamadali, H., et al., 2016. Metabolic analysis of the response of Pseudomonas putida DOT-T1E strains to toluene using Fourier transform infrared spectroscopy and gas chromatography mass spectrometry.

Metabolomics: Official Journal of the Metabolomic Society 12 (7). Available at https://doi.org/10.1007/s11306-016-1054-1, http://www.kluweronline.com/issn/1573-3890/.

Segura, A., Godoy, P., Van Dillewijn, P., Hurtado, A., Arroyo, N., Santacruz, S., et al., 2005. Proteomic analysis reveals the participation of energy- and stress-related proteins in the response of Pseudomonas putida DOT-T1E to toluene. Journal of Bacteriology 187 (17), 5937–5945. Available at https://doi.org/10.1128/JB.187.17.5937-5945.2005.

Serrano, Peters, Wagner, Graham, Chen, Feng, et al., 2025. Progress and New Challenges in Image-based Profiling. arXiv.

Sharma, A.K., Pandit, J., 2023. Advanced bioremediation strategies for organophosphorus compounds. Microbiology and Biotechnology Letters 51 (4), 374–389. Available at https://doi.org/10.48022/mbl.2308.08011.

Sharma, M., Sidhu, A.K., Samota, M.K., Gupta, M., Koli, P., Choudhary, M., 2023. Post-translational modifications in histones and their role in abiotic stress tolerance in plants. Proteomes 11 (4). Available at https://doi.org/10.3390/proteomes11040038, www.mdpi.com/journal/proteomes.

Shekhar, S.K., Godheja, J., Modi, D.R., 2019. Molecular technologies for assessment of bioremediation and characterization of microbial communities at pollutant-contaminated sites. Bioremediation of Industrial Waste for Environmental Safety: Volume II: Biological Agents and Methods for Industrial Waste Management. Springer Singapore, Singapore, pp. 437–474.

Singh, O.V., 2006. Proteomics and metabolomics: the molecular make-up of toxic aromatic pollutant bioremediation. Proteomics 6 (20), 5481–5492. Available at https://doi.org/10.1002/pmic.200600200.

Singh, D., Geat, N., Mehriya, M., Rajawat, M.V.S., Prasanna, R., Kumar, A., et al., 2021. Omics (genomics, proteomics, metabolomics, etc.) tools to study the environmental microbiome and bioremediation. Waste to Energy: Prospects and Applications. Springer Singapore, India, pp. 235–260. Available at https://doi.org/10.1007/978-981-33-4347-4_10, https://link.springer.com/book/10.1007/978-981-33-4347-4.

Sivanich, M.K., Gu, T.J., Tabang, D.N., Li, L., 2022. Recent advances in isobaric labeling and applications in quantitative proteomics. Proteomics 22 (19-20). Available at https://doi.org/10.1002/pmic.202100256, http://onlinelibrary.wiley.com/journal/10.1002/(ISSN)1615-9861.

M.R. Somee, M.A. Amoozegar, S.M.M. Dastgheib, M. Shavandi, L.G. Maman, S. Bertilsson, et al., 2022. Genome-resolved Analyses Show an Extensive Diversification in Key Aerobic Hydrocarbon-degrading Enzymes Across Bacteria and Archaea. bioRxiv, Iran bioRxiv. Available from: https://www.biorxiv.org. doi: 10.1101/2022.06.22.496985.

Sun, S., Xie, S., Chen, H., Cheng, Y., Shi, Y., Qin, X., et al., 2016. Genomic and molecular mechanisms for efficient biodegradation of aromatic dye. Journal of Hazardous Materials 302, 286–295. Available at https://doi.org/10.1016/j.jhazmat.2015.09.071, www.elsevier.com/locate/jhazmat.

Talukdar, T., 2015. Thiolomics: molecular mechanisms of thiol-cascade in plant growth and nutrition. PlantOmics: The Omics of Plant Science. Springer India, India, pp. 491–533. Available at http://doi.org/10.1007/978-81-322-2172-2, https://doi.org/10.1007/978-81-322-2172-2_17.

Tamura, K., Dudley, J., Nei, M., Kumar, S., 2007. MEGA4: molecular evolutionary genetics analysis (MEGA) software version 4.0. Molecular Biology and Evolution 24 (8), 1596–1599. Available at https://doi.org/10.1093/molbev/msm092.

Tarazona, S., Balzano-Nogueira, L., Gómez-Cabrero, D., Schmidt, A., Imhof, A., Hankemeier, T., et al., 2020. Harmonization of quality metrics and power calculation in multi-omic studies. Nature Communications 11 (1). Available at https://doi.org/10.1038/s41467-020-16937-8.

Thakur, S., Chandra, A., Kumar, V., Bharti, S., 2025. Environmental Pollutants: Endocrine Disruptors/Pesticides/Reactive Dyes and Inorganic Toxic Compounds Metals, Radionuclides, and Metalloids and Their Impact on the Ecosystem. Springer Science and Business Media LLC, pp. 55–100. Available at https://doi.org/10.1007/978-981-97-7221-6_3.

Timofeeva, A.M., Galyamova, M.R., Sedykh, S.E., 2023. Plant growth-promoting bacteria of soil: designing of consortia beneficial for crop production. Microorganisms 11 (12), 2864.

Trivedi, P., Kumar, A., Gupta, N., Patel, C.N., 2025. Artificial intelligence and machine learning in microbial degradation of pollutants and toxins. Microbial Metabolomics: Recent Developments, Challenges and Future Opportunities. Springer Nature Singapore, Singapore, pp. 377–400.

Tyagi, S., Kapoor, R.T., Singh, R., Shah, M.P., 2025. Insights on microbial enzymes mediated biodegradation of azo dyes: a sustainable strategy for environment clean up. Bioremediation Journal. Available at https://doi.org/10.1080/10889868.2025.2498695, www.tandf.co.uk/journals/titles/10889868.asp.

Urbance, J.W., Cole, J., Saxman, P., Tiedje, J.M., 2003. BSD: the biodegradative strain database. Nucleic Acids Research 31 (1), 152–155. Available at https://doi.org/10.1093/nar/gkg032.

Vahabi, N., Michailidis, G., 2022. Unsupervised multi-omics data integration methods: a comprehensive review. Frontiers in Genetics 13, 854752.

Varshney, R., Singh, P., Yadav, D., 2022. Hazardous wastes treatment, storage, and disposal facilities. Hazardous Waste Management. Elsevier, pp. 33–64.

Vincenzo, M.D. 2024. Review on Multi-modal AI Models to Integrate Imaging and Omics Data (Master's Thesis).

O. Vince, P. Oldach, V. Pereno, M.H.Y. Leung, C. Greco, G. Minto-Cowcher, et al., 2025. Breaking Through Biology's Data Wall: Expanding the Known Tree of Life by Over 10x using a Global Biodiscovery Pipeline. bioRxiv, United Kingdom bioRxiv. Available from: https://www.biorxiv.org. https://doi.org/10.1101/2025.06.11.658620.

Volkova, S., Matos, M.R.A., Mattanovich, M., Marín de Mas, I., 2020. Metabolic modelling as a framework for metabolomics data integration and analysis. Metabolites 10 (8), 303. Available at https://doi.org/10.3390/metabo10080303.

Wang, X., Cho, J.H., Poudel, S., Li, Y., Jones, D.R., Shaw, T.I., et al., 2020. JUMPm: A tool for large-scale identification of metabolites in untargeted metabolomics. Metabolites 10 (5). Available at https://doi.org/10.3390/metabo10050190, https://www.mdpi.com/2218-1989/10/5/190/pdf.

Wanichthanarak, K., Fahrmann, J.F., Grapov, D., 2015. Genomic, proteomic, and metabolomic data integration strategies. Biomarker Insights 10, BMI-S29511.

Wani, A.K., Dhanjal, D.S., Akhtar, N., Chopra, C., Goyal, A., Singh, R., 2022. Role of genomics, metagenomics, and other meta-omics approaches for expunging the environmental contaminants by bioremediation. Omics for Environmental Engineering and Microbiology Systems. CRC Press, India, pp. 19–51. Available at https://doi.org/10.1201/9781003247883-2, https://www.routledge.com/Omics-for-Environmental-Engineering-and-Microbiology-Systems/Kumar-Garg-Kumar-Biswas/p/book/9781032162836.

Wibberg, D., Bremges, A., Dammann-Kalinowski, T., Maus, I., Igeño, M.I., Vogelsang, R., et al., 2016. Finished genome sequence and methylome of the cyanide-degrading *Pseudomonas pseudoalcaligenes* strain CECT5344 as resolved by single-molecule real-time sequencing. Journal of Biotechnology 232, 61–68.

Xue, L., Ren, H., Li, S., Gao, M., Shi, S., Chang, E., et al., 2015. Comparative proteomic analysis in Miscanthus sinensis exposed to antimony stress. Environmental Pollution 201, 150–160. Available at https://doi.org/10.1016/j.envpol.2015.03.004.

Yunusa, Y.R., Umar, Z.D., 2022. Effective microbial bioremediation via the multi-omics approach: an overview of trends, problems and prospects. UMYU Journal of Microbiology Research 6 (1), 127–145.

Zampolli, J. 2017. Assessment of Microbial Degradative Functions for the Evaluation of the Environmental Quality of Hydrocarbon Contaminated Soils.

Zerbino, D.R., Birney, E., 2008). Velvet: algorithms for de novo short read assembly using de Bruijn graphs. Genome Research 18 (5), 821–829.

Zhao, B., Poh, C.L., 2008. Insights into environmental bioremediation by microorganisms through functional genomics and proteomics. Proteomics 8 (4), 874–881. Available at https://doi.org/10.1002/pmic.200701005.

Zhao, D., Kumar, S., Zhou, J., Wang, R., Li, M., Xiang, H., 2017. Isolation and complete genome sequence of Halorientalis hydrocarbonoclasticus sp. nov., a hydrocarbon-degrading haloarchaeon. Extremophiles: Life Under Extreme Conditions 21 (6), 1081–1090. Available at https://doi.org/10.1007/s00792-017-0968-5.

Zhou, G., Li, S., Xia, J., 2020. Network-based approaches for multi-omics integration. Computational Methods and Data Analysis for Metabolomics 469–487.

Ziganshina, E.E., Mohammed, W.S., Doijad, S.P., Shagimardanova, E.I., Gogoleva, N.E., Ziganshin, A.M., 2018. Draft genome sequence of Brevibacterium epidermidis EZ-K02 isolated from nitrocellulose-contaminated wastewater environments. Federation Data in Brief 17, 119–123. Available at https://doi.org/10.1016/j.dib.2017.12.053, https://www.journals.elsevier.com/data-in-brief.

Chapter 6

Proteomic engineering for sustainability of agroecosystem: an innovative tool

Suhail Ashraf[1], Khalid E. Hamed[2], Mona Saleh Al Tami[3], Mohammed Aladhadh[4], and S. Nakkeeran[1]

[1]*Department of Plant Biotechnology, Tamil Nadu Agricultural University, Coimbatore, Tamil Nadu, India,* [2]*Department of Plant Protection, College of Agriculture and Food, Qassim University, Buraydah, Saudi Arabia,* [3]*Department of Biology, College of Science, Qassim University, Buraydah, Saudi Arabia,* [4]*Department of Food Science and Human Nutrition, College of Agriculture and Food, Qassim University, Buraydah, Saudi Arabia*

6.1 Introduction

Global agriculture is currently confronted with unprecedented challenges driven by climate change, environmental degradation, and rapidly growing food demand (Borah et al., 2024). Increasing global temperatures, irregular precipitation patterns, soil degradation, and loss of biodiversity are significantly affecting crop productivity and agricultural sustainability (Yuan et al., 2024). According to recent assessments, climate change alone could reduce global crop yields by 10%–25% by mid-century if adaptive strategies are not implemented (Lee et al., 2023; Ortiz et al., 2008). At the same time, the global population is projected to reach nearly 9.7 billion by 2050, placing immense pressure on food production systems (FAO, 2024).

Historically, the Green Revolution significantly increased agricultural productivity through the development of high-yielding crop varieties, expanded irrigation systems, and intensive use of fertilizers and pesticides (Borlaug, 1971; Gaud, 1968). Although these innovations dramatically improved food availability, they also led to unintended environmental consequences, including soil degradation, nutrient runoff, groundwater depletion, and ecological imbalance (Tilman et al., 2017; Hassan et al., 2025). Consequently, modern agriculture must transition toward sustainable intensification, a concept that aims to increase crop productivity while minimizing environmental impacts and conserving ecosystem services.

Proteomics Driven Bioremediation. DOI: https://doi.org/10.1016/B978-0-443-45294-9.00014-X

Advances in molecular biology and biotechnology have opened new opportunities for achieving sustainable agricultural development. Among these, proteomic engineering has emerged as a powerful approach for understanding plant responses to environmental stress and improving crop resilience. Proteomics focuses on the large-scale identification, quantification, and functional characterization of proteins, which represent the final products of gene expression and are directly responsible for cellular processes and phenotypic traits (Kosová et al., 2011; Jorrin Novo, 2021).

Unlike genomics, which identifies genetic potential, proteomics provides direct insights into the dynamic biochemical responses of plants under real environmental conditions. Proteins regulate essential physiological processes, including metabolism, signal transduction, stress defense, and nutrient assimilation. Therefore, proteomic studies enable researchers to identify stress-responsive proteins, regulatory pathways, and biomarkers associated with crop performance under adverse environmental conditions (Agrawal et al., 2010; Komatsu and Mustafa, 2022).

Recent technological advances have dramatically expanded the capabilities of proteomics in agricultural research. High-resolution mass spectrometry, label-free quantitative proteomics, and data-independent acquisition technologies allow researchers to identify thousands of proteins simultaneously with high sensitivity and accuracy (Aebersold and Mann, 2016). Additionally, integration of proteomics with genomics, transcriptomics, metabolomics, and phenomics, collectively known as multiomics approaches, has enabled systems-level understanding of plant stress responses and crop productivity.

Artificial intelligence and machine learning are increasingly being applied to analyze large proteomic datasets, predict protein structures, and identify functional protein networks associated with stress tolerance and crop productivity (Jumper et al., 2021; Baek et al., 2021; Dubey et al., 2026). Furthermore, genome editing technologies such as CRISPR-Cas systems enable precise modification of protein-coding genes, facilitating the development of crops with improved tolerance to drought, salinity, temperature stress, and pathogens (Liang et al., 2025).

Proteomic engineering also contributes to the development of climate-smart agriculture, an approach aimed at enhancing productivity, increasing resilience to climate change, and reducing greenhouse gas emissions. Through the identification of protein biomarkers associated with stress tolerance, nutrient efficiency, and photosynthetic performance, proteomic tools support the breeding of climate-resilient crop varieties capable of maintaining yield stability under environmental stress (Kausar et al., 2022; Thingujam et al., 2025).

In addition to crop improvement, proteomics plays a crucial role in understanding plant–microbe interactions in the rhizosphere. Beneficial microorganisms such as plant growth-promoting rhizobacteria (PGPR) and mycorrhizal fungi influence plant nutrition, stress tolerance, and disease resistance through complex molecular signaling networks. Proteomic studies have revealed key proteins involved in these interactions, enabling the development of biofertilizers and biostimulants that reduce reliance on chemical inputs (Backer et al., 2018; Chandran et al., 2021).

Furthermore, the integration of proteomic technologies with precision agriculture and biosensor systems is enabling real-time monitoring of crop health. Nanobiosensors capable of detecting protein biomarkers associated with stress responses can facilitate early detection of disease, nutrient deficiency, or environmental stress, allowing farmers to implement targeted management strategies (Khan et al., 2024).

Overall, proteomic engineering represents a transformative approach for sustainable agriculture by bridging the gap between genotype and phenotype. By identifying key proteins involved in plant growth, metabolism, and stress adaptation, proteomic technologies provide valuable tools for developing resilient crop systems capable of sustaining global food security while preserving environmental resources.

6.2 Concept and scope of proteomic engineering

Proteomic engineering involves large-scale identification, characterization, and modification of proteins to improve plant performance and stress adaptation. Advances in proteomics technologies allow simultaneous identification of thousands of proteins, enabling researchers to understand plant responses to environmental stress at a systems level. Proteomic analysis includes protein identification, quantification, post-translational modification analysis, protein–-protein interactions, and functional characterization (Fig. 6.1).

Proteomic engineering extends beyond protein discovery to include targeted modification and redesign. Integration with genomics and metabolomics bridges genotype and phenotype, enabling identification of biomarkers associated with stress resilience, nutrient efficiency, and crop quality traits (Kosová et al., 2011). This system-level understanding provides a molecular basis for climate-smart breeding programs.

6.3 Technological advances driving proteomic engineering

Recent technological advancements have transformed proteomic engineering into a practical tool for sustainable agriculture. High-resolution mass spectrometry enables precise identification and quantification of proteins, including low-abundance stress-responsive proteins. Label-free quantitative proteomics has improved comparative protein analysis and biomarker discovery (Mirzaei et al., 2016).

CRISPR-based genome editing now enables targeted modification of protein-coding genes to improve enzyme efficiency and stress tolerance. For example, modification of ion transport proteins enhances salinity tolerance in plants (Apse et al., 1999). Artificial intelligence and machine learning are increasingly used to predict protein structures and functional interactions, accelerating the design of stress-tolerant enzymes and improved photosynthetic proteins.

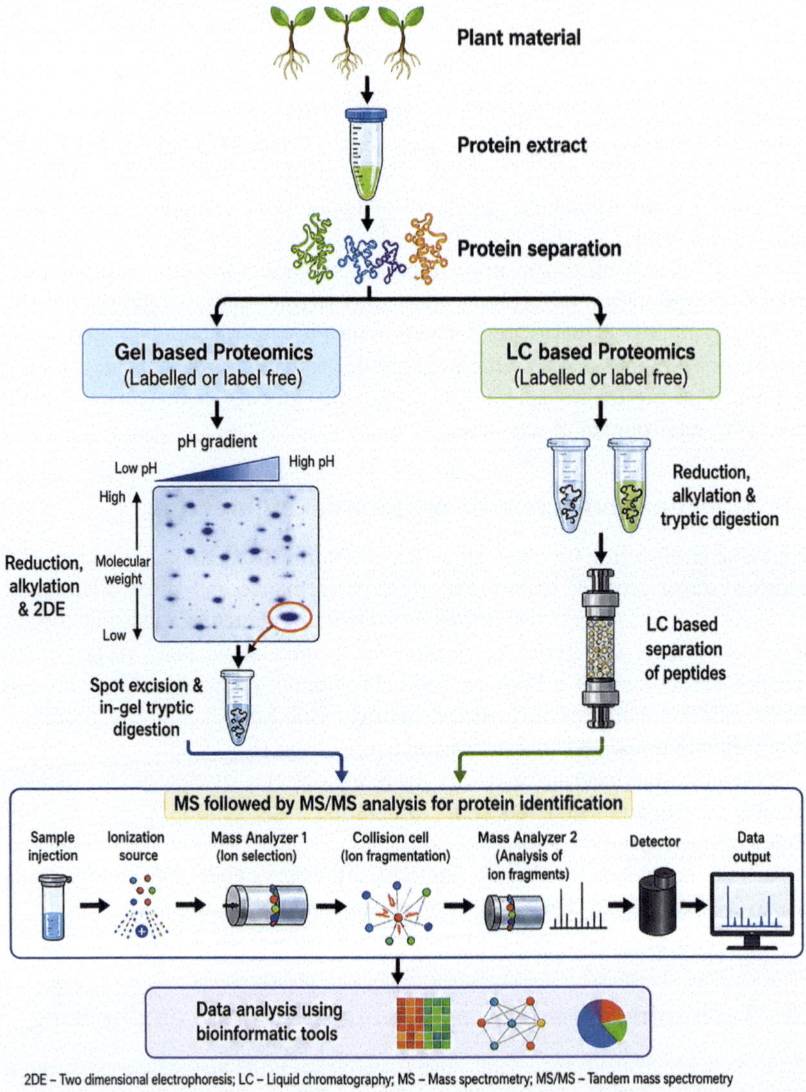

FIGURE 6.1 Workflow of plant proteomics analysis. Plant tissue proteins are extracted and separated, followed by analysis using gel-based or liquid chromatography (LC)-based proteomics approaches. Proteins or peptides undergo reduction, alkylation, and enzymatic digestion before mass spectrometry (MS/MS) for identification and quantification. The resulting data are processed using bioinformatic tools for protein characterization and functional interpretation.
Schematic workflow of plant proteomic analysis.

Synthetic biology complements proteomic engineering by enabling the redesign of metabolic pathways for improved nutrient assimilation and photosynthetic efficiency. Meanwhile, nanobiosensors and portable protein

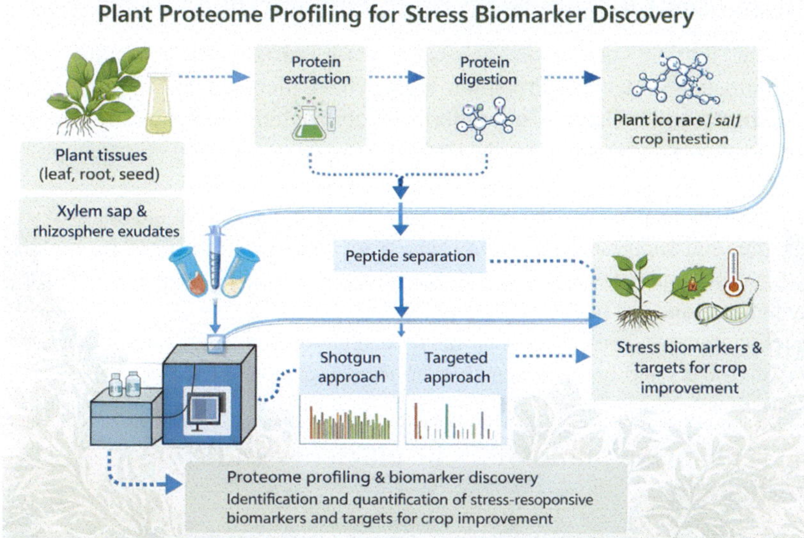

FIGURE 6.2 Plant tissues (leaf, root, and seed) and xylem sap or rhizosphere exudates are subjected to protein extraction and enzymatic digestion, followed by peptide separation and LC-MS analysis. Shotgun and targeted proteomic approaches enable proteome profiling, leading to the identification and quantification of stress-responsive proteins and biomarkers associated with climate resilience, stress tolerance, and targets for crop improvement.
LC-MS-based proteomic profiling of plant tissues for biomarker discovery.

detection systems are emerging as tools for real-time crop health monitoring, bridging molecular biology with precision agriculture (Fig. 6.2).

6.4 Proteomic engineering and stress resilience

Abiotic stresses such as drought, salinity, temperature extremes, and flooding represent major constraints on crop productivity. Proteomic studies reveal extensive protein reprogramming in response to environmental stress, including induction of protective proteins, detoxification enzymes, and osmoprotectants (Kosová et al., 2011; Hasegawa et al., 2000).

Drought stress induces the accumulation of late embryogenesis abundant proteins and dehydrins that stabilize cellular structures during water deficit (Shinozaki et al., 2003). Salinity stress activates ion transporters and vacuolar Na^+/H^+ antiporters that regulate ionic balance and improve salt tolerance (Apse et al., 1999). Heat stress activates heat shock proteins that protect cellular proteins from denaturation, while cold stress responses involve antifreeze proteins and membrane stabilizers.

Proteomic analysis also identifies defense proteins involved in plant immune responses. Pathogenesis-related proteins and defense enzymes play essential roles in resistance against pathogens, reducing reliance on chemical pesticides (Haas and Défago, 2005).

6.5 Nutrient use efficiency and plant–microbe interactions

Efficient nutrient utilization is essential for sustainable agriculture and environmental protection. Proteomic studies have identified enzymes involved in nitrogen assimilation, phosphorus mobilization, and nutrient uptake mechanisms. Phosphatases produced by rhizosphere microorganisms improve phosphorus availability and plant uptake (Abd-Alla, 1994).

Beneficial plant-associated microbes enhance nutrient acquisition, improve soil structure, and suppress pathogens. These plant probiotics contribute to nitrogen fixation, phosphorus solubilization, and phytohormone production, supporting plant growth and soil health (Islam and Hossain, 2012). Proteomic research reveals signaling proteins involved in plant-microbe symbiosis, enabling the development of biofertilizers and biopesticides that reduce agrochemical dependency.

6.6 Enhancing photosynthetic efficiency and crop productivity

Improving photosynthetic efficiency remains essential for increasing crop productivity. Rubisco, the key enzyme in carbon fixation, exhibits limited catalytic efficiency and sensitivity to temperature fluctuations. Proteomic engineering and synthetic biology aim to redesign Rubisco and associated proteins to enhance carbon assimilation and reduce photorespiration losses.

Proteomic research has also enabled modification of storage proteins to improve nutritional quality, digestibility, and amino acid composition. These improvements contribute to enhanced food quality while maintaining yield performance.

6.7 Integration with precision agriculture

Proteomic engineering is increasingly integrated with precision agriculture technologies to enable data-driven crop management. Protein biomarkers serve as early indicators of stress, enabling timely interventions. Nanobiosensors capable of detecting protein markers allow real-time monitoring of crop health, supporting precision irrigation, nutrient management, and disease control.

This integration reduces input waste, minimizes environmental impacts, and enhances resource efficiency while maintaining productivity.

Proteomic engineering contributes to sustainable agriculture by enabling the identification of stress-responsive proteins, improving nutrient use efficiency, enhancing photosynthetic performance, and strengthening plant defense mechanisms. These molecular insights facilitate the development of resilient crop varieties capable of maintaining productivity under adverse environmental conditions. The major agricultural challenges addressed through proteomic engineering and their corresponding molecular targets and sustainability outcomes are summarized in Table 6.1.

TABLE 6.1 Major contributions of proteomic engineering to sustainable agroecosystems.

Agricultural challenge	Proteomic insight	Key molecular targets	Agricultural outcome	Sustainability impact	References
Drought stress	Identification of stress-responsive proteins and osmoprotectants	LEA proteins, dehydrins, and aquaporins	Improved drought tolerance	Yield stability under water scarcity	Kausar et al. (2022)
Salinity stress	Regulation of ion transport and osmotic balance	Na+/H+ antiporters, ion transporters	Development of salt-tolerant crops	Cultivation in saline soils	Bhatt et al. (2024)
Heat stress	Activation of protein folding and stabilization mechanisms	Heat shock proteins (HSP70, HSP90)	Improved thermotolerance	Climate resilience	Zhao et al. (2017)
Cold stress	Membrane protection and antifreeze protein synthesis	Antifreeze proteins, lipid-modifying enzymes	Cold-tolerant cultivars	Expansion of cultivation areas	Jorrin Novo (2021)
Plant diseases	Activation of immune signaling and defense proteins	Pathogenesis-related proteins, chitinases	Enhanced pathogen resistance	Reduced pesticide use	Torres et al. (2006)
Nutrient deficiency	Identification of enzymes involved in nutrient assimilation	Nitrate reductase, phosphatases	Improved nutrient use efficiency	Reduced fertilizer dependency	Backer et al. (2018)

(*Continued*)

TABLE 6.1 (Continued)

Agricultural challenge	Proteomic insight	Key molecular targets	Agricultural outcome	Sustainability impact	References
Soil degradation	Proteins involved in plant–microbe symbiosis	Symbiotic signaling proteins	Improved soil health	Sustainable soil ecosystems	Chandran et al. (2021); Hassan et al. (2025)
Low photosynthetic efficiency	Optimization of carbon fixation pathways	Rubisco, Rubisco activase	Increased biomass production	Higher productivity with fewer inputs	Aebersold and Mann (2016)
Poor nutritional quality	Engineering storage proteins and metabolic pathways	Seed storage proteins	Improved food nutritional value	Nutritional security	Agrawal et al. (2010); Uday et al. (2024)

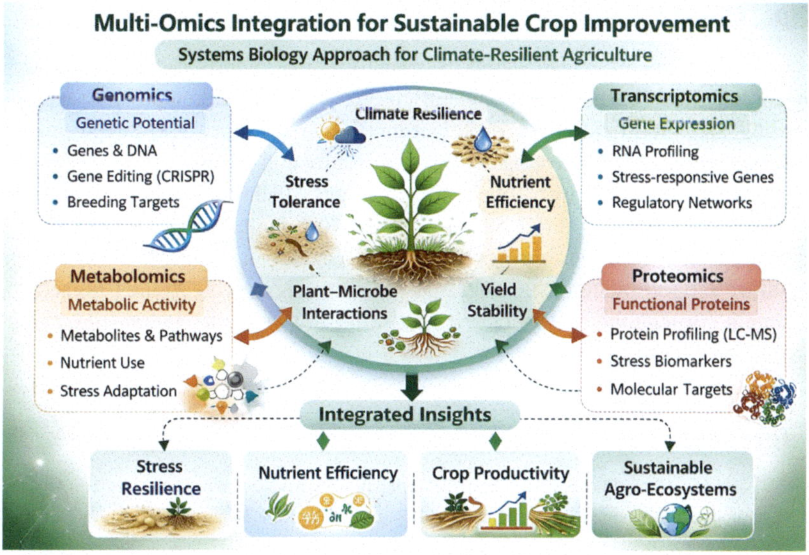

FIGURE 6.3 Multiomics integration supporting climate-resilient and sustainable crop improvement. Multiomics integration for climate-resilient crop improvement.

6.8 Climate-smart breeding and environmental benefits

Integration of proteomics with genomics, metabolomics, and phenomics supports climate-adaptive crop breeding. Proteomic biomarkers associated with stress resilience and yield stability enable multi-trait selection strategies for developing climate-resilient cultivars.

Proteomic engineering contributes to environmental sustainability by reducing fertilizer and pesticide use, improving soil health, and lowering greenhouse gas emissions. Enhanced nutrient efficiency reduces nutrient runoff and water pollution, while improved stress tolerance ensures yield stability under climate stress (Fig. 6.3).

6.9 Challenges and future prospects

Despite its transformative potential, proteomic engineering faces challenges including high instrumentation costs, complex data interpretation, limited infrastructure in developing regions, and regulatory concerns regarding engineered crops. Addressing these barriers requires capacity building, interdisciplinary collaboration, and supportive policy frameworks.

Future innovations are expected to include AI-guided protein design, synthetic photosynthesis optimization, smart biosensors for field diagnostics, and integration with regenerative agriculture practices. As technological advances reduce costs and improve accessibility, proteomic engineering will play a central role in sustainable agricultural transformation.

6.10 Conclusion

Proteomic engineering represents a transformative approach for achieving sustainable and climate-resilient agriculture. By enabling precise understanding and modification of protein functions, it supports the development of crops capable of thriving under environmental stress while reducing agrochemical inputs. Integration with multiomics technologies, plant probiotics, nano-biosensors, and precision agriculture tools provides a holistic framework for resilient agro-ecosystems. As global agriculture confronts mounting environmental challenges, proteomic engineering stands poised to become a cornerstone technology for ensuring food security while preserving ecosystem integrity.

References

Abd-Alla, M.H., 1994. Phosphatases and the utilization of organic phosphorus by Rhizobium leguminosarum biovar viceae. Letters in Applied Microbiology 18 (5), 294–296. Available at https://doi.org/10.1111/j.1472-765x.1994.tb00873.x.

Aebersold, R., Mann, M., 2016. Mass-spectrometric exploration of proteome structure and function. Nature 537 (7620), 347–355. Available at https://doi.org/10.1038/nature19949.

Agrawal, G.K., Jwa, N.S., Lebrun, M.H., Job, D., Rakwal, R., 2010. Plant secretome: unlocking secrets of the secreted proteins. Proteomics 10 (4), 799–827. Available from. Available at https://doi.org/10.1002/pmic.200900514Nepal, http://www3.interscience.wiley.com/cgi-bin/fulltext/123195818/PDFSTART.

Apse, M.P., Aharon, G.S., Snedden, W.A., Blumwald, E., 1999. Salt tolerance conferred by overexpression of a vacuolar Na+/H+ antiport in Arabidopsis. Science (New York, N.Y.) 285 (5431), 1256–1258. Available at https://doi.org/10.1126/science.285.5431.1256.

Backer, R., Rokem, J.S., Ilangumaran, G., Lamont, J., Praslickova, D., Ricci, E., et al., 2018. Plant growth-promoting rhizobacteria: context, mechanisms of action, and roadmap to commercialization of biostimulants for sustainable agriculture. Frontiers in Plant Science 871 Available from. Available at https://doi.org/10.3389/fpls.2018.01473, https://www.frontiersin.org/articles/10.3389/fpls.2018.01473/pdf.

Baek, M., DiMaio, F., Anishchenko, I., Dauparas, J., Ovchinnikov, S., Lee, G.R., et al., 2021. Accurate prediction of protein structures and interactions using a three-track neural network. Science (New York, N.Y.) 373 (6557), 871–876. Available from. Available at https://doi.org/10.1126/science.abj8754, https://science.sciencemag.org/content/373/6557/871.full.

Bhatt, D., Nath, M., Badoni, S., Joshi, R. (Eds.), 2024. Current Omics Advancement in Plant Abiotic Stress Biology. Elsevier.

Borah, A., Sahu, S., Srivastava, R.P., Singh, M., Tyagi, D.B., 2024. Exploring the economic challenges threatening global agriculture and food security. Ecology, Environment and Conservation 30 (05), S193. Available at https://doi.org/10.53550/eec.2024.v30i05s.031.

Borlaug, N.E. 1971. The green revolution, peace, and humanity.

Chandran, H., Meena, M., Swapnil, P., 2021. Plant growth-promoting rhizobacteria as a green alternative for sustainable agriculture. Sustainability 13 (19), 10986. Available at https://doi.org/10.3390/su131910986.

Dubey, P.K.A., Vijayarani, K., Manoj, M., Gavaskar, S., Robert, A.A., Prabhu, J.P.A., et al., 2026. Artificial intelligence approaches for constructing digital plant proteomics. AI-Enhanced Plant Omics CABI International, India, pp. 43–66. Available at http://doi.org/10.1079/9781800628649.0003, https://www.cabidigitallibrary.org/doi/book/10.1079/9781800628649.0000.

FAO. 2024. The state of food and agriculture 2024. Food and Agriculture Organization of the United Nations.

Gaud, W.S., 1968. The Green Revolution: Accomplishments and Apprehensions. The Society for International Development.

Haas, D., Défago, G., 2005. Biological control of soil-borne pathogens by fluorescent pseudomonads. Nature Reviews Microbiology 3 (4), 307–319. Available at https://doi.org/10.1038/nrmicro1129.

Hasegawa, P.M., Bressan, R.A., Zhu, J.K., Bohnert, H.J., 2000. Plant cellular and molecular responses to high salinity. Annual Review of Plant Biology 51, 463–499. Available from. Available at https://doi.org/10.1146/annurev.arplant.51.1.463, http://www.annualreviews.org/journal/arplant.

Hassan, F.U., Islam, M.S., Basit, A., Ahmad, R., Dhakane, A.A., Toriman, M.E.H., et al., 2025. Soil health assessment strategies and future perspectives towards a sustainable agriculture. Soils and Sustainable Agriculture: Interplay of Soil, Plant, Water and Environmental Systems for Sustainable Agriculture. Springer Nature Switzerland, Cham, pp. 295–319.

Islam, M.T., Hossain, M.M., 2012. Plant probiotics in phosphorus nutrition in crops, with special reference to rice. Bacteria in Agrobiology: Plant Probiotics 325 363. Available at https://doi.org/10.1007/978-3-642-27515-9_18, http://doi.org/10.1007/978-3-642-27515-9 Available from.

Jorrin Novo, J.V., 2021. Proteomics and plant biology: contributions to date and a look towards the next decade. Expert Review of Proteomics 18 (2), 93–103. Available at https://doi.org/10.1080/14789450.2021.1910028.

Jumper, J., Evans, R., Pritzel, A., Green, T., Figurnov, M., Ronneberger, O., et al., 2021. Highly accurate protein structure prediction with AlphaFold. Nature 596 (7873), 583–589. Available at https://doi.org/10.1038/s41586-021-03819-2.

Kausar, R., Wang, X., Komatsu, S., 2022. Crop proteomics under abiotic stress: from data to insights. Plants 11 (21), 2877. Available at https://doi.org/10.3390/plants11212877.

Khan, M. Q., Khan, J., Alvi, M. A. H., Nawaz, H., Fahad, M., & Umar, M. (2024). Discover Nano.

Komatsu, S., Mustafa, G., 2022. Plant proteomics. Plant Omics: Advances in Big Data Biology. CABI, GB, pp. 30–49.

Kosová, K., Vítámvás, P., Prášil, I.T., Renaut, J., 2011. Plant proteome changes under abiotic stress - contribution of proteomics studies to understanding plant stress response. Journal of Proteomics 74 (8), 1301–1322. Available at https://doi.org/10.1016/j.jprot.2011.02.006.

Lee, H., Calvin, K., Dasgupta, D., Krinner, G., Mukherji, A., Thorne, P., & Park, Y. (2023). IPCC, 2023: Climate change 2023: Synthesis report, summary for policymakers. Contribution of working groups i, II and III to the sixth assessment report of the intergovernmental panel on climate change [core writing team, h. Lee and j. Romero (eds.)]. IPCC, geneva, Switzerland (pp. 1-34). Intergovernmental Panel on Climate Change (IPCC).

Liang, X., Yu, S., Ju, Y., Wang, Y., Yin, D., 2025. Multi-scale remote-sensing phenomics integrated with multi-omics: advances in crop drought–heat stress tolerance mechanisms and perspectives for climate-smart agriculture. Plants 14 (18), 2829. Available at https://doi.org/10.3390/plants14182829.

Mirzaei, M., Wu, Y., Handler, D., Maher, T., Pascovici, D., Ravishankar, P., et al., 2016. Applications of quantitative proteomics in plant research. Agricultural Proteomics Volume 1: Crops, Horticulture, Farm Animals, Food, Insect and Microorganisms. Springer International Publishing, Australia, pp. 1–29. Available at http://doi.org/10.1007/978-3-319-43275-5, http://doi.org/10.1007/978-3-319-43275-5_1.

Ortiz, R., Sayre, K.D., Govaerts, B., Gupta, R., Subbarao, G.V., Ban, T., et al., 2008. Climate change: can wheat beat the heat? Agriculture, Ecosystems and Environment 126 (1-2), 46–58. Available at https://doi.org/10.1016/j.agee.2008.01.019.

Shinozaki, K., Yamaguchi-Shinozaki, K., Seki, M., 2003. Regulatory network of gene expression in the drought and cold stress responses. Current Opinion in Plant Biology 6 (5), 410–417. Available at https://doi.org/10.1016/s1369-5266(03)00092-x.

Thingujam, D., Gouli, S., Cooray, S.P., Chandran, K.B., Givens, S.B., Gandhimeyyan, R.V., et al., 2025. Climate-resilient crops: integrating AI, multi-omics, and advanced phenotyping to address global agricultural and societal challenges. Plants 14 (17) Available from. Available at https://doi.org/10.3390/plants14172699, www.mdpi.com/journal/plants.

Tilman, D., Clark, M., Williams, D.R., Kimmel, K., Polasky, S., Packer, C., 2017. Future threats to biodiversity and pathways to their prevention. Nature 546 (7656), 73–81. Available from. Available at https://doi.org/10.1038/nature22900, http://www.nature.com/nature/index.html.

Torres, M.A., Jones, J.D.G., Dangl, J.L., 2006. Reactive oxygen species signaling in response to pathogens. Plant Physiology 141 (2), 373–378. Available from. Available at https://doi.org/10.1104/pp.106.079467, http://www.plantphysiol.org/.

Uday, G., Shardha, H.B., Priyanka, K., Jagirdhar, S., 2024. Advances in Plant Proteomics toward Improvement of Crop Productivity and Stress Resistance. Informa UK Limited, pp. 11–29. Available at http://doi.org/10.1201/b23255-2.

Yuan, X., Li, S., Chen, J., Yu, H., Yang, T., Wang, C., et al., 2024. Impacts of global climate change on agricultural production: a comprehensive review. Agronomy 14 (7), 1360. Available at https://doi.org/10.3390/agronomy14071360.

Zhao, C., Liu, B., Piao, S., Wang, X., Lobell, D.B., Huang, Y., et al., 2017. Temperature increase reduces global yields of major crops in four independent estimates. Proceedings of the National Academy of Sciences 114 (35), 9326–9331. Available at https://doi.org/10.1073/pnas.1701762114.

Chapter 7

Yeast proteomic strategies for bioremediation of oil contaminated soils

Pritam Acharya[1], Tapas Kumar Ghosh[2,3], Saurabh Acharyya[1], Dipankar Chattopadhyay[4], and Burhan Hamid[5]

[1]*Department of General Science, Calcutta Institute of Technology, Howrah, West Bengal, India,* [2]*Department of Polymer Science and Technology and Center for Research in Nanoscience and Nanotechnology, University of Calcutta, Kolkata, West Bengal, India,* [3]*Applied Science and Humanities, Global Institute of Management and Technology, Krishnanagr, West Bengal, India,* [4]*Department of Polymer Science and Technology, University of Calcutta, Kolkata, West Bengal, India,* [5]*Centre of Research for Development, University of Kashmir, Srinagar, Jammu and Kashmir, India*

7.1 Introduction

Global industrialisation and urbanisation have led to considerable environmental and social degradation. Industrial sectors are substantial contributors to the national economies of some emerging nations; yet, they also represent considerable sources of environmental pollution. Numerous poisons are released by industry, including wastewater, which constitutes a major contaminant of soil and water. A variety of chemical compounds are utilised to process raw materials, producing high-quality, user-friendly products at a low cost and in a reduced timeframe. Typically, low-quality, inexpensive, or non-biodegradable substances are employed, and their toxicity is frequently disregarded, leading to detrimental impacts on the environment and human health (Smart and Smart, 2003). The industrial revolution has led to significant health issues exacerbated by extensive environmental contamination. Nearly every second species on Earth is presently under a significant threat to its survival due to heightened environmental pollution. Numerous publicly available studies indicate that industrial effluents contain a wide array of highly harmful or dangerous substances. Industrial effluents contain numerous organic and inorganic pollutants that significantly impair natural ecosystems and present health risks to humans and other organisms. Microorganisms, including bacteria, yeast, and fungi, serve as effective, simple, and ecofriendly

Proteomics Driven Bioremediation. DOI: https://doi.org/10.1016/B978-0-443-45294-9.00001-1

agents for disinfection and can be utilized for the remediation of heavy metals and other environmental contaminants (Tarfeen et al., 2022; Hamid et al., 2023; Kuddus et al., 2024). Microbes may easily adjust to unfavourable environmental conditions and possess varied dietary needs. Furthermore, they have numerous intracellular and extracellular enzymatic systems that utilise complex contaminants as an energy source; yeast proteases have been utilised as an efficient remediation tool (Adessi et al., 2000; Mohiddin and Hamid, 2018; Hamid et al., 2022).

Moreover, bacteria experience a gradual genetic transformation that enables them to develop new metabolic pathways for the degradation of xenobiotics (Thakur et al., 2019). The technology of phytoremediation is crucial for the elimination of pollutants from the environment (Sharma et al., 2023; Singh and Santal, 2015). Additionally, bioremediation employing microorganisms is an appealing method for the removal of pollutants (Santal et al., 2011). Recent advancements in novel remediation through genetic modification may enhance the efficacy of addressing environmental challenges, while conventional methods entail high costs, yield minimal impacts, and pose hazards to ecosystems (Santal et al., 2023). Genetically modified microorganisms, utilising techniques such as 16S rRNA, DNA microarray, and next-generation sequencing, demonstrate enhanced bioremediation efficacy compared to their naturally occurring equivalents. The genome-enabled experimental and modelling organism substantially facilitates the evaluation of physiology and enhances the efficacy of organisms employed for bioremediation. The application of proteomics in bioremediation research provides an in-depth understanding of the protein composition of microbial cells and serves as an effective approach to clarify the molecular mechanisms underlying the elimination of dangerous compounds from the environment. Advanced technologies, in conjunction with genomic analyses of soil microbes, have enabled the identification of catabolic pathways crucial for the degradation of organic and inorganic contaminants in soil. The detected species have proven essential in enhancing bioremediation methods and in the creation of novel biosensors for environmental contaminant detection (Khan et al., 2025). Ex situ and in situ represent the two categories of bioremediation methodologies. This chapter analyses the importance of microbes in the degradation of environmental toxins and their mechanisms, alongside the efficacy of recombinant DNA technology in creating genetically modified microorganisms for bioremediation purposes. In situ treatment approaches involve the biodegradation of pollutants at the contaminated site, whereas ex situ remediation techniques necessitate the transfer of pollutants from contaminated areas to a different place. Due to its constraints, several academics do not regard ex situ cleanup as a viable option for remediation. Profitability may vary by region, and while microbes might facilitate toxin remediation in vitro, their efficacy in vivo is uncertain (Barupal et al., 2019a, b). *Bacillus* sp. AS2 was examined for its ability to degrade textile dyes sourced from a contaminated environment (Santal et al., 2022).

Comprehensive research on the fundamental and applied dimensions of microbial complementarity is essential to evaluate the biodegradation potential of microorganisms found in contaminated environments and to examine alterations in their structure and metabolic functions. Microorganisms have been utilised for the conversion of food waste into valuable compost and as a potential tool for bioremediation (Hamid et al., 2023). Therefore, a multidisciplinary approach is essential to elucidate the undiscovered routes and chemistries, allowing bioremediation to evolve into a versatile environmental management and monitoring strategy. This chapter explores proteomics-driven bioremediation as a transformative strategy for precision environmental restoration. Unlike genomics, which provides a static map of genetic potential, proteomics offers a real-time snapshot of the active enzymes and catabolic pathways involved in pollutant degradation. Fig. 7.1 demonstrates different sources of environmental pollution.

By integrating advanced tools such as mass spectrometry and bioinformatics, researchers can elucidate the molecular mechanisms governing the breakdown of xenobiotics and heavy metals. This systems-level approach facilitates the identification of stress-response biomarkers and the development of genetically modified microorganisms with enhanced metabolic efficiency.

FIGURE 7.1 **Diverse origins of environmental.** The figure shows multiple sources of pollution that lead to contamination of the environment.

Ultimately, leveraging the proteome allows for the optimisation of in situ and ex situ strategies, bridging the gap between laboratory research and real-world application to ensure a cleaner, more sustainable environment (Kumar et al., 2024).

7.2 The shift from genomics to proteomics

Genomics indicates the potential capabilities of an organism, whereas proteomics reveals its actual responses to environmental stimuli. Proteins are the principal agents of cellular function, acting as enzymes that facilitate the degradation of complicated contaminants. Proteomics reveals alterations such as phosphorylation or acetylation that modulate the activity of degradative enzymes, which are undetectable by DNA sequencing (Malla et al., 2018). It records the dynamic alterations in microbial protein expression upon exposure to heavy metals, hydrocarbons, or pesticides. Certain proteins can function as biomarkers to assess the health and advancement of a bioremediation site.

7.3 Development of proteomics

The approach for analysing protein profiles is termed 'proteomics', which encompasses the examination of an organism's proteins (Anderson and Anderson, 1998). The names "proteomics" and "proteome" were introduced in 1995, marking a significant post-genomic advancement that emerged from the development of extensive and intricate genome sequencing datasets (Wasinger et al., 1995). Proteins are essential components of living organisms, possessing diverse functions. The word 'proteomics' originated in 1997, in relation to genomics, the study of the genome. The proteome encompasses the complete array of proteins produced or modified by an organism or system (James, 1997). Proteomics has facilitated the discovery of an ever-increasing array of proteins. This varies over time and is influenced by specific requirements or pressures that a cell or organism encounters. Proteomics is an interdisciplinary field that has greatly benefited from the genetic information obtained from several genome projects, particularly the Human Genome Project (Hood and Rowen, 2013). It encompasses the examination of proteomes concerning the comprehensive aspects of protein composition, structure, and function. It constitutes a substantial component of utilitarian genomics. Proteomics pertains to the exploratory analysis of proteins and proteomes but is often specifically associated with protein purification and mass spectrometry. Proteomics-based analyses have been important in determining alterations in the organisation and abundance of proteins, as well as in identifying critical proteins involved in the physiological response of organisms when exposed to anthropogenic pollutants. Proteomic studies are particularly essential, since the observed phenotype is a direct consequence of protein function rather than of genomic sequencing. Several studies have indicated that the proteome plays a critical role in the bioremediation of polycyclic aromatic hydrocarbons (PAHs) (Singh and Nagaraj, 2006). Modifications in 2-DE for compartmental proteomics have been implemented by demonstrating a selective approach for multidimensional protein identification technology (Mud PIT) (Paoletti et al., 2004). Mass spectrometry has transformed ecological proteomics in the context of bioremediation (Aebersold and Mann, 2003). Matrix-assisted laser desorption/ionisation time-of-flight mass spectrometry (MALDI-TOF-MS) is the predominant mass spectrometry technique for identifying proteins from two-dimensional electrophoresis (2-DE) gels (Aitken and Learmonth, 2002). Surface-enhanced laser desorption-ionisation time-of-flight mass spectrometry (SELDI-TOF-MS) technology is an alternative method for identifying proteins associated with PAH degradation pathways (Merchant and Weinberger, 2000). Scientists have reported on the use of liquid chromatography-mass spectrometry (LC-MS) for the direct localisation and identification of potential pollutants in water.

7.3.1 The role of yeasts in hydrocarbon degradation

7.3.1.1 Diversity of hydrocarbonoclastic yeasts: Yarrowia, Candida, *and* Pichia *sp.*

Hydrocarbonoclastic yeasts are polyphyletic groupings that may utilise petroleum hydrocarbons as their exclusive carbon and energy source. In contrast to several bacteria, these yeasts exhibit remarkable resilience to low pH and elevated osmotic pressure commonly present in contaminated industrial soils. *Yarrowia lipolytica* is commonly recognised as the "model" hydrocarbonoclastic yeast, possessing a comprehensive array of alkane hydroxylases (ALK) and lipases. Its capacity to break down n-alkanes and effectively collect lipids positions it as a leading contender for soil bioaugmentation (Bankar et al., 2009; Beopoulos et al., 2009). Species such as *C. tropicalis* and *C. viswanathii* are recognised for their fast degradation of PAHs. They are frequently segregated from marine oil spills and tropical soils owing to their elevated tolerance for aromatic toxicity (Ameen et al., 2021; Hassanshahian et al., 2012). Furthermore, *Pichia* sp. and *Debaryomyces* sp. are commonly detected in hypersaline, oil-contaminated settings. *Pichia guilliermondii* has demonstrated considerable capability in degrading crude oil while concurrently generating elevated levels of riboflavin, which serves as a redox mediator in the degradation process (Amaral et al., 2010). Moreover, reports on psychrophilic yeasts utilized in the remediation of soil contaminants have been reported (Hamid et al., 2014; Mohiddin and Hamid, 2018)

7.4 Mechanism of the bioremediation process

Bioremediation involves the utilisation of microorganisms to decompose environmental contaminants. This is an environmentally sustainable and cutting-edge approach that employs microbial processes to entirely eliminate hazardous pollutants. Microorganisms are crucial in regulating biogeochemical cycles, sustaining the atmosphere, maintaining human health, suppressing plant diseases, promoting growth, and facilitating the remediation of environmental toxins. Microbial-mediated bioremediation is highly significant, as it offers a more cost-effective, straightforward, and environmentally sustainable alternative to conventional nonbiological remediation methods, such as "muck, suck, and truck", where contaminants are merely extracted and transported elsewhere (Griggs et al., 2013). Nonetheless, the potential of bioremediation remains unfulfilled. One explanation for this is that bioremediation methods that prove efficient in one location may not yield the same results in different locations. Moreover, microbial systems that effectively remediate contaminants in laboratory settings may not function satisfactorily in real-world environments. The causes of these failures remain ambiguous, leading many managers to hesitate about utilising bioremediation for environmental remediation (Morris et al., 2011). The mechanisms governing the development and activity of microorganisms in contaminated environments remain poorly understood, consequently constraining the use of bioremediation (Lovley, 2003; Tarfeen et al., 2022). The dynamic behaviour, dietary versatility, and ability to adapt to harsh environmental conditions render microorganisms the most ideal life forms for endurance. The microbial capability is favourable and beneficial to humanity, especially in the removal of pollutants and other dangerous substances from the environment. A complete comparison of conventional and microbial bioremediation is shown in Fig. 7.2.

Feature	Conventional	Microbial Bioremediation
Action	Physical extraction and transportation of waste	Complete elimination/decomposition of waste
Sustainability	Low; simply moves the problem elsewhere	High, environmentally sustainable, and cutting-edge
Cost	Generally, more expensive	More cost-effective and straightforward
Outcome	Contaminants remain hazardous in a new site	Contaminants are converted to harmless substances

FIGURE 7.2 **Comparison of conventional and microbial.** The figure depicts a comparative analysis of conventional methods and microbial methods of bioremediation and demonstrates how effective and ecofriendly microbial bioremediation is as compared to conventional methods.

Microbes possess the ability to digest environmental toxins through various enzymatic processes, hence alleviating or eliminating pollution. A comprehensive list of bacteria involved in bioremediation processes is accessible (Satyanarayana et al., 2012; Abou Seeda et al., 2017). Microorganisms can facilitate environmental restoration through many activities, including binding, oxidation, volatilisation, immobilisation, and chemical change of contaminants. A prevalent bioremediation strategy involves the oxidation of harmful organic contaminants into innocuous byproducts. Oxygen is the predominant electron acceptor in microbial respiration and serves as the agent for the aerobic destruction of a diverse array of organic contaminants, including arenes like benzene and xenobiotics such as pesticides, which have been extensively researched (Wackett et al., 2001). Although a wide taxonomic diversity can digest aerobic pollutants, Pseudomonas species and their near relatives are the most extensively studied organisms due to their capacity to degrade various contaminants. Bioremediation strategies should be formulated based on an understanding of the specific microorganisms present in contaminated sites, encompassing their metabolic pathways and responses to alterations in environmental circumstances. Regrettably, this specialist information is not readily accessible in practice, and the application of bacteria in bioremediation remains highly experimental rather than grounded in established knowledge (Malla et al., 2018). While the science of bioremediation is currently not optimal, it appears to be attainable. The mechanisms of hydrocarbon uptake include bioremediation through yeast proteomics, specifically involving cell surface hydrophobicity and biosurfactants. The principal obstacle in oil bioremediation is the substrate's hydrophobicity, which restricts its bioavailability. Yeasts utilise a bifurcated technique to surmount this challenge, as outlined below.

7.4.1 Cell surface hydrophobicity

Yeasts alter their cell wall composition, particularly the concentrations of mannoproteins and hydrophobins, to enhance their attraction for oil droplets. The "direct contact" technique enables cells to attach to and "coat" oil droplets, thus promoting passive or active transport across the membrane.

7.4.2 Biosurfactant production

Numerous yeasts, especially *Y. lipolytica*, excrete extracellular glycolipids and lipopeptides (e.g., Yarriolipins). These surfactants diminish the interfacial tension between the aqueous phase and the oil, emulsifying the hydrocarbons into smaller micelles that are more readily internalised (Amaral et al., 2010; Fontes et al., 2008). A comparison of yeast versus bacterial efficacy in acidic and saline soils may be illustrated. Although bacteria like *Alcanivorax* are abundant degraders in neutral sea habitats, yeasts frequently surpass them in stressed terrestrial ecosystems. The majority of oil-degrading bacteria are suppressed at pH levels below 5.0. Nonetheless, yeasts such as *Rhodotorula* and *Candida* flourish in acidic mine drainage or acidic forest soils, sustaining elevated metabolic rates in conditions when bacterial activity halts (Margesin and Schinner, 2001). Conversely, elevated salinity induces osmotic stress that results in the lysis of many bacterial cells. Yeasts preserve intracellular osmotic equilibrium by accumulating polyols (glycerol and erythritol), enabling their activity in saline oil-contaminated coastal soils (Hassanshahian et al., 2012). Yeasts typically exhibit reduced nitrogen and phosphorus demands compared to bacteria, rendering them more appropriate for "low-input" bioremediation in nutrient-deficient industrial soils (Brar et al., 2006).

7.5 Proteomics in bioremediation

Protein analysis poses significant challenges due to the ubiquitous presence of proteins in every intracellular compartment within the envelope or membrane, unlike other biological macromolecules. In contrast, the majority of soluble proteins are found in the cytoplasm or periplasm of an organism. 'Extracellular protein' refers to proteins generated outside of cells. Given the varied roles and structures of proteins, only rational combinations of several analytical techniques can yield a thorough comprehension of a cell's true state. The dimensions of the microbial genome, the number of proteins in the cell, and the effectiveness of its response to environmental changes are the only criteria for evaluating the proteome complexity of a microbial community. In recent decades, dominant proteomics techniques have depended on the cost-effectiveness and accessibility of mass spectrometers, as well as the availability and reliability of procedures (Sankpal, 2022). The proteomics study offers an extensive insight into the protein composition of organisms. The proteomics approach offers an extensive insight into the protein composition of biological systems. When combined with other metagenomic techniques, it can clarify the molecular mechanisms driving these biological activities, hence aiding in the identification of innovative, environmentally friendly solutions. Examination of protein sequences from bacteria in in vitro cultures and ambient samples demonstrates discrepancies in protein production and composition (Lacerda and Reardon, 2009). We may also identify proteins that provide critical functions and responses to pollutants in microbial strains (Dangi et al., 2019). The analysis of membrane proteins is crucial in bioremediation. PAHs, widespread environmental pollutants, must be eradicated from the environment through the use of both natural and genetically modified microorganisms. Specific PAHs have been remediated both in situ and ex situ. Physiological changes in an organism during bioremediation can elucidate bioremediation-related genes and their regulatory mechanisms (Hivrale et al., 2015). Employing proteomics technology, researchers have discovered previously unidentified genes and proteins implicated in the degradation of toluene and ethylbenzene in anaerobic conditions. A multitude of genes and proteins related to toluene are expressed during its anaerobic biodegradation. Additionally, two operons related to toluene were notably active in the transformed cells. In a study of the ethylbenzene pathway, Ebd protein was produced in cells exposed to ethylbenzene, but not in those treated with acetophenone. In contrast, the APC protein was synthesised in both instances (Kühner et al., 2005). The proteomics methodology revealed new pathways for aerobic and anaerobic biodegradation of toxic waste, offering useful insights for the discovery of novel proteins. The majority of regulated proteins are involved in several categories, such as oxidative stress response, general stress response, energy metabolism, transcription regulation, transport molecules, and nucleotide biosynthesis (Santos et al., 2004). *Pycnoporus sanguineus* is a fungus exhibiting a potent and promising metabolic route for the bioenhancement and biodegradation of triphenyl phosphate (TPhP)-contaminated aquatic sediment. This is essential for enhancing the utilisation of white rot fungi in the bioremediation of TPhP contamination. It can degrade TPhP under conditions favourable to fungal growth. Proteomic analysis reveals the participation of aromatic molecule dioxygenase, enzymes that produce oxidative species, methyltransferase hydroxylase, and methylating agents in the metabolism of TPhP (Feng et al., 2021). Metaproteomics has been utilised to clarify protein expression in ambient microbial communities without the need for laboratory cultivation. This method illustrates the cooperative operation of microorganisms within a designated environment (Herbst et al., 2016). The amalgamation of proteomics techniques with developments in protein structure databases augments the efficacy of this approach. Metaproteomics analyses encounter hurdles like the varied distribution of species, substantial heterogeneity in protein expression levels among microorganisms, and considerable genetic diversity within microbial communities. Notwithstanding these challenges, metaproteomics holds considerable potential to link the genetic diversity and activities of microbial communities to their influence on ecosystem dynamics (Bharagava et al., 2019). Genomics can ascertain a yeast strain's potential for oil degradation, but it cannot verify the activity of those genes in field circumstances. Proteomics elucidates the functional phenotype by identifying the enzymes produced in response to contaminants. This is crucial as mRNA levels frequently do not align with protein abundance due to post-translational controls (Vido et al., 2001). The two-dimensional electrophoresis method is usually employed to separate proteins based on isoelectric point and molecular weight. It is proficient at viewing isoforms but constrained in identifying low-abundance or extremely hydrophobic membrane proteins (Santos et al., 2004). A method called shotgun proteomics (LC-MS/MS) cleaves whole protein mixtures into peptides. It provides enhanced throughput and sensitivity, enabling the detection of thousands of proteins in a single analysis, rendering it suitable for intricate soil-yeast systems (Sabirova et al., 2011). Yeasts play a significant role in the remediation of environmental pollutants (Fig. 7.3).

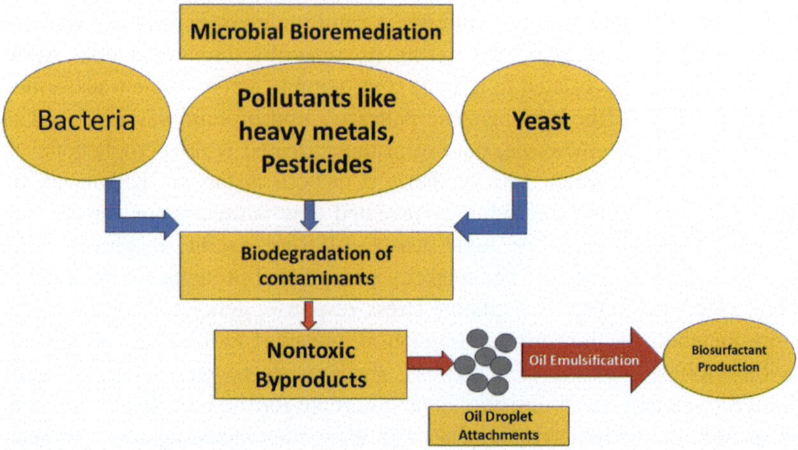

FIGURE 7.3 **Bioremediation process at a glance by.** The figure demonstrates the role of yeast in the bioremediation of environmental pollutants.

7.6 Key strategies in proteomic bioremediation

The incorporation of proteomics into environmental science entails many strategic procedures aimed at enhancing contaminant removal.

7.6.1 Identification and extraction of innovative biocatalysts

The most effective technique entails shotgun proteomics to identify enzymes capable of degrading intricate man-made compounds. Researchers can utilise mass spectrometry to discover specific proteins that are "upregulated" (made in greater numbers) in specialised microbial strains exposed to contaminants such as crude oil or pesticides to mitigate stress. The impact of this process is the isolation of uncommon enzymes, including dioxygenases and hydrolases, which can then be mass-produced or utilised to develop "superdegraders" specifically designed for contaminated locations (Yang and Ding, 2014).

7.6.2 Metaproteomic: administration of microbial consortia

Ecological sustainability frequently relies on collaboration. Metaproteomics is the examination of the complete protein repertoire from a diverse assemblage of microorganisms (a consortium) present in soil or water (Nieto et al., 2023).

7.6.2.1 In situ insight

It uncovers "metabolic cross-talk", when one species degrades a toxin into a simpler intermediary, subsequently completed by a second species.

7.6.2.2 Optimisation

By comprehending these collaborations, researchers can engineer synthetic microbial consortia that exhibit greater stability and efficiency compared to single-strain cultures, thereby safeguarding the enduring health of the treated ecosystem.

7.6.3 Engineering resilience via stress response profiling

A significant obstacle in bioremediation is that contaminants frequently exterminate the microorganisms designated for their degradation. A fundamental proteomic approach involves the profiling of stress-response proteins, including:

Efflux pumps: Proteins that expel harmful heavy metals (e.g., Lead or Cadmium) before cellular injury.

Chaperones: Proteins that rectify other impaired proteins during toxic stress. By understanding these survival mechanisms, researchers can choose or create strains that effectively break down contaminants while enduring the most extreme industrial conditions, hence enhancing the reliability of the cleanup process (Misu et al., 2025).

7.6.4 Real-time proteomic fingerprinting

To ensure sustainability, we must monitor the repair process without disturbing the site. Targeted proteomics serves as a diagnostic tool. Instead of merely measuring chemical quantities, researchers aim to identify "proteomic fingerprints"—unique biomarker proteins that appear solely during the active breakdown of a pollutant.

Enduring competitive advantage: This enables real-time adjustments, including the addition of specific nutrients or biostimulants, to sustain the microbial population at peak efficiency, thus preventing the stagnation of the remediation process. Real-time monitoring through protein biomarkers involves proteomics, enabling researchers to find "reporter proteins" that indicate ongoing degradation. The presence of elevated alkane hydroxylase proteins in a soil sample indicates that the yeast population is actively metabolising the oil, offering a more precise measure than mere microbial numbers. Soils polluted with oil frequently contain heavy metals. Proteomics facilitates the identification of yeast proteins associated with metal chelation (metallothioneins), elucidating the survival strategies yeasts employ in response to dual toxicity (Vido et al., 2001).

7.6.5 Incorporating the "multiomics" pipeline

The fundamental strategy for an ecologically sustainable environment is the amalgamation of proteomics with genomics (potential outcomes) and metabolomics (current conditions). This comprehensive perspective guarantees that the by-products of bioremediation are genuinely non-toxic. For example, proteomics verifies enzyme activity, whereas metabolomics ascertains that the final product is a benign chemical such as water or CO_2 (Bastas and Kumar, 2023).

7.7 Proteomic mapping of yeast metabolic pathways under oil stress

7.7.1 Activation of the cytochrome P450 monooxygenase system

The initial phase of alkane degradation involves the terminal hydroxylation of the hydrocarbon chain. Proteomic analyses in *Y. lipolytica* demonstrate a significant elevation of alkane-induced cytochrome P450 (CYP52 family) and their associated reductases during exposure to crude oil (Beopoulos et al., 2009).

7.7.2 Regulation of peroxisomal β-oxidation and glyoxylate cycle

Upon oxidation to fatty acids, hydrocarbons are transported into the peroxisome. Proteomics indicates elevated levels of Acyl-CoA oxidase and glyoxylate cycle enzymes (isocitrate lyase and malate synthase), which circumvent the CO$_2$-releasing phases of the TCA cycle to preserve carbon for biomass synthesis (Bankar et al., 2009).

7.7.3 Proteomic signatures of oxidative stress and membrane adaptation

Exposure to hydrocarbons generates reactive oxygen species. Yeast proteomes react by upregulating superoxide dismutase and catalase. Additionally, proteins associated with ergosterol production are regulated to preserve membrane fluidity in response to the solvent effects of oil (Santos et al., 2004).

7.8 The future of proteomics-driven bioremediation, challenges, ethics, and future frontiers

The future of proteomics-based bioremediation depends on transitioning from descriptive research to precise environmental engineering. By pinpointing the precise proteins and enzymes (the "catalytic units") involved in pollutant degradation, researchers can advance from generic bioaugmentation to tailored metabolic optimisation. The primary challenge is moving from "lab-pure" cultures to "dirty" field proteomics. Proteogenomics—the integration of genomic and proteomic data is essential to assign functions to the many "hypothetical proteins" found in wild yeast strains (Sabirova et al., 2011). The release of genetically modified yeast (GMOs) requires strict ethical and regulatory oversight. Proteomics serves as a safety tool to ensure that engineered yeasts do not produce unexpected, toxic secondary metabolites in the environment (Khan et al., 2025).

7.8.1 Artificial intelligence integration

Artificial intelligence and machine learning are utilised to forecast protein activities and simulate degradation kinetics, facilitating the "rational design" of resilient, field-deployable microorganisms.

7.8.2 Metaproteomic

This technology facilitates the examination of complete microbial communities in situ, elucidating how complex ecosystems, rather than isolated strains, cooperate to detoxify heavy metals and organic contaminants.

7.8.3 Real-time monitoring

Future innovations in mass spectrometry and protein microarrays will enable instantaneous surveillance of wastewater and soil health, employing proteins as bioindicators for prompt intervention. This "multiomics" technique converts bioremediation from a trial-and-error methodology into a precise, environmentally acceptable instrument for sustainable restoration.

7.9 Conclusion

The use of yeast proteomic strategies signifies a clear transition from descriptive to functional environmental research in the fight against hydrocarbon pollution. Although conventional techniques and genomic mapping offer the foundational "blueprints" for possible repair, proteomics supplies the essential "operational data" necessary for the effective management of oil-contaminated soils. The efficacy of bioremediation in contemporary times relies on our capacity to elucidate the intricate metabolic interactions between hydrocarbonoclastic yeasts and their progressively hazardous surroundings. This chapter has illustrated that yeasts like *Yarrowia lipolytica* and *Candida* sp. exhibit distinctive enzymatic resilience, notably via the activation of Cytochrome P450 systems and specialized beta oxidation pathways, enabling them to surpass bacterial counterparts in acidic or saline oil-stressed soils. By utilizing modern LC-MS/MS proteomic methodologies, researchers may now pinpoint specific protein biomarkers indicative of ongoing degradation in real time. This skill converts bioremediation from an unpredictable "trial-and-error" approach into a well-designed method. The shift from laboratory-scale proteomics to field-scale application necessitates a multidisciplinary strategy that integrates bioinformatics, geochemistry, and synthetic biology. As we approach 2030, the objective is not just to bring microbes to a location, but to monitor and manage their proteome flux to guarantee the complete mineralization of xenobiotics. The integration of recombinant DNA technology and proteome validation provides a sustainable and scientifically sound framework for rehabilitating oil-affected terrestrial ecosystems. The future of environmental management depends on our capacity to interpret the "protein language" of this extraordinary fungus, transforming industrial waste into ecological vitality.

References

Abou Seeda, M.A., Yassen, A.A., Abou El-Nour, E.Z.A.A., 2017. Microorganism as a tool of bioremediation technology for cleaning waste and industrial water. Bioscience Research 14 (3), 633–644. Available at http://www.isisn.org/BR-14-2017/633-644-14(3)2017BR-1506.pdf.

Adessi, C., Matton, G., Ayala, G., Turcatti, G., Mermod, J.J., Mayer, P., et al., 2000. Solid phase DNA amplification: characterisation of primer attachment and amplification mechanisms. Nucleic Acids Research 28 (20), 87. Available at https://doi.org/10.1093/nar/28.20.e87.

Aebersold, R., Mann, M., 2003. Mass spectrometry-based proteomics. Nature 422 (6928), 198–207. Available at https://doi.org/10.1038/nature01511.

Aitken, A., Learmonth, M., 2002. Protein identification by in-gel digestion and mass spectrometric analysis. Molecular Biotechnology 20 (1), 095–098. Available at https://doi.org/10.1385/MB:20:1:095.

Amaral, P.F., et al., 2010. Yarrowia lipolytica and its biotechnological applications. Critical Reviews in Biotechnology.

Ameen, F., et al., 2021. Proximate and ultimate analysis of petroleum hydrocarbon degradation by tropical yeast species. Journal of King Saud University - Science.

Anderson, N.L., Anderson, N.G., 1998. Proteome and proteomics: new technologies, new concepts, and new words. Electrophoresis 19 (11), 1853–1861. Available at https://doi.org/10.1002/elps.1150191103, http://onlinelibrary.wiley.com/journal/10.1002/(ISSN)1522-2683.

Bankar, A.V., Kumar, A.R., Zinjarde, S.S., 2009. Environmental and industrial applications of Yarrowia lipolytica. Applied Microbiology and Biotechnology 84 (5), 847–865. Available at https://doi.org/10.1007/s00253-009-2156-8.

Barupal, T., Meena, M., Sharma, K., 2019. Inhibitory effects of leaf extract of Lawsonia inermis on Curvularia lunata and characterization of novel inhibitory compounds by GC–MS analysis. Biotechnology Reports 23, e00335. Available at https://doi.org/10.1016/j.btre.2019.e00335.

Barupal, T., Chittora, D., Meena, M., 2019b. Solid waste: characterization, assessment, monitoring, and remediation. Advances in Waste-to-Energy Technologies. CRC Press, pp. 1–19.

Beopoulos, A., Cescut, J., Haddouche, R., Uribelarrea, J.-L., Molina-Jouve, C., Nicaud, J.-M., 2009. Yarrowia lipolytica as a model for bio-oil production. Progress in Lipid Research 48 (6), 375–387. Available at https://doi.org/10.1016/j.plipres.2009.08.005.

Bharagava, R.N., Saxena, G., Purchase, D., Mulla, S.I., 2019. Applications of Metagenomics in Microbial Bioremediation of Pollutants Microbial Diversity in the Genomic Era. Elsevier, India, pp. 459–477. Available at https://www.sciencedirect.com/book/9780128148495, 10.1016/B978-0-12-814849-5.00026-5.

Dangi, A.K., Sharma, B., Hill, R.T., Shukla, P., 2019. Bioremediation through microbes: systems biology and metabolic engineering approach. Critical Reviews in Biotechnology 39 (1), 79–98. Available at https://doi.org/10.1080/07388551.2018.1500997, http://www.tandfonline.com/loi/ibty20.

Fontes, G.C., et al., 2008. Production of biosurfactant by Yarrowia lipolytica. Brazilian Archives of Biology and Technology.

Griggs, D., Stafford-Smith, M., Gaffney, O., Rockström, J., Öhman, M.C., Shyamsundar, P., et al., 2013. Policy: Sustainable development goals for people and planet. Nature 495 (7441), 305–307. Available at https://doi.org/10.1038/495305a.

Hamid, B., Rana, R.S., Chauhan, D., Singh, P., Mohiddin, F.A., Sahay, S., 2014. Psychrophilic yeasts and their biotechnological applications: a review. African Journal of Biotechnology 13 (22), 2188–2197.

Hamid, B., Bashir, Z., Yatoo, A.M., Mohiddin, F., Majeed, N., Bansal, M., et al., 2022. Cold-active enzymes and their potential industrial applications—a review. Molecules (Basel, Switzerland) 27 (18). Available at https://doi.org/10.3390/molecules27185885, http://www.mdpi.com/journal/molecules.

Hamid, B., Majeed, N., Ganai, B.A., Hassan, S., Bashir, Z., Wani, P.A., et al., 2023. Heavy-metal tolerant bacterial strains isolated from industrial sites and scrap yards in Kashmir, India. Journal of Basic Microbiology 63 (12), 1361–1372. Available at https://doi.org/10.1002/jobm.202300349, http://onlinelibrary.wiley.com/journal/10.1002/(ISSN)1521-4028.

Herbst, F.A., Lünsmann, V., Kjeldal, H., Jehmlich, N., Tholey, A., von Bergen, M., et al., 2016. Enhancing metaproteomics the value of models and defined environmental microbial systems. Proteomics 16 (5), 783–798. Available at https://doi.org/10.1002/pmic.201500305, http://onlinelibrary.wiley.com/journal/10.1002/(ISSN)1615-9861.

Hivrale, A.U., Pawar, P.K., Rane, N.R., Govindwar, S.P., 2015. Application of genomics and proteomics in bioremediation. Toxicity and Waste Management Using Bioremediation. IGI Global, India, pp. 97–112. Available at http://www.igi-global.com/chapter/application-of-genomics-and-proteomics-in-bioremediation/141795 10.4018/978-1-4666-9734-8.ch005.

Hood, L., Rowen, L., 2013. The human genome project: big science transforms biology and medicine. Genome Medicine 5 (9), 79. Available at https://doi.org/10.1186/gm483.

James, P., 1997. Protein identification in the post-genome era: the rapid rise of proteomics. Quarterly Reviews of Biophysics 30 (4), 279–331. Available at https://doi.org/10.1017/s0033583597003399.

Khan, M., Khan, T., Tabassum, B., Hashim, M., 2025. Advancement in genomics and proteomics to enhance bioremediation strategies for environmental restoration. Sustainable Green Technologies: Bioremediation of Environmental Pollution. Springer Nature, Switzerland, pp. 275–295.

Kuddus, M., Roohi, Bano, N., Sheik, G.B., Joseph, B., Hamid, B., et al., 2024. Cold-active microbial enzymes and their biotechnological applications. Microbial Biotechnology 17 (4). Available at https://doi.org/10.1111/1751-7915.14467, http://onlinelibrary.wiley.com/journal/10.1111/(ISSN)1751-7915.

Kühner, S., Wöhlbrand, L., Fritz, I., Wruck, W., Hultschig, C., Hufnagel, P., et al., 2005. Substrate-dependent regulation of anaerobic degradation pathways for toluene and ethylbenzene in a denitrifying bacterium, strain EbN1. Journal of Bacteriology 187 (4), 1493–1503. Available at https://doi.org/10.1128/jb.187.4.1493-1503.2005.

Kumar, V., Sharma, J.K., Kumar, S., Gulia, J., Singh, N.P., Santal, A.R., 2024. Metagenomic and proteomic approach for bioremediation of environmental pollutants. Development in Waste Water Treatment Research and Processes: Applied Technologies for Clean up of

Environmental Contaminants. Elsevier, India, pp. 635–647. Available at https://www.sciencedirect.com/book/9780443136153 10.1016/B978-0-443-13615-3.00020-0.

Lacerda, C.M.R., Reardon, K.F., 2009. Environmental proteomics: applications of proteome profiling in environmental microbiology and biotechnology. Briefings in Functional Genomics and Proteomics 8 (1), 75–87. Available at https://doi.org/10.1093/bfgp/elp005.

Lovley, D.R., 2003. Cleaning up with genomics: applying molecular biology to bioremediation. Nature Reviews Microbiology 1 (1), 35–44. Available at https://doi.org/10.1038/nrmicro731.

Malla, M.A., Dubey, A., Yadav, S., Kumar, A., Hashem, A., Abd-Allah, E.F., 2018. Understanding and designing the strategies for the microbe-mediated remediation of environmental contaminants using omics approaches. Frontiers in Microbiology 9. Available at https://doi.org/10.3389/fmicb.2018.01132, https://www.frontiersin.org/articles/10.3389/fmicb.2018.01132/full.

Margesin, R., Schinner, F., 2001. Biodegradation and bioremediation of hydrocarbons in extreme environments. Applied Microbiology and Biotechnology 56 (5-6), 650–663. Available at https://doi.org/10.1007/s002530100701.

Misu, I.J., Kayess, M.O., Siddiqui, M.N., Gupta, D.R., Islam, M.N., Islam, T., 2025. Microbiome engineering for sustainable rice production: strategies for biofertilization, stress tolerance, and climate resilience. Microorganisms 13 (2). Available at https://doi.org/10.3390/microorganisms13020233, www.mdpi.com/journal/microorganisms.

Mohiddin, F.A., Hamid, B., 2018. Cold-active yeast proteases as a versatile tool for industry: a review. SKUAST Journal of Research 20 (2), 127–134.

Nieto, E.E., Macchi, M., Valacco, M.P., Festa, S., Morelli, I.S., Coppotelli, B.M., 2023. Metaproteomic and gene expression analysis of interspecies interactions in a PAH-degrading synthetic microbial consortium constructed with the key microbes of a natural consortium. Biodegradation 34 (2), 181–197. Available at https://doi.org/10.1007/s10532-022-10012-3, https://www.springer.com/journal/10532.

Paoletti, A.C., Zybailov, B., Washburn, M.P., 2004. Principles and applications of multidimensional protein identification technology. Expert Review of Proteomics 1 (3), 275–282. Available at https://doi.org/10.1586/14789450.1.3.275.

Sabirova, J.S., et al., 2011. Proteomic insights into the hydrocarbon-degrading abilities of yeast species. Journal of Proteomics.

Sankpal, N., 2022. Importance of proteomics in bioremediation. International Journal for Research in Applied Science and Engineering Technology 10 (3), 650–653. Available at https://doi.org/10.22214/ijraset.2022.40706.

Santal, A.R., Singh, N.P., Saharan, B.S., 2011. Biodegradation and detoxification of melanoidin from distillery effluent using an aerobic bacterial strain SAG 5 of Alcaligenes faecalis. Journal of Hazardous Materials 193, 319–324. Available at https://doi.org/10.1016/j.jhazmat.2011.07.068.

Santal, A.R., Rani, R., Kumar, A., Sharma, J.K., Singh, N.P., 2022. Biodegradation and detoxification of textile dyes using a novel bacterium Bacillus sp. AS2 for sustainable environmental cleanup. Biocatalysis and Biotransformation 1–15. Available at https://doi.org/10.1080/10242422.2022.2136128.

Santal, A.R., et al., 2023. Genetic engineering of microbes for environmental cleanup. Journal of Environmental Management.

Santos, P.M., et al., 2004. Insights into the yeast response to organic solvents: a proteomics approach. Molecular & Cellular Proteomics.

Satyanarayana, T., Prakash, A., Johri, B.N., 2012. Microorganisms in environmental management: microbes and environment. Microorganisms in Environmental Management: Microbes and

Environment 1–819. Available at https://doi.org/10.1007/978-94-007-2229-3, http://doi.org/10.1007/978-94-007-2229-3.

Sharma, J.K., Kumar, N., Singh, N.P., Santal, A.R., 2023. Phytoremediation technologies and their mechanisms for removal of heavy metals from contaminated soil: an approach for a sustainable environment. Frontiers in Plant Science 14, 78.

Singh, O.V., Nagaraj, N.S., 2006. Transcriptomics, proteomics and interactomics: Unique approaches to track the insights of bioremediation. Briefings in Functional Genomics and Proteomics 4 (4), 355–362. Available at https://doi.org/10.1093/bfgp/eli006.

Singh, N.P., Santal, A.R., 2015. Phytoremediation of heavy metals: the use of green approaches to clean the environment. In: Ansari, A.A., Gill, S.S., Lanza, G.R., Newman, L. (Eds.), Phytoremediation: Management of Environmental Contaminants. Springer, Cham.

Smart, A., Smart, J., 2003. Urbanization and the global perspective. Annual Review of Anthropology 32 (1), 263–285. Available at https://doi.org/10.1146/annurev.anthro.32.061002.093445.

Tarfeen, N., Nisa, K.U., Hamid, B., Bashir, Z., Yatoo, A.M., Dar, M.A., et al., 2022. Microbial remediation: a promising tool for reclamation of contaminated sites with special emphasis on heavy metal and pesticide pollution: a review. Processes 10 (7). Available at https://doi.org/10.3390/pr10071358, http://www.mdpi.com/journal/processes.

Thakur, M., Medintz, I.L., Walper, S.A., 2019. Enzymatic bioremediation of organophosphate compounds—progress and remaining challenges. Frontiers in Bioengineering and Biotechnology 7. Available at https://doi.org/10.3389/fbioe.2019.00289, http://journal.frontiersin.org/journal/bioengineering-and-biotechnology#archive.

Wasinger, V.C., Cordwell, S.J., Poljak, A., Yan, J.X., Gooley, A.A., Wilkins, M.R., et al., 1995. Progress with gene-product mapping of the Mollicutes: Mycoplasma genitalium. Electrophoresis 16 (1), 1090–1094. Available at https://doi.org/10.1002/elps.11501601185.

Vido, K., et al., 2001. A proteome analysis of sulphur metabolism in Saccharomyces cerevisiae. Microbiology (Reading, England).

Yang, G., Ding, Y., 2014. Recent advances in biocatalyst discovery, development and applications. Bioorganic & Medicinal Chemistry 22 (20), 5604–5612. Available at https://doi.org/10.1016/j.bmc.2014.06.033.

Chapter 8

Proteomics as a versatile tool for addressing environmental pollution

Somna Oberoi[1], Balwinder Singh[1,2], and Amritpal Kaur[1]
[1]*Department of Food Science and Technology, Guru Nanak Dev University, Amritsar, Punjab, India,* [2]*P.G. Department of Botany, Khalsa College Amritsar, Amritsar, Punjab, India*

8.1 Introduction

Anthropogenic activities are major contributors to the release of heavy metals into aquatic ecosystems. Industrial operations, including metal smelting, petrochemical production, mining, and wastewater discharge, account for substantial inputs of these elements into natural waters (Vareda et al., 2019). Agricultural practices also represent an important source of contamination, as fertilizers and pesticides containing zinc, arsenic, and copper are transported into rivers and coastal systems through surface runoff (Angon et al., 2024). Furthermore, the combustion of fossil fuels, particularly in coal-fired power plants and vehicular emissions, releases airborne mercury and lead, which subsequently enter marine environments through atmospheric deposition processes (Maciejczyk et al., 2021). In addition to heavy metals, a wide range of emerging contaminants such as pharmaceuticals, pesticides, nanomaterials, cyanotoxins, microplastics and associated plasticizers, flame retardants, per- and polyfluoroalkyl substances (PFAS), and numerous other industrial chemicals and by-products are being continuously introduced into the environment as a result of agricultural intensification, industrial development, transportation, and other human activities as illustrated in Fig. 8.1(Guo et al., 2021; Meher and Zarouri, 2025). Collectively, they represent significant threats to ecological integrity and human health. Owing to their non-biodegradable nature and pronounced toxicity, many of these pollutants persist in the environment and exert deleterious effects even at low concentrations. Exposure to such contaminants has been shown to interfere with essential physiological and biochemical processes across diverse taxa (Saravanan et al., 2022).

Proteomics Driven Bioremediation. DOI: https://doi.org/10.1016/B978-0-443-45294-9.00019-9

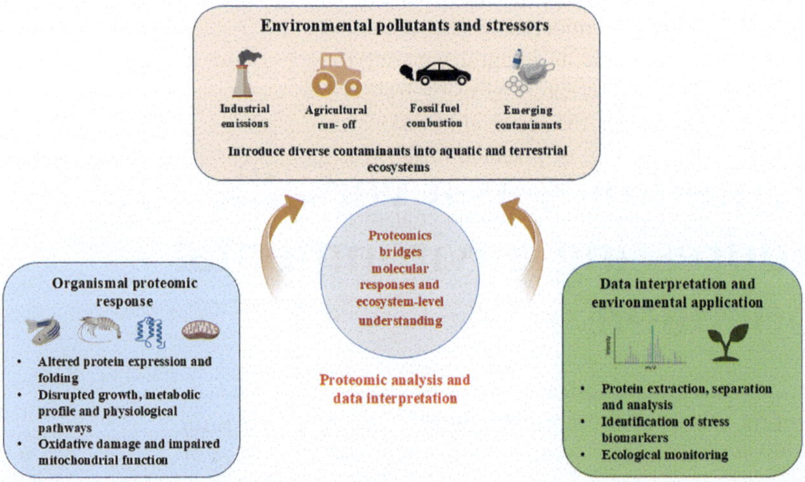

FIGURE 8.1 Overview of proteomics as a bridge between emerging contaminants and ecological monitoring.

For effective evaluation, systematic toxicity assessment methods are essential to identify the harmful effects of pollutants as well as their degradation by-products. Such evaluations also facilitate the comparison of treatment strategies aimed at reducing pollutant toxicity. In this context, proteomics-based approaches offer a powerful platform, providing scientific insights for early diagnosis of ecological risks and enabling the efficacy of pollution mitigation techniques, by examining alterations at the level of cellular metabolic pathways and functional protein interaction networks (Guo et al., 2021). As demonstrated in Fig. 8.1, proteomics serves as a bridge linking molecular-level responses to ecosystem-level understanding by elucidating how environmental pollutants trigger organismal proteomic responses and how this information can be applied in environmental monitoring, protection, and mitigation strategies. However, proteomics remains underutilized in eco-immunology despite its potential to enhance biomarker discovery, host–pathogen interaction studies, and environmental health assessments (Vicente-Santos et al., 2025).

8.2 Environmental pollution as a global challenge

Environmental pollution has emerged as one of the most pressing global challenges, given its widespread ubiquity, escalating economic costs, and profound health impacts. It is broadly defined as the introduction of harmful substances into the environment in gaseous, liquid, or solid forms (Saxena, 2025). Pollutants, on the other hand, are specifically characterized as substances that enter the environment in concentrations exceeding their natural background levels. When present in excess, these substances disrupt the delicate balance of ecosystems, leading to adverse consequences for both the environment and the species that rely upon it (Briffa et al., 2020). Human activities have markedly increased nitrogen deposition in terrestrial ecosystems beyond preindustrial times, triggering a chain of ecological consequences. Elevated nitrogen levels accelerate soil acidification and eutrophication, while altering species composition and leading to biodiversity losses (Clark et al., 2017). The anthropogenic activities, particularly industrialization, urbanization, and intensive agriculture, have significantly accelerated the release of chemical pollutants into the air, water, and soil (Weldeslassie et al., 2017). Recent estimates indicate that air pollution alone accounted for approximately 8.1 million fatalities worldwide in 2021, with over 700,000 deaths among children below 5 years of age (State of Global Air Report 2024, 2024). Similarly, exposure to hazardous chemicals such as lead, pesticides, and occupational carcinogens was responsible for 2 million deaths and 53 million disability-adjusted life years globally in 2019 (Chemical safety – Fact sheet: Exposure to selected chemicals, 2021). Apart from human health, pollutants such as nitrogen compounds, heavy metals, organic and inorganic chemicals, and gases result in long-lasting ecological effects, including biodiversity loss, reduced oxygen levels, soil and water acidification, destruction of habitat, and disruption of ecosystems (Clark et al., 2017; Saxena, 2025). Chemicals such as PFAS illustrate the persistent and bioaccumulative nature of many modern pollutants. Recent studies have shown that PFAS build up extensively in both terrestrial and marine organisms, raising concerns about their widespread ecological and health impacts (Khan et al., 2023; Gang et al., 2024).

Environmental pollution remains a truly global challenge, transcending geographical and economic boundaries by affecting both developed and developing nations alike. Given its far-reaching and long-term consequences, the scientific community continues to explore novel approaches aimed at lessening environmental harm, including reducing industrial and agricultural emissions, promoting cleaner production processes, and advancing the development of environmental-friendly alternatives to conventional chemicals and materials. Combating pollution effectively requires more than isolated interventions; coordinated global efforts, long-term strategies, and sustained investment in research and technology (Mondschein et al., 2021; Judge et al., 2022). Such measures are critical to safeguarding ecosystems and ensuring that future generations inherit a healthier and more sustainable planet.

8.2.1 Conventional approaches for pollution monitoring

The presence of both natural and synthetic toxic compounds in the environment necessitates the development of reliable detection and monitoring methods. Traditionally, pollution monitoring and toxicity screening have relied mainly on animal-based models, which have provided valuable insights into pollutant presence and their toxicological outcomes. While these approaches remain important, they are increasingly recognized as limited as they are time-consuming, costly, and raise ethical concerns regarding animal welfare. Moreover, such methods often fail to capture the mechanistic details of toxicity at the molecular and cellular levels (Hur et al., 2018). Conventional approaches for toxicity evaluation of compounds and their transformation products have primarily focused on phenotypic changes in target cells or organisms, which arise through complex multistep genetic processes and regulatory networks (Lundberg and Borner, 2019). However, the link between proteins and their regulated phenotypes often remains unclear, making it difficult for traditional methods to elucidate the molecular mechanisms underlying toxicity induced by pollutants or stressors. In some cases, these methods may even lead to incorrect conclusions, particularly when changes in gene expression do not translate into measurable phenotypic outcomes (Guo et al., 2021).

Biomonitoring of aquatic environments and the health of ecosystems is central to developing effective strategies for environmental protection, sustainable development, and human well-being. In the last few decades, pollution exposure biomarkers have been widely utilized to evaluate the health of organisms and, in turn, the ecological conditions of their habitats. However, the use of single biomarkers in response to biotic or abiotic stressors often suffers from limited specificity and sensitivity (Campos et al., 2012). Moreover, conventional biomarker-based monitoring approaches depend heavily on prior knowledge of pollutant mechanisms, thereby restricting assessments to a narrow set of well-characterized biomolecules while overlooking many others whose links to contamination remain poorly understood. Since environmental contaminants can elicit multiple and sometimes uncorrelated responses in organisms, biomonitoring often requires the combined use of several biomarkers. Despite this, the approach remains inherently biased and restricted in scope (Carmona et al., 2017).

Consequently, there is a growing demand to transition from traditional animal-based toxicity tests toward molecular-level approaches such as omics. This shift aligns with the Tox21 vision, emphasizing the development and application of high-throughput in vitro screening methods designed to enhance precision, efficiency, and reduce reliance on animal models (Jin et al., 2025). Within this framework, emerging omics technologies represent a powerful alternative, as they allow high-throughput and quantitative monitoring of a vast array of biomolecules simultaneously, thereby providing a more comprehensive understanding of biological responses to environmental contaminants (Carmona et al., 2017). The application of omics tools—including genomics, transcriptomics, proteomics, and metabolomics—has become increasingly significant in environmental management, where they serve not only to evaluate ecosystem health but also to reinforce risk assessment strategies

(Simões et al., 2018). Whereas genomics and transcriptomics reveal the genetic potential and transcriptional activity of cells, proteomics focuses on the actual functional machinery inside and outside the cell (Guo et al., 2025). While metabolomics emphasizes the identification, characterization, and quantification of metabolites arising from various biochemical pathways within living systems, proteomic analysis offers a more mechanistic understanding of organismal responses to environmental stressors. By examining alterations in protein abundance and post-translational modifications, proteomics provides valuable insights into the functional adjustments organisms undergo when exposed to environmental challenges (Diniz et al., 2017).

8.3 Proteomics: an overview

Proteins are highly dynamic molecules within an organism and often undergo post-transcriptional modifications, resulting in the formation of multiple isoforms after mRNA is transcribed from DNA. Unlike nucleotides, proteins display far greater diversity, with a single gene capable of generating more than 100 distinct protein variants. Owing to this diversity and functional significance, proteins have become valuable tools for monitoring and assessment across multiple fields, including medicine, forensics, aquaculture, agriculture, and environmental monitoring (Gajahin Gamage et al., 2022). Environmental sciences have greatly benefited from proteomics, as it enables a deeper exploration of ecosystems through the comprehensive study of interactions among multiple proteomes across species (Marshall et al., 2025). Toxicity in aquatic environments imposes significant stress on aquatic organisms, and these stress responses can be effectively detected and characterized through proteomic analyses (Gajahin Gamage et al., 2022).

Unlike conventional toxicity screening technologies, proteomics offers a sensitive approach for detecting sublethal effects of environmental contaminants by profiling protein expression, interactions, and posttranslational modifications. Recent research has demonstrated that the accumulation of human and veterinary pharmaceutical residues significantly alters protein expression profiles in aquatic phytoplankton, with advanced proteomic technologies offering detailed insights into these molecular perturbations (Chia et al., 2021). Through the identification of proteins involved in toxicological responses, it captures dynamic molecular changes that occur shortly after exposure, often preceding visible ecological effects (Li et al., 2021; Xue et al., 2023; Geng et al., 2024). Latest findings further demonstrate its utility, showing that proteomic alterations can be detected within hours of exposure, long before quantifiable reductions in organismal performance are observed (Guzmán-Fierro et al., 2024). By capturing these molecular responses, proteomic approaches serve as novel tools for systematically assessing the ecological risks posed by pollutants and pollutant-processing technologies. Such analyses extend beyond individual protein changes to the level of cellular metabolism and functional protein–protein interaction networks, offering a systems-level understanding of how contaminants disrupt biological pathways and compromise organismal health (Guo et al., 2021). A recent longitudinal nationwide study based on the UK Biobank cohort explored the protein-mediated effects of five major air pollutants on mental health, specifically depression and anxiety. The study identified 23 proteins playing a pivot role in mediating the risk of developing depression associated with air pollution exposure, while no mediating proteins were detected for anxiety onset. The findings also revealed 38 proteins associated with the severity of anxiety symptoms and nine proteins linked to the severity of depressive symptoms, providing robust proteomic evidence supporting the role of protein-mediated pathways in the relationship between air pollution and the development and progression of depression, thereby offering new molecular insights into the environmental roots of mental health (Wang et al., 2025). Faria et al. (2018) investigated the toxicity of acrylamide in zebrafish to better understand the mechanisms underlying its neurotoxic effects. Acrylamide is a widespread environmental contaminant, commonly generated as a by-product of food processing, water treatment, and various industrial manufacturing processes. Its ubiquity in both aquatic and terrestrial ecosystems raises significant concerns about potential risks to human and ecological health.

Proteomic investigations have revealed previously unrecognized cellular effects of environmental stressors, such as those induced by thermal shifts, anaerobic conditions, and osmotic imbalance. Advances in proteomic technologies have facilitated the identification of both broad and pollutant-specific biological markers, including those reflecting critical post-translational modifications. Moreover, investigations into proteomic alterations during early developmental stages of marine organisms highlight the importance of signaling pathways essential for survival and adaptation in rapidly changing environments. Commonly observed responses involve changes in protein regulating energy metabolism, cytoskeletal organization, protein stabilization and degradation, oxidative stress responses, and intracellular signaling (Tomanek, 2011). A study demonstrated that anthropogenic pressure markedly altered the proteome of *Daphnia pulex*, leading to the differential expression of 17 significantly upregulated and 181 downregulated proteins. Among the 198 altered proteins, only 13 could be fully characterized, providing evidence of strong human-induced impacts in Lianquihue compared to the relatively pristine conditions of Icalma. The pattern of up- and downregulation suggests pronounced cellular stress, with consequences for key physiological processes, particularly those linked to cellular metabolism and exoskeleton composition. These findings highlight the significance of integrating molecular–ecological approaches into environmental monitoring and management strategies and offer valuable insight into how anthropogenic stressors alter freshwater ecosystems (Norambuena et al., 2025).

Overall, proteomics represents a cost-effective and versatile approach that enables comprehensive data acquisition while relying on minimal invasive sampling, small sample volumes, and reduced use of experimental animals, whichare especially advantageous when studying vulnerable or underrepresented species (Vicente-Santos et al., 2025).

8.4 Approaches and techniques in proteomics

8.4.1 Gel-based approaches

Gel-based techniques represent some of the earliest approaches in proteomics, valued for their ability to resolve differentially expressed proteins with relatively high resolution while requiring only basic instrumentation and minimal technical expertise (Henke et al., 2024). Conventional two-dimensional electrophoresis (2-DE) separates proteins by isoelectric point (pI) in the first dimension, and by molecular weight using sodium dodecyl sulfate–polyacrylamide gel electrophoresis (SDS-PAGE) in the second (Alharbi, 2020). Proteins are typically visualized using binding dyes such as Coomassie Brilliant Blue or silver staining, which allows for wide protein detection within gels. Resolved proteins can also be transferred for sequencing by Edman degradation, or excised to nitrocellulose membranes for immunodetection utilizing techniques such as Western blotting. Despite being valuable, this approach is limited by variability in electrophoretic resolution, making comparative analyses and accurate quantification challenging (Marshall et al., 2025). To address these constraints, a new approach, two-dimensional difference gel electrophoresis (2D-DIGE), continues to hold value for resolving complex protein mixtures. This technique enables the direct visualization of hundreds to thousands of protein species within a single gel, offering critical insights into alterations in isoelectric point and molecular weight. Such changes may result from processes including protein truncation, degradation, genetic variations, alternative splicing, post-translational processing, and diverse post-translational modifications (Arentz et al., 2015). The 2D-DIGE approach optimizes the 2-DE workflow by labeling protein samples with distinct fluorescent dyes before electrophoresis, allowing simultaneous resolution and comparison of multiple samples on the same gel without changing the intrinsic pI or MW of proteins (Alharbi, 2020). By lowering inter-gel variation, 2D-DIGE enhances both the accuracy and reproducibility of quantification, making it particularly useful in comparative environmental proteomics (López-Barea and Abril, 2017). Although its use in large-scale field-based environmental monitoring remains limited due to high costs (Campos et al., 2012).

To improve analytical efficiency and broaden its applicability in ecotoxicological research, researchers frequently employ 2-DE followed by mass spectrometry (MS)-based protein identification to achieve precise characterization of differentially expressed proteins (Tubatsi et al., 2022). For example, red swamp crayfish (*Procambarus clarkii*) collected from aquatic ecosystems in Doñana National Park and surrounding areas with different levels of pollution were subjected to a proteome analysis using a 2D-DIGE + MS technique. The study demonstrated that *P. clarkii* accumulates metals primarily in the digestive glands and gills, indicative of the degree of sediment contamination levels. Moreover, the species exhibited a metabolic shift towards increased aerobic glycolysis, presumably to sustain ATP synthesis and supply reducing equivalents under oxidative stress conditions attributed to impaired mitochondrial function. The proteins identified as differentially expressed reveal the physiological functions disrupted by pollutants and offer insight into the molecular mechanisms underlying the toxicity of pollutants. Notably, these proteins hold potential as biomarkers of exposure for environmental risk assessment (Fernández-Cisnal et al., 2017). Overall, gel-based techniques alone suffer from limited reproducibility and are less effective for insoluble proteins or those expressed at low abundance, as spot detection largely depends on protein concentration and staining sensitivity (Henke et al., 2024).

8.4.2 Mass spectrometry-based proteomics

Mass spectrometry (gel-free technique) has emerged as an indispensable analytical platform across diverse scientific disciplines due to its precision, sensitivity, and capacity for structural elucidation (Singh et al., 2025). With the growing complexity of modern scientific challenges, ranging from unraveling the intricacies of biological systems to detecting trace levels of contaminants in environmental matrices, the role of versatile and robust techniques such as MS has become increasingly vital (Adams and Adriaens, 2021). Unlike conventional spectroscopic approaches, which depend on the interaction of matter with electromagnetic radiation, MS generates direct, high-resolution data on the molecular composition of complex mixtures, often enabling the detection and characterization of analytes present at trace concentrations (Chen et al., 2021).

In proteomics, MS offers unparalleled insights into the composition, structure, function, and regulation of the proteome (Aebersold and Mann, 2016). By accurately measuring the mass-to-charge (m/z) ratio of ionized peptides, MS thereby facilitates precise identification of proteins along with their post-translational modifications (Jin et al., 2025). Unlike traditional targeted biochemical methods, MS-based proteomics directly identifies and quantifies amino acid sequences, enabling an unbiased and systematic analysis of protein expression, modification, and interaction profiles across different biological and environmental contexts (Aebersold and Mann, 2016). In the field of environmental sciences, MS has emerged as a cornerstone tool for the detection and quantification of emerging contaminants. Its exceptional selectivity and adaptability make it highly effective in monitoring trace pollutants across water, soil, and air (Meher and Zarouri, 2025). By enabling the detection of pollutants and their metabolites at ultratrace levels, MS enables precise identification and quantification of contaminants such as radioactive isotopes, industrial toxins, and emerging pollutants. This analytical accuracy underpins its vital role in environmental monitoring, regulatory compliance, and safeguarding ecosystem and public health (Tamara et al., 2022).

The general workflow of MS-based proteomics typically follows a bottom-up approach, as shown in Fig. 8.2. Proteins are first extracted from the biological sample and subsequently digested into smaller fragments (known as peptides) through the action of one or more enzymes. These peptides, typically 7–30 amino acids in length, are separated by chromatography and ionized via electrospray before introduction into the mass spectrometer. In data-dependent acquisition (DDA) mode, peptides are first detected and then selectively fragmented, enabling sequence information, which is later decoded through bioinformatic analysis (Shuken, 2023). Despite being commonly used for peptide sequencing, DDA can be constrained by stochastic sampling, which limits coverage and reproducibility in intricate samples. Alternative acquisition techniques, such as data-independent acquisition (DIA), have been developed to overcome these obstacles. DIA is especially helpful for examining proteome changes in response to altering ecological and environmental conditions, as it renders high-throughput, comprehensive, and reproducible proteomic investigations feasible (Krasny and Huang, 2021; Root et al., 2021). In contrast, top-down proteomics involves the direct loading of intact proteins into the mass spectrometer, allowing in-depth examination of posttranslational modifications while circumventing various inference-related challenges inherent to

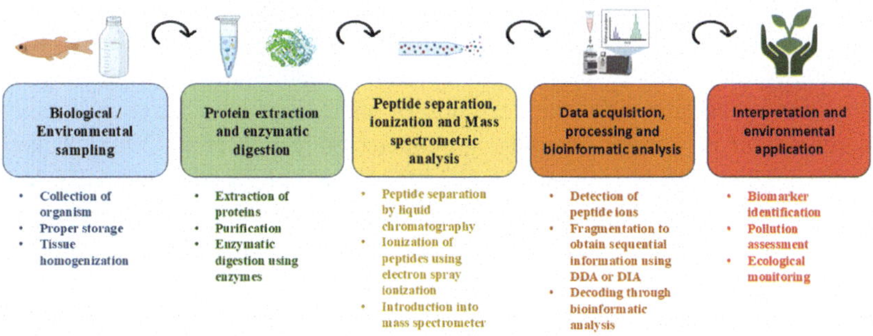

FIGURE 8.2 Typical workflow of environmental proteomics, illustrating the sequential stages from organism sampling and protein extraction to proteomic analysis, bioinformatics interpretation, and environmental sampling. *DDA*, Data-dependent acquisition; *DIA*, Data-independent acquisition.

bottom-up proteomics. This approach provides the advantage of achieving total sequence coverage and allows for the accurate characterization of proteoforms—the distinct molecular variants of a protein resulting from genetic variation, alternative splicing, and diverse post-translational modifications (Catherman et al., 2014).

Multiple MS-based techniques have been employed in environmental research, each offering distinct advantages tailored to the physicochemical properties of different pollutants. Gas chromatography–mass spectrometry (GC-MS) remains one of the most established methods for assessing volatile and semi-volatile organic contaminants. Separating analytes based on their volatility before they enter the MS detector, it allows high accuracy identification and quantification of chemical compounds such as pesticides, solvents, and industrial chemicals. This ability has rendered it a desirable tool for measuring air quality and identifying soil pollution. In contrast, liquid chromatography–mass spectrometry (LC-MS) is particularly advantageous for the analysis of nonvolatile, polar, or thermally unstable compounds that cannot be readily explored by GC-MS. By exploiting differences in solubility across various solvent systems, LC-MS is capable of handling complex environmental matrices such as wastewater or surface water. Its sensitivity makes it particularly well-suited for detecting pollutants present in very low concentrations in aquatic systems. For pollutants that are volatile and vulnerable to rapid fluctuations in concentration, real-time monitoring techniques are beneficial. Proton transfer reaction mass spectrometry (PTR-MS) provides direct detection of volatile organic compounds in air without the requirement for prior chromatographic separation. The rapid, high-throughput performance allows continuous monitoring of volatile compounds, making it useful for studying emissions from industrial processes, traffic, and other dynamic sources (Meher and Zarouri, 2025). A related method, selected ion flow tube mass spectrometry (SIFT-MS), also enables direct and real-time detection of trace gases through controlled chemical ionization instead of chromatographic separation to identify compounds with high sensitivity and specificity, making it particularly well-suited for the detection of short-lived or reactive pollutants in ambient air that might otherwise go undetected (Smith et al., 2025). Taken together, the analytical strengths of GC-MS, LC-MS, PTR-MS, and SIFT-MS are complementary. Upon collective application, these techniques enable researchers to document the diversity of emerging contaminants across air, water, and soil systems. Beyond simply detecting pollutants, they also assist in the determination of trace contamination pathways, assessing ecological risks, and providing information for remediation efforts (Meher and Zarouri, 2025).

8.4.3 Array-based approaches

The array-based approaches, particularly protein microarrays (often referred to as protein chips), have emerged as valuable tools in proteomics for high-throughput screening and functional characterization of proteins. Unlike gel- or mass spectrometry-based techniques, protein arrays enable the simultaneous analysis of thousands of proteins or peptides immobilized on a chip surface (Jin et al., 2025). Protein microarrays are broadly categorized into three types: analytical protein arrays, functional protein arrays, and reverse-phase protein arrays. Analytical arrays are mainly designed for the detection and quantification of proteins or antigens. Typically, antibodies are immobilized onto glass slides and probed with target analytes, facilitating applications such as protein expression profiling, biomarker identification, cell surface marker detection, and glycosylation analysis (Atak et al., 2016). Leveraging the high specificity of antigen–antibody interactions, these arrays serve as sensitive platforms for quantifying protein abundance and investigating binding affinities (Jin et al., 2025). Functional protein microarrays are typically constructed using full-length proteins or functional domains, making them highly effective for large-scale investigations of protein activities and interactions. By immobilizing purified proteins derived from an organism, these arrays enable systematic analyses of protein–protein interactions, enzymatic activities, and immune responses. In addition to interaction mapping, they offer valuable insights into biochemical pathways disrupted by toxicants, thereby advancing the understanding of molecular mechanisms underlying environmental and biological stress conditions (Zhu et al., 2022). Reverse-phase protein microarrays (RPPAs) differ from conventional protein arrays by immobilizing proteins directly from biological samples, such as cell or tissue lysates, onto the array surface. Detection is then achieved through the application of specific antibodies, enabling high-throughput quantification of protein expression across a large number of samples. This format is particularly advantageous for comparative analyses, as it allows the simultaneous assessment of proteomic alterations between different experimental groups, including those influenced by environmental exposures or pathological conditions (Coarfa et al., 2021).

8.5 Proteomics approaches in environmental biomonitoring: case studies

Proteomics offers a robust framework for understanding organismal and ecosystem responses to environmental stressors at the molecular level. By evaluating the effects of pollutants on protein expression, structure, and function, proteomic analyses enable early diagnosis of ecological risks and assessment of remediation effectiveness. These techniques also provide insights into the evolutionary adaptations of species under prolonged environmental stress and help elucidate the complex interactions between pollutants and biomolecules. Furthermore, systematic screening of functional protein biomarkers allows for the identification of biochemical pathways involved in pollutant accumulation, transport, and transformation, thereby enhancing our capacity to monitor and predict environmental health (Guo et al., 2021). A comparative summary of representative proteomic approaches linking environmental pollutants to molecular stress responses in diverse model organisms investigated is presented in Table 8.1.

Kwan et al. (2019) conducted an in situ deep-sea proteomic investigation on the amphipod *Abyssorchomene distinctus* to assess copper-induced contamination in the South China Sea. Using tandem mass tag (TMT) labeling combined with LC-MS/MS, the research identified 2937 proteins, of which 1918 were reliably quantified. Comparative proteomic analysis revealed significant changes in proteins associated with ion regulation (e.g., Na^+/K^+-ATPase), cuticle and exoskeleton remodeling, and chitin metabolism. Conducting experiments directly under natural deep-sea pressure and temperature regimes ensured ecological relevance. The findings demonstrate that proteomics can detect subtle yet biologically relevant stress responses in organisms within their native habitats, underscoring its value in environmental monitoring.

Similarly, Bjørnstad et al. (2006) investigated the applicability of proteomics in ecotoxicoproteomics research using the filter-feeding mussel *Mytilus edulis*. Mussels were exposed to a three-week continuous exposure to oil, or to oil supplemented with alkylphenols and additional polycyclic aromatic hydrocarbons. The plasma protein expression profiles were analyzed using protein chip array technology coupled with surface-enhanced laser desorption/ionization time-of-flight mass spectrometry. The spiked oil treatment produced stronger proteomic alterations than oil alone, with 83 protein peaks significantly affected compared to 49 peaks under crude oil alone. The majority of the peaks were upregulated, and shared expression changes were more pronounced under the spiked oil exposure. This investigation demonstrated the potential of proteomic profiling for detecting pollutant-induced molecular perturbations, supporting its utility in environmental risk assessment.

Further evidence supporting the value of proteomics in pollution assessment was provided by Rivaldi et al. (2025), who examined *Sinanodonta woodiana* as a sentinel species for evaluating biochemical alterations along pollution gradients in the Winongo stream. By examining sites with low, moderate, and high contamination, distinct metabolomic and proteomic shifts were observed. Proteomic profiling via LC-MS revealed a progressive increase in β-actin protein coverage from 17% in unexposed mussels to 23% and 34% under moderate and high pollution levels, respectively, indicating enhanced cellular stress and cytoskeletal remodeling in response to pollutant exposure. Consequently, β-actin and related proteins were proposed as early biomarkers of chronic environmental stress in freshwater ecosystems. In a related study, Nugroho et al. (2025) employed combined metabolomic and proteomic analyses to investigate biochemical responses in the gills and digestive glands of *S. woodiana* collected from environments with differing pollution intensities. Significant alterations, particularly in

TABLE 8.1 Proteomic studies linking environmental contaminants to molecular stress responses in model organisms.

Geographic area	Source of stress	Pollutant	Model species investigated	Proteomic approach used	Key findings (stress/protein alterations)	Reference
North Sea, Europe	Oil contamination	Crude oil + polycyclic aromatic hydrocarbons + alkylphenols	Mytilus edulis (blue mussel)	Protein chip array + SELDI-TOF MS	83 protein peaks were altered in spiked oil exposure; the majority were upregulated	Bjørnstad et al. (2006)
Doñana National Park and surrounding areas, Spain	Aquatic pollution	Various metals and co-contaminants	Procambarus clarkii (Red swamp crayfish)	2D-DIGE + MS	Differential expression of proteins involved in oxidative stress; impaired mitochondrial function; protein folding; metabolic shift to aerobic glycolysis	Fernández-Cisnal et al. (2017)
Barcelona, Spain	Waterborne chemical exposure	Acrylamide	Danio rerio (Zebrafish)	μLC-MS/MS	57 proteins downregulated and 65 upregulated in the brain; disruptions in presynaptic vesicle cycling, neurotransmitter release, and cysteine adduct formation	Faria et al. (2018)
Natural spring, Southern Germany	Thermal warming	Increasing temperature and microplastics exposure	Crunoecia irrorata (Aquatic insect)	LC-MS/MS	Identified 1,358 differentially abundant proteins linked to trehalose metabolism; heat-shock protein regulation; evidence of molecular acclimation mechanisms to warming	Ebner et al. (2019)
Deep-sea habitats, South	Metal exposure due to deep-sea mining	Copper	Abyssorchomene distinctus (Deep-sea amphipods)	TMT labeling + LC-MS/MS	Identified 2,937 proteins and 1,918 reliably quantified proteins; differential expression of proteins linked to ion	Kwan et al. (2019)

(Continued)

TABLE 8.1 (Continued)

Geographic area	Source of stress	Pollutant	Model species investigated	Proteomic approach used	Key findings (stress/protein alterations)	Reference
China Sea (In situ)					regulation (Na^+/K^+-ATPase), cuticle structure, and chitin metabolism	
Parentis Biscarrosse lake, France	Hydrocarbons contamination	Crude oil + polycyclic aromatic hydrocarbons	Corbicula fluminea (Asian clams)	LC-MS/MS	Proteome alterations in gills, muscles, and ganglia	Miserazzi et al. (2020)
Laboratory study, University of Aveiro, Portugal	Pesticide exposure	Spinosad (insecticide) exposure	Chironomus riparius (non-biting midge larvae)	LC-MS/MS	Altered expression of globins, actin, cuticle proteins, arginine kinase, and elongation factor 1-alpha; disruptions in physiological pathways; changes in protein profile identified as early-warning biomarkers of pesticide exposure	Monteiro et al. (2020)
Singapore aquatic systems	Pharmaceutical residues from agriculture, hospitals, and industries	Human and veterinary drugs	Aquatic phytoplankton	LC-MS	Disruption in growth, photosynthetic activities, metabolite profiles, and oxidative damage	Chia et al. (2021)
Augsburg, Germany	Microplastic pollution	Polystyrene microplastics	Daphnia magna (Water flea)	MS	Significant alteration of 41 proteins in abundance	Trotter et al. (2021)
Lake Aesake, Papua New Guinea	Metal pollution	Copper	Chlorella sp. (green alga)	LC-MS/MS	Modulation of proteins involved in photosynthesis, carbon fixation, glycolysis, and oxidative phosphorylation	Shakya et al. (2022)

United Kingdom biobank	Air pollution	Particulate matter 2.5, nitrogen dioxide, sulfur dioxide	Human cohort study	Proteomic mediation analysis	Identified 23 proteins mediating depression onset and 38 proteins linked to anxiety severity	Wang et al. (2025)
Winongo Stream, Yogyakarta Province, Indonesia	Aquatic pollution	Contaminants	Sinanodonta woodiana (Mussels)	LC-MS	Increased β-actin coverage (from 17% to 34%) across pollution gradients	Rivaldi et al. (2025)
Code Stream, Yogyakarta, Indonesia	Aquatic pollution	Pollutants	Sinanodonta woodiana (Mussels)	LC-MS	Increased β-actin coverage (from 13% to 38%) across pollution gradients	Nugroho et al. (2025)
Northern Patagonia, Chile (Llanquihue and Icalma lakes)	Economic and human activities	Total dissolved solids, electrical conductivity, heavy metals	Daphnia pulex (Zooplankton)	MS	198 proteins were differentially expressed in response to environmental stressors	Norambuena et al. (2025)

β-actin peptide expression, were detected, reflecting complex cellular adjustments to pollutant stress. Several metabolites and peptides—such as 7-oxocholesterol, glycol oleate, and hydroxy eicosapentaenoic acid were identified as potential biomarkers indicative of contaminant exposure. Collectively, these studies underscore the application of proteomic approaches in aquatic biomonitoring, providing molecular-level insights that complement conventional ecological assessments.

Ebner et al. (2019) further expanded the application of proteomics in environmental biomonitoring by exploring the molecular acclimation strategies of the stenotopic caddisfly *Crunoecia irrorata* subjected to elevated water temperatures in freshwater springs of southern Germany. Using an LC-MS/MS-based proteomics workflow, over 1300 proteins were successfully identified despite the complexity of whole-larvae extracts, revealing significant alterations in proteins associated with trehalose metabolism, cuticle and tracheal system modification, and heat-shock protein regulation. This approach demonstrated its capability to simultaneously capture multiple physiological responses to environmental variations, allowing comprehensive assessments of adaptive capacity across diverse taxa.

In a seminatural mesocosm study, Miserazzi et al. (2020) employed a proteomics approach to elucidate the molecular mechanisms underlying behavioral disturbances in the Asian clam (*Corbicula fluminea*) exposed to crude oil contamination in France. The LC-MS/MS-based quantitative proteomics revealed no significant polycyclic aromatic hydrocarbon accumulation in the combined muscle and nervous ganglia tissues, in contrast to the gills, where moderate levels were detected. The proteomic analysis identified distinct tissue-specific responses: a slowdown in metabolic and/or cellular processes in muscles and ganglia associated with minor toxicological effects; and an increase in metabolic and/or cellular processes in gills associated with a greater toxicological impact. These proteomic alterations provide mechanistic insight into how differential physiological stress contributes to observed behavioral changes following oil exposure.

Shakya et al. (2022) explored the biochemical effects of copper exposure on the amino acid profile and proteome of the tropical freshwater microalga *Chlorella* sp., aiming to elucidate molecular responses under environmentally relevant copper concentrations. Amino acid profiling and LC-MS/MS analyses revealed that sublethal toxicity could be detected before any measurable decline in cell division, even at very low copper levels (1–2 μg/L). Significant alterations were observed in amino acids such as glycine, arginine, and branched-chain amino acids, alongside changes in proteins involved in their metabolism, including branched-chain amino acid transaminase, carbamoyl-phosphate synthase, ornithine aminotransferase, and thiosulfate sulfurtransferase. These molecular responses indicate the potential of such amino acids and proteins to serve as sensitive biomarkers of copper-induced toxicity in aquatic systems.

Trotter et al. (2021) examined the long-term impacts of polystyrene microplastic exposure on the freshwater cladoceran *Daphnia magna*, focusing on proteomic, morphological, and life-history alterations. Employing high-resolution MS-based quantitative proteomics, the researchers identified 3,784 proteins, of which 41 exhibited significant changes in abundance ($q < 0.05$) following polystyrene exposure. Upregulated proteins included sulfotransferases and γ-aminobutyric acid transaminase, associated with metabolic and neurotransmission pathways, while downregulated proteins encompassed DNA-directed RNA polymerase subunits and digestive enzymes linked to nutrient assimilation and reproductive processes. These proteomic alterations correlated with observed morphological and life-history impairments, suggesting that prolonged microplastics exposure can compromise metabolic efficiency and overall organismal fitness in *D. magna*.

Monteiro et al. (2020) investigated the sublethal molecular effects of spinosad exposure in *Chironomus riparius* larvae using LC-MS/MS. The study revealed marked alterations in the expression of key proteins, including globins, actin, cuticle proteins, arginine kinase, and elongation factor 1-alpha, indicating disruptions in essential physiological pathways. These proteomic changes elucidate the molecular mechanisms underlying the toxic action of spinosad and its influence on higher biological processes. Furthermore, the results suggest that alterations in protein expression may serve as early-warning biomarkers of pesticide exposure, providing mechanistic insights into organismal stress responses and supporting ecological risk assessment of environmental contaminants. Together, these studies affirm the growing role of proteomics in environmental biomonitoring and ecotoxicological assessment.

8.6 Metaproteomics

Metaproteomics provides comprehensive insights into the composition, diversity, and functional potential of microbial communities across a wide range of ecosystems (Kumar et al., 2024). Examining the proteins expressed within a community, it enables the detection of environmentally driven shifts in microbial function and clarifies how these alterations influence pollutant degradation processes. This community-level perspective offers a more effective framework for addressing environmental pollution than traditional approaches that focus solely on individual microorganisms or isolated enzymatic mechanisms (Zahra et al., 2025). A major strength of metaproteomic analysis lies in its capacity to identify which microorganisms are actively contributing to ecosystem functions by assigning proteins to specific biological pathways. The relative abundance of these proteins provides a direct measure of metabolic activity within distinct microbial groups. Analysis of extracellular or secreted proteins further reveals patterns of microbe–microbe and microbe–environment interactions, enabling a deeper understanding of community responses to environmental stressors and contaminants. Moreover, peptide sequences obtained from environmental samples can be taxonomically traced to their source organisms, allowing detailed molecular characterization of active microbial phenotypes in complex habitats (Fernández-Carro et al., 2025). Applications of metaproteomics in pollution monitoring continue to expand. The comparative metaproteomics was used to investigate microbial biofilms colonizing marine plastic debris (Messer et al., 2024). The analysis revealed active expression of xenobiotic-degrading enzymes, including depolymerases and hydrolases from genera such as *Pseudomonas*, *Psychrobacter*, *Planococcus*, and *Pseudoalteromonas*. These microbes also exhibited elevated levels of stress-response proteins and pathways associated with oxidative phosphorylation and the citric acid cycle, indicating functional adaptation to the plastic environment.

Ongoing methodological advancements have further strengthened the utility of metaproteomics for decoding microbial functional dynamics. Early groundwork was established through improvements in large-scale protein separation using 2-DE. The subsequent development of high-throughput LC-MS/MS has markedly enhanced analytical depth, sensitivity, and precision, transforming the entire workflow of metaproteomic investigations (Nebauer et al., 2024; Chaurasia et al., 2025)

8.7 Future perspectives

Historically, proteomics faced challenges in extending its application beyond a few model organisms due to limited genomic resources. However, recent advancements have facilitated the development of pan-proteomics strategies, enabling the characterization of proteomes and their dynamic regulation across a broad spectrum of species using draft or incomplete genomes. Proteomic approaches possess distinct strengths and serve pivotal roles in regulating growth and developmental processes, deciphering stress-response mechanisms, mapping protein–protein interaction networks, and uncovering biosynthetic pathways of bioactive compounds. However, proteomics alone cannot provide a complete understanding of biological systems, as it lacks coverage at the genomic, transcriptomic, and metabolic levels. Additionally, it faces inherent challenges such as the detection of low-abundance proteins, the complexity of post-translational modifications, and the computational intensity of data processing. To address these limitations, the integration of proteomics with complementary omics approaches, such as genomics, transcriptomics, and metabolomics is essential to achieve a more holistic understanding of the biological complexity and functional regulatory networks.

8.8 Conclusion

Proteomics has emerged as a revolutionary technique in environmental sciences, offering molecular-level insights that transcend far beyond the capabilities of conventional monitoring approaches. Proteomic analyses establish a clear connection between pollutant exposure and biological effects by interpreting changes in protein abundance, structure, and post-translational modifications. The integration of gel-based, mass spectrometry-based, and array-based proteomic technologies has enabled researchers to uncover both broad and pollutant-specific biomarkers, revealing intricate pathways involved in oxidative stress, energy metabolism, detoxification, and cellular repair mechanisms across diverse species and ecosystems. The versatility of proteomics in detecting early molecular signs of stress in species exposed to heavy metals, hydrocarbons, pesticides, nanomaterials, and microplastics is illustrated by the case studies addressed throughout this chapter. These outcomes suggest the ability of proteomic biomarkers to function as sensitive markers of contaminant exposure, even before overt physiological or ecological damage becomes apparent. Such molecular evidence supports ecological risk assessment frameworks, aids in environmental management decisions, and deepens current understanding of pollutant toxicity and adaptive responses.

Overall, proteomics stands as a vital and evolving tool for modern biomonitoring. It bridges the gap between chemical detection and biological consequence, translating complex molecular information into actionable knowledge for environmental protection. As methodologies continue to advance, future research should focus on expanding proteomic databases for nonmodel species and improving field-based workflows to build a holistic understanding of ecosystem health under escalating industrial, agricultural, and anthropogenic stress.

Acknowledgments

Amritpal Kaur acknowledges SERB for the Science and Technology Award for Research (STAR award, STR/2021/000019). Somna Oberoi sincerely acknowledges the Department of Science and Technology (DST), Government of India, for financial support under the INSPIRE Fellowship program [IF200213].

References

Adams, F., Adriaens, M., 2021. Methodological aspects of analytical chemistry. Journal of Analytical Chemistry 76 (5), 671–673. Available at https://doi.org/10.1134/s1061934821050038.

Aebersold, R., Mann, M., 2016. Mass-spectrometric exploration of proteome structure and function. Nature 537 (7620), 347–355. Available at https://doi.org/10.1038/nature19949.

Alharbi, R.A., 2020. Proteomics approach and techniques in identification of reliable biomarkers for diseases. Saudi Journal of Biological Sciences 27 (3), 968–974. Available at https://doi.org/10.1016/j.sjbs.2020.01.020, http://www.ksu.edu.sa/sites/Colleges/CollegeofScinces/BotanyDepartment/SJBS/Default.aspx.

Angon, P.B., Islam, M.S., KC, S., Das, A., Anjum, N., Poudel, A., et al., 2024. Sources, effects and present perspectives of heavy metals contamination: soil, plants and human food chain. Heliyon 10 (7). Available at https://doi.org/10.1016/j.heliyon.2024.e28357, https://www.sciencedirect.com/science/journal/24058440.

Arentz, G., Weiland, F., Oehler, M.K., Hoffmann, P., 2015. State of the art of 2D DIGE. Proteomics - Clinical Applications 9 (3-4), 277–288. Available at https://doi.org/10.1002/prca.201400119, http://onlinelibrary.wiley.com/journal/10.1002/(ISSN)1862-8354.

Atak, A, Mukherjee, S, Jain, R, Gupta, S, Singh, V.A, Gahoi, N, Manubhai, K.P, Srivastava, S, 2016. Protein microarray applications: Autoantibody detection and posttranslational modification. Proteomics 16, 2557–2569. Available at https://doi.org/10.1002/pmic.201600104.

Bjørnstad, A., Larsen, B.K., Skadsheim, A., Jones, M.B., Andersen, O.K., 2006. The potential of ecotoxicoproteomics in environmental monitoring: biomarker profiling in mussel plasma using proteinchip array technology. Journal of Toxicology and Environmental Health, Part A 69 (1-2), 77–96. Available at https://doi.org/10.1080/15287390500259277.

Briffa, J., Sinagra, E., Blundell, R., 2020. Heavy metal pollution in the environment and their toxicological effects on humans. Heliyon 6 (9), e04691. Available at https://doi.org/10.1016/j.heliyon.2020.e04691.

Campos, A., Tedesco, S., Vasconcelos, V., Cristobal, S., 2012. Proteomic research in bivalves. towards the identification of molecular markers of aquatic pollution. Journal of Proteomics 75 (14), 4346–4359. Available at https://doi.org/10.1016/j.jprot.2012.04.027.

Carmona, J.A., Doña, C.M., López-Barea, J., 2017. New trends in aquatic pollution monitoring: from conventional biomarkers to environmental proteomics. Environmental Problems in Marine Biology: Methodological Aspects and Applications. CRC Press, Spain, pp. 150–171. Available at https://doi.org/10.1201/9781315119113, http://www.tandfebooks.com/doi/book/10.1201/9781315119113.

Catherman, A.D., Skinner, O.S., Kelleher, N.L., 2014. Top down proteomics: facts and perspectives. Biochemical and Biophysical Research Communications 445 (4), 683–693. Available at https://doi.org/10.1016/j.bbrc.2014.02.041, http://www.elsevier.com/inca/publications/store/6/2/2/7/9/0/index.htt.

Chaurasia, P., Kumar, A., Sarim, K.M., 2025. Metaproteomics: recent advances and applications. Meta-omics in Crop Improvement: Volume I: Methods and Approaches. Springer, pp. 77–91.

Chemical Safety – Fact Sheet: Exposure to Selected Chemicals. 2021.

Chen, R., Brown, H.M., Cooks, R.G., 2021. Metabolic profiles of human brain parenchyma and glioma for rapid tissue diagnosis by targeted desorption electrospray ionization mass spectrometry. Analytical and Bioanalytical Chemistry 413 (25), 6213–6224. Available at https://doi.org/10.1007/s00216-021-03593-0, https://link.springer.com/journal/216/volumes-and-issues.

Chia, M.A., Lorenzi, A.S., Ameh, I., Dauda, S., Cordeiro-Araújo, M.K., Agee, J.T., et al., 2021. Susceptibility of phytoplankton to the increasing presence of active pharmaceutical ingredients (APIs) in the aquatic environment: a review. Aquatic Toxicology 234. Available at https://doi.org/10.1016/j.aquatox.2021.105809, http://www.elsevier.com/wps/find/journaldescription.cws_home/505509/description#description.

Clark, C.M., Bell, M.D., Boyd, J.W., Compton, J.E., Davidson, E.A., Davis, C., et al., 2017. Nitrogen-induced terrestrial eutrophication: cascading effects and impacts on ecosystem services. Ecosphere 8 (7). Available at https://doi.org/10.1002/ecs2.1877.

Coarfa, C., Grimm, S.L., Rajapakshe, K., Perera, D., Lu, H.Y., Wang, X., et al., 2021. Reverse-phase protein array: technology, application, data processing, and integration. Journal of Biomolecular Techniques 32 (1), 15–29. Available at https://doi.org/10.7171/jbt.21-3202-001, https://www.ncbi.nlm.nih.gov/pmc/articles/PMC8121080/pdf/jbt.32-15.pdf.

Diniz, M., Madeira, D., Araújo, E., 2017. Effects of climate change on marine organisms: a proteomic approach. Environmental Problems in Marine Biology: Methodological Aspects and Applications. CRC Press, Portugal, pp. 190–212. Available at https://doi.org/10.1201/9781315119113, http://www.tandfebooks.com/doi/book/10.1201/9781315119113.

Ebner, J.N., Ritz, D., von Fumetti, S., 2019. Comparative proteomics of stenotopic caddisfly Crunoecia irrorata identifies acclimation strategies to warming. Molecular Ecology 28 (19), 4453–4469. Available at https://doi.org/10.1111/mec.15225, http://onlinelibrary.wiley.com/journal/10.1111/(ISSN)1365-294X.

Faria, M., Ziv, T., Gómez-Canela, C., Ben-Lulu, S., Prats, E., Novoa-Luna, K.A., et al., 2018. Acrylamide acute neurotoxicity in adult zebrafish. Scientific Reports 8 (1). Available at https://doi.org/10.1038/s41598-018-26343-2, www.nature.com/srep/index.html.

Fernández-Carro, E., Letsiou, S., Tsironi, S., Chaniotis, D., Ciriza, J., Beloukas, A., 2025. Alternatives integrating omics approaches for the advancement of human skin models: a focus on metagenomics, metatranscriptomics, and metaproteomics. Microorganisms 13 (8), 1771. Available at https://doi.org/10.3390/microorganisms13081771.

Fernández-Cisnal, R., García-Sevillano, M.A., Gómez-Ariza, J.L., Pueyo, C., López-Barea, J., Abril, N., 2017. 2D-DIGE as a proteomic biomarker discovery tool in environmental studies with Procambarus clarkii. Science of the Total Environment 584-585, 813–827. Available at https://doi.org/10.1016/j.scitotenv.2017.01.125, www.elsevier.com/locate/scitotenv.

Gajahin Gamage, N.T., Miyashita, R., Takahashi, K., Asakawa, S., Senevirathna, J.D.M., 2022. Proteomic applications in aquatic environment studies. Proteomes 10 (3). Available at https://doi.org/10.3390/proteomes10030032, www.mdpi.com/journal/proteomes.

Gang, D., Jia, H., Ji, H., Li, J., Yu, H., Hu, C., et al., 2024. Ecological risk of per-and polyfluorinated alkyl substances in the phytoremediation process: a case study for ecologically keystone species across two generations. Science of The Total Environment 949, 174961. Available at https://doi.org/10.1016/j.scitotenv.2024.174961.

Geng, Q., Liu, B., Fan, D., Cao, Z., Li, L., Lu, P., et al., 2024. Strictosamide ameliorates LPS-induced acute lung injury by targeting ERK2 and mediating NF-κB signaling pathway. Journal of Ethnopharmacology 322. Available at https://doi.org/10.1016/j.jep.2023.117593, https://www.sciencedirect.com/science/journal/03788741.

Guo, H., Wang, L., Deng, Y., Ye, J., 2021. Novel perspectives of environmental proteomics. Science of The Total Environment 788, 147588. Available at https://doi.org/10.1016/j.scitotenv.2021.147588.

Guo, T., Steen, J.A., Mann, M., 2025. Mass-spectrometry-based proteomics: from single cells to clinical applications. Nature 638 (8052), 901–911. Available at https://doi.org/10.1038/s41586-025-08584-0, https://www.nature.com/nature/.

Guzmán-Fierro, V., Dieguez-Seoane, A., Roeckel, M., Lema, J.M., Trueba-Santiso, A., 2024. Environmental proteomics as a useful methodology for early-stage detection of stress in anammox engineered systems. Science of the Total Environment 912. Available at https://doi.org/10.1016/j.scitotenv.2023.169349, https://www.sciencedirect.com/science/journal/00489697.

Henke, A.N., Chilukuri, S., Langan, L.M., Brooks, B.W., 2024. Reporting and reproducibility: proteomics of fish models in environmental toxicology and ecotoxicology. Science of the Total Environment 912. Available at https://doi.org/10.1016/j.scitotenv.2023.168455, https://www.sciencedirect.com/science/journal/00489697.

Jin, Y., Sun, Z., Gu, A.Z., Zhou, X., 2025. Proteomics technologies in toxicity screening: a review. Environmental Chemistry Letters 23 (1), 67–80. Available at https://doi.org/10.1007/s10311-024-01816-1, https://www.springer.com/journal/10311.

Judge, M., Fernando, J.W., Begeny, C.T., 2022. Dietary behaviour as a form of collective action: a social identity model of vegan activism. Appetite 168. Available at https://doi.org/10.1016/j.appet.2021.105730, https://www.sciencedirect.com/science/journal/01956663.

Hur, J., Danes, L., Hsieh, J.H., McGregor, B., Krout, D., Auerbach, S., 2018. Tox21 enricher: web-based chemical/biological functional annotation analysis tool based on Tox21 toxicity screening platform. Molecular Informatics 37 (5). Available at https://doi.org/10.1002/minf.201700129.

Khan, B., Burgess, R.M., Cantwell, M.G., 2023. Occurrence and bioaccumulation patterns of per- and polyfluoroalkyl substances (PFAS) in the marine environment. ACS ES and T Water 3 (5), 1243–1259. Available at https://doi.org/10.1021/acsestwater.2c00296, https://pubs.acs.org/page/aewcaa/about.html.

Krasny, L., Huang, P.H., 2021. Data-independent acquisition mass spectrometry (DIA-MS) for proteomic applications in oncology. Molecular Omics 17 (1), 29–42. Available at https://doi.org/10.1039/d0mo00072h, http://www.rsc.org/journals-books-databases/about-journals/molecular-omics/.

Kumar, V., Sharma, J.K., Kumar, S., Gulia, J., Singh, N.P., Santal, A.R., 2024. Metagenomic and proteomic approach for bioremediation of environmental pollutants. Development in Waste Water Treatment Research and Processes: Applied Technologies for Clean up of Environmental Contaminants. Elsevier, India, pp. 635–647. Available at https://doi.org/10.1016/B978-0-443-13615-3.00020-0, https://www.sciencedirect.com/book/9780443136153.

Kwan, Y.H., Zhang, D., Mestre, N.C., Wong, W.C., Wang, X., Lu, B., et al., 2019. Comparative proteomics on deep-sea amphipods after in situ copper exposure. Environmental Science and Technology 53 (23), 13981–13991. Available at https://doi.org/10.1021/acs.est.9b04503, http://pubs.acs.org/journal/esthag.

Li, D., Jiang, L., Hong, Y., Cai, Z., 2021. Multilayered glycoproteomic analysis reveals the hepatotoxic mechanism in perfluorooctane sulfonate (PFOS) exposure mice. Environmental Pollution 268 (Pt A), 115774. Available at https://doi.org/10.1016/j.envpol.2020.115774.

Lundberg, E., Borner, G.H.H., 2019. Spatial proteomics: a powerful discovery tool for cell biology. Nature Reviews Molecular Cell Biology 20 (5), 285–302. Available at https://doi.org/10.1038/s41580-018-0094-y.

López-Barea, J., Abril, N., 2017. 2D-DIGE as a proteomic biomarker discovery tool in environmental studies with Procambarus clarkii. Science of the Total Environment 584.

Maciejczyk, P., Chen, L.C., Thurston, G., 2021. The role of fossil fuel combustion metals in PM2.5 air pollution health associations. Atmosphere 12 (9). Available at https://doi.org/10.3390/atmos12091086, https://www.mdpi.com/2073-4433/12/9/1086/pdf.

Marshall, K., Twum, Y., Li, Y., Gao, W., 2025. Spotting targets with 2D-DIGE proteomics. Advances in Clinical Chemistry 125, 1–22. Available at https://doi.org/10.1016/bs.acc.2024.11.003, https://www.sciencedirect.com/science/bookseries/00652423.

Meher, A.K., Zarouri, A., 2025. Environmental applications of mass spectrometry for emerging contaminants. Molecules (Basel, Switzerland) 30 (2). Available at https://doi.org/10.3390/molecules30020364, http://www.mdpi.com/journal/molecules.

Messer, L.F., Lee, C.E., Wattiez, R., Matallana-Surget, S., 2024. Novel functional insights into the microbiome inhabiting marine plastic debris: critical considerations to counteract the challenges of thin biofilms using multi-omics and comparative metaproteomics. Microbiome 12 (1). Available at https://doi.org/10.1186/s40168-024-01751-x, https://microbiomejournal.biomedcentral.com/.

Miserazzi, A., Perrigault, M., Sow, M., Gelber, C., Ciret, P., Lomenech, A.M., et al., 2020. Proteome changes in muscles, ganglia, and gills in Corbicula fluminea clams exposed to crude oil: relationship with behavioural disturbances. Aquatic Toxicology 223, 105482. Available at https://doi.org/10.1016/j.aquatox.2020.105482.

Mondschein, J., Clark-Ginsberg, A., Kuehn, A., 2021. Smart cities as large technological systems: overcoming organizational challenges in smart cities through collective action. Sustainable Cities and Society 67, 102730. Available at https://doi.org/10.1016/j.scs.2021.102730.

Monteiro, H.R., Pestana, J.L.T., Soares, A.M.V.M., Devreese, B., Lemos, M.F.L., 2020. Chironomus riparius proteome responses to spinosad exposure. Toxics 8 (4), 117. Available at https://doi.org/10.3390/toxics8040117.

Nebauer, D.J., Pearson, L.A., Neilan, B.A., 2024. Critical steps in an environmental metaproteomics workflow. Environmental Microbiology 26 (5). Available at https://doi.org/10.1111/1462-2920.16637, http://onlinelibrary.wiley.com/journal/10.1111/(ISSN)1462-2920.

Norambuena, J.A., Poblete-Grant, P., Beltrán, J.F., De los Ríos-Escalante, P., Aranzaez-Ríos, C., Farías, J.G., 2025. Proteomic Profile of Daphnia pulex in response to heavy metal pollution in lakes of Northern Patagonia. International Journal of Molecular Sciences 26 (1). Available at https://doi.org/10.3390/ijms26010417, http://www.mdpi.com/journal/ijms.

Nugroho, A.P., Rivaldi, M., Yudha, D.S., Ukasha, T., Hadisusanto, S., 2025. Active biomonitoring using mussels: an evaluation of the untargeted metabolomics and proteomics responses as biomarkers of stream ecosystem health. Environmental Science and Pollution Research 32 (34), 20375–20394. Available at https://doi.org/10.1007/s11356-025-36771-z, https://www.springer.com/journal/11356.

Rivaldi, M., Frediansyah, A., Aziz, S.A.A., Nugroho, A.P., 2025. Active biomonitoring of stream ecosystems: untargeted metabolomic and proteomic responses and free radical scavenging activities in mussels. Ecotoxicology (London, England) 34 (3), 425–443. Available at https://doi.org/10.1007/s10646-024-02846-9, https://www.springer.com/journal/10646.

Root, L., Campo, A., MacNiven, L., Con, P., Cnaani, A., Kültz, D., 2021. A data-independent acquisition (DIA) assay library for quantitation of environmental effects on the kidney proteome of Oreochromis niloticus. Molecular Ecology Resources 21 (7), 2486–2503. Available at https://doi.org/10.1111/1755-0998.13445.

Saravanan, A., Kumar, P.S., Hemavathy, R.V., Jeevanantham, S., Harikumar, P., Priyanka, G., et al., 2022. A comprehensive review on sources, analysis and toxicity of environmental pollutants and its removal methods from water environment. Science of the Total Environment 812 Available from: Available at https://doi.org/10.1016/j.scitotenv.2021.152456, www.elsevier.com/locate/scitotenv.

Saxena, V., 2025. Water quality, air pollution, and climate change: investigating the environmental impacts of industrialization and urbanization. Water, Air, & Soil Pollution 236 (2). Available at https://doi.org/10.1007/s11270-024-07702-4.

Shakya, M., Silvester, E., Rees, G., Rajapaksha, K.H., Faou, P., Holland, A., 2022. Changes to the amino acid profile and proteome of the tropical freshwater microalga Chlorella sp. in response to copper stress. Ecotoxicology and Environmental Safety 233, 113336. Available at https://doi.org/10.1016/j.ecoenv.2022.113336.

Shuken, S.R., 2023. An introduction to mass spectrometry-based proteomics. Journal of Proteome Research 22 (7), 2151–2171. Available at https://doi.org/10.1021/acs.jproteome.2c00838, http://pubs.acs.org/journal/jprobs.

Simões, T., Novais, S.C., Natal-da-Luz, T., Devreese, B., de Boer, T., Roelofs, D., et al., 2018. An integrative omics approach to unravel toxicity mechanisms of environmental chemicals: effects of a formulated herbicide. Scientific Reports 8 (1). Available at https://doi.org/10.1038/s41598-018-29662-6, www.nature.com/srep/index.html.

Singh, N., Das, S., Barman, K., Das Kurmi, B., Patel, P., 2025. Analytical mass spectrometry as a versatile tool in omics, diagnostics, and environmental monitoring. International Journal of Medical, Pharmaceutical and Health Sciences 2 (2), 275–285. Available at https://doi.org/10.62946/ijmphs/2.2.275-285.

Smith, D., Španěl, P., Demarais, N., Langford, V.S., McEwan, M.J., 2025. Recent developments and applications of selected ion flow tube mass spectrometry (SIFT-MS). Mass Spectrometry

Reviews 44 (2), 101–134. Available at https://doi.org/10.1002/mas.21835, http://onlinelibrary.wiley.com/journal/10.1002/(ISSN)1098-2787.

State of Global Air Report 2024. 2024.

Tamara, S., Den Boer, M.A., Heck, A.J.R., 2022. High-resolution native mass spectrometry. Chemical Reviews 122 (8), 7269–7326. Available at https://doi.org/10.1021/acs.chemrev.1c00212, http://pubs.acs.org/journal/chreay.

Tomanek, L., 2011. Environmental proteomics: changes in the proteome of marine organisms in response to environmental stress, pollutants, infection, symbiosis, and development. Annual Review of Marine Science 3 (1), 373–399. Available at https://doi.org/10.1146/annurev-marine-120709-142729.

Trotter, B., Wilde, M.V., Brehm, J., Dafni, E., Aliu, A., Arnold, G.J., et al., 2021. Long-term exposure of Daphnia magna to polystyrene microplastic (PS-MP) leads to alterations of the proteome, morphology and life-history. Science of the Total Environment 795 Available from: Available at https://doi.org/10.1016/j.scitotenv.2021.148822, www.elsevier.com/locate/scitotenv.

Tubatsi, G., Kebaabetswe, L.P., Musee, N., 2022. Proteomic evaluation of nanotoxicity in aquatic organisms: a review. Proteomics 22 (21). Available at https://doi.org/10.1002/pmic.202200008, http://onlinelibrary.wiley.com/journal/10.1002/(ISSN)1615-9861.

Vareda, J.P., Valente, A.J.M., Durães, L., 2019. Assessment of heavy metal pollution from anthropogenic activities and remediation strategies: a review. Journal of Environmental Management 246, 101–118. Available at https://doi.org/10.1016/j.jenvman.2019.05.126, https://www.sciencedirect.com/journal/journal-of-environmental-management.

Vicente-Santos, A., Sandoval-Herrera, N., Czirják, G.Á., Neely, B.A., Becker, D.J., 2025. Proteomics approaches to ecoimmunology: new insights into wildlife immunity and disease. Integrative and Comparative Biology. Available at https://doi.org/10.1093/icb/icaf044.

Wang, M., Ma, Y., Zheng, Y., Tao, Y., Mu, W., Yang, L., et al., 2025. Proteomic biomarkers as mediators in the association between air pollution and incident depression and anxiety: evidence from the UK biobank. Depression and Anxiety 2025 (1). Available at https://doi.org/10.1155/da/3984517.

Weldeslassie, T., Naz, H., Singh, B., Oves, M., 2017. Chemical contaminants for soil, air and aquatic ecosystem. Modern Age Environmental Problems and their Remediation. Springer International Publishing, Eritrea, pp. 1–22. Available at https://doi.org/10.1007/978-3-319-64501-8, 10.1007/978-3-319-64501-8_1.

Xue, B., Yang, Q., Jin, Y., Zhu, Q., Lan, J., Lin, Y., et al., 2023. Genotoxicity assessment of haloacetaldehyde disinfection byproducts via a simplified yeast-based toxicogenomics assay. Environmental Science & Technology 57 (44), 16823–16833. Available at https://doi.org/10.1021/acs.est.3c04956.

Zahra, A.F.T., Tabassum, M., Perumal, S., 2025. Novel approaches to plastic pollution: leveraging machine learning and metaproteomics for advanced plastic degradation. Malaysian Journal of Science Health & Technology 11 (1), 78–85. Available at https://doi.org/10.33102/mjosht.v11i1.451.

Zhu, D., Fang, C., Yang, Z., Ren, Y., Yang, F., Zheng, S., et al., 2022. Tubulin-binding peptide RR-171 derived from human umbilical cord serum displays antitumor activity against hepatocellular carcinoma via inducing apoptosis and activating the NF-kappa B pathway. Cell Proliferation 55 (5), e13241.

Chapter 9

Proteomics: novel and potential strategies for pollution assessment

Jaspreet Singh[1], Balwinder Singh[2,3], and Amritpal Kaur[3]
[1]*P.G. Department of Biotechnology, Khalsa College Amritsar, Amritsar, Punjab, India,* [2]*P.G. Department of Botany, Khalsa College Amritsar, Amritsar, Punjab, India,* [3]*Department of Food Science and Technology, Guru Nanak Dev University, Amritsar, Punjab, India*

9.1 Introduction

Environmental pollutants present a major global risk to both ecosystems and human health (Kumar et al., 2024). Environmental pollutant exposure contributes significantly to global ailments (Carberry and Rager, 2023). Fine particulate contaminants are suspected as important contributors to inflammatory oxidative damages and progression of disease, but the related biological mechanisms remain poorly understood (Tang et al., 2025). Recently, screening of potential health risks due to environmental pollutants has advanced to a new paradigm shift involving the application of high-throughput in silico screening methods (Krewski et al., 2020). Pollutant–biomolecule interactions regulate enzymes and cellular pathways, making protein target identification crucial for understanding pollutant-induced biotransformation, as proteins are the key elements of molecular recognition in biology (Caro et al., 2017). Prediction of quantitative information related to molecular features, viz. transcripts, proteins, and metabolites, to understand the toxicological effects of environmental pollutants has become an integral innovative tool in contemporary pollutant-induced toxicology research by using transcriptomic biomarkers and pattern-matching approaches (Meier et al., 2025).

Recently, screening of potential health risks due to environmental pollutants has advanced to a new paradigm shift involving the application of high-throughput in silico screening methods (Krewski et al., 2020). Among in silico methods, that is, genomics and metabolomics, the proteomic exploration offers a robust means to decipher molecular pathways associated with contaminant detection. Proteomics in bioremediation research offers a comprehensive perspective on the cellular

Proteomics Driven Bioremediation. DOI: https://doi.org/10.1016/B978-0-443-45294-9.00018-7

protein composition and molecular mechanisms. Proteomics-based toxicity assessments deliver valuable insights for early identification of environmental hazards of pollutants. Large-scale plasma proteomics has identified novel molecular targets of ambient pollutants (Pan et al., 2025). Across a number of disciplines, proteomics serves as a foundational step toward systems biology in understanding the organism-level cellular mechanisms (Tomanek, 2011). It offers a high-throughput strategy for identifying the biochemical networks associated with stress adaptation. Proteomics is a set of computational biology and empirical verification techniques that provide critical insight into the ecological risks of pollutants. Exceedingly, proteomic analysis is also utilized in assessing the pollutant biodegradation. As a data-driven technology, its comprehensive approach involves the identification of efficient protein biomarkers, functional enzymes, etc., for pollution assessment and remediation. As a leveraging bioinformatical and experimental validation technique, nowadays it appears as a potent environmental technology referred to as environmental proteomics. Proteomics propounds a powerful approach to decipher biomarkers, molecular mechanisms, as well as metabolic networks related to stress responses, pollutant interactions, and ecological risk evaluation.

9.2 Proteins as molecular effectors

A diverse array of pollutant discharge from industries, farming practices, etc., has substantial accumulative threats to ecological and biological systems (Thakur et al., 2025). The functional proteins are valuable in revealing the pollutant-induced transformations and adversaries related to environmental stressors. Under stress conditions, UAs become pivotal in the regulation of phenotypic expressions of physiological traits (Poblete-Naredo and Albores, 2016). The interactions between pollutants and their target proteins involve various biochemical and molecular mechanisms that influence enzymatic performance, signaling mechanisms, thereby shifting normal cellular dynamics (Caro et al., 2017). Various experimental techniques have been employed to identify pollutant-protein interactions, viz., microbe identification, cultivation, extraction, and gene cloning, and many more. But these conventional methodologies have disadvantages like being resource-intensive and requiring significant labor, time and energy. A new approach of combining computational biology with experimental validation techniques works well to mitigate these constraints. This integrated OMIC data-based approach is rapid and cost-effective (Vitorino, 2024).

9.3 OMICs technology

OMICs technology is a general platform used nowadays to study the cellular metabolites (Yuan et al., 2012). OMICs approaches targeting proteome, genome, and metabolome in living cells are expanded specifically as proteomics, genomics, and metabolomics (Lee et al., 2017). The advent of OMICs technologies has further raised the scientific expectations, leading to the swift evolution of diverse tools for in silico studies. Similar continuous improvements and application of Omics technologies in pollutant toxicity studies have given rise to a set of techniques grouped as toxicogenomics, toxicoproteomics, and toxicometabolomics (Bernhard et al., 2023). Bioinformatical analysis of OMIC data on experimental studies, when modeled with toxicokinetic and dynamic thinking, results in the conceptual framework used to address the questions related to environmental pollution (Altenburger et al., 2012). By-passing the animal-intensive testing, the OMICs data analysis, viz., proteomics, transcriptomics, or metabolomics, is an attractive novel approach towards comprehensive molecular insights into hazardous pollutant-exposed systems (Sprenger et al., 2022). Their application as the new approach methodologies (NAMs) anchored with cellular toxicity pathways is based on the presumption that any induced health change must be underpinned by some alterations at the OMIC levels (Li et al., 2023b). Omics, as a high-throughput processing technique, is characterized by 4Vs representing volume, variety, velocity, and veracity of Big Data (Souza et al., 2020).

9.4 Proteomics as an OMIC tool

OMIC technologies are now coming up as incredible risk assessment tools (Harrill et al., 2021). In pollution assessment proteomics potential is focused not only on evaluating the drastic effects of chemical pollutants but also concerns with safety as well. (Hudson et al., 2021). Proteomics allows phenotypic insights into the molecular mechanisms triggered by pollutant exposure (Liang et al., 2020). Proteomics is a dedicated field focused on examining the fluctuating in protein structures and peptidal alterations. It also estimates the dynamic changes in the composition and quantifies the modifications in the amino acid contents (Bantscheff et al., 2012). Its objective is to uncover the functions of biological processes by deciphering how proteins engage and associate with one another, thereby revealing the fundamental mechanisms underlying the principles of biological systems. Nevertheless, owing to the intricate and ever-changing compositions of the proteome, the proteomic investigations are complex (Veenstra, 2021). Therefore proteomic studies present significantly greater challenges in comparison to studies in other omics branches, viz., genomics and transcriptomics, which offer less formidable obstacles. A standard proteomic scheme encompasses multiple sequential stages and is outlined in Fig. 9.1. However, in contrast to the static frameworks of genomics and transcriptomics, proteomics, as a transformative discipline, elucidates more of the dynamic behavior of proteins and offers deeper insights (Palabiyik and Palabiyik, 2025). Proteomics provides better insights into the plant response towards environmental pollutants (toxic gases and heavy metals) that pose critical abiotic stress affecting both human and

Initial protein isolation
↓
Separation and characterization
↓
Quantitative profiling
↓
Interpretive data analysis
↓
Computational bioinformatics integration
↓
Functional validation
↓
Identification proteins

FIGURE 9.1 Sequential stages in proteomic analysis.

plant health, through protein modifications or de novo protein synthesis to combat stress (Harish et al., 2022). Environmental proteomics, focused on posttranslational modifications (PTMs), is a powerful tool for understanding the impact of environmental stressors through the identification of pollutant-specific biomarkers and their cellular effects, immune response profiling in infected organisms, and developmental signaling insights during changing environments (Tomanek, 2011).

9.5 Posttranslational modification proteomic profiling

Proteins are both structural and functional units of a living cell. Amino acid sequences dictate the formation of secondary conformation (α-helical and β-pleated strands) of proteins (Yang et al., 2018), thereby influencing the physical and functional properties, viz., solubility, stability, and ligand affinity (Starr and Thornton, 2016). PTMs play a crucial role in cellular activity, as they modulate amino acid composition and interactions (Nobeli et al., 2009). Structural and functional mutations reflect the responses of organisms towards environmental changes (Zhang et al., 2017). Comparative proteomic analyses of PTM profiles across different stages of environmental stress reveal dynamic changes in protein regulation. Under prolonged stress conditions, alterations in protein folding stability, abundance, and PTMs serve as direct evidence of transformation processes (Modi et al., 2021). For example, Timothy (*Phleum pratense*) grass pollen, polluted with heavy metals and phenolics, undergoes PTM due to oxidative stress that promotes protein unfolding via deamidation, thereby enhancing allergic responses by altering allergen structure (Smiljanic et al., 2019). PTM profiling enables the study of enzymatic activity in cells and tissues and, in combination with proteomic offers valuable insights that help in more understanding of regulatory networks (Eisenberg-Lerner et al., 2016). PTM profiling is a better tool for cellular landscape mapping for large-scale studies, as PTMs enhance protein diversity by altering structure and function in

TABLE 9.1 Frequently reported protein oxidative modifications.

Protein oxidative modifications	Chemical basis	References
Direct amino acid oxidation	α-amidation pathway generating carbonyl groups	Madian and Regnier (2010a)
Indirect advanced glycation	Lipid peroxidation adducts formation	Madian and Regnier (2010b)
Disulfide bonding, glutathione covalent binding	Methionine sulfoxide, glutathione adduct formation	Shacter (2000)

a site-specific, dynamic manner that generates cellular signaling, adaptive responses, and contributes to phenotypic diversity across organisms (Salovska and Liu, 2023).

9.6 Redox proteomics

Cytoskeleton structure and function happen to be the soft targets of oxidative stress related to pollution exposure (Amelina et al., 2007). The environmental pollutants induce oxidative stress-related protein inactivation through PTMs, leading to an imbalanced cellular metabolism. Further refinement of proteomic data by incorporating PTMs signatures can be utilized in pollution impact assessment (Li et al., 2023a). Oxidative PTM methods given in Table 9.1 cause proteasome- and lysosome-mediated degradation, leading to reduced protein functionality. Redox proteomics provides new molecular markers of pollutants for revealing the xenobiotic action mechanisms, for example, hydrogen peroxide (H_2O_2) induced disulfide bond alterations (McDonagh et al., 2006). Redox proteomics is commonly employed to study biochemical alterations induced by environmental contaminants (Madian and Regnier, 2010b). Redox proteomics for profiling pollutant-induced oxidative modifications in biological systems enables the identification of oxidative stress biomarkers. The systematic cataloging of oxidized proteins across diverse biological contexts has application in environmental impact assessment via molecular-level insights through improved analytical tools and software (Braconi et al., 2011).

9.7 Techniques in proteomic workflow

General techniques utilized in proteomic workflow comprise (1) extraction (sample preparation), (2) 2D gel electrophoresis (protein separation); (3) Mass spectrometry-based techniques, viz., LC-MS/MS and MALDI-TOF (peptide sequencing and protein identification); (4) iTRAQ, TMT (quantification); and

(5) SRM, PRM targeted proteomics (biomarker monitoring). Complex proteome requires extensive fractionation as a prerequisite (Li et al., 2005); that involves comprehensive analysis like diagonal gel electrophoresis for S—S modifications, Western blot analysis for modified protein, for example, s actin, 2-dimensional gel electrophoresis (2-DE) for protein carbonylation and glutathionylation, and extensive liquid chromatography coupled to tandem mass spectroscopy for modified proteins. Mass spectrometry (MS)-based assays grasp conserved motifs and construct biomarkers using partial sequences. Phylogenetically related multi-species protein sequence alignments confirm peptide selection (Jubeaux et al., 2012). Algorithmic guided matching after LC-MS/MS detection design potential peptides enabling functional verification of proteins.

Functional proteins from the same family as experimentally validated biomarkers are screened using public sequence databases analysis across taxa, for identification of prototypic peptides through targeted proteomics using SRM/MRM modes (Simon et al., 2010). Homology search using SWISS-PROT & TrEMBL databases coupled to peptide mass fingerprinting and LC-MS/MS dramatically improves the proteome analytical capability among specific subsets of proteins, for example, cytoskeletal proteins alterations. Table 9.2 mentions some of the important proteomics augmented cytoskeletal protein biomarkers. Scalable proteomic method using blue mussel (*Mytilus edulis*) presents a reproducible protocol for assessing marine pollution by combining liquid chromatography with 2-DE, along with cell fractionation and ion-exchange chromatography of digestive gland tissue that identified a 13-protein expression signature (PES) involved in β-oxidation, amino acid metabolism, detoxification,

TABLE 9.2 Proteomics assessed xenobiotic toxicity biomarkers.

Xenobiotic contaminant	Biomarkers	References
Diallylphtalate, bisphenol A	Carbonic anhydrase, hydroxyacid oxidase, Mn-superoxide dismutase, glutathione transferase	Apraiz et al. (2006)
PBDE-47, bisphenol A	Enoyl-CoA hydratase, Peroxin 10, Phospholipase A2, Cytochrome P450	
Benzo[a]pyrene	Aspartate aminotransferase, N-acetyltransferase, dual-specific phosphatase, Alcohol dehydrogenase	Riva et al. (2011)
Aroclor 1254	Actin, Tropomyosin	Rodríguez-Ortega et al. (2003)

protein degradation, organelle biogenesis, and protein folding. The principal component analysis and hierarchical clustering validated spatial discrimination of pollution gradients, offering a robust platform for eco-toxicological research with potential for automated, high-throughput pollution monitoring in marine environments (Amelina et al., 2007).

9.8 Shotgun proteomics

Numerous biological processes operate simultaneously and in a coordinated manner within every living cell; as a result, the integrated and dynamic understanding of the proteome as a system remains challenging (Bhushan and Nita-Lazar, 2024). Evolution of shotgun proteomics technology relies upon the procurement and processing of data from diverse sources. It has now emerged as a method of choice in broad-spectrum proteome mapping and characterization (Aebersold and Mann, 2016). As a high-throughput bottom-up approach, the shotgun proteomics bypasses the gel-based separation and directly identifies and quantifies the digested peptides following multiple sequential stages as outlined in Fig. 9.2. It comprehensively investigates the complete repertoire of expressed proteins to provide critical insights into protein expression, structure, and function, deciphering the molecular impact of hazardous toxicants (Katsavelis et al., 2025). The implementation of nanoflow reversed phase-high performance liquid chromatography coupled online to electrospray ionization mass spectrometry (ESI-MS) provides access to the

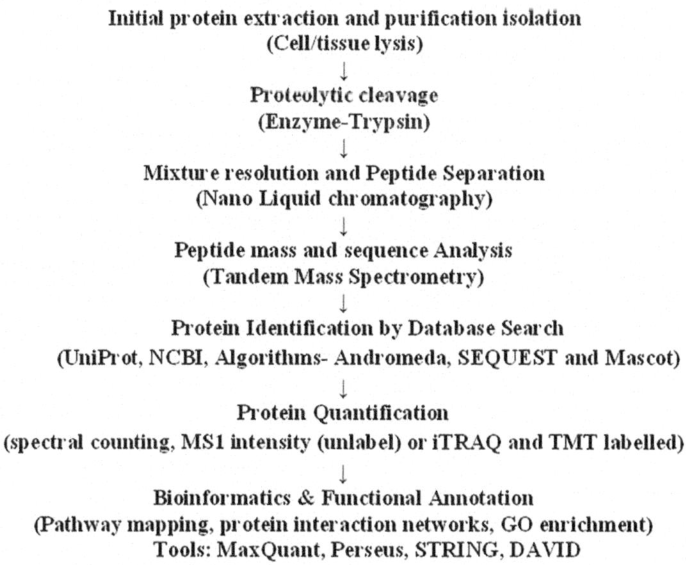

FIGURE 9.2 Sequential stages in shotgun proteomics.

big datasets yielding high-throughput protein identification, which has made shotgun proteomics a standard methodology of choice (Gorbach et al., 2025). Sydney rock oysters are useful as biomonitoring species in proteomic studies for detecting heavy metal pollution (Pb, Cu, Zn) in coastal waters due to their filter-feeding behavior and high bioaccumulation capacity. Researchers used label-free shotgun proteomics (SDS-PAGE-based), gas phase fractionation, and the Bivalvia protein database to identify 56 potential biomarker proteins in pollution monitoring bioassays, as the heavy metal stress significantly altered protein biosynthetic pathways (Muralidharan et al., 2012).

9.9 Proteomics over conventional assessment methods

Conventional methods evaluating pollutant toxicity rely upon downstream gene expression into phenotypic characteristics (Lundberg and Borner, 2019). The conventional pollutant evaluation methods (Table 9.3) remain inconclusive for

TABLE 9.3 Conventional techniques of protein analysis.

Conventional techniques	Analysis	method	References
Tandem mass spectrometry	Peptide sequence	Peptidal mass and amino acid composition detection	Medzihradszky and Chalkley (2015)
Nuclear magnetic resonance	Small protein structural conformations	Measured chemical shifts, detect atomic nuclei within the protein	Cavalli et al. (2007)
Cryo-electron microscopy	Frozen protein complex tertiary structure	Electron beam imaging	Gong et al. (2016)
Serial femtosecond crystallography	Protein microcrystal high-resolution structure	X-ray crystallography and X-ray free-electron laser (X-ray pulses)	Boutet et al. (2012)
Deep mutational scanning technology	Three-dimensional protein folding	Combination analysis of protein structures with mutations in amino sequences	Rollins et al. (2019)

subtle pollutants (e.g., organic pollutants), as phenotypic shifts remain unobserved and mechanisms driving toxicity are tough to understand. Proteomics, being a comprehensive analytical technique, offers a promising alternative to track functional protein profiles at the molecular level and reveal the toxic potential of pollutants that might mistakenly be considered as nontoxic by conventional assessment methods. For example, proteomic analysis of DINCH (a plasticizer and potential metabolic disrupting chemical linked to adipogenesis) revealed its molecular effects on metabolic disruption, highlighting the power of proteomics in evaluating the metabolic disrupting chemicals (Aldehoff et al., 2025). Proteomics has emerged as an effective tool in monitoring aquatic pollution in organisms like fish, crustaceans, and bivalves, especially mussels, due to an understanding of contamination-induced pathways in these species (López-Pedrouso et al., 2020). The biochemical responses towards complex pollutant mixtures using green crab *Carcinus maenas* as bioindicators across Tunisian coastal sites were studied (Ghedira et al., 2016). The crabs' biochemical responses involving 700 and 900 protein spots in gills and digestive glands, respectively, revealed that metal content and metallothionein levels were highest in Bizerte, followed by Teboulba, Gargour, and Mahres. The Bizerte crabs showed the highest number of altered protein spots (associated with molting, oxidative stress, inflammation, and proteolysis) in gills.

9.10 Quantitative proteomics

Proteins are tracked down (selection and identification) as targeting biomarkers in pollution monitoring. The proteomic investigations often rely on informatics analysis and lack the experimental process of validation. The combination of both systematic and integrative approaches involving high-throughput screening with functional verification is often advocated to overcome this limitation, making proteomics a more effective tool for tracking pollutants (Trapp et al., 2014). Moreover, integration of artificial intelligence and machine learning with proteomics can improve data interpretation, biomarker prioritization, and pollution assessments. Assessing toxicological risks of by-products produced through pollutant transformation is also important for environmental safety (Yang et al., 2016). Effective toxicity assessment of environmental pollutants requires a systematic evaluation that not only quantifies toxicity but is also aimed at mitigating the toxic effects. Identification of novel biomarkers that can detect environmental stress due to pollutants, their accumulation, and transformation is required (Chatterjee and Roy, 2022). Quantitative proteomics enables the isolation of individual by-products for the targeted analysis on model organisms. It offers a systems-level view of affected proteins and metabolic pathways by making use of advanced separation techniques like preparative HPLC, high-speed counter-current chromatography (Shi et al., 2013), and molecular imprinting (Song et al., 2012). The 4D-label-free proteomic analysis showed that *Enterobacter* sp.

FM-1 responds to cadmium toxicity under high cadmium stress by triggering 1148 differentially expressed proteins (DEPs), with 815 proteins involved in oxidative phosphorylation, citrate cycle, peroxisome function, pyruvate and glyoxylate metabolisms (Li et al., 2023a). Li et al. (2016) used relative and absolute quantification (iTRAQ) to study the response of nonhyperaccumulator maize line 9782 roots to lead pollution. They identified differentially expressed proteins (DEPs) involved in PTM, protein turnover, and chaperone activity. The study highlighted the role of transcriptional and post-transcriptional regulation in maize root adaptation towards lead stress using integrated proteomic and transcriptomic insights (Li et al., 2016).

9.11 Proteomics target validation

For understanding the modification of proteins and phenotypic expression, the temporal proteome dynamic characterization over biological processes is necessary. Temporal proteomics and validation strategies link pollutant-induced protein modification to the phenotypic changes (Guo et al., 2021). As the proteome changes very often within a defined observation window, the incorporation of the time dimension improves cognizance towards cellular ways and actions. Temporal proteomics helps link these changes to phenotypic effects, providing insights into cellular responses (Sobsey et al., 2020).

Validation strategies include confirming mRNA and protein expression levels, assessing protein network functions, and analyzing protein–protein interactions (Ding and Kihara, 2019). Proteomic (MALDI-TOF mass spectrometry) has revealed how microalgae (primary producers) respond to pesticide stress, as persistent organic pollutants revealed adaptive molecular pathways involved in microalgal resilience. The mercury-induced neurotoxicity in chronically exposed marine medaka (*Oryzias melastigma*) showed 16 significantly altered protein spots in proteomic analysis related to oxidative stress, cytoskeletal assembly dysfunction, and metabolic disturbances (Wang et al., 2015).

9.12 Molecular dynamics simulation in proteomics

Molecular dynamics simulation (MDS) (computational technique) models the movement of atoms in proteins over time to understand the dynamic behavior and flexibility, thereby providing insights into crucial biological processes like protein folding, conformational changes, and protein interactions with other molecules (Jez, 2017). Pollutants induce alteration in structural and binding properties necessary for protein functioning, including ligand recognition and enzymatic catalysis (Diezmann and Seitz, 2011). For example, a decrease in arginine and proline content in psychrophilic proteins results in the formation of salt bridges and multiple hydrogen bonds, which reduces protein flexibility. Molecular dynamics simulations provide the motion trajectory of each interaction unit and help in analyzing the structural changes induced by pollutants. Amber, Gromacs, and Namd, etc., are the commonly used software for MDS proteomic studies (Lee et al., 2017). The modeling tools like STITCH 4.0 and STRING 9.0 mapped protein–chemical and protein–protein interactions affected by air pollutants (carbon monoxide and hydrogen sulfide). These tools identified six key proteins (FECH, HMOX1, ALB, CTH, CBS, and CBSL) involved in Reactome and KEGG pathways. Functional analysis conducted via OmicsBox v3.0 revealed pollutant impacts on key pathways like erythropoiesis, oxidative stress, amino acid metabolism, respiratory, cardiovascular, and neurological signaling. Molecular Docking (AutoDock Vina) showed CO has a strong binding affinity with ADI1 (-1.9 kcal/mL), while H_2S has a weaker binding (below −0.9 kcal/mL), highlighting their influence on critical biological processes and protein networks (Sheikh et al., 2025).

9.13 Molecular docking and homology modeling

Computerized in silico structural modeling programmes serve as a powerful tool to predict and verify experimental outcomes, particularly in understanding protein-ligand interactions. Pollutants can be computationally placed at protein active sites, with iterative optimization of molecular positions and conformation. Molecular docking and homology modeling (MDHM) optimizes protein conformations and binding affinities to near natural conformations for pollutant recognition, using score functions (Pagadala et al., 2017). For example, docking and modeling of rat nuclear hormone receptors to predict organic pollutant binding (Wu et al., 2009), specific binding of 4,5-dichloro-2-n-octyl-4-isothiazolin-3-one (Chen et al., 2017) helps in understanding pollutant-protein interactions. In addition to MDHM analysis, the protein-ligand interactions prediction also includes the chemical and proteomic data based integrated statistical approaches (CPDISA) (Yamanishi et al., 2011). The CPDISA clusters various protein domains that bind similar chemical motifs to improve specificity assessments and forecast novel protein-pollutant interactions.

9.14 Pollutant docking

Structurally similar pollutants dock with the same target protein, allowing researchers to validate the binding capabilities against a shared target biomarker (Liu et al., 2017). The tools, like the maximum common substructure algorithm (MCSA), screen structural resemblance and evaluate structural similarity among related compounds for the identification of pollutants with shared substructures (Cao et al., 2008). MCSA narrows down the specific peptides/domains within a protein that recognize structurally similar pollutants. For example, acetylcholinesterase enzyme can be used to monitor a wide range of organic pollutants for broad-spectrum monitoring and biomarker discovery (Fu et al., 2018). The ligand-based virtual screening specifies the binding potential of structurally overlapping compounds (Eckert and Bajorath, 2007). The heat shock protein 70 can be used as a biomarker responsive to benzene ring-containing pollutants, viz., bisphenol A, 3-benzylidene camphor, xylene, and toluene (Schirling et al., 2006).

9.15 Proteomics integrated biomonitoring

The direct quantification method based upon the expression of specific biomolecules is used for measuring the toxic potential of bioaccumulated xenobiotic pollutants within exposed organisms (Sarkar et al., 2006). Environmental proteomics identifies protein markers of pollutants to generate protein patterns that indicate xenobiotics-induced protein variations. By integrating precise proteomic markers, biomonitoring offers sharper insights into ecosystem health, structural integrity as well as sustainability, thus providing a direct way to quantify the toxic potential of bioaccumulated pollutants (Knigge et al., 2004). The integrated biomarker proteomic (IBP) index for pollution monitoring in European Eel (*Anguilla anguilla*) explored the use of proteomic data as early warning biomarkers for environmental pollution (Roland et al., 2016). The authors identified 34 protein spots as pollutant-specific protein expression signatures, supporting IBP as a promising tool for ecological risk assessment and offering a global index of proteome alteration applicable to endangered species. Biomonitoring is now being recognized as an important component in global programmes, particularly in assessing pollutants in oceans and river basins, like in biomonitoring programmes initiated by European member states in the Water Framework Directive.

9.16 Microbial proteomics for assessment of environmental pollution

Microbial proteomics is a large-scale study of the entire protein repertoire directly extracted from complex environmental microbial communities, offering critical insights into microbe–environment interactions. It has significant potential in revealing the functional dynamics across diverse aquatic and soil ecosystems (Wang et al., 2016). Genome decoding and high-throughput sequencing innovations are rapidly establishing microbial community proteomics as a powerful tool in chemical hazard assessment and health risk evaluation (Ge et al., 2013). Aquatic habitat comprising Earth's largest ecosystem harbors diverse microbial communities deciphering metabolic functioning in regulating biogeochemical cycling of biogenic elements (Azam and Malfatti, 2007). Similarly, soil, as a vast and diverse microbiota ecosystem that drives decomposition and remediation, offers insights into microbe–environment interactions (Bastida et al., 2010). Microbial encoded proteins that show resemblance to known reference catalytic proteins can be potentially evaluated for their specific pollutant degradation activity. Selected microbial strains are assayed for biodegradation capabilities, and the short-listed ones are further analyzed for functionality (Ufarté et al., 2015). Functional verification techniques confirm homologous proteins that recognize and actively transform the target pollutants. Subsequently, the comprehensive studies that include proteomics, metabolomics, and related regulatory networks analysis reveal the degradation mechanisms (Mishra et al., 2021).

Bioinformatic tools based on protein-pollutant recognition are gaining attraction for predicting the pollutant degradation by specific microbial enzymes. These tools leverage databases like GenBank and UniProt for elucidating the ligand-protein interactions and identification of target pollutants recognized by functional proteins (Nagarajan et al., 2022). After identifying target proteins, the sequence alignment is performed for homologous sequences in various microbes. The conserved sequences are essential for predicting and selecting functional proteins. The sequence identity in enzyme superfamilies, maintaining three-dimensional structural similarity with conserved active site residues, is critical for substrate recognition, binding, and catalysis (Gerlt and Babbitt, 2001). Notably, most of the amino acid substitutions occurring outside the active sites are responsible for preserving the structural integrity and protein folding (Redfern et al., 2008). Proteomic analysis follows the gene editing techniques to isolate and identify the functional proteins. The RNA interference (RNAi) and CRISPR/Cas9 help in knocking out genes encoding potential functional proteins. In silico predictions and experimental protein-pollutant affinity validations are then performed for protein recognition. Additionally, the protein function verification includes metabolomic analysis that encompasses the extraction, identification, pathway mapping, and bioinformatics interpretation steps. These systematic and comprehensive approaches help investigate the sequence similarity thresholds for identifying effective proteins.

Some reported pioneering metaproteomic studies on the microbial communities devoted to environmental pollution assessment have improved our understanding of the natural microbial community systems under pollutant stress. *Halobacterium* and related halophiles that thrive in hypersaline environments through unique metabolic adaptations, revealing their physiological versatility and novel enzyme potential (Kennedy et al., 2001). Highly adaptable, easily cultured pseudomonads with diverse metabolic modes of aerobic decomposition and pollutant biodegradation showed toxin-degrading capacity due to flexible metabolism (Krayl et al., 2003). Gram-negative *Shewanella oneidensis* bacterium with respiratory versatility that enables metals and radionuclide reduction is valuable in bioremediation (Elias et al., 2006). Proteomic studies on Methanogens in engineered ecosystems reveal that *Methylococcus capsulatus* regulates over 100 proteins in metal (copper) stress as specialized stress enzymes (Kao et al., 2004). Similarly, proteomic analysis on *Desulfovibrio vulgaris* (Sulfate-reducing bacteria) reveals nearly

1000 proteins linked to bioremediation, along with many whose functions are being resolved through non-homology approaches (Zhang et al., 2006). Though still emerging, the metaproteome analysis links the microbial protein functions to pollutant-induced alterations in the ecosystems despite limitations (Schneider and Riedel, 2010).

9.17 Future perspectives

Environmental study in the coming times is predicted to be a multidisciplinary consortium of proteomics, genomics, and metabolomics branches of bioinformatics. KEGG (Kyoto encyclopedia of genes and genomes) is an important resource in multi-omics research and integrating data from various OMICS branches and categorizing further into areas like environmental information processing, organismal systems, and cellular processes. However, processing such big data from OMICS databases requires more advanced bioinformatics modeling tools that can offer the holistic and accurate assessment of environmental stressors and their possible biological effects. Proteomic investigations are mere informatics analyses that need experimental validation. Therefore to overcome this snag, the combination of both systematic and integrative approaches involving high-throughput screening with functional verification is often advocated. Integrating artificial intelligence and machine learning with proteomics can also enhance data interpretation, biomarker prioritization, and pollution assessments.

9.18 Conclusion

Proteins as proteomic tools are tracked down (selection and identification) as targeting biomarkers in pollution monitoring. Uncovering the critical molecular and systemic mechanisms provides insights into the ecological risks of environmental pollutants. Understanding of pollutant–protein interactions and systematic screening of biomarkers reflects the impact of environmental pollutants and helps in exploring stressor degradation technology. In silico approaches for environmental pollution screening are gaining attention for predicting, identifying, and mitigating environmental xenobiotics that disrupt ecological balance. These methods are efficient, cost-effective, and less time-consuming and resource-intensive than traditional remediation methods.

Acknowledgment

AK acknowledges SERB for the Science and Technology Award for Research (STAR award, STR/2021/000019) and ANRF for research project (ANRF/IRG/2025/000260/LS).

References

Aebersold, R., Mann, M., 2016. Mass-spectrometric exploration of proteome structure and function. Nature 537 (7620), 347–355. Available at https://doi.org/10.1038/nature19949.

Aldehoff, A.S., Karkossa, I., Broghammer, H., Krupka, S., Weiner, J., Goerdeler, C., et al., 2025. Advanced proteomics approaches hold potential for the risk assessment of metabolism-disrupting chemicals as omics-based NAM: a case study using the phthalate substitute DINCH. Environmental Science & Technology 59 (31), 16193–16216. Available at https://doi.org/10.1021/acs.est.5c01206.

Altenburger, R., Scholz, S., Schmitt-Jansen, M., Busch, W., Escher, B.I., 2012. Mixture toxicity revisited from a toxicogenomic perspective. Environmental Science and Technology 46 (5), 2508–2522. Available at https://doi.org/10.1021/es2038036.

Amelina, H., Apraiz, I., Sun, W., Cristobal, S., 2007. Proteomics-based method for the assessment of marine pollution using liquid chromatography coupled with two-dimensional electrophoresis. Journal of Proteome Research 6 (6), 2094–2104. Available at https://doi.org/10.1021/pr060689s.

Apraiz, I., Mi, J., Cristobal, S., 2006. Identification of proteomic signatures of exposure to marine pollutants in Mussels (Mytilus edulis). Molecular & Cellular Proteomics 5 (7), 1274–1285. Available at https://doi.org/10.1074/mcp.m500333-mcp200.

Azam, F., Malfatti, F., 2007. Microbial structuring of marine ecosystems. Nature Reviews Microbiology 5 (10), 782–791. Available at https://doi.org/10.1038/nrmicro1747.

Bantscheff, M., Lemeer, S., Savitski, M.M., Kuster, B., 2012. Quantitative mass spectrometry in proteomics: critical review update from 2007 to the present. Analytical and Bioanalytical Chemistry 404 (4), 939–965. Available at https://doi.org/10.1007/s00216-012-6203-4.

Bastida, F., Nicolás, C., Moreno, J.L., Hernández, T., García, C., 2010. Tracing changes in the microbial community of a hydrocarbon-polluted soil by culture-dependent proteomics. Pedosphere 20 (4), 479–485. Available at https://doi.org/10.1016/s1002-0160(10)60037-9.

Bernhard, A., Poulsen, R., Brun Hansen, A.M., Hansen, M., 2023. Toxicometabolomics as a tool for next generation environmental risk assessment. EFSA Journal 21 (1). Available at https://doi.org/10.2903/j.efsa.2023.e211005, efsa.onlinelibrary.wiley.com/journal/18314732/.

Bhushan, V., Nita-Lazar, A., 2024. Recent advancements in subcellular proteomics: growing impact of organellar protein niches on the understanding of cell biology. Journal of Proteome Research 23 (8), 2700–2722. Available at https://doi.org/10.1021/acs.jproteome.3c00839.

Boutet, S., Lomb, L., Williams, G.J., Barends, T.R.M., Aquila, A., Doak, R.B., et al., 2012. High-resolution protein structure determination by serial femtosecond crystallography. Science (New York, N.Y.) 337 (6092), 362–364. Available at https://doi.org/10.1126/science.1217737, http://www.sciencemag.org/content/337/6092/362.full.pdf.

Braconi, D., Bernardini, G., Santucci, A., 2011. Linking protein oxidation to environmental pollutants: redox proteomic approaches. Journal of Proteomics 74 (11), 2324–2337. Available at https://doi.org/10.1016/j.jprot.2011.06.029.

Cao, Y., Jiang, T., Girke, T., 2008. A maximum common substructure-based algorithm for searching and predicting drug-like compounds. Bioinformatics (Oxford, England) 24 (13), i366. Available at https://doi.org/10.1093/bioinformatics/btn186.

Carberry, C.K., Rager, J.E., 2023. The impact of environmental contaminants on extracellular vesicles and their key molecular regulators: a literature and database-driven review. Environmental and Molecular Mutagenesis 64 (1), 50–66. Available at https://doi.org/10.1002/em.22522, http://onlinelibrary.wiley.com/journal/10.1002/(ISSN)1098-2280.

Caro, J.A., Harpole, K.W., Kasinath, V., Lim, J., Granja, J., Valentine, K.G., et al., 2017. Entropy in molecular recognition by proteins. Proceedings of the National Academy of Sciences of the United States of America 114 (25), 6563–6568. Available at https://doi.org/10.1073/pnas.1621154114, http://www.pnas.org/content/114/25/6563.full.pdf.

Cavalli, A., Salvatella, X., Dobson, C.M., Vendruscolo, M., 2007. Protein structure determination from NMR chemical shifts. Proceedings of the National Academy of Sciences 104 (23), 9615–9620. Available at https://doi.org/10.1073/pnas.0610313104.

Chatterjee, M., Roy, K., 2022. Recent advances on modelling the toxicity of environmental pollutants for risk assessment: from single pollutants to mixtures. Current Pollution Reports 8 (2), 81–97. Available at https://doi.org/10.1007/s40726-022-00219-6.

Chen, L., Au, D.W.T., Hu, C., Peterson, D.R., Zhou, B., Qian, P.Y., 2017. Identification of molecular targets for 4,5-dichloro-2-n-octyl-4-isothiazolin-3-one (DCOIT) in teleosts: new insight into mechanism of toxicity. Environmental Science and Technology 51 (3), 1840–1847. Available at https://doi.org/10.1021/acs.est.6b05523, http://pubs.acs.org/journal/esthag.

Diezmann, F., Seitz, O., 2011. DNA-guided display of proteins and protein ligands for the interrogation of biology. Chemical Society Reviews 40 (12), 5789–5801. Available at https://doi.org/10.1039/c1cs15054e.

Ding, Z., Kihara, D., 2019. Computational identification of protein-protein interactions in model plant proteomes. Scientific Reports 9 (1). Available at https://doi.org/10.1038/s41598-019-45072-8.

Eckert, H., Bajorath, J., 2007. Molecular similarity analysis in virtual screening: foundations, limitations and novel approaches. Drug Discovery Today 12 (5-6), 225–233. Available at https://doi.org/10.1016/j.drudis.2007.01.011.

Eisenberg-Lerner, A., Ciechanover, A., Merbl, Y., 2016. Post-translational modification profiling – a novel tool for mapping the protein modification landscape in cancer. Experimental Biology and Medicine 241 (14), 1475–1482. Available at https://doi.org/10.1177/1535370216651732.

Elias, D.A., Monroe, M.E., Smith, R.D., Fredrickson, J.K., Lipton, M.S., 2006. Confirmation of the expression of a large set of conserved hypothetical proteins in Shewanella oneidensis MR-1. Journal of Microbiological Methods 66 (2), 223–233. Available at https://doi.org/10.1016/j.mimet.2005.11.009.

Fu, H., Xia, Y., Chen, Y., Xu, T., Xu, L., Guo, Z., et al., 2018. Acetylcholinesterase is a potential biomarker for a broad spectrum of organic environmental pollutants. Environmental Science & Technology 52 (15), 8065–8074. Available at https://doi.org/10.1021/acs.est.7b04004.

Ge, Y., Wang, D.Z., Chiu, J.F., Cristobal, S., Sheehan, D., Silvestre, F., et al., 2013. Environmental OMICS: current status and future directions. Journal of Integrated OMICS 3 (2), 75–87. Available at https://doi.org/10.5584/jiomics.v3i2.141, www.jiomics.com/index.php/jio/index.

Gerlt, J.A., Babbitt, P.C., 2001. Divergent evolution of enzymatic function: mechanistically diverse superfamilies and functionally distinct suprafamilies. Annual Review of Biochemistry 70, 209–246. Available at https://doi.org/10.1146/annurev.biochem.70.1.209.

Ghedira, J., Chicano-Gálvez, E., Fernández-Cisnal, R., Jebali, J., Banni, M., Chouba, L., et al., 2016. Using environmental proteomics to assess pollutant response of Carcinus maenas along the Tunisian coast. Science of The Total Environment 541, 109–118. Available at https://doi.org/10.1016/j.scitotenv.2015.09.032.

Gong, X., Qian, H., Zhou, X., Wu, J., Wan, T., Cao, P., et al., 2016. Structural insights into the niemann-pick C1 (NPC1)-mediated cholesterol transfer and ebola infection. Cell 165 (6), 1467–1478. Available at https://doi.org/10.1016/j.cell.2016.05.022.

Gorbach, D.P., Leonova, T.S., Orlova, A.A., Frolova, N.V., Gurina, A.K., Danko, K.V., et al., 2025. Shotgun proteomics in the study of adaptive stress responses of plant proteome. Ecological Genetics 23 (1), 39–63. Available at https://doi.org/10.17816/ecogen642716, https://journals.eco-vector.com/ecolgenet/article/view/636450/194044.

Guo, H., Wang, L., Deng, Y., Ye, J., 2021. Novel perspectives of environmental proteomics. Science of The Total Environment 788, 147588. Available at https://doi.org/10.1016/j.scitotenv.2021.147588.

Harish, B.M., Sharma, S., Choudhary, D.R., Kumar, A., Choudhary, A., Kaur, H., et al., 2022. Plant proteomics and environmental pollution. Plants and their Interaction to Environmental Pollution: Damage Detection, Adaptation, Tolerance, Physiological and Molecular Responses. Elsevier, India, pp. 411–428. Available at http://doi.org/10.1016/B978-0-323-99978-6.00023-6, https://www.sciencedirect.com/book/9780323999786.

Harrill, J.A., Viant, M.R., Yauk, C.L., Sachana, M., Gant, T.W., Auerbach, S.S., et al., 2021. Progress towards an OECD reporting framework for transcriptomics and metabolomics in regulatory toxicology. Regulatory Toxicology and Pharmacology 125. Available at https://doi.org/10.1016/j.yrtph.2021.105020, http://www.elsevier.com/inca/publications/store/6/2/2/9/3/9/index.htt.

Hudson, K.M., Shiver, E., Yu, J., Mehta, S., Jima, D.D., Kane, M.A., et al., 2021. Transcriptomic, proteomic, and metabolomic analyses identify candidate pathways linking maternal cadmium exposure to altered neurodevelopment and behavior. Scientific Reports 11 (1). Available at https://doi.org/10.1038/s41598-021-95630-2, www.nature.com/srep/index.html.

Jez, J.M., 2017. Revisiting protein structure, function, and evolution in the genomic era. Journal of Invertebrate Pathology 142, 11–15. Available at https://doi.org/10.1016/j.jip.2016.07.013, http://www.elsevier.com/inca/publications/store/6/2/2/8/8/3/index.htt.

Jubeaux, G., Audouard-Combe, F., Simon, R., Tutundjian, R., Salvador, A., Geffard, O., et al., 2012. Vitellogenin-like proteins among invertebrate species diversity: potential of proteomic mass spectrometry for biomarker development. Environmental Science & Technology 46 (11), 6315–6323. Available at https://doi.org/10.1021/es300550h.

Kao, W.C., Chen, Y.R., Yi, E.C., Lee, H., Tian, Q., Wu, K.M., et al., 2004. Quantitative proteomic analysis of metabolic regulation by copper ions in Methylococcus capsulatus (Bath). Journal of Biological Chemistry 279 (49), 51554–51560. Available at https://doi.org/10.1074/jbc.M408013200.

Katsavelis, D., van der Hart, M.G.C., Wolters, J.C., Permentier, H.P., Horvatovich, P., Cremers, T.I.F.H., 2025. Impact of sample preparation strategies on the quantitative accuracy of low-abundance serum proteins in shotgun proteomics. Journal of Proteome Research 24 (9), 4792–4803. Available at https://doi.org/10.1021/acs.jproteome.5c00572, http://pubs.acs.org/journal/jprobs.

Kennedy, S.P., Ng, W.V., Salzberg, S.L., Hood, L., DasSarma, S., 2001. Understanding the adaptation of Halobacterium species NRC-1 to its extreme environment through computational analysis of its genome sequence. Genome Research 11 (10), 1641–1650. Available at https://doi.org/10.1101/gr.190201.

Knigge, T., Monsinjon, T., Andersen, O.K., 2004. Surface-enhanced laser desorption/ionization-time of flight-mass spectrometry approach to biomarker discovery in blue mussels (Mytilus edulis) exposed to polyaromatic hydrocarbons and heavy metals under field conditions. Proteomics 4 (9), 2722–2727. Available at https://doi.org/10.1002/pmic.200300828.

Krayl, M., Benndorf, D., Loffhagen, N., Babel, W., 2003. Use of proteomics and physiological characteristics to elucidate ecotoxic effects of methyl tert -butyl ether in Pseudomonas putida KT2440. Proteomics 3 (8), 1544–1552. Available at https://doi.org/10.1002/pmic.200300477.

Krewski, D., Andersen, M.E., Tyshenko, M.G., Krishnan, K., Hartung, T., Boekelheide, K., et al., 2020. Toxicity testing in the 21st century: progress in the past decade and future perspectives. Archives of Toxicology 94 (1), 1–58. Available at https://doi.org/10.1007/s00204-019-02613-4.

Kumar, V., Sharma, J.K., Kumar, S., Gulia, J., Singh, N.P., Santal, A.R., 2024. Metagenomic and proteomic approach for bioremediation of environmental pollutants. Development in Waste Water Treatment Research and Processes: Applied Technologies for Clean up of Environmental Contaminants. Elsevier, India, pp. 635–647. Available at http://doi.org/10.1016/B978-0-443-13615-3.00020-0, https://www.sciencedirect.com/book/9780443136153.

Lee, H.J., Jedrychowski, M.P., Vinayagam, A., Wu, N., Shyh-Chang, N., Hu, Y., et al., 2017. Proteomic and metabolomic characterization of a mammalian cellular transition from quiescence to proliferation. Cell Reports 20 (3), 721–736. Available at https://doi.org/10.1016/j.celrep.2017.06.074, http://www.sciencedirect.com/science/journal/22111247.

Li, X., Gong, Y., Wang, Y., Wu, S., Cai, Y., He, P., et al., 2005. Comparison of alternative analytical techniques for the characterisation of the human serum proteome in HUPO Plasma Proteome Project. Proteomics 5 (13), 3423–3441. Available at https://doi.org/10.1002/pmic.200401226.

Li, G.K., Gao, J., Peng, H., Shen, Y.O., Ding, H.P., Zhang, Z.M., et al., 2016. Proteomic changes in maize as a response to heavy metal (lead) stress revealed by iTRAQ quantitative proteomics. Genetics and Molecular Research 15 (1). Available at https://doi.org/10.4238/gmr.15017254.

Li, Y., Shi, X., Chen, Y., Luo, S., Qin, Z., Chen, S., et al., 2023a. Quantitative proteomic analysis of the mechanism of Cd toxicity in Enterobacter sp. FM-1: comparison of different growth stages. Environmental Pollution 336, 122513. Available at https://doi.org/10.1016/j.envpol.2023.122513.

Li, Y., Zhang, Z., Jiang, S., Xu, F., Tulum, L., Li, K., et al., 2023b. Using transcriptomics, proteomics and phosphoproteomics as new approach methodology (NAM) to define biological responses for chemical safety assessment. Chemosphere 313, 137359. Available at https://doi.org/10.1016/j.chemosphere.2022.137359.

Liang, X., Martyniuk, C.J., Simmons, D.B.D., 2020. Are we forgetting the "proteomics" in multi-omics ecotoxicology? Comparative Biochemistry and Physiology - Part D: Genomics and Proteomics 36. Available at https://doi.org/10.1016/j.cbd.2020.100751, https://www.journals.elsevier.com/comparative-biochemistry-and-physiology-part-d-genomics-and-proteomics.

Liu, J.Q., Dai, S.X., Zheng, J.J., Guo, Y.C., Li, W.X., Li, G.H., et al., 2017. The identification and molecular mechanism of anti-stroke traditional Chinese medicinal compounds. Scientific Reports 7. Available at https://doi.org/10.1038/srep41406, www.nature.com/srep/index.html.

López-Pedrouso, M., Varela, Z., Franco, D., Fernández, J.A., Aboal, J.R., 2020. Can proteomics contribute to biomonitoring of aquatic pollution? A critical review. Environmental Pollution 267, 115473. Available at https://doi.org/10.1016/j.envpol.2020.115473.

Lundberg, E., Borner, G.H.H., 2019. Spatial proteomics: a powerful discovery tool for cell biology. Nature Reviews. Molecular Cell Biology 20 (5), 285–302. Available at https://doi.org/10.1038/s41580-018-0094-y.

Madian, A.G., Regnier, F.E., 2010a. Profiling carbonylated proteins in human plasma. Journal of Proteome Research 9 (3), 1330–1343. Available at https://doi.org/10.1021/pr900890k.

Madian, A.G., Regnier, F.E., 2010b. Proteomic identification of carbonylated proteins and their oxidation sites. Journal of Proteome Research 9 (8), 3766–3780. Available at https://doi.org/10.1021/pr1002609.

McDonagh, B., Tyther, R., Sheehan, D., 2006. Redox proteomics in the mussel, Mytilus edulis. Marine Environmental Research 62 (1), S101. Available at https://doi.org/10.1016/j.marenvres.2006.04.001.

Medzihradszky, K.F., Chalkley, R.J., 2015. Lessons in de novo peptide sequencing by tandem mass spectrometry. Mass Spectrometry Reviews 34 (1), 43–63. Available at https://doi.org/10.1002/mas.21406, http://onlinelibrary.wiley.com/journal/10.1002/(ISSN)1098-2787.

Meier, M.J., Harrill, J., Johnson, K., Thomas, R.S., Tong, W., Rager, J.E., et al., 2025. Progress in toxicogenomics to protect human health. Nature Reviews Genetics 26 (2), 105–122. Available at https://doi.org/10.1038/s41576-024-00767-1, https://www.nature.com/nrg/.

Mishra, S., Lin, Z., Pang, S., Zhang, W., Bhatt, P., Chen, S., 2021. Recent advanced technologies for the characterization of xenobiotic-degrading microorganisms and microbial communities. Frontiers in Bioengineering and Biotechnology 9. Available at https://doi.org/10.3389/fbioe.2021.632059.

Modi, T., Campitelli, P., Kazan, I.C., Ozkan, S.B., 2021. Protein folding stability and binding interactions through the lens of evolution: a dynamical perspective. Current Opinion in Structural Biology 66, 207–215. Available at https://doi.org/10.1016/j.sbi.2020.11.007, http://www.elsevier.com/locate/sbi.

Muralidharan, S., Thompson, E., Raftos, D., Birch, G., Haynes, P.A., 2012. Quantitative proteomics of heavy metal stress responses in Sydney rock oysters. Proteomics 12 (6), 906–921. Available at https://doi.org/10.1002/pmic.201100417.

Nagarajan, M., Maadurshni, G.B., Manivannan, J., 2022. Systems toxicology approach explores target-pathway relationship and adverse health impacts of ubiquitous environmental pollutant bisphenol A. Journal of Toxicology and Environmental Health, Part A 85 (6), 217–229. Available at https://doi.org/10.1080/15287394.2021.1994492.

Nobeli, I., Favia, A.D., Thornton, J.M., 2009. Protein promiscuity and its implications for biotechnology. Nature Biotechnology 27 (2), 157–167. Available at https://doi.org/10.1038/nbt1519, http://www.nature.com/nbt/index.html.

Pagadala, N.S., Syed, K., Tuszynski, J., 2017. Software for molecular docking: a review. Biophysical Reviews 9 (2), 91–102. Available at https://doi.org/10.1007/s12551-016-0247-1, http://www.springer.com/life+sciences/biochemistry+%26+biophysics/journal/12551.

Palabiyik, A.A., Palabiyik, E., 2025. Proteomics in biomarker discovery: uncovering disease-specific interactome and post-translational modification networks. Journal of Proteins and Proteomics 16 (3), 237–251. Available at https://doi.org/10.1007/s42485-025-00190-y.

Pan, C., Qi, X., Yang, X., Cheng, B., Cheng, S., Liu, L., et al., 2025. Large-scale plasma proteomics uncovers novel targets linking ambient air pollution and depression. Molecular Psychiatry 30 (8), 3650–3659. Available at https://doi.org/10.1038/s41380-025-02953-x.

Poblete-Naredo, I., Albores, A., 2016. Molecular biomarkers to assess health risks due to environmental contaminants exposure. Biomédica 36 (2), 309. Available at https://doi.org/10.7705/biomedica.v36i3.2998.

Redfern, O.C., Dessailly, B., Orengo, C.A., 2008. Exploring the structure and function paradigm. Current Opinion in Structural Biology 18 (3), 394–402. Available at https://doi.org/10.1016/j.sbi.2008.05.007.

Riva, C., Binelli, A., Rusconi, F., Colombo, G., Pedriali, A., Zippel, R., et al., 2011. A proteomic study using zebra mussels (D. polymorpha) exposed to benzo(α)pyrene: the role of gender and exposure concentrations. Aquatic Toxicology 104 (1-2), 14–22. Available at https://doi.org/10.1016/j.aquatox.2011.03.008.

Rodríguez-Ortega, M.J., Grøsvik, B.E., Rodríguez-Ariza, A., Goksøyr, A., López-Barea, J., 2003. Changes in protein expression profiles in bivalve molluscs (Chamaelea gallina) exposed to four model environmental pollutants. Proteomics 3 (8), 1535–1543. Available at https://doi.org/10.1002/pmic.200300491.

Roland, K., Kestemont, P., Dieu, M., Raes, M., Silvestre, F., 2016. Using a novel "Integrated Biomarker Proteomic" index to assess the effects of freshwater pollutants in European eel peripheral blood mononuclear cells. Journal of Proteomics 137, 83–96. Available at https://doi.org/10.1016/j.jprot.2016.01.007.

Rollins, N.J., Brock, K.P., Poelwijk, F.J., Stiffler, M.A., Gauthier, N.P., Sander, C., et al., 2019. Inferring protein 3D structure from deep mutation scans. Nature Genetics 51 (7), 1170–1176. Available at https://doi.org/10.1038/s41588-019-0432-9, http://www.nature.com/ng/index.html.

Salovska, B., Liu, Y., 2023. Post-translational modification and phenotype. Proteomics 23 (3-4). Available at https://doi.org/10.1002/pmic.202200535.

Sarkar, A., Ray, D., Shrivastava, A.N., Sarker, S., 2006. Molecular biomarkers: their significance and application in marine pollution monitoring. Ecotoxicology (London, England) 15 (4), 333–340. Available at https://doi.org/10.1007/s10646-006-0069-1.

Schirling, M., Jungmann, D., Ladewig, V., Ludwichowski, K.U., Nagel, R., Köhler, H.R., et al., 2006. Bisphenol A in artificial indoor streams: II. Stress response and gonad histology in Gammarus fossarum (Amphipoda). Ecotoxicology (London, England) 15 (2), 143–156. Available at https://doi.org/10.1007/s10646-005-0044-2.

Schneider, T., Riedel, K., 2010. Environmental proteomics: analysis of structure and function of microbial communities. Proteomics 10 (4), 785–798. Available at https://doi.org/10.1002/pmic.200900450.

Shacter, E., 2000. Quantification and significance of protein oxidation in biological samples. Drug Metabolism Reviews 32 (3-4), 307–326. Available at https://doi.org/10.1081/dmr-100102336.

Sheikh, M.R., Mahmud, H.H., Hossen, M.S., Saha, D., Uddin, M.E., Hossain, M.F., et al., 2025. Modeling the interactions between chemicals and proteins to predict the health consequences of air pollution. International Journal of Environmental Research and Public Health 22 (3). Available at https://doi.org/10.3390/ijerph22030418, http://www.mdpi.com/journal/ijerph.

Shi, S.Y., Peng, M.J., Zhang, Y.P., Peng, S., 2013. Combination of preparative HPLC and HSCCC methods to separate phosphodiesterase inhibitors from Eucommia ulmoides bark guided by ultrafiltration-based ligand screening. Analytical and Bioanalytical Chemistry 405 (12), 4213–4223. Available at https://doi.org/10.1007/s00216-013-6806-4, https://link.springer.com/journal/216/volumes-and-issues.

Simon, R., Jubeaux, G., Chaumot, A., Lemoine, J., Geffard, O., Salvador, A., 2010. Mass spectrometry assay as an alternative to the enzyme-linked immunosorbent assay test for biomarker quantitation in ecotoxicology: application to vitellogenin in Crustacea (Gammarus fossarum). Journal of Chromatography. A 1217 (31), 5109–5115. Available at https://doi.org/10.1016/j.chroma.2010.06.015.

Smiljanic, K., Prodic, I., Apostolovic, D., Cvetkovic, A., Veljovic, D., Mutic, J., et al., 2019. In-depth quantitative profiling of post-translational modifications of Timothy grass pollen allergome in relation to environmental oxidative stress. Environment International 126, 644–658. Available at https://doi.org/10.1016/j.envint.2019.03.001, www.elsevier.com/locate/envint.

Sobsey, C.A., Ibrahim, S., Richard, V.R., Gaspar, V., Mitsa, G., Lacasse, V., et al., 2020. Targeted and untargeted proteomics approaches in biomarker development. Proteomics 20 (9).

Available at https://doi.org/10.1002/pmic.201900029, http://onlinelibrary.wiley.com/journal/10.1002/(ISSN)1615-9861.

Song, X., Li, J., Xu, S., Ying, R., Ma, J., Liao, C., et al., 2012. Determination of 16 polycyclic aromatic hydrocarbons in seawater using molecularly imprinted solid-phase extraction coupled with gas chromatography-mass spectrometry. Talanta 99, 75–82. Available at https://doi.org/10.1016/j.talanta.2012.04.065, https://www.journals.elsevier.com/talanta.

Souza, T.M., Kleinjans, J.C.S., Jennen, D.G.J., 2020. CHAPTER 7: Toxicogenomics and toxicoinformatics: supporting systems biology in the big data era. Issues in Toxicology 214–241. Available at https://doi.org/10.1039/9781782623656-00214, http://www.rsc.org/shop/books/series/17.asp.

Sprenger, H., Kreuzer, K., Alarcan, J., Herrmann, K., Buchmüller, J., Marx-Stoelting, P., et al., 2022. Use of transcriptomics in hazard identification and next generation risk assessment: a case study with clothianidin. Food and Chemical Toxicology 166. Available at https://doi.org/10.1016/j.fct.2022.113212, www.elsevier.com/locate/foodchemtox.

Starr, T.N., Thornton, J.W., 2016. Epistasis in protein evolution. Protein Science 1204–1218. Available at https://doi.org/10.1002/pro.2897, http://onlinelibrary.wiley.com/journal/10.1002/(ISSN)1469-896X.

Tang, H., Huang, J., Lin, H., Zhang, X., Yang, Q., Luo, N., et al., 2025. The global burden and biomarkers of cardiovascular disease attributable to ambient particulate matter pollution. Journal of Translational Medicine 23 (1). Available at https://doi.org/10.1186/s12967-025-06375-9.

Thakur, S., Chandra, A., Kumar, V., Bharti, S., 2025. Environmental Pollutants: Endocrine Disruptors/Pesticides/Reactive Dyes and Inorganic Toxic Compounds Metals, Radionuclides, and Metalloids and Their Impact on the Ecosystem. Springer Science and Business Media LLC, pp. 55–100. Available at http://doi.org/10.1007/978-981-97-7221-6_3.

Tomanek, L., 2011. Environmental proteomics: changes in the proteome of marine organisms in response to environmental stress, pollutants, infection, symbiosis, and development. Annual Review of Marine Science 3 (1), 373–399. Available at https://doi.org/10.1146/annurev-marine-120709-142729.

Trapp, J., Armengaud, J., Salvador, A., Chaumot, A., Geffard, O., 2014. Next-generation proteomics: toward customized biomarkers for environmental biomonitoring. Environmental Science & Technology 48 (23), 13560–13572. Available at https://doi.org/10.1021/es501673s.

Ufarté, L., Laville, É., Duquesne, S., Potocki-Veronese, G., 2015. Metagenomics for the discovery of pollutant degrading enzymes. Biotechnology Advances 33 (8), 1845–1854. Available at https://doi.org/10.1016/j.biotechadv.2015.10.009.

Veenstra, T.D., 2021. Omics in systems biology: current progress and future outlook. Proteomics 21 (3-4). Available at https://doi.org/10.1002/pmic.202000235, http://onlinelibrary.wiley.com/journal/10.1002/(ISSN)1615-9861.

Vitorino, R., 2024. Transforming clinical research: the power of high-throughput omics integration. Proteomes 12 (3), 25. Available at https://doi.org/10.3390/proteomes12030025.

Wang, Y., Wang, D., Lin, L., Wang, M., 2015. Quantitative proteomic analysis reveals proteins involved in the neurotoxicity of marine medaka Oryzias melastigma chronically exposed to inorganic mercury. Chemosphere 119, 1126–1133. Available at https://doi.org/10.1016/j.chemosphere.2014.09.053, www.elsevier.com/locate/chemosphere.

Wang, D.Z., Kong, L.F., Li, Y.Y., Xie, Z.X., 2016. Environmental microbial community proteomics: status, challenges and perspectives. International Journal of Molecular Sciences 17 (8). Available at https://doi.org/10.3390/ijms17081275, http://www.mdpi.com/1422-0067/17/8/1275/pdf.

Wu, B., Zhang, Y., Kong, J., Zhang, X., Cheng, S., 2009. In silico predication of nuclear hormone receptors for organic pollutants by homology modeling and molecular docking. Toxicology Letters 191 (1), 69–73. Available at https://doi.org/10.1016/j.toxlet.2009.08.005.

Yamanishi, Y., Pauwels, E., Saigo, H., Stoven, V., 2011. Extracting sets of chemical substructures and protein domains governing drug-target interactions. Journal of Chemical Information and Modeling 51 (5), 1183–1194. Available at https://doi.org/10.1021/ci100476q.

Yang, B., Kookana, R.S., Williams, M., Ying, G.G., Du, J., Doan, H., et al., 2016. Oxidation of ciprofloxacin and enrofloxacin by ferrate(VI): products identification, and toxicity evaluation. Journal of Hazardous Materials 320, 296–303. Available at https://doi.org/10.1016/j.jhazmat.2016.08.040, www.elsevier.com/locate/jhazmat.

Yang, M., Qin, H., Wang, W., Zhang, H., Long, Y., Ye, J., 2018. Global proteomic responses of Escherichia coli and evolution of biomarkers under tetracycline stress at acid and alkaline conditions. Science of The Total Environment 627, 1315–1326. Available at https://doi.org/10.1016/j.scitotenv.2018.01.342.

Yuan, M., Breitkopf, S.B., Yang, X., Asara, J.M., 2012. A positive/negative ion-switching, targeted mass spectrometry-based metabolomics platform for bodily fluids, cells, and fresh and fixed tissue. Nature Protocols 7 (5), 872–881. Available at https://doi.org/10.1038/nprot.2012.024.

Zhang, W., Culley, D.E., Gritsenko, M.A., Moore, R.J., Nie, L., Scholten, J.C.M., et al., 2006. LC-MS/MS based proteomic analysis and functional inference of hypothetical proteins in Desulfovibrio vulgaris. Biochemical and Biophysical Research Communications 349 (4), 1412–1419. Available at https://doi.org/10.1016/j.bbrc.2006.09.019.

Zhang, Y., Sun, J., Chen, C., Watanabe, H.K., Feng, D., Zhang, Y., et al., 2017. Adaptation and evolution of deep-sea scale worms (Annelida: Polynoidae): insights from transcriptome comparison with a shallow-water species. Scientific Reports 7. Available at https://doi.org/10.1038/srep46205, www.nature.com/srep/index.html.

Chapter 10

Exploring the role of immobilized microbial enzyme for bioremediation of dyes: a proteomic approach

Rimsha Amin[1], Burhan Hamid[1], Parvaze Ahmad Wani[1], Asfa Rizvi[2], and Bilal Ahmed[3]

[1]*Centre of Research for Development, University of Kashmir, Srinagar, Jammu and Kashmir, India,* [2]*Department of Botany, School of Chemical and Life Sciences, Jamia Hamdard University, New Delhi, India,* [3]*Agricultural and Biological Engineering, Purdue University, West Lafayette, IN, United States*

10.1 Introduction

Various pollutants, such as heavy metals, pesticides, dyes, and many more, have exerted negative impacts on the environment, human health, and other organisms. Dyes pose a major risk to the environment by affecting pH, temperature, oxygen content of water, contaminating soil, and public health (Al-Tohamy et al., 2022). Nowadays, synthetic dyes are widely used in industries, laboratories, and schools that leech out to the nearby streams, causing an increase in pollution as they are recalcitrant, toxic, mutagenic, and carcinogenic in nature. Even at low concentrations, dyes cause harmful effects on the environment, such as a decrease in the passage of sunlight, an increase in BOD and COD, and inhibition of photosynthesis and plant growth (Ardila-Leal et al., 2021). Although many physical and chemical methods like filtration, coagulation, and oxidizers have been used for treating wastewater, they require large amounts of energy and resources, can be toxic to the environment, and do not completely mineralize pollutants (Periyasamy, 2025). Thus an alternative method is required for treating these pollutants that can be eco-friendly, less toxic, and reduce waste generation (Yamaguchi and Miyazaki, 2024). For treatment, bioremediation using microorganisms such as bacteria, fungi, yeast, and algae has been used, as they can discolor and even mineralize dyes completely (Ardila-Leal et al., 2021). Microbial enzymes,

Proteomics Driven Bioremediation. DOI: https://doi.org/10.1016/B978-0-443-45294-9.00016-3

including laccases, peroxidases, and azoreductases, have the ability to cleave chromophoric groups (e.g., azo, anthraquinone, nitro) in dyes and thus decolorize and detoxify toxic dyes (Yadav et al., 2025). However, the use of free enzymes for treating dye-containing industrial wastewater has some disadvantages, including low stability, rapid deactivation, and limited reusability (Al-Sakkaf et al., 2023). Therefore, enzymes are immobilized on various carriers by employing various methods, including adsorption, covalent binding, entrapment, and cross-linking, etc., to achieve one or more goals such as increased catalytic efficiency, stability, and reusability in bioremediation systems (Tadesse and Liu, 2025). For example, immobilized fungal and bacterial enzymes have been used for decolorization of dyes like crystal violet, malachite green, and reactive dyes, showing better activity and stability than free enzymes (Abdelhamid et al., 2024). Many studies include exploration of various effective methods for immobilization of enzymes and versatile carriers such as metal-organic frameworks (MOFs), porous nanostructures, hydrogels, and advanced magnetic carriers to improve the performance of enzymes. These immobilized enzymes and microbial systems show a broad range of applications in industrial wastewater treatment (Periyasamy, 2025), dye bioremediation, and protecting human health and the environment. This has driven increasing interest in biological and enzyme-based approaches for dye remediation (Fig. 10.1).

10.2 Dyes: types, sources, and fate

Dyes are organic compounds highly soluble in water, natural or synthetic, that are added to the substrate, imparting different colors. Two main components, such as chromophores and auxochromes, give the dye a characteristic color, also increasing its affinity for the fibers to be colored (Kumar et al., 2025). Dyes have polar auxotrophic groups that bind to polar groups of fibers (Lellis et al., 2019) and absorb certain wavelengths of light and reflect a visible color (Souza et al., 2022). Structurally, dyes are classified into basic dyes, acidic dyes, azo dyes, disperse dyes, vat dyes, soluble, insoluble dyes, anthraquinone-based dyes, and metal complex dyes (Benkhaya et al., 2021). Dyes play an important role in textile manufacturing, paper, leather industry, as well as in medicine and biology, but cannot be degraded and thus are very toxic to the environment and human health. Among dyes, synthetic dyes are toxic compounds obtained from petrochemicals that are stable and complex aromatic compounds incapable of degradation by conventional methods. For example, azo dye, acridine dye, anthraquinone dye, crystal violet, methylene blue, and safranin. Out of all these, azo-derived dyes are most commonly used in dye production industries, but due to their loose binding to fabric, they go into streams, rivers, and other water bodies, causing trouble to humans and the ecosystem. For the degradation of azo dyes, different aerobic and anaerobic methods have been used for transferring their genotoxic forms into less-toxic forms (Ihsanullah et al., 2020).

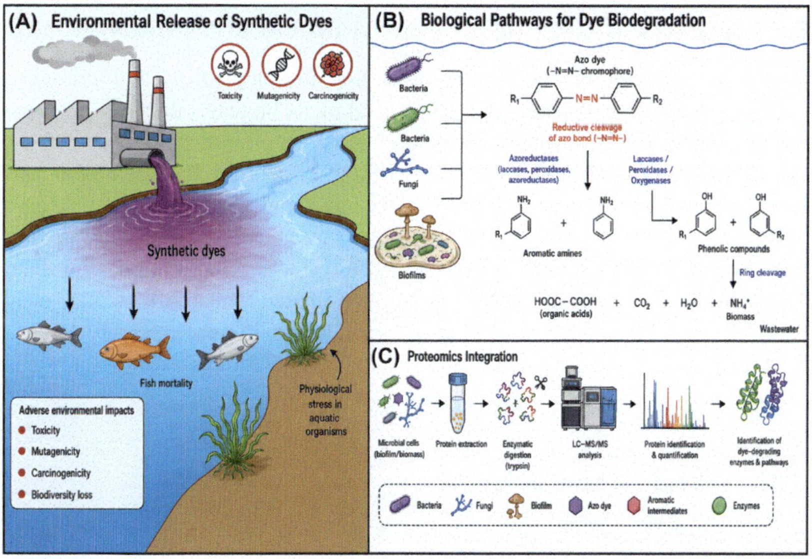

FIGURE 10.1 **Environmental impact of synthetic dyes and biological pathways for dye biodegradation.** (A) Synthetic dyes released from textile and related industries enter aquatic systems, where they persist due to their chemical stability and exert toxic effects on ecosystems. (B) Microbial cells and their enzymes, including laccases, peroxidases, and azoreductases, can cleave dye chromophores, leading to decolourization and, in some cases, detoxification of dye molecules. (C) LC-MS/MS-based proteomics integrated with bioremediation systems can identify key enzymes and pathways, which enables mechanistic understanding and guides enzyme selection, immobilization design, and process optimization.

The main source of these dyes is residuals released along with wastewater during dyeing, washing (Slama et al., 2021), containing unused dyes and other toxic intermediates from textile, leather, tannery, paper, cosmetics, drugs, paint, and other industrial sectors. Industries use these dyes for various purposes, and the wastewater effluent from these industries, along with other toxic compounds, severely affects the environment (Kayani et al., 2025). Natural dyes have been used for coloring textiles, food, and imparting color to various materials. Natural dyes have originated from numerous natural resources, and based on their origin, they are classified into animal, plant, microbial, and mineral dyes; however, plants are the major sources of natural dyes. For example, Indigo is obtained from the leaves of the plant *Indigofera tinctoria*, and *Polygonum tinctorium* is used in coloring different fabrics (Saxena and Raja, 2014). Synthetic dyes are organic compounds that are produced from raw materials of petrochemical origin (Souza et al., 2022). Azo dyes, a xenobiotic pollutant, are found in various foods and drugs. Anthraquinone dye is used in many industries. In schools for different experimental purposes, dyes like Crystal violet, methylene blue, and safranin have been frequently used for staining and visualization purposes.

10.3 Impacts of dyes on the environment and human health

The textile industry is one of the most potential industries that produces large amounts of industrial releases, including a huge production of dyes. Although these dyes are potent colorants in various food, textile, and paper industries, these dyes, however, pose serious effects on human health and the ecosystem due to their persistence, toxicity, and release of carcinogenic byproducts. Even at low concentrations, these dyes affect the aquatic ecosystem and its inhabitants; some persist in water and bio-magnify, produce toxic aromatic amines, while other dye-contaminated waters affect skin, eyes (Lellis et al., 2019), liver, and kidney, are neurotoxic, and intensify human health risks.

10.3.1 Effect on human health

The discovery of synthetic dyes allowed better production of dyes, generating industrial development, but increasing environmental pollution levels. These dyes are poorly biodegradable, harmful to the environment, cause mutations, metabolic alterations, have carcinogenic effects, and bioaccumulate, and all of this affects human beings. For example, anthraquinoid dye is very toxic and persists longer in wastewater, leading to many health issues (Ardila-Leal et al., 2021). Dyestuffs used as colorants are a major threat to human beings, particularly those working at these industrial workplaces (Muñoz et al., 2023), and cause serious respiratory issues and other side effects like watery eyes, skin irritation, vertigo, and asthma (Ozkurt et al., 2012) when these color particles are inhaled. Most of the health problems in humans and animals have been associated with textile dyes, as they are extremely harmful and may even cause cancer, cardiovascular problems, respiratory issues, and central nervous system disorders (Kumar et al., 2025). Genetic toxicity of textile dyes is the greatest potential long-term hazard to human health. A mutagenic and cytotoxic dye, Azure-B, is widely used in the textile industry, which intercalates with the DNA helix (Lellis et al., 2019) and inhibits nervous system enzymes involved in human behavior. Due to the contact of contaminated water with aquatic animals, these toxic compounds transfer through the food chain to humans, causing health disorders such as nausea, hypertension, bleeding, ulceration of the skin, and mucous membranes. Long exposure to these dyes can even damage the liver, kidney, reproductive system, brain, and central nervous system (CNS) (Ardila-Leal et al., 2021). Studies have found many dyes, such as Disperse Red 1 dye, Crystal Violet dye, and Basic Red 9 dye, used by many textile industries, and are highly mutagenic, leading to DNA damage, and their disposal in water bodies causes skin, lung irritations, and even cancer (Souza et al., 2022).

10.3.2 Effect on the environment

The consequences of the xenobiotic and recalcitrant nature of the dyes end up impacting the structure and functioning of the ecosystems. During the dyeing process, it was observed that nearly 50% of the colorant is released into the environment (Lellis et al., 2019). Some dyes persist for a very long time, and their degradation products can be carcinogenic or mutagenic, posing long-term health risks to aquatic organisms, compromising photosynthesis, and impacting aquatic flora and fauna (Kumar et al., 2025). Synthetic dyes in soil can significantly affect plant health and development due to chemical toxicity and changes in soil parameters. These dyes disrupt the absorption of essential nutrients and water by plants, affecting their growth and causing stunted development, chlorosis, and reduced biomass production (Kumar et al., 2025). Lowering the photosynthetic activity due to limited light penetration in water, one of the effects of these dyes, leads to oxygen depletion and disruption of natural aquatic life processes (Dutta et al., 2022). The textile industry releases various waste products that ultimately enter the environment, mainly into the water supply (Saratale et al., 2019). This discharge of wastewater containing dyes from different textile industries leads to contamination of the environment, which is a major concern nowadays (Sudarshan et al., 2023). The effects of the xenobiotic and persistent nature of the dyes impact the structure and functioning of the ecosystems (Lellis et al., 2019). The direct or indirect discharge of colored wastewater from industries into rivers, streams, and other water bodies also affects the surrounding ecosystem. These effects include toxicity to soil microbial populations and obstruction to enzymatic activities that are essential for nitrogen fixation and organic matter breakdown. It diminishes soil resilience and capacity to recover from contamination. Many dyes have hazardous components that cannot degrade easily, polluting water and soil (Kumar et al., 2025). These dyes are composed of heavy metals like nickel, copper, chromium, etc., which are even worse for the aquatic environment and plant growth (Lellis et al., 2019). The discharge of wastewater containing dyes while irrigation contaminates soil, reducing yield, affecting microbial population performing various metabolic activities, and leading to genotoxic effects on plants, and when discharged into water bodies, affects fish and phytoplankton.

10.4 Bioremediation pathways and enzyme systems

Bioremediation is a simple, versatile, and eco-friendly method of degradation of harmful pollutants present in soil, water, sludge, waste, or effluent. Bioremediation uses biological agents such as plants, bacteria, fungi, and yeast, where microorganisms are mostly preferred due to rapid growth and easy manipulation. Many heavy metals released from industries like antimony, chromium, mercury, zinc, copper, and cadmium are toxic to aquatic and terrestrial habitats and their inhabitants. Therefore degradation or removal of these pollutants is necessary to prevent their harmful effects on the environment (Ayilara and Babalola, 2023). In this regard, bioremediation is a preferred technique that causes biological modifications of pollutants by several processes that majorly include detoxification and mineralization. Several microorganisms belonging to different genera like *Bacillus*, *Micrococcus*, *Pseudomonas*, *Aspergillus*, etc. (Kour et al., 2022) and their enzymes have been used for degrading pollutants like hydrocarbons, heavy metals, dyes, and xenobiotics. Such microorganisms possess an extraordinary property of surviving in extreme environmental conditions and produce metabolites that can transform environmental pollutants from toxic to less-toxic forms. Thus microbial activity helps in reviving the contaminated sites naturally (Kour et al., 2022). Microorganisms naturally compete for a preferred carbon (C) source for energy requirements and also look for nitrogen (N) and phosphorus (P) through pollutants, metabolize them either partially or fully, and biodegrade (Ayilara and Babalola, 2023). Microbial bioremediation has thus proven effective in reclaiming the contaminated sites (Kour et al., 2022).

In textile industries, many dyes have been used for making fabrics, for instance, Azo dyes for coloring of various materials in cosmetics, food, and textiles. Such dyes eventually contaminate water bodies. For the degradation of such dyes, numerous microbial species have been utilized, including but not limited to *Sphingomonas paucimobilis*, *Pichia occidentalis, Shewanella indica*, *Rheinheimera* sp., *Paenibacillus polymyxa*, *Micrococcus luteus*, *Marinobacterium* sp., *Halomonas* sp., *Brevibacillus* sp., *Bacillus fusiformis*, and *Acinetobacter junii* (Ngo and Tischler, 2022; Pinheiro et al., 2022). Microbial enzymes have been used for bioremediation and detoxification of dyes and other harmful substances, which include cytochrome P450s, azoreductases (flavoenzymes) (Al-Sakkaf et al., 2023), laccases, tyrosinases, dehalogenases, dehydrogenases, proteases, and lipases (Desai et al., 2021; Sen and Nigam, 2022; Njiki et al., 2020; Ramaswamy et al., 2021). The remediation of these pollutants using microbial enzymes is considered environment-friendly, cost-effective, sustainable, energy-saving, and innovative, and sustainable.

However, due to slow processing and less obtainability of useful microorganisms, genetically engineered microorganisms have been used for bioremediation to produce specific enzymes at optimized conditions to metabolize the xenobiotic pollutants to obtain energy and biomass and transform them into harmless molecules or even to novel products without using dangerous chemicals (Kumar et al., 2025). For example, *Pseudomonas entomophila BS1* has been used to break down Reactive Black 5 (RB5). In another study, *Pseudomonas* sp. SUK1 and *Providencia rettgeri* (formerly *Pseudomonas rettgeri*) strain HSL1 has been used to degrade four azo dyes, including Direct Red 81, Reactive Orange 16, Reactive Black 5, and Disperse Red 78 (Kumar et al., 2025). As an example, the degradation pathway of an azo dye, Acid Black 210 (AB210), is shown in Fig. 10.2. It was proposed based on GC-MS analysis of metabolic intermediates and the involvement of specific enzymatic systems. AB210 is a structurally complex azo dye comprising three azo linkages and exhibiting a molecular weight of 859.8. The decolorization of AB210 by a bacterial consortium under microaerophilic conditions has previously been investigated, demonstrating the ability of microbial systems to initiate reductive cleavage of the azo bonds and facilitate subsequent breakdown of the dye molecule (Agrawal et al., 2014).

These microbes used in bioremediation can be used at optimum temperatures and genetically modified to produce more enzymes, which will enhance their biodegrading ability (Ayilara and Babalola, 2023). For degradation of azo dyes, many bacterial enzymes are used, including azoreductase, peroxidase, laccase, and tyrosinase, etc. (Lellis et al., 2019). Microorganisms found in extreme conditions, called extremophiles, have also been used for bioremediation of textile dyes found in unfavorable conditions, including *Bacillus pallidus*, Pseudomonas, *Anoxybacillus flavithermus, Clostridiales*, and *Streptomyces* (Lellis et al., 2019).

10.4.1 Techniques used for bioremediation of pollutants

Microbes can remove pollutants from the environment using different mechanisms: enzymatic oxidation, enzymatic reduction, bioleaching, biostimulation, bioaugmentation, bioaccumulation, biosorption, and precipitation. They can be done either at the native site of pollution (in situ) or taken from the pollution site to another location (ex situ), like land farming, composting, and soil biopile to immobilize the contaminants (Ayilara and Babalola, 2023). An overview of the major bioremediation techniques and their underlying mechanisms discussed in this section is illustrated in Fig. 10.3.

Those different microbial mechanisms are briefly discussed here.

1. **Enzymatic oxidation**
   Enzymatic oxidation is the process of oxidizing pollutant compounds, including dyes, heavy metals, etc., from a higher oxidation state to a lower one, making them less toxic using an enzyme (oxidoreductase), released by the microbes involved. An example of an oxidoreductase enzyme is laccase, which catalyzes the oxidation of aromatic amines (Ayilara and Babalola, 2023).
2. **Bioaugmentation**
   Microorganisms that are already selected or genetically modified are specially added to polluted sites to feed on toxic pollutants and enhance the process of remediation (Kour et al., 2022) in a process referred to as bioaugmentation. External microbes are added to polluted sites to augment the resident microbes, or the microorganisms present at the polluted site unable to do bioremediation are genetically modified and then returned to the original site for degradation of pollutants. It is a very effective, rapid, and cost-effective method of bioremediation (Ayilara and Babalola, 2023). *Aeromonas punctata* and *Shewanella putrefaciens* can degrade azo textile

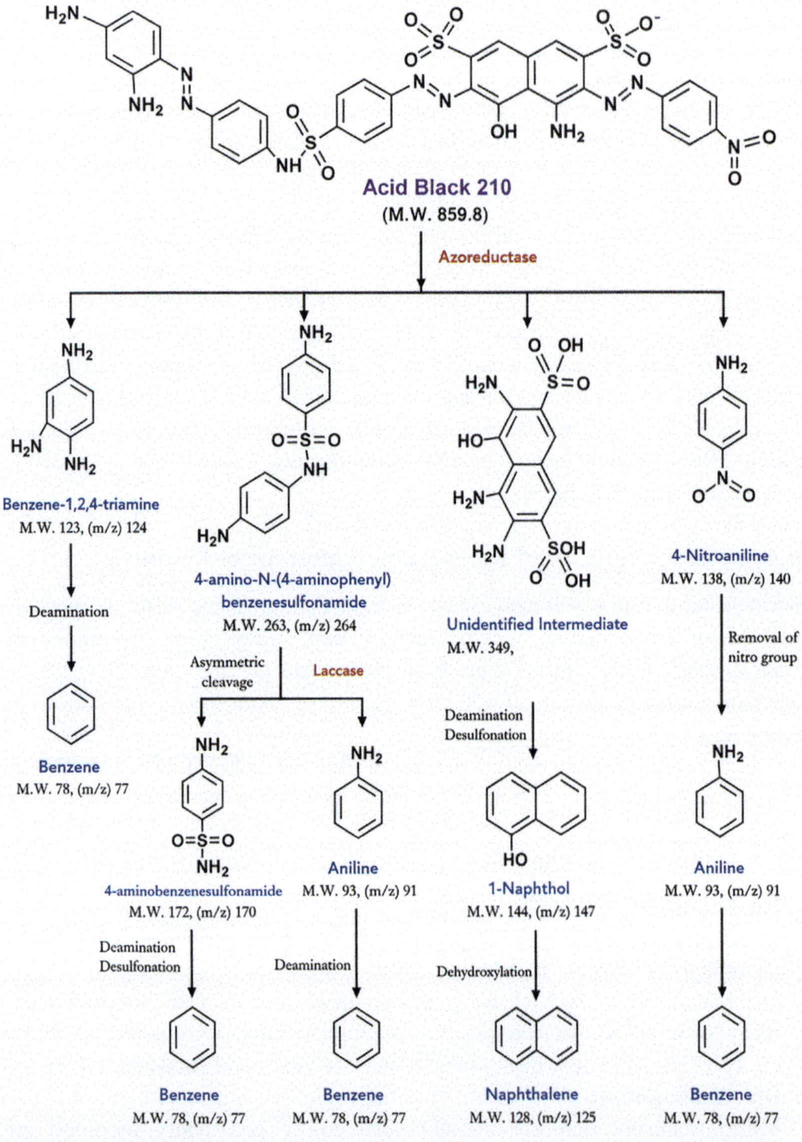

FIGURE 10.2 **Proposed bacterial degradation pathway of the azo dye Acid Black 210 by *Providencia* sp. SRS82.** The pathway illustrates the initial reductive cleavage of azo bonds, followed by the formation of aromatic amine intermediates and their subsequent transformation through enzymatic reactions such as deamination, desulfonation, and oxidative cleavage, ultimately yielding simpler aromatic compounds. The proposed pathway is based on GC-MS identification of metabolites and the involvement of key bacterial enzyme systems.
Adapted from Agrawal, S., Tipre, D., Patel, B., Dave, S., 2014. Optimization of triazo Acid Black 210 dye degradation by Providencia sp. SRS82 and elucidation of degradation pathway. Process Biochemistry, 49 (1), 110–119. https://doi.org/10.1016/j.procbio.2013.10.006.

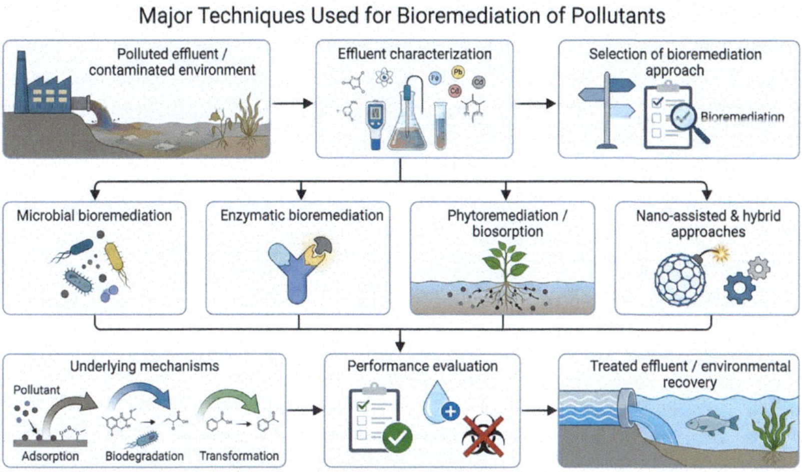

FIGURE 10.3 Overview of major bioremediation techniques for pollutant removal. Schematic overview of major techniques used for bioremediation of pollutants. Following effluent characterization, appropriate bioremediation strategies, including microbial, enzymatic, phytoremediation/biosorption, and nano-assisted approaches, are selected. These techniques operate through adsorption, biodegradation, and transformation mechanisms, leading to reduced pollutant toxicity and environmental recovery.

dyes Acid Red 88, Reactive Black 5, Direct Red 81, and Disperse Orange 3, followed by bioaugmentation (Lellis et al., 2019).

3. **Biostimulation**
Biostimulation is yet another eco-friendly and efficient process of stimulation of native microorganisms by the addition of nutrients, including phosphorus and nitrogen, O_2 (Kour et al., 2022), metabolites, and enzymes, etc. (Ayilara and Babalola, 2023), while maintaining the integrity of the microbial diversity. For example, an azo dye, Navitan Fast Blue S5R, was degraded in the presence of glucose by *Pseudomonas aeruginosa*. Also, *Aeromonas hydrophila* can degrade the Crystal Violet dye by laccase enzymes due to the addition of sucrose and yeast extract (Lellis et al., 2019).
4. **Bioaccumulation**
In microbial dye bioremediation systems, bioaccumulation can be defined as the active uptake and intracellular retention of dye molecules or their metabolites, which enables a subsequent enzymatic transformation and contributes significantly to effective decolorization and detoxification (Ayilara and Babalola, 2023).

10.5 Enzyme immobilization approaches

Immobilization refers to the technique of entrapping or anchoring biocatalysts in or on an inert carrier or support, rendering them more stable and functionally reusable (Saha and Rao, 2021). Enzymes have low operational stability and catalytic efficiency and thus need to be immobilized for better activity (Yamaguchi and Miyazaki, 2024), cost reduction, reusability, and resistance towards harsh environmental conditions (Saha and Rao, 2021). To increase catalyst concentration without disrupting their conformational structures, enzymes from different microorganisms are immobilized on a durable, reusable, and stable support material like microspheres, 3D printing, beads, and hydrogels. These immobilized enzymes are cost-effective and can be reused.

Different types of enzyme immobilization methods have been used to acquire different amounts and catalytic activity of enzymes, which include carrier binding, cross-linking, and entrapment.

10.5.1 Adsorption

Physical adsorption is a simple, economical, cost-effective, reversible, and reusable method. It involves hydrophobic interactions, hydrogen bonding, and van der Waals forces between enzymes and specific active groups of the support material, like alumina, activated carbon (Maghraby et al., 2023), silica, microcrystalline cellulose, Sepabeads, and resins (Al-Sakkaf et al., 2023). Physical adsorption involves a wide range of solid supports or carriers but has some disadvantages, like cell desorption due to weak forces, non-specific binding, shorter operational lifespan, and less tolerance to harsh conditions (Ogundolie et al., 2024).

10.5.2 Entrapment

Entrapment is an irreversible method where enzymes are placed in gels using covalent or noncovalent bonds, maintaining their viability. It involves entrapment materials like pristine materials, nanofibers, and mesoporous silica having high mechanical stability, low mass transfer (Maghraby et al., 2023), high adsorption capacity, large surface area, and adjustable pore size (Al-Sakkaf et al., 2023). The entrapment technique involves the placement of immobilized cells in fibers or a support matrix and then encasing the entire structure in a matrix of either collagen, alginate, agar, carrageenan, cellulose, gelatin, polyacrylamide, polystyrene, epoxy resin, polyester, polyurethane, or resins (Ogundolie et al., 2024). Alginate is widely used in the entrapment of enzymes because of its ability to rapidly form polymerized hydrogels with cations, such as Ca^{2+} (Al-Sakkaf et al., 2023). For example, a bacterial mixture was immobilized on alginate beads to treat the effluent of a textile and dyeing industry (Saha and Rao, 2021). Another example using the entrapment immobilization method for increasing their thermal stability includes Manganese peroxidase (MnP), which was entrapped in an agarose gel matrix for effluent decolorization and industrial pollutant degradation (Bilal et al., 2017).

10.5.3 Encapsulation

Encapsulation is also an irreversible method of immobilization. Enzymes are restricted within semi-permeable capsules and allow movement of small substrates, nutrients, or products, providing protection from the external environment, reducing enzyme leakage (Al-Sakkaf et al., 2023), controlled release of enzymes, and enhancing bioremediation efficiency (Ogundolie et al., 2024). Studies have found that enzymes like HRP and laccases have the ability to degrade dyes after encapsulation (Saha and Rao, 2021).

10.5.4 Covalent binding

Covalent bonding is a stable, irreversible technique where highly stable and strong covalent bonds are formed using amino acids in an enzyme's side chain, such as arginine, aspartic acid, and histidine, and a reactive group of various supports, such as hydrophilic polysaccharide polymers (Maghraby et al., 2023), porous silica, agarose and chitosan, increasing specific activity, reusability, and stability of the enzymes (Al-Sakkaf et al., 2023). Some of the enzymes covalently immobilized for dye degradation include laccase, azoreductase, and HRP (Saha and Rao, 2021).

10.5.5 Crosslinking

Cross-linking is an irreversible carrier-free immobilization method for immobilizing enzymes that do not require a support material, and reduces enzyme leakage. For example, glutaraldehyde, diazonium salt, and hexamethylene diisocyanate are some of the cross-linking agents used to cross-link enzymes by forming a chemical bond between the enzyme and this compound and thus forming an insoluble aggregate that retains their enzymatic activity (Al-Sakkaf et al., 2023). Several enzymes, such as laccase, tyrosinase, and peroxidases, have been immobilized by crosslinking and used for dye degradation and wastewater treatment (Ogundolie et al., 2024). Immobilized HRP using adsorption and cross-linking was used to degrade dyes like Direct Black 38 and Direct Blue 53 (Saha and Rao, 2021).

Fig. 10.4 provides an integrated overview of dye pollution, the advantages of immobilized bioremediation systems over free systems, and the role of proteomics (proteomics is discussed in the latter part of this chapter) in guiding enzyme selection, immobilization design, and process optimization.

10.6 Immobilized microbial enzymes used in bioremediation

Microbial cells are immobilized to a specific position to limit their movement to increase their stability in extreme conditions, reusability, efficient degradation of pollutants (Ogundolie et al., 2024), and provide thermostability.

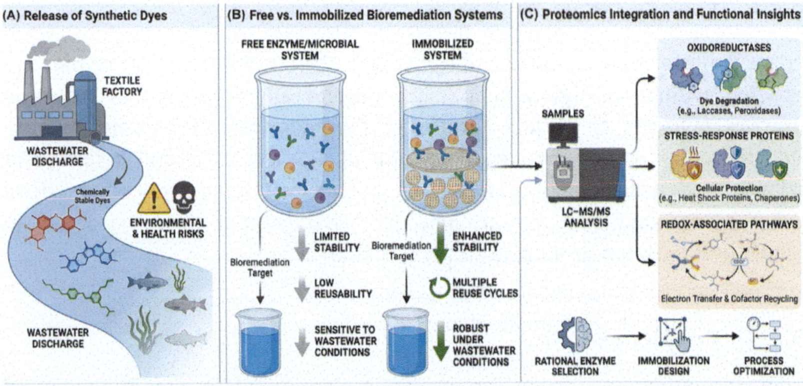

FIGURE 10.4 **Proteomics-guided immobilized enzyme systems for dye bioremediation.** (A) Release of synthetic dyes from textile and related industries into wastewater streams, where chemically stable dye molecules pose environmental and health risks. (B) Comparison between free and immobilized bioremediation systems, which highlights the improved stability, reusability, and operational strength of immobilized enzymes and microbial cells under wastewater conditions. (C) Integration of LC-MS/MS-based proteomics to compare free versus immobilized systems. This enables identification of key oxidoreductases, stress-response proteins, and redox-associated pathways. Proteomic insights in this chapter support rational enzyme selection, immobilization design, and process optimization for efficient and durable dye remediation.

Nowadays instead of whole microbial cells, only enzymes like laccases, peroxidases, lignin peroxidase, azoreductases, hydrolase (Souza et al., 2022), tyrosinase, manganese, and HRP (Bilal et al., 2017) are immobilized on different support systems using physical or chemical methods to increase enzyme stability and reusability while maintaining integrity of the enzyme (Al-Sakkaf et al., 2023) lacking in free enzymes (Table 10.1). Many studies involved microbial enzymes immobilized using different techniques for various pollutant degradation. Immobilized enzymes from *Bacillus* sp., *Pseudomonas luteola*, and *Enterobacter agglomerans* have been used for the degradation of dyes (Saha and Rao, 2021).

Enzymatic bioremediation is an innovative technique as these enzymes can function at a wide range of pH and temperature. Oxidoreductases have been adsorbed on carriers for dye-decolorization. Azoreductases and peroxidases from microbes degrade harmful azo dyes by polymerizing the phenolic constituents of dyes and cleaving the azo bond (-N=N-) through electron transfer and releasing the aromatic amines that will be degraded into CO_2 and H_2O (Lellis et al., 2019). For example, immobilized azoreductase from *Brevibacillus laterosporus* showed efficient degradation of Methyl Orange (Lang et al., 2013). Peroxidases have been widely used for the treatment of wastewater containing colored dyestuffs (Saha and Rao, 2021). Many peroxidases have been used for decolorization of textile dyes, including lignin peroxidase (LiP) and Mn peroxidase (MnP) (Lellis et al., 2019). LiP and

TABLE 10.1 Enzymes, their microbial sources, and immobilization techniques.

S. no.	Enzymes	Microbial sources	Immobilization approach	Dyes degraded	References
1.	Azoreductase	*Bacillus subtilis*	Entrapment in polyacrylamide gel	Methyl Orange	Ikram et al. (2022)
2.	Laccase	*Pichia manshurica*	Immobilization on sodium alginate	Malachite Green	Wakil et al. (2019)
3.	Manganese peroxidase (MnP)	*Phaerochaete chrysosporium*	Immobilization on silica gel	Azo dyes	Urek and Pazarlioglu (2008)
4.	Lignin peroxidase (LiP)	*Trametes versicolor*	Adsorption on agave fiber	Basic Green	Castillo-Carvajal et al. (2013)
5.	Horse-radish peroxidase (HRP)	Recombinant microbial expression	Immobilization on polypropylene films	Basic Red 29	Kumar et al. (2016)
6.	Tyrosinase	*Aspergillus nidulans*	Immobilized on chitosan beads	Congo Red	El-Shora et al. (2025)
7.	Dye-decolorizing peroxidase	*Pseudomonas* sp.	Entrapment in a sol-gel matrix	Azo dyes	Tuttolomondo et al. (2014)

MnP have been immobilized in chitosan beads, glutaraldehyde-activated gelatin matrix, Ca-alginate beads, carbon nanotubes, agar-agar, etc., for better decolorization, thermostability, reusability, and catalytic activity (Saha and Rao, 2021). For example, MnP has been immobilized using crosslinking in chitosan beads for the degradation of textile effluent (Bilal et al., 2016).

Laccases are another example of dye-decolorizing enzymes that have been found to efficiently degrade dyes using a wide range of carriers, including alumina, silica beads, polypropylene chloride film, porous glass beads, etc. For example, laccase enzymes from *Kluyveromyces dobzhanskii* and *Pichia manshurica* were immobilized using the sodium alginate calcium chloride method for decolorization of dyes (Wakil et al., 2019), including malachite green and methyl red (Al-Sakkaf et al., 2023). Researchers proposed a transformation pathway for the anthraquinone dye Remazol Brilliant Blue R (RBBR) based on its enzymatic treatment with immobilized laccase derived from *Trametes pubescens*. The laccase-mediated oxidation resulted in the cleavage of the chromophoric structure that led to the formation of lower molecular weight transformation products. These degradation products were anticipated to exhibit reduced environmental persistence and toxicity compared to the parent dye molecule (Lellis et al., 2019). Also, laccase was immobilized on cross-linked polymethacrylate (PMMA)/(CNTs) that effectively decolorized both Methylene Blue and Orange II and improved the removal of dyes from wastewater samples (Al-Sakkaf et al., 2023). Fungal laccase from *T. pubescens* was also immobilized by cross-linking, which increased the degradation of synthetic dyes (Zheng et al., 2016). In one study, it was found that the gene encoding laccase enzyme from *Bacillus subtilis* could be transferred to *E. coli* for decolorizing azo dyes (Lellis et al., 2019).

In a recent study, horse-radish peroxidase (HRP) has been used for the detoxification of Methyl Orange (Ogundolie et al., 2024), Methylene Blue, Rhodamine B, and Congo Red (Saha and Rao, 2021). In another study, the HRP enzyme was immobilized on carriers like hydroxyl-terminated bisphenol A, bisphenol AF-based polysulfones (Celebi et al., 2013), and polypropylene (PP) films (Kumar et al., 2016) for decolorization of dyes, Acid Black 1, Reactive Blue 19, and Basic Red 29 (Saha and Rao, 2021). In an attempt to achieve more efficient dye decolourization, HRP was immobilized on graphene oxide-SiO_2 (Al-Sakkaf et al., 2023). Tyrosinases have been found to degrade synthetic dyes better when immobilized onto cross-linked polyacrylonitrile (PAN) or swollen chitosan beads (Saha and Rao, 2021). In one study, PersiManXyn1, a model enzyme, was immobilized on magnetic graphene oxide (GO) for improved dye removal, such as Methylene Blue (Al-Sakkaf et al., 2023). In some cases, use of these enzymes releases degradation products that are mutagenic and carcinogenic, which impact human and animal health (Souza et al., 2022). The degradation products can be further degraded by aerobic treatment (Kumar et al., 2025).

10.7 Applications of immobilized enzymes in the remediation of dyes

10.7.1 Dye remediation and wastewater treatment

Enzymes break down chromophore structures of dye molecules that are responsible for color of the substance and prevent the toxic effects of dyes. Several immobilized enzymes like laccase, peroxidases, and azoreductases have been used to decolorize textile dyes. Studies have found that the presence of a wide variety of carriers for immobilization of enzymes allows effective degradation and decolorization of textile dyes. For example, immobilized laccase degrades Cibacron D-Blue SGL dye, methyl orange and reactive black 5, immobilized manganese peroxidase degrades methylene blue, and immobilized horseradish peroxidase enzyme degrades direct black 38 and anthraquinone acid violet (Maghraby et al., 2023). During dyeing, these harmful textile dyes, including azo, anthraquinone, and triarylmethane dyes, are released into water bodies along with wastewater and can affect the whole ecosystem, thereby requiring a proper wastewater treatment (Zdarta et al., 2021). Wastewater treatment is the removal of industrial, agricultural, and other wastes from water bodies. Immobilization techniques have been preferred over physical or chemical treatment (Ogundolie et al., 2024) for proper wastewater treatment, immobilized manganese peroxidase (Maghraby et al., 2023), immobilized oxidoreductases, and immobilized laccase beads, which have been used for decolorizing various dyes used in laboratories, preventing draining into the water, and thus, a sustainable wastewater management method. For example, a support material, polyacrylamide–alginate cryogel (PAG) was used for immobilizing laccase from *Trametes versicolor* to decolorize textile dye-containing wastewater (Zdarta et al., 2021).

10.7.2 Protection of human health and the environment

Dyes used in many industries, such as textiles, food, paper production, and pharmaceuticals, are resistant to biodegradation, carcinogenic, mutagenic, and thus lethal to human health and the environment. These dyes are a source of deadly intermediates that are poisonous and can even be cancerous (Othman and Flaifil, 2025). Thus various methods have been used for enzyme immobilization to detoxify harmful effluents. Immobilized enzymes are an effective means to protect the ecosystem from colorant pollutants. For example, laccases immobilized on various support materials/carriers like alginate beads (Tuncay and Yagar, 2020) were used for degradation of dyes including malachite green (MG), Reactive Blue, Brilliant Green (BG), and Crystal Violet (CV), that pose serious risks to human health and environment like affecting phytoplanktons, eutrophication, genotoxicity in human genome, respiratory disorders and various skin diseases. These immobilized enzymes, including laccase, oxidoreductase, Manganese Peroxidase (MnP), and Lignin Peroxidase (LiP), decrease BOD, cytotoxicity, mutagenicity, and other toxic effects, protecting aquatic life, human health, and the environment (Bilal and Asgher, 2015). Peroxidases are a potential eco-friendly method of degrading synthetic dyes and thus reducing the use of harsh chemicals and their harmful impacts on the ecosystem (Khan, 2025).

10.8 Proteomics as a tool to understand dye biodegradation and enzyme immobilization

10.8.1 Why proteomics matters in dye bioremediation?

Omics-based approaches, like proteomics, provide a comprehensive view of the intracellular processes for microbial decolorization of toxic dyes, for example, textile dyes. Also, multiomics is highlighted as providing an integrative understanding across biomolecular levels that can explain the molecular basis of dye biodegradation and pave the way for the development of effective bioremediation methods (Jasińska et al., 2024). This is useful with regard to dye biodegradation because multiple enzyme classes (that include azo reductases and oxidoreductases, such as laccases and peroxidases, as discussed in previous sections) can participate in dye degradation, depending on dye structure and redox conditions. This creates a complex and condition-dependent enzymatic landscape that proteomics has helped to resolve at the protein-expression level (Desai et al., 2021; Sen and Nigam, 2022; Njiki et al., 2020). As we know, the industrially relevant operations often occur in saline conditions and at variable pH. Therefore, the ability of extremotolerant microorganisms to withstand dyes and sustain biodegradation under such stresses encourages comparative proteomic designs to identify stress-resilient biodegradation systems (Al-Tohamy et al., 2020; Jasińska et al., 2024; Yuan et al., 2021). Since there is always a chance that multiple enzymes may be implicated in dye degradation, depending upon dye class and process conditions, proteomics-based profiling of enzymes supports sensible selection of "which enzyme(s)" to immobilize. And this can feasibly be achieved by identifying highly expressed and condition-robust catalysts during the course of biodegradation (Desai et al., 2021; Jasińska et al., 2024; Njiki et al., 2020). In particular, where biodegradation involves combinations such as laccase working with reductases (e.g., NADH-DCIP reductase and azo reductase) on different dyes, proteomics can guide whether immobilization should focus on a single broad-spectrum oxidoreductase (e.g., laccase) or a targeted reductase for cleaving the azo bond (-N=N-) (Njiki et al., 2020; Ramaswamy et al., 2021). Sometimes, the detoxification outcome depends on the metabolite profile that may cause incomplete decolorization of a dye, resulting in problematic aromatic intermediates. In this regard, the proteomics-informed immobilization strategies should be evaluated together with metabolite/toxicity endpoints to ensure that immobilized catalysts drive transformations toward nontoxic product distributions (Manogaran et al., 2021; Sarkar et al., 2021).

10.8.2 Experimental design for proteomics in dye-remediation systems

Studies show and advocate that proteomics can be used to identify enzymes that actively degrade dyes. These enzymes can then be immobilized to improve their stability and practical application. Lastly, selecting suitable reactors or treatment processes must ensure efficient and consistent dye removal under realistic wastewater conditions (Alsaiari et al., 2021; Yang et al., 2020). Based on the literature, a stepwise and comprehensible workflow consistent with the studies can be as follows.

1. First, screen and select dye-degrading microorganisms that can tolerate real wastewater conditions, such as high salinity and extreme pH. The choice can be extreme-tolerant strains that are inherently tolerant to such stresses and an appropriately designed microbial consortia. These seem effective because synergistic interactions have proven to enhance dye degradation efficiency (Al-Tohamy et al., 2020; Baena-Baldiris et al., 2020; Yuan et al., 2021).
2. The strains or consortia should be assessed for their treatment performance beyond color removal. This can be done by identifying degradation products using widely used analytical tools such as MS, NMR, or FTIR. Toxicity testing is also essential to confirm that the resulting by-products are stable and environmentally safe (Kaur et al., 2023; Sarkar et al., 2021)
3. Apply proteomics as a part of an integrated omics or multi-omics strategy to understand dye biodegradation at the molecular level. This approach helps identify key enzymes and pathways that are actively expressed when the microorganism is in the high-performance phase (Desai et al., 2021; Jasińska et al., 2024).
4. Enzymes should be prioritized for immobilization based on their well-known roles in dye degradation, for example, laccases, oxidoreductases, and azoreductases. Proteomics data should guide enzyme selection by revealing condition-specific expression patterns (Sen and Nigam, 2022; Njiki et al., 2020; Ramaswamy et al., 2021).
5. Immobilized enzymes can be implemented as reusable treatment modules on proven platforms, such as magnetic nanoparticle-immobilized laccase systems. These systems have demonstrated sustained dye removal efficiency even after multiple reuse cycles (Alsaiari et al., 2021; Desai et al., 2021).
6. Integrate enzyme modules with suitable wastewater treatment processes that may include high-efficiency biological reactors. Systems can go hybrid as per the need. Hybrid systems combine pretreatments with biological

detoxification and further improve overall treatment performance (Yang et al., 2020; Osorio et al., 2020; Zhang et al., 2021).

7. Dye bioremediation should be validated at both microbial community and process scales. This can be achieved by combining proteomics with metagenomic and microbial diversity analyses. This is of importance because it links microbial composition and redox conditions to functional roles in dye degradation (Isah et al., 2024; Zhuang et al., 2020; Santhanarajan et al., 2021).

Moreover, proteomics sampling conditions should also be defined carefully, with an experimental contrast that is likely to influence dye biodegradation and enzyme performance. This allows proteomic data to be interpreted in a meaningful way. By doing so, the data can be linked to biological mechanisms involved in dye removal. One important consideration here is the textile effluent stress, mainly salinity and pH, which are generally high. These conditions remarkably influence the microbial population and the functions of microbial enzymes. Proteomic analysis under such stresses, for example, salinity levels up to 10% NaCl at a variable range of pH 4–10, can help find stress-tolerant proteins and pathways involved in azo-dye biodegradation (Jasińska et al., 2024). Process parameters such as pH, temperature, dye concentration, and incubation time should also be considered through experimental design. Factorial or response surface methodology-based approaches have commonly been used in dye-removal research. These approaches provide a structured way to capture proteomic changes associated with optimized as well as suboptimal conditions (Ayed et al., 2022; Uppala and Muthukumaran, 2021). In addition to this, redox conditions play a key role in dye degradation. During the use of a mixed culture, whole community, or microbial consortia, sampling from paired aerobic and anaerobic systems or reactor zones allows identification of distinctive proteins. Among these, some are specifically involved in reductive and some in oxidative steps of dye transformation (Patel et al., 2022).

In hybrid setups, advanced oxidation or photocatalytic pretreatments are applied before biological treatment. Proteomic sampling can also be extended to these systems. Such pretreatments usually generate smaller dye intermediates. Studies report that these intermediates can enhance subsequent microbial degradation. Therefore, the proteomes before and after pretreatment are compared, which provides plentiful insights into how microbial metabolism adapts to these transformed substrates (Payathuparambil and Aly Hassan, 2025; Jasińska et al., 2024). Apart from the proteome, comparisons between free (i.e., soluble) and immobilized enzyme systems can be included in experimental designs. Proteomic analysis of soluble/free enzymes and immobilized laccase systems at their optimal conditions, for example, pH 4 and 25°C, revealed changes that were related to stability, accessibility, and enzyme orientation (Kołodziejczak-Radzimska et al., 2021). In that scenario, consideration must also be given to sample digestion and protein recovery strategies. This ensures compatibility with downstream proteomic workflows (Bolivar and López-Gallego, 2020; Nagy et al., 2022).

10.8.3 Proteomics platforms and approaches

Mass spectrometry (MS) based proteomics, which is generally performed using liquid chromatography-tandem mass spectrometry (LC-MS/MS), has recently seen an uptrend and is being used to study protein expression in complex biological systems (Yan et al., 2023). This approach has become routine in many laboratories and is also suitable for environmental and microbial studies. From the perspective of dye-remediation research, LC-MS/MS is useful because microbial communities are rarely simple or stable. The communities are often mixed, stressed, and constantly adapting. Use of proteomics-based analyses makes it possible to identify and quantify large numbers of proteins under these variable conditions. This used to be difficult to accomplish with targeted assays alone. In most such studies, there have been either two choices: (1) to capture broad proteome-level responses to dyes and associated environmental stress, and (2) to compare protein expression across different conditions, that is, salinity, redox state, or dye concentration (Mani et al., 2021). Another consideration is to detect extracellular enzymes, belonging to the group of laccases and peroxidases, since they directly participate in dye decolourization (Mani et al., 2021). When the quality control is appropriate, along with data-processing workflows, LC MS/MS-based proteomics can generate reproducible and interpretable datasets (Patterson et al., 2023).

10.8.3.1 Shotgun (bottom-up) LC-MS/MS proteomics for complex remediation samples

In shotgun, or bottom-up, proteomics, proteins are first digested into peptides and then analyzed by LC-MS/MS. This technique is widely used for proteome analysis because a single shotgun run can identify and quantify thousands of relevant peptides and proteins. In the case of dye remediation research, this is important for assessing global proteome changes under different conditions. This approach can easily distinguish between dye exposure and control systems, or stressed versus unstressed microbial communities (Zhao et al., 2023). In fact, remediation samples are often complex. They have mixed microbial consortia, biofilms of various stages, and immobilized systems. All these typically contain proteins with different abundances and physicochemical properties. In such cases, deeper proteome coverage is usually achieved by combining optimized sample-preparation strategies with high-resolution MS (Houston et al., 2024). Studies have shown that this combination allows the detection of a substantial fraction of predicted proteins, highlighting the importance of error-free sample preparation (Du et al., 2022; Houston et al., 2023). When an even higher degree of separation is required, multidimensional approaches can be applied. This includes Modified MudPIT workflows, which are, for example, used to increase proteome depth in complex samples. Therefore, applying these techniques consistently across experiments helps ensure reliable identification and comparison in dye-remediation studies.

10.8.3.2 Quantitative LC-MS/MS strategies: label-free and data-independent acquisition-based approaches

Remediation studies require comparisons across multiple experimental conditions because they report different dye loads, time points, or reactor stages. Due to these, maintaining quantitative consistency across LC-MS/MS runs becomes a central concern (Patterson et al., 2023; Yan et al., 2023). Label-free quantification and data-independent acquisition (DIA) approaches are commonly used in such studies. DIA is often preferred because it enhances run-to-run reproducibility and enables more reliable comparisons across repeated sampling operations (Yan et al., 2023; Jacome et al., 2020). This is of importance in long-term experiments or multi-reactor designs, where small technical variations can make the real biological trends doubtful. In practice, quantitative consistency depends not only on acquisition strategy but also on data processing. Computational tools that support cross-run peak alignment and ion selection, therefore, are essential when analyzing DIA datasets. Automated DIA analysis platforms also help solve the challenges associated with signal interference and peptide identification in complex environmental samples (Jacome et al., 2020). Experimental variability can arise at multiple stages, such as at sample preparation and instrument performance. Because of this, quality-control procedures originally developed for large-scale proteomics studies are directly applicable to remediation-focused experiments. These remediation-focused experiments generally involve many samples or batches. When applied consistently, these approaches help ensure that observed differences in protein abundance reflect biological responses rather than technical noise (Patterson et al., 2023).

10.8.3.3 Isobaric labeling (TMT/iTRAQ) for multiplexed comparative designs

Isobaric labeling approaches include tandem mass tags (TMT) and iTRAQ. These techniques enable the simultaneous analysis of multiple samples within a single LC-MS/MS run. By analyzing many conditions (multiple dyes, stresses, and time points) together (multiplex runs), these approaches reduce instrument time and improve cross-condition normalization (Wu et al., 2022). TMT-based workflows are commonly used in structured comparative studies to quantify protein-expression differences across defined experimental contrasts (Liang et al., 2021; Zimmermann et al., 2024). On the other hand, iTRAQ remains a popular option for high-throughput screening. However, the successful application of these methods depends on robust computational pipelines for protein identification and quantification (Liang et al., 2021; Wu et al., 2022). Such pipelines can be adapted to remediation datasets, where batch effects and experimental covariates are common.

10.8.3.4 Secretome and extracellular proteomics for dye decolourization enzymes

Secretome proteomics focuses on proteins that are released into the extracellular environment (Suh et al., 2024). This approach is relevant for dye-remediation studies because many dye-transforming reactions occur outside the cell. Enzymes such as laccases and peroxidases are often secreted and act directly on dye molecules present in the surrounding medium. For this reason, LC-MS/MS analysis of conditioned media or extracellular fractions can provide direct insight into the enzymes responsible for dye decolourization (Mani et al., 2021). Differences in decolourization efficiency among dyes are frequently linked to enzyme-substrate compatibility. Factors such as redox potential and accessibility at the enzyme active site play an important role. Secretome proteomics makes it possible to examine these relationships at the level of protein expression rather than contingent activity alone (Mani et al., 2021). From a methodological perspective, careful handling of extracellular fractions is essential. Intracellular proteins released during cell lysis can easily contaminate secretome samples and complicate interpretation. Established quality-control practices developed for large-scale proteomics, therefore, provide useful guidance for improving reproducibility and confidence in extracellular proteomic analyses applied to dye-remediation systems (Patterson et al., 2023; Suh et al., 2024).

10.8.3.5 Targeted LC-MS/MS proteomics for monitoring key enzymes

As dye-remediation studies progress from discovery toward application, the focus often shifts to a smaller set of enzymes. At this stage, it becomes important to monitor these enzymes across different operational conditions or reactor systems. Targeted LC-MS/MS approaches, such as selected reaction monitoring or parallel reaction monitoring (SRM/PRM), are well suited for this purpose. These methods allow sensitive and reproducible quantification of predefined protein targets. However, they also require careful assay design and method development. Moreover, they are generally more resource-intensive than discovery-based workflows (Batruch et al., 2020). Due to these, targeted proteomics is rarely used in isolation. Instead, it is most effective when applied as a follow-up to shotgun LC-MS/MS. As a matter of fact, enzymes identified during discovery studies can be tracked over time using targeted assays. This helps confirm whether key enzymes remain stable, consistently expressed, and functionally relevant under different remediation conditions. When combined in this way, discovery and targeted proteomics help bridge the gap between mechanistic understanding and routine monitoring in applied dye-remediation systems (Mai et al., 2021; Patterson et al., 2023).

10.9 Comparative proteomics of free versus immobilized systems

Comparative proteomics, most commonly based on LC-MS/MS analysis, offers a way to examine how immobilization alters protein expression in cells, enzymes, or biofilm-supported microbial consortia when compared with free or planktonic systems. These comparisons are useful for understanding why immobilized systems usually perform better and show greater operational stability during dye removal in wastewater (Al-Tohamy et al., 2020; Ledakowicz and Paździor, 2021). This line of investigation is very justified. Free-cell dye biodegradation is frequently inhibited by dye toxicity, process instability, and inefficient substrate transfer. As a result, immobilization strategies are commonly adopted to address these limitations and improve overall treatment performance in applied remediation systems (Asranudin et al., 2024; Pete et al., 2021). Thus proteomics provides a direct way to examine how these design choices influence functional protein expression.

10.9.1 Enzyme stability and operational durability in free versus immobilized systems

Improving operational stability is one of the main reasons for immobilizing enzymes or microbial cells. Advantages of immobilization have been discussed in previous sections of this chapter. However, from a proteomics standpoint, this leads to a straightforward expectation: under comparable operating conditions, immobilized systems should retain functional dye-degrading enzymes for longer periods than free systems (Gamage et al., 2022). This is well justified in practice. With regard to stability, dye-active oxidoreductases have been immobilized on a wide range of supports. These supports include chitosan beads, gelatin or alginate matrices, covalent supports, hydrogels, and nanomaterial-based carriers. Studies have reported optimized immobilization conditions and improved dye decolourization relative to free enzymes. Together, these studies support the idea that immobilization alters not only enzyme performance but also the functional state and durability of catalytic proteins in water-treatment settings (Aslam et al., 2022; Harguindeguy et al., 2021; Kumar et al., 2022). It is evident from the sustained dye decolourization and detoxification over multiple reuse cycles that stability and durability are key comparative endpoints. Proteomics can help interpret the study outcomes in many ways, at the protein level by tracking the persistence of catalytic isoforms, chaperones, or oxidative stress-related proteins over time (Chakravarthi et al., 2021).

Furthermore, it is important to recognize that immobilization performance depends on more than the choice of support material. Two other aspects (1) enzyme orientation and (2) the local microenvironment, play a significant role in determining the immobilization performance. Immobilization chemistries such as covalent attachment, crosslinking, and surface functionalization can influence active-site accessibility and protein conformation (Zeballos et al., 2023; Cutiño-Avila et al., 2022). This, in turn, affects functional proteome readouts. Alternative strategies, including crosslinked enzyme aggregates or peroxidases immobilized on iron oxyhydroxide or photocatalytic supports, further broaden the range of systems. Proteomics can be applied to these systems to link immobilization microenvironments with protein stability and persistence under wastewater treatment conditions (Zerva et al., 2020; Silva Tavares et al., 2020).

10.9.2 Stress-response remodeling in immobilized microbial systems

One of the practical reasons for immobilizing microorganisms is the expectation that immobilization can reduce exposure to environmental stress. This is especially relevant in textile wastewater systems, where organisms are exposed to fluctuating conditions and inhibitory compounds. For this reason, stress-response pathways receive particular attention in comparative proteomics studies of free and immobilized systems (Singh et al., 2025). Maintaining stable dye biodegradation under such conditions requires more than efficient enzyme activity. Microbial cells must also remain viable and resilient over time (Ledakowicz and Paździor, 2021). Immobilization is often assumed to protect toxic dyes, salinity, or pH variation; however, these benefits are not always obvious from process performance alone. Comparative proteomics offers a way to look more closely. Changes in general stress proteins, envelope-associated systems, and metabolic regulators can provide insight into how microorganisms adjust to immobilized growth and continue to function under stress (Basharat et al., 2024). With this logic, proteomics helps translate the assumed advantages of immobilization into measurable biological responses.

10.9.3 Redox metabolism, electron-transfer processes, and oxidoreductase expression

In many treatment designs, immobilization is combined with strategies that intentionally modify redox conditions. For example, bioelectrochemical systems and hybrid processes that integrate advanced oxidation with biological treatment. This makes redox metabolism an important aspect of comparative proteomics, as far as dye-remediation research is concerned. Advanced oxidation pretreatments can generate smaller and more biodegradable dye intermediates. They can also influence electron availability for downstream biological reactions. Collectively, these effects provide a clear rationale for examining redox-associated enzymes and cofactor-dependent pathways using proteomics under different treatment configurations (Ledakowicz and Paździor, 2021)

Evidence from studies conducted with systems operating under electron-rich conditions further highlights the importance of electron transfer. For example, enhanced dye decolourization has been reported in constructed wetland-microbial fuel cell systems, where electron availability plays a crucial role. Comparative proteomics can help identify which redox pathways and catalytic proteins are preferentially expressed under these conditions (Xie et al., 2024).

Amongst textile wastewater studies, dye biodegradation is consistently linked to oxidoreductase enzymes, including laccases, peroxidases, and azoreductases. Changes in the abundance or isoform composition of these enzymes, therefore, represent a key focus for comparative proteomics in free versus immobilized systems (Desai et al., 2021). Numerous studies report effective dye decolourization and detoxification by immobilized laccases and peroxidases, often with improved stability during extended operation. Proteomics can also help clarify whether these benefits arise from altered enzyme expression, stabilization of specific isoforms, or more effective enzyme deployment over time (Gamage et al., 2022; Silva Tavares et al., 2020; Chakravarthi et al., 2021).

10.9.4 Transport proteins and cell envelope adaptation

Free and immobilized growth modes differ in how substrates and metabolites move between the cell and its environment. In free-cell systems, limitation in substrate transport is considered a reason for poor or unstable dye biodegradation. This provides a reasonable basis for expecting differences in membrane, envelope, and transport-related proteins between free and immobilized states (Asranudin et al., 2024). Although transporter-related proteomic patterns can vary from system to system. Comparative proteomics has repeatedly identified changes in protein groups associated with the cell wall, membrane, and envelope in response to environmental conditions. These observations support the idea that immobilization can drive measurable adaptations at the cell surface that are relevant to wastewater treatment performance (Heo et al., 2023). From this perspective, proteomics can be used to examine whether immobilization supports altered expression of membrane-associated and transport-related proteins that influence dye uptake, metabolite export, and tolerance to wastewater-related stress. Such analyses are consistent with the broader use of proteomics to resolve pathway-level responses to environmental perturbations rather than isolated enzymatic effects (Shiny Matilda et al., 2020).

10.10 Challenges, gaps, and future directions

Despite its growing application, the use of proteomics in dye biodegradation and wastewater treatment faces several practical challenges. One of them is the lack of practicality in many experimental designs. Textile dye effluents are generally highly saline and exhibit wide pH variability (usually alkaline), yet proteomic studies are frequently performed under simplified laboratory conditions. As a result, protein-expression patterns observed in the laboratory may not accurately represent those operating in real treatment systems (Vincent et al., 2022; Yuan et al., 2021). Therefore, incorporating realistic stress conditions is essential to avoid biases and misinterpretation of condition-specific responses as treatment-relevant mechanisms.

Another critical bottleneck is the sample preparation. Wastewater matrices contain biofilms, immobilized carriers, and extracellular polymers that complicate the process of protein extraction and digestion. Studies have used different methods. These differences in methodology can strongly influence LC-MS/MS outcomes. This, in turn, limits the comparability across studies unless workflows are carefully optimized and transparently reported (Araújo et al., 2021; Ogata and Ishihama, 2021). Reproducibility also remains a concern in multi-condition studies because variability can arise during both sample handling and MS analysis. Although data-independent acquisition and micro-flow LC-MS/MS improve robustness of the process, consistency must be deliberately engineered through quality-control strategies rather than assumed (Bian et al., 2020).

Another major concern is that the dye decolourization is sometimes equated with detoxification, which is not always true. Such conclusions must be supported by proteomic evidence of enzyme induction, by chemical identification of degradation products, and toxicity assessments to confirm environmentally safe transformation (Zdarta et al., 2021).

Future work should focus on the following aspects: (1) standardization of workflows; (2) integration of proteomics with complementary meta-omics and targeted approaches; and (3) emphasizing translating protein-level insights into durable and scalable immobilized systems. Mechanistic understanding with operational feasibility and cost considerations must be aligned for advancing proteomics-guided dye remediation from laboratory studies to practical wastewater treatment applications.

10.11 Conclusion

This chapter highlights the growing potential of microbial enzymes for the bioremediation of dye-contaminated wastewaters, with emphasis on the advantages offered by enzyme and cell immobilization strategies. Limitations of conventional treatments in terms of energy demand, secondary pollution, and incomplete detoxification underscore the need for more sustainable alternatives.

Immobilized microbial enzymes, especially oxidoreductases such as laccases, peroxidases, and azoreductases, offer clear benefits in terms of stability, reusability, and operational robustness under conditions relevant to textile effluents. A central theme of this chapter is the role of proteomics in strengthening mechanistic understanding and guiding logical process design. Proteomic analyses provide direct insight into which enzymes and stress-response pathways are actively involved in dye biodegradation, and how these profiles differ between free and immobilized systems. By linking protein expression to operational conditions, immobilization chemistry, and treatment performance, proteomics helps bridge the gap between laboratory-scale observations and practical wastewater treatment applications. However, the chapter also emphasizes that proteomics must be applied with careful experimental design and interpreted alongside chemical and toxicity data to avoid overestimating decolourization as detoxification.

Largely, integrating proteomics with enzyme immobilization and process engineering offers a promising framework for developing durable and efficient dye-remediation systems. Future progress will depend on realistic experimental conditions, standardized analytical workflows, and a stronger focus on translating protein-level insights into scalable and economically viable treatment technologies.

References

Abdelhamid, M.A.A., Khalifa, H.O., Yoon, H.J., Ki, M.R., Pack, S.P., 2024. Microbial immobilized enzyme biocatalysts for multipollutant mitigation: harnessing nature's toolkit for environmental sustainability. International Journal of Molecular Sciences 25 (16). Available at https://doi.org/10.3390/ijms25168616, http://www.mdpi.com/journal/ijms.

Agrawal, S., Tipre, D., Patel, B., Dave, S., 2014. Optimization of triazo Acid Black 210 dye degradation by Providencia sp. SRS82 and elucidation of degradation pathway. Process Biochemistry 49 (1), 110–119. Available at https://doi.org/10.1016/j.procbio.2013.10.006.

Al-Sakkaf, M.K., Basfer, I., Iddrisu, M., Bahadi, S.A., Nasser, M.S., Abussaud, B., et al., 2023. An up-to-date review on the remediation of dyes and phenolic compounds from wastewaters using enzymes immobilized on emerging and nanostructured materials: promises and challenges. Nanomaterials 13 (15). Available at https://doi.org/10.3390/nano13152152, http://www.mdpi.com/journal/nanomaterials.

Al-Tohamy, R., Kenawy, E.R., Sun, J., Ali, S.S., 2020. Performance of a newly isolated salt-tolerant yeast strain Sterigmatomyces halophilus SSA-1575 for azo dye decolorization and detoxification. Frontiers in Microbiology 11. Available at https://doi.org/10.3389/fmicb.2020.01163, https://www.frontiersin.org/journals/microbiology#.

Al-Tohamy, R., Ali, S.S., Li, F., Okasha, K.M., Mahmoud, Y.A.G., Elsamahy, T., et al., 2022. A critical review on the treatment of dye-containing wastewater: ecotoxicological and health concerns of textile dyes and possible remediation approaches for environmental safety. Ecotoxicology and Environmental Safety 231. Available at https://doi.org/10.1016/j.ecoenv.2021.113160, http://www.elsevier.com/inca/publications/store/6/2/2/8/1/9/index.htt.

Alsaiari, N.S., Amari, A., Katubi, K.M., Alzahrani, F.M., Harharah, H.N., Rebah, F.B., et al., 2021. The biocatalytic degradation of organic dyes using laccase immobilized magnetic nanoparticles. Applied Sciences (Switzerland) 11 (17). Available at https://doi.org/10.3390/app11178216, https://www.mdpi.com/2076-3417/11/17/8216/pdf.

Araújo, M.J., Sousa, M.L., Felpeto, A.B., Turkina, M.V., Fonseca, E., Martins, J.C., et al., 2021. Comparison of sample preparation methods for shotgun proteomic studies in aquaculture species. Proteomes 9 (4). Available at https://doi.org/10.3390/proteomes9040046, www.mdpi.com/journal/proteomes.

Ardila-Leal, L.D., Poutou-Piñales, R.A., Pedroza-Rodríguez, A.M., Quevedo-Hidalgo, B.E., 2021. A brief history of colour, the environmental impact of synthetic dyes and removal by using accases. Molecules (Basel, Switzerland) 26 (13), 3813. Available at https://doi.org/10.3390/molecules26133813 PMID: 34206669; PMCID: PMC8270347.

Aslam, S., Ali, A., Asgher, M., Farah, N., Iqbal, H.M.N., Bilal, M., 2022. Fabrication and catalytic characterization of laccase-loaded calcium-alginate beads for enhanced degradation of dye-contaminated aqueous solutions. Catalysis Letters 152 (6), 1729–1741. Available at https://doi.org/10.1007/s10562-021-03765-8, www.springer.com/journal/10562.

Asranudin, N., Purnomo, A.S., Holilah, N., Prasetyoko, D., El Messaoudi, N., Rohmah, A.A., et al., 2024. Adsorption and biodegradation of the azo dye methyl orange using Ralstonia pickettii immobilized in polyvinyl alcohol (PVA)-alginate-hectorite beads (BHec-RP). RSC Advances 14 (26), 18277–18290. Available at https://doi.org/10.1039/d3ra08692e, http://pubs.rsc.org/en/journals/journal/ra.

Ayed, L., Bekir, K., Jabeur, C., 2022. Modeling and optimization of biodegradation of methylene blue by Staphylococcus aureus through a statistical optimization process: a sustainable approach for waste management. Water Science and Technology 86 (2), 380–394. Available at https://doi.org/10.2166/wst.2022.211, https://watermark.silverchair.com/wst086020380.pdf?.

Ayilara, M.S., Babalola, O.O., 2023. Bioremediation of environmental wastes: the role of microorganisms. Frontiers in Agronomy 5. Available at https://doi.org/10.3389/fagro.2023.1183691, www.frontiersin.org/journals/agronomy#.

Baena-Baldiris, D., Montes-Robledo, A., Baldiris-Avila, R., 2020. Franconibacter sp., 1MS: a new strain in decolorization and degradation of azo dyes ponceau S red and methyl orange. ACS Omega 5 (43), 28146–28157. Available at https://doi.org/10.1021/acsomega.0c03786, pubs.acs.org/journal/acsodf.

Basharat, Z., Foster, L.J., Abbas, S., Yasmin, A., 2024. Comparative proteomics of bacteria under stress conditions. Proteogenomics: Methods and Protocols. Springer, pp. 129–162.

Batruch, I., Lim, B., Soosaipillai, A., Brinc, D., Fiala, C., Diamandis, E.P., 2020. Mass spectrometry-based assay for targeting fifty-two proteins of brain origin in cerebrospinal fluid. Journal of Proteome Research 19 (8), 3060–3071. Available at https://doi.org/10.1021/acs.jproteome.0c00087, http://pubs.acs.org/journal/jprobs.

Benkhaya, S., M'rabet, S., Lgaz, H., El Bachiri, A., El Harfi, A., 2021. Dyes: classification, pollution, and environmental effects. Dye Biodegradation, Mechanisms and Techniques: Recent Advances. Springer, pp. 1–50.

Bian, Y., Zheng, R., Bayer, F.P., Wong, C., Chang, Y.C., Meng, C., et al., 2020. Robust, reproducible and quantitative analysis of thousands of proteomes by micro-flow LC–MS/MS. Nature Communications 11 (1). Available at https://doi.org/10.1038/s41467-019-13973-x, http://www.nature.com/ncomms/index.html.

Bilal, M., Asgher, M., 2015. Sandal reactive dyes decolorization and cytotoxicity reduction using manganese peroxidase immobilized onto polyvinyl alcohol-alginate beads. Chemistry Central Journal 9 (1). Available at https://doi.org/10.1186/s13065-015-0125-0.

Bilal, M., Asgher, M., Iqbal, M., Hu, H., Zhang, X., 2016. Chitosan beads immobilized manganese peroxidase catalytic potential for detoxification and decolorization of textile effluent.

International Journal of Biological Macromolecules 89, 181–189. Available at https://doi.org/10.1016/j.ijbiomac.2016.04.075.

Bilal, M., Asgher, M., Iqbal, H.M.N., Hu, H., Wang, W., Zhang, X., 2017. Bio-catalytic performance and dye-based industrial pollutants degradation potential of agarose-immobilized MnP using a Packed Bed Reactor System. International Journal of Biological Macromolecules 102, 582–590. Available at https://doi.org/10.1016/j.ijbiomac.2017.04.065, www.elsevier.com/locate/ijbiomac.

Bolivar, J.M., López-Gallego, F., 2020. Characterization and evaluation of immobilized enzymes for applications in flow reactors. Current Opinion in Green and Sustainable Chemistry 25, 100349. Available at https://doi.org/10.1016/j.cogsc.2020.04.010.

Castillo-Carvajal, L.C., Pedroza-Rodríguez, A.M., Barragán-Huerta, B.E., 2013. Adsorption and biological removal of basic green 4 dye using white-rot fungi immobilized on agave tequilana weber waste. Fresenius Environmental Bulletin 22 (8), 2334–2343.

Celebi, M., Kaya, M.A., Altikatoglu, M., Yildirim, H., 2013. Enzymatic decolorization of anthraquinone and diazo dyes using horseradish peroxidase enzyme immobilized onto various polysulfone supports. Applied Biochemistry and Biotechnology 171 (3), 716–730. Available at https://doi.org/10.1007/s12010-013-0377-x.

Chakravarthi, B., Mathkala, V., Palempalli, U.M.D., 2021. Degradation and detoxification of congo red azo dye by immobilized laccase of *Streptomyces sviceus*. Journal of Pure & Applied Microbiology 15 (2).

Cutiño-Avila, B.V., Sánchez-López, M.I., Cárdenas-Moreno, Y., González-Durruthy, M., Ramos-Leal, M., Guerra-Rivera, G., et al., 2022. Modeling and experimental validation of covalent immobilization of Trametes maxima laccase on glyoxyl and MANA-Sepharose CL 4B supports, for the use in bioconversion of residual colorants. Biotechnology and Applied Biochemistry 69 (2), 479–491.

Desai, C., Jain, K.R., Boopathy, R., van Hullebusch, E.D., Madamwar, D., 2021. Editorial: Eco-sustainable bioremediation of textile dye wastewaters: innovative microbial treatment technologies and mechanistic insights of textile dye biodegradation. Frontiers in Microbiology 12. Available at https://doi.org/10.3389/fmicb.2021.707083, https://www.frontiersin.org/journals/microbiology#.

Dutta, P., Rabbi, M., Sufian, M., Mahjebin, S., 2022. Effects of textile dyeing effluent on the environment and its treatment: a review. Engineering and Applied Science Letters (EASL 5, 1–17.

Du, X., Zhou, D., Zhou, J., Xue, J., Cheng, Z., 2022. Marek's disease virus and reticuloendotheliosis virus coinfection enhances viral replication and alters cellular protein profiles. Frontiers in Veterinary Science 9. Available at https://doi.org/10.3389/fvets.2022.854007, journal.frontiersin.org/journal/veterinary-science.

El-Shora, H.M., Elazab, N.T., Al-Anazi, A., El-Sayyad, G.S., Ibrahim, M.E., Alfakharany, M.W., 2025. Fungal tyrosinase immobilized on chitosan, calcium alginate, and silica gel for phenol elimination and dye decolorization. Enzyme and Microbial Technology 189. Available at https://doi.org/10.1016/j.enzmictec.2025.110655, https://www.sciencedirect.com/science/journal/01410229.

Gamage, A., Thiviya, P., Mani, S., Ponnusamy, P.G., Manamperi, A., Evon, P., et al., 2022. Environmental properties and applications of biodegradable starch-based nanocomposites. Polymers 14 (21), 4578. Available at https://doi.org/10.3390/polym14214578.

Harguindeguy, M., Antonelli, C., Belleville, M.-P., Sanchez-Marcano, J., Pochat-Bohatier, C., 2021. Gelatin supports with immobilized laccase as sustainable biocatalysts for water treatment. Journal of Applied Polymer Science 138 (2). Available at https://doi.org/10.1002/app.49669.

Heo, L., Han, Y., Cho, Y., Choi, J., Lee, J., Han, S.W., 2023. A putative glucose 6-phosphate isomerase has pleiotropic functions on virulence and other mechanisms in Acidovorax citrulli. Frontiers in Plant Science 14. Available at https://doi.org/10.3389/fpls.2023.1275438, https://www.frontiersin.org/journals/plant-science.

Houston, S., Gomez, A., Geppert, A., Eshghi, A., Smith, D.S., Waugh, S., et al., 2023. Deep proteome coverage advances knowledge of Treponema pallidum protein expression profiles during infection. Scientific Reports 13 (1). Available at https://doi.org/10.1038/s41598-023-45219-8, https://www.nature.com/srep/.

Houston, S., Gomez, A., Geppert, A., Goodyear, M.C., Cameron, C.E., 2024. In-depth proteome coverage of in vitro-cultured Treponema pallidumand quantitative comparison analyses with in vivo-grown treponemes. Journal of Proteome Research 23 (5), 1725–1743. Available at https://doi.org/10.1021/acs.jproteome.3c00891.

Ihsanullah, I., Jamal, A., Ilyas, M., Zubair, M., Khan, G., Atieh, M.A., 2020. Bioremediation of dyes: current status and prospects. Journal of Water Process Engineering 38. Available at https://doi.org/10.1016/j.jwpe.2020.101680, http://www.journals.elsevier.com/journal-of-water-process-engineering/.

Ikram, M., Naeem, M., Zahoor, M., Hanafiah, M.M., Oyekanmi, A.A., Islam, N.U., et al., 2022. Bacillus subtilis: as an efficient bacterial strain for the reclamation of water loaded with textile azo dye, orange II. International Journal of Molecular Sciences 23 (18). Available at https://doi.org/10.3390/ijms231810637, http://www.mdpi.com/journal/ijms.

Isah, M.M., Taining, Z., Abd Jalil, N.K., Ibrahim, Z., Abdul-Wahab, M.F., 2024. Insights into the microbiota of a textile wastewater treatment system towards sustainable bioremediation of industrial wastes. Key Engineering Materials 1002, 119–131. Available at https://doi.org/10.4028/p-4E4xtI, https://www.scientific.net/KEM.

Jacome, A.S.V., Peckner, R., Shulman, N., Krug, K., DeRuff, K.C., Officer, A., et al., 2020. Avant-garde: an automated data-driven DIA data curation tool. Nature Chemical Biology 17 (12), 1237–1244. Available at https://doi.org/10.1038/s41592-020-00986-4.

Jasińska, A., Walaszczyk, A., Paraszkiewicz, K., 2024. Omics-based approaches in research on textile dye microbial decolorization. Molecules (Basel, Switzerland) 29 (12), 2771. Available at https://doi.org/10.3390/molecules29122771.

Kaur, J., Mudgal, G., Negi, A., Tamang, J., Singh, S., Singh, G.B., et al., 2023. Reactive black-5, congo red and methyl orange: chemical degradation of azo-dyes by agrobacterium. Water (Switzerland) 15 (9). Available at https://doi.org/10.3390/w15091664, http://www.mdpi.com/journal/water.

Kayani, K.F., Mohammed, S.J., Mustafa, M.S., Aziz, S.B., 2025. Dyes and their toxicity: removal from wastewater using carbon dots/metal oxides as hybrid materials: a review. Materials Advances 6 (16), 5391–5409. Available at https://doi.org/10.1039/d5ma00572h, https://www.rsc.org/journals-books-databases/about-journals/materials-advances/.

Khan, M.F., 2025. Recent advances in microbial enzyme applications for sustainable textile processing and waste management. Sci 7 (2). Available at https://doi.org/10.3390/sci7020046, www.mdpi.com/journal/sci.

Kour, D., Khan, S.S., Kour, H., Kaur, T., Devi, R., Rai, P.K., et al., 2022. Microbe-mediated bioremediation: current research and future challenges. Journal of Applied Biology and Biotechnology 10, 6–24. Available at https://doi.org/10.7324/JABB.2022.10s202, jabonline.in/index.php.

Kołodziejczak-Radzimska, A., Nghiem, L.D., Jesionowski, T., 2021. Functionalized materials as a versatile platform for enzyme immobilization in wastewater treatment. Current Pollution Reports 7 (3), 263–276. Available at https://doi.org/10.1007/s40726-021-00193-5.

Kumar, V., Misra, N., Goel, N.K., Thakar, R., Gupta, J., Varshney, L., 2016. A horseradish peroxidase immobilized radiation grafted polymer matrix: a biocatalytic system for dye waste water treatment. RSC Advances 6 (4), 2974–2981. Available at https://doi.org/10.1039/c5ra20513a, http://pubs.rsc.org/en/journals/journalissues.

Kumar, D., Khullar, A., Sharma, N., Gupta, N., 2022. Immobilization of laccase from Bacillus subtilis DS on chitosan beads and standardization of process for biodegradation of textile dyes. Journal of Chemical Technology & Biotechnology 97 (2), 466–473. Available at https://doi.org/10.1002/jctb.6688.

Kumar, M., Mishra, A., Patel, S.K., Kushwaha, J., Singh, S., Mishra, V., et al., 2025. Environmental impacts and strategies for bioremediation of dye-containing wastewater. Bioengineering 12 (10). Available at https://doi.org/10.3390/bioengineering12101043, www.mdpi.com/journal/bioengineering.

Lang, W., Sirisansaneeyakul, S., Ngiwsara, L., Mendes, S., Martins, L.O., Okuyama, M., et al., 2013. Characterization of a new oxygen-insensitive azoreductase from Brevibacillus laterosporus TISTR1911: toward dye decolorization using a packed-bed metal affinity reactor. Bioresource Technology 150, 298–306. Available at https://doi.org/10.1016/j.biortech.2013.09.124, www.elsevier.com/locate/biortech.

Ledakowicz, S., Paździor, K., 2021. Recent achievements in dyes removal focused on advanced oxidation processes integrated with biological methods. Molecules (Basel, Switzerland) 26 (4), 870. Available at https://doi.org/10.3390/molecules26040870.

Lellis, B., Fávaro-Polonio, C.Z., Pamphile, J.A., Polonio, J.C., 2019. Effects of textile dyes on health and the environment and bioremediation potential of living organisms. Biotechnology Research and Innovation 3 (2), 275–290. Available at https://doi.org/10.1016/j.biori.2019.09.001.

Liang, X., He, M., Zhu, B., Zhu, Y., He, X., Liu, D., et al., 2021. TMT-based proteomic explores the influence of DHEA on the osteogenic differentiation of hBMSCs. Frontiers in Cell and Developmental Biology 9. Available at https://doi.org/10.3389/fcell.2021.726549.

Maghraby, Y.R., El-Shabasy, R.M., Ibrahim, A.H., Azzazy, H.M.E.S., 2023. Enzyme immobilization technologies and industrial applications. ACS Omega 8 (6), 5184–5196. Available at https://doi.org/10.1021/acsomega.2c07560, pubs.acs.org/journal/acsodf.

Mai, Y., Dou, L., Yao, Z., Madla, C.M., Gavins, F.K.H., Taherali, F., et al., 2021. Quantification of P-glycoprotein in the gastrointestinal tract of humans and rodents: methodology, gut region, sex, and species matter. Molecular Pharmaceutics 18 (5), 1895–1904. Available at https://doi.org/10.1021/acs.molpharmaceut.0c00574, http://pubs.acs.org/journal/mpohbp.

Mani, P., Fidal, V.T., Keshavarz, T., Chandra, T.S., Kyazze, G., 2021. Laccase immobilization strategies for application as a cathode catalyst in microbial fuel cells for azo dye decolourization. Frontiers in Microbiology 11. Available at https://doi.org/10.3389/fmicb.2020.620075, https://www.frontiersin.org/journals/microbiology#.

Manogaran, M., Yasid, N.A., Othman, A.R., Gunasekaran, B., Halmi, M.I.E., Shukor, M.Y.A., 2021. Biodecolourisation of reactive red 120 as a sole carbon source by a bacterial consortium—toxicity assessment and statistical optimisation. International Journal of Environmental Research and Public Health 18 (5), 1–26. Available at https://doi.org/10.3390/ijerph18052424, https://www.mdpi.com/1660-4601/18/5/2424/pdf.

Muñoz, X., Clofent, D., Cruz, M.J., 2023. Occupational respiratory allergy to reactive dyes. Current Opinion in Allergy and Clinical Immunology 23 (2), 70–75. Available at https://doi.org/10.1097/ACI.0000000000000885, http://journals.lww.com/co-allergy/pages/default.aspx.

Nagy, C., Szabo, R., Gaspar, A., 2022. Microfluidic immobilized enzymatic reactors for proteomic analyses—recent developments and trends (2017–2021). Micromachines 13 (2), 311. Available at https://doi.org/10.3390/mi13020311.

Ngo, A.C.R., Tischler, D., 2022. Microbial degradation of azo dyes: approaches and prospects for a hazard-free conversion by microorganisms. International Journal of Environmental Research and Public Health 19 (8). Available at https://doi.org/10.3390/ijerph19084740, https://www.mdpi.com/1660-4601/19/8/4740/pdf.

Njiki, A., Kamgang-Youbi, G., Nola, M., Laminsi, S., 2020. Biodegradation kinetic studies and optimization for the elimination of azoic and triphenylmethane dyes using an integrated process combining biological treatment and gliding arc plasma. Journal of Chemical Technology & Biotechnology 96 (1), 273–282. Available at https://doi.org/10.1002/jctb.6539.

Ogata, K., Ishihama, Y., 2021. CoolTip: low-temperature solid-phase extraction microcolumn for capturing hydrophilic peptides and phosphopeptides. Molecular & Cellular Proteomics 20, 100170. Available at https://doi.org/10.1016/j.mcpro.2021.100170.

Ogundolie, F.A., Babalola, O.O., Adetunji, C.O., Aruwa, C.E., Manjia, J.N., Muftaudeen, T.K., 2024. A review on bioremediation by microbial immobilization-an effective alternative for wastewater treatment. AIMS Environmental Science 11 (6), 918–939. Available at https://doi.org/10.3934/environsci.2024046, https://www.aimspress.com/aimspress-data/aimses/2024/6/PDF/Environ-11-06-046.pdf.

Osorio, G.P., Hernández-Gómez, F.D.R., Arriola-Morales, J., Castillo-Morales, M., Mendoza-Hernández, J.C., 2020. Blue dye degradation in an aqueous medium by a combined photocatalytic and bacterial biodegradation process. Turkish Journal of Chemistry 44 (1), 180–193. Available at https://doi.org/10.3906/kim-1902-33.

Ozkurt, S., Kargi, B.A., Kavas, M., Evyapan, F., Kiter, G., Baser, S., 2012. Respiratory symptoms and pulmonary functions of workers employed in Turkish textile dyeing factories. International Journal of Environmental Research and Public Health 9 (4), 1068–1076. Available at https://doi.org/10.3390/ijerph9041068, http://www.mdpi.com/1660-4601/9/4/1068/pdf.

Patel, H., Yadav, V.K., Yadav, K.K., Choudhary, N., Kalasariya, H., Alam, M.M., et al., 2022. A recent and systemic approach towards microbial biodegradation of dyes from textile industries. Water (Switzerland) 14 (19). Available at https://doi.org/10.3390/w14193163, http://www.mdpi.com/journal/water.

Patterson, K.L., Arul, A.B., Choi, M.J., Oliver, N.C., Whitaker, M.D., Bodrick, A.C., et al., 2023. Establishing quality control procedures for large scale plasma proteomics analyses. Journal of the American Society for Mass Spectrometry 34 (6), 1105–1116. Available at https://doi.org/10.1021/jasms.3c00050, https://pubs.acs.org/loi/jamsef.

Payathuparambil, A.M., Aly Hassan, A., 2025. Hybrid biofiltration–advanced oxidation process system for removing emerging contaminants from actual wastewater treatment plant effluent. Journal of Water Process Engineering 78, 108830. Available at https://doi.org/10.1016/j.jwpe.2025.108830.

Periyasamy, A.P., 2025. A review of bioremediation of textile dye containing wastewater. Cleaner Water 4, 100092. Available at https://doi.org/10.1016/j.clwat.2025.100092.

Pete, A.J., Bharti, B., Benton, M.G., 2021. Nano-enhanced bioremediation for oil spills: a review. ACS ES&T Engineering 1 (6), 928–946.

Pinheiro, L.R.S., Gradíssimo, D.G., Xavier, L.P., Santos, A.V., 2022. Degradation of azo dyes: bacterial potential for bioremediation. Sustainability (Switzerland) 14 (3). Available at https://doi.org/10.3390/su14031510, https://www.mdpi.com/2071-1050/14/3/1510/pdf.

Ramaswamy, R., Sundaravadivel, K., Prakash, S., Mohan, A., 2021. Biodecolorization of azo dye mixture (Remazol Brilliant Violet 5R and Reactive Red 120) by indigenous bacterial consortium isolated from dye contaminated soil. Malaysian Journal of Microbiology 18 (1). Available at https://doi.org/10.21161/mjm.211202.

Saha, P., B Rao, K.V., 2021. Immobilization as a powerful bioremediation tool for abatement of dye pollution: a review. Environmental Reviews 29 (2), 277–299.

Santhanarajan, A.-E., Sul, W.J., Yoo, K., Seong, H.J., Kim, H.-G., Koh, S., 2021. Metagenomic insight of a full scale eco-friendly treatment system of textile dye wastewater using bioaugmentation of the composite culture CES-1. Microorganisms 9 (7), 1503. Available at https://doi.org/10.3390/microorganisms9071503.

Saratale, R.G., Rajesh Banu, J., Shin, H.-S., Bharagava, R.N., Saratale, G.D., 2019. Textile industry wastewaters as major sources of environmental contamination: bioremediation approaches for its degradation and detoxification. Bioremediation of Industrial Waste for Environmental Safety: Volume I: Industrial Waste and Its Management. Springer, pp. 135–167.

Sarkar, S., Echeverría-Vega, A., Banerjee, A., Bandopadhyay, R., 2021. Decolourisation and biodegradation of textile di-azo dye congo red by Chryseobacterium geocarposphaerae DD3. Sustainability 13 (19), 10850. Available at https://doi.org/10.3390/su131910850.

Saxena, S., Raja, A.S.M., 2014. Natural dyes: sources, chemistry, application and sustainability issues. Roadmap to Sustainable Textiles and Clothing: Eco-friendly Raw Materials, Technologies, and Processing Methods. Springer, pp. 37–80.

Sen, A., Nigam, A., 2022. Bioengineering for decolorization of synthetic dyes in textile effluents using microbial enzymes. Journal of Scientific Research 66 (03), 66–78. Available at https://doi.org/10.37398/jsr.2022.660310.

Shiny Matilda, C., Madhusudan, I., Gaurav Isola, R., Shanthi, C., 2020. Potential of proteomics to probe microbes. Journal of Basic Microbiology 60 (6), 471–483. Available at https://doi.org/10.1002/jobm.201900628.

Silva Tavares, T., da Rocha, E.P., Esteves Nogueira, F.G., Torres, J.A., Silva, M.C., Kuca, K., et al., 2020. Δ-FeOOH as support for immobilization peroxidase: optimization via a chemometric approach. Molecules (Basel, Switzerland) 25 (2), 259. Available at https://doi.org/10.3390/molecules25020259.

Singh, P.K., Parikh, H., Saxena, A., Mishra, B., Tyagi, R., Awasthi, M.K., et al., 2025. Managing municipal wastewater remediation employing alginate-immobilized marine diatoms and silver nanoparticles. Energy and Environment 36 (5), 2333–2344. Available at https://doi.org/10.1177/0958305X241246185, https://journals.sagepub.com/home/EAE.

Slama, H., Ben, C., Bouket, A., Pourhassan, Z., Alenezi, F.N., Silini, A., et al., 2021. Diversity of synthetic dyes from textile industries, discharge impacts and treatment methods. Applied Sciences 11 (14), 6255.

Souza, A.G.V., Maria, T.C., Saran, L.M., Alves, L.M.C., 2022. Enzymatic bioremediation of dyes from textile industry effluents. The Toxicity of Environmental Pollutants 167.

Sudarshan, S., Harikrishnan, S., RathiBhuvaneswari, G., Alamelu, V., Aanand, S., Rajasekar, A., et al., 2023. Impact of textile dyes on human health and bioremediation of textile industry effluent using microorganisms: current status and future prospects. Journal of Applied Microbiology 134 (2). Available at https://doi.org/10.1093/jambio/lxac064.

Suh, H.N., Ji, J.Y., Heo, J.S., 2024. Translating proteome and transcriptome dynamics of periodontal ligament stem cell-derived secretome/conditioned medium in an in vitro model of periodontitis. BMC Oral Health 24 (1). Available at https://doi.org/10.1186/s12903-024-04167-z, https://bmcoralhealth.biomedcentral.com/.

Tadesse, M., Liu, Y., 2025. Recent advances in enzyme immobilization: the role of artificial intelligence, novel nanomaterials, and dynamic carrier systems. Catalysts 15 (6), 571. Available at https://doi.org/10.3390/catal15060571.

Tuncay, D., Yagar, H., 2020. Decolorization of reactive blue-19 textile dye by Boletus edulis laccase immobilized onto rice husks. International Journal of Environmental Science and Technology 17 (6), 3177–3188. Available at https://doi.org/10.1007/s13762-020-02641-z.

Tuttolomondo, M.V., Alvarez, G.S., Desimone, M.F., Diaz, L.E., 2014. Removal of azo dyes from water by sol-gel immobilized Pseudomonas sp. Journal of Environmental Chemical Engineering 2 (1), 131–136. Available at https://doi.org/10.1016/j.jece.2013.12.003.

Uppala, R., Muthukumaran, A., 2021. Optimization of media components and process parameters for microbial mediated remediation of azo dyes: a review. Journal of Microbiology Biotechnology and Food Sciences 11 (3), e3549. Available at https://doi.org/10.15414/jmbfs.3549.

Urek, R.O., Pazarlioglu, N.K., 2008. Manganese peroxidase production by immobilized Phanerochaete chrysosporium BKM-F-1767 in a cell bioreactor. Preparative Biochemistry and Biotechnology 38 (1), 1–12. Available at https://doi.org/10.1080/10826060701774296.

Vincent, D., Bui, A.D., Ram, D., Ezernieks, V., Bedon, F., Panozzo, J., et al., 2022. Mining the wheat grain proteome. International Journal of Molecular Sciences 23 (2), 713. Available at https://doi.org/10.3390/ijms23020713.

Wakil, S.M., Eyiolawi, S.A., Salawu, K.O., Onilude, A.A., 2019. Decolourization of synthetic dyes by laccase enzyme produced by Kluyveromyces dobzhanskii DW1 and Pichia manshurica DW2. African Journal of Biotechnology 18 (1), 1–11.

Wu, Z., Shen, Y., Zhang, X., 2022. TAG-TMTpro, a hyperplexing quantitative approach for high-throughput proteomic studies. Analytical Chemistry 94 (37), 12565–12569. Available at https://doi.org/10.1021/acs.analchem.2c02099.

Xie, X., Hu, J., Cao, X., Zhang, S., Sakamaki, T., Li, X., 2024. In situ utilization of electron-enhanced degradation of azo dyes in a constructed wetland–microbial fuel cell coupling system. Sustainability 16 (8), 3181. Available at https://doi.org/10.3390/su16083181.

Yadav, S., Kundu, N., Bhattacharya, A., Aseri, G.K., Jain, N., 2025. Biocatalysts for change: microbial enzymes as a sustainable solution to industrial pollution. Journal of Pure and Applied Microbiology 19 (3), 1627–1647. Available at https://doi.org/10.22207/jpam.19.3.34.

Yamaguchi, H., Miyazaki, M., 2024. Bioremediation of hazardous pollutants using enzyme-immobilized reactors. Molecules (Basel, Switzerland) 29 (9), 2021. Available at https://doi.org/10.3390/molecules29092021.

Yang, X., López-Grimau, V., Vilaseca, M., Crespi, M., 2020. Treatment of textile wastewater by CAS, MBR, and MBBR: a comparative study from technical, economic, and environmental perspectives. Water 12 (5), 1306. Available at https://doi.org/10.3390/w12051306.

Yan, B., Shi, M., Cai, S., Su, Y., Chen, R., Huang, C., et al., 2023. Data-driven tool for cross-run ion selection and peak-picking in quantitative proteomics with data-independent acquisition LC-MS/MS. Analytical Chemistry 95 (45), 16558–16566. Available at https://doi.org/10.1021/acs.analchem.3c02689, http://pubs.acs.org/journal/ancham.

Yuan, T., Zhang, S., Chen, Y., Zhang, R., Chen, L., Ruan, X., et al., 2021. Enhanced reactive blue 4 biodegradation performance of newly isolated white rot fungus antrodia P5 by the synergistic effect of herbal extraction residue. Frontiers in Microbiology 12. Available at https://doi.org/10.3389/fmicb.2021.644679.

Zdarta, J., Jankowska, K., Bachosz, K., Degórska, O., Kaźmierczak, K., Nguyen, L.N., et al., 2021. Enhanced wastewater treatment by immobilized enzymes. Current Pollution Reports 7 (2), 167–179. Available at https://doi.org/10.1007/s40726-021-00183-7.

Zeballos, N., Comino, N., Andrés-Sanz, D., Santiago-Arcos, J., Azkargorta, M., Elortza, F., et al., 2023. Region-directed enzyme immobilization through engineering protein surface with

histidine clusters. Acs Applied Materials & Interfaces 16 (1), 833–846. Available at https://doi.org/10.1021/acsami.3c15993.
Zerva, A., Pentari, C., Topakas, E., 2020. Crosslinked enzyme aggregates (CLEAs) of laccases from Pleurotus citrinopileatus Induced in olive oil mill wastewater (OOMW). Molecules (Basel, Switzerland) 25 (9), 2221. Available at https://doi.org/10.3390/molecules25092221.
Zhang, C., Peng, Q., Xuan, R., Zhang, Y., Xue, G., Sun, W., 2021. Accelerating biodegradation of an azo dye by using its inorganic intermediates. Journal of Chemical Technology & Biotechnology 96 (8), 2336–2342. Available at https://doi.org/10.1002/jctb.6761.
Zhao, Q., Zhao, Z., Zhang, J., Ni, Y., Ouyang, S., Liu, X., et al., 2023. Fn-HMGB1 adsorption behavior initiates early immune recognition and subsequent osteoinduction of biomaterials. Advanced Healthcare Materials 13 (2). Available at https://doi.org/10.1002/adhm.202301808.
Zheng, F., Cui, B.-K., Wu, X.-J., Meng, G., Liu, H.-X., Si, J., 2016. Immobilization of laccase onto chitosan beads to enhance its capability to degrade synthetic dyes. International Biodeterioration & Biodegradation 110, 69–78. Available at https://doi.org/10.1016/j.ibiod.2016.03.004.
Zhuang, M., Sanganyado, E., Xu, L., Zhu, J., Li, P., Liu, W., 2020. High throughput sediment DNA sequencing reveals azo dye degrading bacteria inhabit nearshore sediments. Microorganisms 8 (2), 233. Available at https://doi.org/10.3390/microorganisms8020233.
Zimmermann, A., Reitnauer, D., Yundung, Y., Menges, A.-L., Meuli, L., Pelisek, J., et al., 2024. Different proteins as biomarkers for sac shrinkage after endovascular aortic repair of abdominal aortic aneurysms. Journal of Cardiovascular Development and Disease 11 (11), 374. Available at https://doi.org/10.3390/jcdd11110374.

Chapter 11

Biodegradation of microplastic through proteomics approach: current and future perspective

Misba Khan[1], Mariyam Rafi[1], Mohd. Rehan Zaheer[2], and Roohi[1]
[1]*Department of Bioengineering, Integral University, Lucknow, Uttar Pradesh, India,* [2]*Department of Chemistry, R.M.P.S.P. Girls Post Graduate College, Basti, Uttar Pradesh, India*

11.1 Introduction

Microplastic (MP) pollution is a growing worldwide concern that has broad effects on both human health and environmental sustainability. MPs are synthetic polymer particles smaller than 5 mm. They can emerge directly from products like personal care items that contain primary MPs or from the fragmentation of larger plastic debris caused by environmental factors like ultraviolet (UV) radiation, temperature changes, and mechanical abrasion. Terrestrial, marine, and atmospheric habitats are now rife with these persistent contaminants. Unsettlingly, fresh research has revealed that groundwater, a crucial supply of drinking and agricultural water, contains MPs, highlighting a new dimension of concern (Syberg et al., 2021; Li et al., 2024b). According to research, only 60% of the 250 million tons of plastic waste produced worldwide each year is gathered. Weathering and UV radiation are two physical and chemical processes that break down the uncollected plastic garbage into tiny plastic particles that spread throughout the environment. Surface runoff, industrial waste discharge, wastewater treatment plant effluents, landfill leachates, and even air deposition are all possible entry points into groundwater. These contaminants are able to evade traditional filtration systems and can build up over time because of their small size and chemical makeup (Hoang et al., 2025). They have the potential to cause chronic illnesses such as oxidative stress, inflammation, and hormone disruption when consumed by either people or animals. They affect the environment by degrading soil quality, lowering plant photosynthetic capability, upsetting microbial equilibrium, and interfering with the global carbon cycle. Additionally, MPs contribute to climate change by affecting greenhouse gas emissions

Proteomics Driven Bioremediation. DOI: https://doi.org/10.1016/B978-0-443-45294-9.00006-0

directly through degradation byproducts and indirectly through the disruption of natural carbon sinks, including microbial ecosystems and vegetation. Remediation of MPs is particularly difficult due to their tenacity and persistence in the environment. Because they are ineffective at eliminating particles smaller than 5 mm or because of secondary pollution, such as greenhouse gas emissions, where conventional methods like incineration, landfill confinement, coagulation-flocculation, and advanced filtering sometimes fail (Goli et al., 2024). As a result, biological degradation, also known as bioremediation, has become a novel and environmentally beneficial substitute. This process breaks down synthetic polymers into less hazardous or biodegradable substances by taking advantage of the innate enzymatic properties of microorganisms such as bacteria, fungi, and algae. Several microbial strains, including *Pseudomonas, Bacillus cereus, Acinetobacter, Aspergillus*, and green algae like *Scenedesmus dimorphus,* have demonstrated the ability to degrade plastics such as polyethylene (PE), polypropylene (PP), and polyethylene terephthalate (PET) (Paladhi et al., 2022). The complicated polymer chains of plastic are broken down into smaller molecules by these organisms' release of extracellular enzymes such as laccase, PETase, cutinase, and esterases. Each microbe's efficiency of degradation is affected by environmental factors like pH, temperature, and nutrient availability, as well as polymer properties like hydrophobicity, surface area, and crystallinity. Since they divide metabolic work and more closely resemble natural ecosystem conditions, microbial consortia or mixed cultures are frequently more effective than single strains.

Proteomics has made tremendous progress in understanding and enhancing these microbial degradation processes. Proteomics, the extensive study of protein expression, provides important information about the enzymatic and metabolic mechanisms that microorganisms employ to break down plastic (Ho et al., 2024). Through the examination of proteins produced in response to plastic exposure, scientists can determine which enzymes break down polymers and how their expression is controlled. Enzymes like hydrolases and oxidases are essential for breaking stable bonds in plastic polymers (Khairul Anuar et al., 2022). Additionally, stress response proteins and biofilm-associated elements that promote microbial colonization on plastic surfaces have also been discovered using proteomic research. Engineering high-efficiency degradation systems requires an understanding of these findings. Microbial strains can be altered to overexpress desired enzymes or show increased resilience to environmental stressors through genetic engineering and synthetic biology. Proteomics-discovered enzymes can also be isolated or synthesized for use in bioreactors or immobilized on filters, offering a very effective but nonliving method of decomposing MPs in municipal or industrial wastewater streams (Muttiah and Hanafiah, 2025).

Proteomics and microbial bioremediation together offer a promising but challenging way to address MP pollution. Despite promising results at the lab level, where certain strains have been able to degrade over 50% of specific polymers in a matter of weeks, bringing these systems to the real world faces regulatory, ecological, and technical difficulties. It's important to carefully analyze variables including fluctuating environmental conditions, interactions with natural microbial populations, and the possible ecological dangers of introducing synthetic strains (Agarwal et al., 2024). Furthermore, a lot of breakdown processes take a long time and cannot completely mineralize the plastic into compounds that are safe for the environment. The advantages of proteomic-guided bioremediation are enormous, notwithstanding these drawbacks. Several United Nations Sustainable Development Goals (SDGs) and other global climate and pollution reduction objectives are in line with these solutions' scalable, low-energy, and sustainable nature. Enzymatic biochemistry, microbial ecology, and system integration research must continue to go from controlled experimental applications to reliable, field-deployable technology. Designing focused, effective, and secure bioremediation techniques that can tackle one of the most prevalent contaminants on Earth is becoming more and more possible as our knowledge of the interactions between microbes and plastics expands due to cutting-edge technologies like proteomics (Khurana et al., 2025).

11.2 Microplastics: properties, sources, and environmental persistence

11.2.1 Classification of microplastics

Plastics are synthetic polymers that are malleable, or flexible, and can change shape. Coal, oil, and natural gas are the sources of the polymers used to produce plastic, which are long chains of carbon, oxygen, hydrogen, silicon, and chlorides. 90% of plastics produced globally are made of synthetic polymers, including PE, polyamide (PA), PP, polyester (PES), polyurethane (PU), acrylic (AC), polystyrene (PS), PET, polyvinyl chloride (PVC), high-density polyethylene (HDPE), low-density polyethylene (LDPE), polyimide (PI), polymethyl methacrylate (PMMA), polytetrafluoroethylene (PFE), and HDPE. Due to its affordability, ease of use, strength, resistance to corrosion, high electrical and thermal conductivity, and advantageous chemical and physical properties, plastic has become an essential and versatile material in all aspects of our lives.

Particles range in size from nm to around 5 mm, but the majority of MP are undetectable to the naked eye. The term "MPs" has been defined differently by a number of academics. It appears that 333 nm is a reasonable lower limit for sampling using neuston nets. However, plastic particles that range in size from a few to roughly 500 nm are frequently found in saltwater. Classification of plastics based on size is shown in Fig. 11.1.

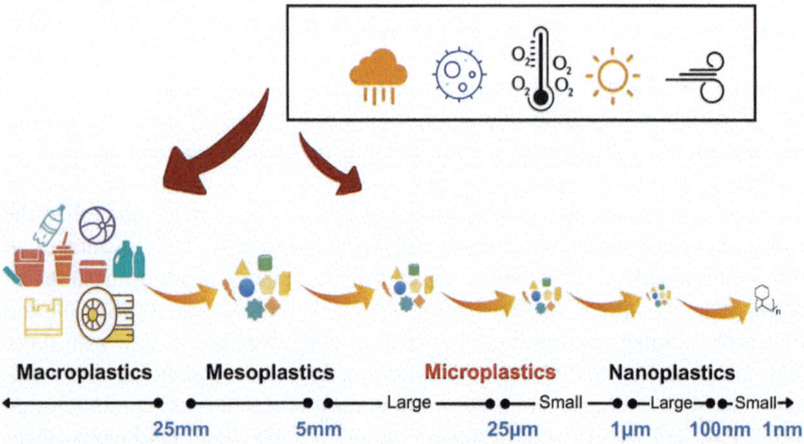

FIGURE 11.1 **Classification of plastics based on size.** For breaking down the plastics, information for its size is highly essential.

Primary resin granules are among the bigger particles referred to as mesoplastics. Additionally, MPs are tiny plastic granules used as scrubbers in cosmetics and as air purifiers. When macroplastics decompose, small plastic particles are produced. They enter aquatic systems gradually due to their small size, where they are forced into drainage channels without being absorbed. Throughout the world, fresh and marine water systems contain trace levels of persistent organic pollutants (POPs), which are captured by these meso- and MPs. POPs' hydrophobicity contributes to their higher concentration in meso-MPs litter compared to the whole fresh and marine ecosystem. The consumption of these contaminated plastics by aquatic creatures poses a serious risk since POPs have the potential to enter the water food chain. Unlike macroplastics, MPs are invisible to the naked eye; even mesoplastic resin grains mixed with sand can be difficult to spot. Naturally, net sampling is not used to gather smaller MPs, and there is no accepted standard technique for counting them in water or sand. MPs are divided into two groups according to their origins: primary MP and secondary MP. Man-made polymers known as primary MP are so tiny that they are discharged into the environment as separate, minuscule particles. They are used as exfoliants in a variety of processes, including chemical formulations, abrasive environments, petrochemical and chemical cleaning, and the manufacturing of synthetic clothing. They can also be included in items like shower gels and other cleaning supplies for hygiene and beauty. Additionally, they could be the consequence of large plastic products being abrasive during production, use, or maintenance, such as tyre erosion while driving or abrasion of synthetic fabrics after washing. Made of polyethylene, polypropylene, and polystyrene granules, microbeads are a type of virgin plastic that is used in cosmetic and medicinal products. They are around 2 mm in size. Recycled MP is created when bigger plastic products break down into smaller secondary plastic pieces in the aquatic environment. Photocatalysis, microbiological degradation, thermal decomposition, high temperature decomposition, hydrolysis, and other weathering processes linked to negligent discharge, such as discarded plastic bags, or unforeseen losses, such as fishing gear, are the causes of this environmental change (Bexeitova et al., 2024). MPs in water can either circulate or sink and be stored in the bed sediments. MPs' irregular shapes, which depend on their morphological properties, may include spherical, rod-like, and cubic forms, and should be taken into account when assessing risks to the environment and human health. Since sharp MPs physically excite human tissue, they may be dangerous. Furthermore, numerous chemicals are employed to make plastic polymers, most of which are detrimental to vital human organs, depending on the end usage. Due to significant environmental and public health problem because of MPs contamination in drinking water, the biodegradable MPs are supposed to show a good promise as a substitute for traditional plastics. Despite their potential, biodegradable MPs also contaminate water, leading to complicated issues. MPs have been detected in drinking water systems all around the world, even biodegradable varieties. Regions that are severely impacted by traditional MPs also typically have the highest levels of contamination from biodegradable MPs. In nations like China, India, Indonesia, and the Philippines, major cities and coastal areas report high levels of MP pollution in their water systems. Significant MP pollution, for instance, has been discovered in tests of the Yangtze River in China and the Ganges River in India, two important drinking water supplies. Unlike traditional plastic, biodegradable MP is made to break down faster. One of the primary varieties is polylactic acid (PLA), which is utilized in packaging and single-use items. Renewable materials like maize starch are used to make PLA. Utilized in packaging and medicine, polyhydroxyalkanoates (PHAs) are created through microbial fermentation. Starch-based plastics, which are made from natural organic sources, are frequently seen in single-use products like bags. Made from plant cellulose, cellulose-based polymers are utilized in coatings and films. Biodegradable MPs have a number of drawbacks, including incomplete degradation in natural conditions, particularly in water, despite their faster rate of breakdown. MP pollution may linger if biodegradable MP doesn't entirely break down. After three years in the maritime environment, biodegradable plastic bags held up well, according to a University of Plymouth study. The breakdown of biodegradable polymers may result in the discharge of hazardous additives or byproducts into the water. For instance, PLA may emit lactic acid, which may change the water's pH and endanger aquatic life. The decomposition of biodegradable plastics may nevertheless produce tiny particles that linger in the environment, much like conventional MPs (Yao et al., 2022).

11.2.1.1 Characteristics of microplastics

(1) Size: MPs are typically plastics that have a particle size of less than 5 mm. The initial research on MP size in the atmosphere revealed information on the range of 100–5000 micrometers in total atmospheric fallout. As of right now, atmospheric MPs have been found to range significantly in size from 4 to 5000 micrometers. The size-measuring investigations conducted between 2015 and 2018 were not accurate enough due to limitations in the methodologies used. Less than 2000 micrometers is now the focus of the research on MP size, thanks to the development of both qualitative and quantitative approaches. Methods of sampling and analysis were frequently used to operationally establish the size limits of MPs. The range of measurable size reported by earlier investigations on air samples was expanded by the identification of MPs as small as 10 micrometers. The overall atmospheric deposition in central London revealed a large number of fluorescent nonfibrous particles with fibers 20 micrometers and smaller, suggesting the presence of smaller MPs in the atmosphere that merit more investigation (Yao et al., 2022). According to these investigations, MPs make up a significant portion of the airborne particulate matter that is swallowed and inhaled. The research also suggests that breathing in MPs from the air may be harmful to people's health.

The amount of indoor and outdoor MPs in the same area varies as well, and different researchers have differing opinions about this subject. MPs were found to be less than 1650 micrometers in outdoor air and no fibers larger than 3250 micrometers in indoor air in research conducted in Paris, France (Gaston et al., 2020). Although Californian interior and outdoor air did not significantly differ in fiber length, their data indicate that indoor fragments are half as large as outdoor ones. According to data from the two cities, the size of MPs in Paris's indoor and outdoor air is greater than that of California, suggesting that MPs may vary from city to city. The size of MPs ranged from 20 to 2000 micrometers, according to a survey of all suspended MP particles in the West Pacific Ocean. The MPs gathered over the Eastern Indian Ocean and South China Sea are less than 2000 micrometers. As a result, larger MPs are challenging to suspend in the air and even transmit over greater distances.

(2) Color: MPs come in a variety of colors, such as translucent, red, black, and white. According to the circumstances as reported, the majority of atmospheric MPs are black, blue, and white. When evaluating atmospheric MP samples visually for the first time, color is usually thought to be beneficial. In the streets of the Tehran metropolitan area, 29.9% of all atmospheric MPs are black, making them the most prevalent type (Yao et al., 2022). This may lead us to believe that a significant amount of street dust is composed of tyre crumbs.

(3) Shape: In atmospheric MPs, the most prevalent morphologies are fibrous and fractured. There are also irregular shapes like spheres, films, and fragments. The largest percentage of fibrous MPs was found in air deposition samples by the Paris and Yantai studies. Additionally, the most common MP shape found in various interior surroundings is fibers. Distinctly shaped MPs have distinct physical characteristics. Films are often thin and flexible, fibers are typically very thin, and flakes are typically thick and asymmetrical. The original morphology of the primary MPs, the processes of erosion and degradation on the plastic particles' surface, and the length of time spent in the environment all influence the shape of MPs in the atmosphere. The forms of atmospheric MPs are frequently varied in different functional regions of the same metropolis. For instance, whereas more fibrous MPs were detected in urban regions, more film-like, spherical, and fragmented MPs were found in industrial locations. MPs were primarily made up of fibrous particles of various sizes and black debris, while the majority of the street dust samples were dominated by spherical and film-like particles (Yao et al., 2022).

11.2.2 Sources and routes of environmental entry

The sources of MP vary widely, as plastics have a broad range of uses and applications. Primary MPs are micrometer-sized particles that are purposefully produced for certain uses or goods, including industrially produced drink bottle pellets and personal care product microbeads, and ultimately end up in the environment. Particles are created when larger plastic waste breaks up, and such fragments are referred to as secondary MP. Due to the small number of primary MPs and the effects of weathering and ageing, these two classes eventually become nearly identical in the environment. Once the material reaches the MP scale, however, its diversity still reflects the complexity and diversity of its sources (Koelmans et al., 2022).

Numerous human activities also release MPs into the environment, and they can enter groundwater systems through a number of interrelated paths. Landfills, wastewater treatment facilities, industrial waste discharge, septic tank overflows, and runoff from cities and farms are important sources. MP pollution is caused by improper disposal of plastic waste, air deposition, and surface water infiltration. Additionally, the usage of compost, sewage sludge, and plastic mulch films—all of which break down over time into tiny plastic particles—makes agricultural areas important entrance locations. Through irrigation or rainfall, these deteriorated materials can enter aquifers as well, making agriculture a vital conduit.

Groundwater contamination is largely caused by landfills, particularly those that are poorly maintained or situated close to aquifers. High concentrations of MPs, which are created when garbage containing plastic breaks down, are present in leachates from these locations. These disposal sites frequently receive microbeads from personal hygiene products and other main microplastics, which are then broken up by physical, chemical, and biological processes. These MPs may then move into groundwater systems through soil layers, especially in areas with permeable soil or heavy rainfall. Leachates from landfills frequently contain secondary MPs, which are produced when commonplace objects like bags and bottles break down (Liu and Zheng, 2025).

Another significant source is wastewater treatment facilities used in homes and businesses. Tiny plastic particles are discharged during product use, cleaning, and manufacturing, and they wind up in wastewater. These particles are frequently released into surface waters or kept in sludge that can be used again in agriculture since they are not entirely eliminated by traditional treatment methods. Laundry-shedding textile fibers and MPs from cleaning and cosmetic goods find their way into sewer systems and have the potential to contaminate groundwater through direct discharge or runoff. Another major factor in the introduction of MPs into aquatic systems is industrial facilities that use plastic pellets or resins.

Stormwater runoff and atmospheric deposition are two indirect but significant pathways for the movement of MPs. Rainfall can cause particles to land on the surfaces of soil or water after being sent into the air by wind, traffic, or industrial activity. After being deposited, they may seep into groundwater through the earth or be transported by runoff into lakes and rivers. It has been determined that urban stormwater and agricultural runoff are significant sources of MPs, especially during the rainy season. The intricacy of MP pollution and the pressing need for all-encompassing mitigation solutions are highlighted by the combined impact of these various sources and transport routes (Hoang et al., 2025).

11.2.3 Environmental fate and challenges in degradation of microplastics

MPs are becoming widely recognized as ubiquitous environmental contaminants that affect freshwater ecosystems as well as marine ones. MPs are very challenging to regulate and remediate because of their complex behavior after they are discharged into freshwater environments. Although they can evade traditional treatment and filtration systems due to their small size (usually less than 5 mm), their varied chemical compositions and shapes make detection and control much more challenging. Though intended to provide sustainable alternatives, even biodegradable versions frequently fail to fully decompose in the natural environment, releasing persistent fragments or secondary pollutants such as acids that stay in the environment. When these particles are introduced into aquatic systems, environmental factors such as wind, precipitation, water currents, and seasonal events have a significant impact on their mobility and permanence, which helps to disperse them widely. There are further long-term hazards from bioaccumulation through food webs, trapping in biofilms, and accumulation in sediments. These elements demonstrate the unique resistance of MPs to natural degradation and the necessity of sophisticated methods for tracking, cleaning, and controlling their ambient presence (Salehi et al., 2024).

Mitigation measures are complicated by the different transit and destiny pathways that MPs go through after being introduced into aquatic systems. Particle mobility and settling behavior are influenced by various factors, including size, shape, density, and hydrophobicity. MPs can become entangled in biological tissues, settle in sediments, or float in the aqueous column. Their dispersal is greatly influenced by environmental factors such as wind, tides, river currents, and precipitation. For example, storms can raise MP concentrations by 40 times by introducing particles into rivers and lakes from terrestrial sources. MP loads are exceptionally high in regions with dense populations, industrial activity, and important freshwater supplies, such as the Yangtze and Ganges rivers. MPs also change their buoyancy when they interact with biological or mineral materials (such as biofilms, feces, or soil aggregates), which frequently makes it easier for them to settle and accumulate in benthic zones, where they can cause long-term problems.

The absence of standardized sampling and analysis methods is one of the primary obstacles to comprehending the fate of MPs. Different mesh sizes, filtration techniques, and pretreatment chemicals are used in a wide range of current systems. Studies that use low-density separation media, like NaCl, frequently overlook high-density MPs, such as PET and PVC. Furthermore, due to visual misinterpretation, commonly utilized mesh sizes may overstate the existence of tiny MP fractions or eliminate them. Analytical techniques like pyrolysis-gas chromatography-mass spectrometry (Py-GC–MS), Raman spectroscopy, and Fourier transform-infrared spectroscopy (FTIR) are used, but each has drawbacks in terms of resolution, time, and expense (Berkel and Özbek, 2024). A thorough worldwide assessment of MP pollution in freshwater ecosystems is hampered by this methodological variability, which also makes cross-study comparisons challenging.

MPs degrade slowly and inefficiently in the environment because synthetic polymers are resistant. While exposure to sunlight (photodegradation), oxygen (oxidation), and microbes (biodegradation) all have a role in fragmentation, environmental factors, including temperature, pH, salt, and UV exposure, have a significant impact on these processes. MP degradation rates, for instance, vary across moving rivers and stagnant freshwater basins because of differences in microbial communities and mechanical stress (Niu et al., 2023). Notwithstanding their plant-based beginnings, these polymers frequently endure in aquatic environments, generating leachates and microfragments that may upset the pH balance and aquatic metabolism.

The toxicological consequences of MPs exacerbate their environmental destiny. The high surface area and hydrophobicity of MPs allow them to easily absorb harmful chemicals such as heavy metals, medicines, and POPs. These pollutants have the ability to enter aquatic life and bioaccumulate throughout the food chain. The exposure of fish, amphibians, and invertebrates to MPs has been associated with immunotoxicity, oxidative stress, endocrine disruption, and developmental abnormalities. Serious public health issues are raised by the fact that studies have also shown that MPs can aid in the spread of infections like *Helicobacter pylori* and even SARS-CoV-2 (Desdouits et al., 2023). The ability of ageing MPs to bind toxicants is increased due to their rough and broken surfaces, which increases their potential as environmental damage vectors.

The distribution of MP is significantly influenced by human and geospatial variables. Typically, high quantities are found close to recreational lakes, wastewater disposal locations, and urban and industrial areas. Poorly managed waste and industrial runoff cause rivers in China's textile centers, as well as in portions of India and Nigeria, to receive significant MP loads. MPs have also been found in high-altitude and isolated lakes, suggesting long-distance hydrological and atmospheric transmission. Older, denser particles are frequently discovered buried in sediment studies, serving as slow-release sources of pollutants as well as historical records. MP redistribution is further influenced by seasonal changes like monsoons or snowmelt, which makes their environmental behavior extremely dynamic and localized. MP sources, transportation, and their environmental fate can be easily understood in Fig. 11.2.

A complicated interaction between physicochemical characteristics, ambient factors, and human activity shapes the environmental fate of MPs in freshwater ecosystems. The durability of the material and the unpredictability of the environment impede gradual, frequently incomplete degradation

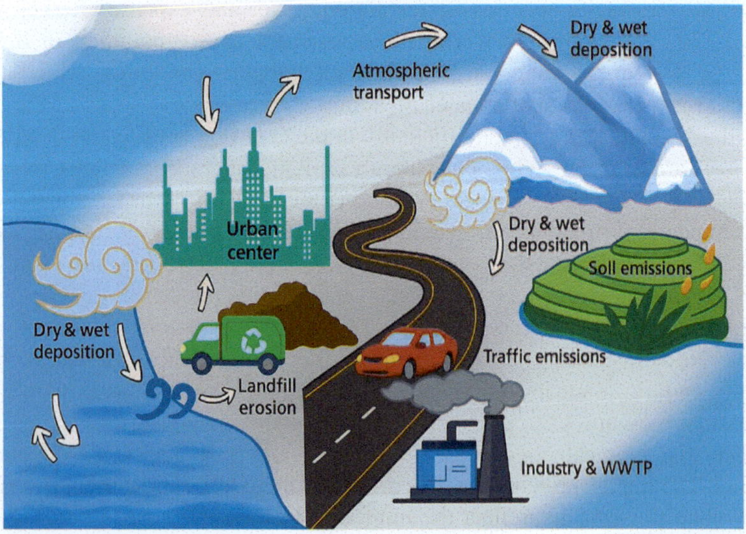

FIGURE 11.2 **Source, transport, and fate of MPs in the environment.** Understanding the microplastics' journey is a key step to reducing pollution.

processes. Despite their potential, biodegradable MPs have many of the same drawbacks as traditional polymers. Global assessments and mitigation measures are made more difficult by the continued inconsistency in detection and analysis. A diverse strategy is needed to address the issues raised by MPs, including public awareness campaigns, technological advancements in waste management, regulatory interventions, and the standardization of detection techniques. The persistence and hazards of MPs in freshwater systems can only be successfully addressed by concerted international efforts (Yao et al., 2022).

11.3 Microbial biodegradation of microplastics

11.3.1 Microbial diversity in plastic-degrading ecosystems

MPs are biodegraded in a variety of ecosystems, including soil, compost, landfills, and marine environments by microbial communities, especially those made up of bacteria, fungi, and algae. *Pseudomonas, Bacillus, Alcanivorax,* and *Actinomycetes* are the most common degraders, with species from these genera often isolated for their capacity to digest a variety of synthetic polymers. The most noteworthy is *Pseudomonas aeruginosa*, which has been identified in sewage-contaminated soil and superworm intestines and has been demonstrated to break down PE, PS, and PP with varying degrees of effectiveness. *P. putida* and *P. stutzeri* are two consortia that have shown notable degradation of LDPE, whereas marine species such as *P. aestusnigri* and *P. rhodesiae* can build biofilms on plastic surfaces. Similar to this, *Bacillus* species that have been found in mangroves and marine environments, such as *B. cereus* and *B. subtilis,* show detectable degrading activity on PE, PET, and PS. Other species, including the hydrocarbon-degrading *Alcanivorax borkumensis,* are frequently found in marine plastic biofilms and have shown the ability to break down polycaprolactone and LDPE (Jebashalomi et al., 2024). Despite being mostly isolated from soil, actinomycetes such as *Rhodococcus ruber* and *Nocardioides marinus* have also been demonstrated to lower the weight of PET and PS in deep-sea and marine environments.

Beyond bacteria, marine fungi have special potential because of their filamentous structure, which allows them to colonize plastic surfaces that are inaccessible to many microorganisms. Species like *Rhodotorula mucilaginosa*, *Penicillium* sp., and *Aspergillus* sp. have been isolated from a variety of marine environments, such as the Bay of Bengal and the North Sea, and have shown effectiveness in breaking down polyethylene by producing degrading enzymes or forming biofilms. Significant mass loss in PE has been demonstrated by *Alternaria alternata* and *Zalerion maritimum*, the latter of which degraded by more than 56% in just 14 days. Extracellular substances that aid in film perforation or surface roughening of the plastic substrate are frequently secreted by these marine fungi. Although terrestrial fungi like *Phanerochaete* and *Trichoderma* are renowned degraders, their marine equivalents exhibit potential activity but have received less research attention (Zeghal et al., 2021). Furthermore, some cyanobacteria and microalgae, such as *Uronema africanum* and *Oscillatoria subbrevis*, have shown the capacity to colonize and destroy plastics, particularly LDPE, even in the absence of additives, despite the fact that algae are under-represented in studies on plastic biodegradation. These microbial communities' enzymatic arsenal, which includes cutinases, esterases, and laccases, among others, allows complicated polymers to be first oxidized and broken down into oligomers, which are subsequently subjected to further metabolism. Therefore, a key element in reducing the environmental load of MPs is the taxonomically rich and functionally diverse microbial diversity found in plastic-degrading ecosystems (Dey et al., 2022).

11.3.2 Enzymatic mechanisms in microplastic degradation

Enzymatic mechanisms in the degradation of MPs include crucial stages in which microorganisms adhere to plastic surfaces and create biofilms that release enzymes such as hydrolases, laccases, cutinases, and esterases. Functional groups like carbonyl, carboxy, and hydroxyl are introduced by these enzymes to depolymerize long-chain polymers into oligomers. Following their absorption, the oligomers undergo β-oxidation and the citric acid cycle, which results in the production of CO_2, CH_4, and H_2O. This enzymatic biodegradation is cheap and environmentally beneficial when compared to traditional techniques (Lv et al., 2024). The particular enzymes and their breakdown products linked to various plastic polymers are summarized in Table 11.1.

11.3.3 Notable bacterial and fungal strains

The microbial biodegradation of MPs is mostly dependent on notable bacterial and fungal strains, some of which exhibit exceptional enzymatic capabilities. *Ideonella sakaiensis*, one of the most researched bacteria, is well-known for generating two essential enzymes, ISF6–4831 and ISF6–0224, which hydrolyze PET into environmentally safe monomers such as ethylene glycol (EG) and terephthalic acid (TPA). *Pseudomonas* species are very well-known, especially for their adaptable enzymatic systems that use alkane hydroxylases to break down high-molecular-weight polymers like polyethylene. These bacteria are particularly effective in waste management applications because they can use plastics as their source of carbon and are frequently found in contaminated soils and aquatic settings (Hayat et al., 2025).

Aspergillus species are the most common and well-researched fungi for the breakdown of MPs. Laccase, esterase, lipase, and peroxidase are among the extracellular enzymes produced by strains like *A. niger, A. flavus,* and *A. tubingensis* that degrade complex plastic polymers into simpler molecules. These fungi have the ability to grow biofilms on plastic surfaces, which speeds

TABLE 11.1 Key enzymes involved in the biodegradation of different types of plastics.

Types of plastics	Microbial species	Sources	Enzymes	Degradation products	References
PHB	*Pseudomonas* sp.	Activated sludge	Depolymerase	3-HB monomer	Di et al. (2019)
	Agrobacterium sp.	Sewage	Depolymerase	3-HB monomer, HB-HB dimers	Lv et al. (2024)
	DSGZ				
PLA	-	Environmental metagenomes	Polyesterases	Lactic acid monomers, dimmers, and longer oligomers	Hajighasemi et al. (2018)
PBS	*Fusarium solani*	Phytopathogenic	Cutinase	SA	Santos-Beneit et al. (2023)
PVC	*Cochliobolus*	Plastic dumped soils	Laccases	Aromatic compounds, polycyclic aromatic hydrocarbons	Sumathi et al. (2016)
PET	*Stenotrophomonas maltophilia* PRS8	Soil of a landfill	Cutinase-like enzyme	TPA, MHET	Jabloune et al. (2020); Bollinger et al. (2020); Carr et al. (2023)
	Streptomyces scabies	Plant	Cutinase	BHET	
	Pseudomonas aestusnigri	Marine	Polyester hydrolase	MHET	
	Streptomyces sp. SM14	Marine sponge-derived	Hydrolyzing enzyme	TPA, MHET, BHET, soluble di-aromatic oligomers	

(*Continued*)

TABLE 11.1 (Continued)

Types of plastics	Microbial species	Sources	Enzymes	Degradation products	References
PBAT	*Fusarium solani*	Phytopathogenic	Cutinase	PBAT: TPA, BD-TPA	Muroi et al. (2017); Santos-Beneit et al. (2023)
	Bacillus pumilus	Soils	Hydrolase	TPA, adipic acid, 1,4-butanediol	
PCL	*Fusarium solani*	Phytopathogenic	Cutinase	6HH	Santos-Beneit et al. (2023)

3-HB, 3-hydroxybutyric acid; *6-HH*, 6-hydroxyhexanoic acid; *BD-TPA*, 1,4-butenediol terephthalic acid; *BHET*, Bis-(2-hydroxyethyl) terephthalate; *MHET*, Mono-(2-hydroxyethyl) terephthalate (MHET); *SA*, Succinic acid; *TPA*, Terephthalic acid.

up degradation and improves enzymatic activity. For example, under controlled settings, *A. niger* has demonstrated a reduction in polyethylene weight of up to 19%–24% in a matter of months. *Aspergillus* is essential for sustainable plastic waste treatment solutions due to its diversity, flexibility, and enzymatic competency (Liu et al., 2025).

11.4 Proteomic technologies in environmental microbiology

The study of environmental microbiology looks at how microbial communities affect pollutant degradation, biogeochemical cycles, and ecosystem function (Kasmawati and Nasution, 2024). However, identifying which proteins are actively produced and functional in particular environmental conditions necessitates going beyond genomic data in order to comprehend microbial activity in complicated ecosystems (Ramond et al., 2025). Proteomics provides a formidable toolkit to reveal the activity and functional potential of microbial communities in many habitats, supporting ecological research, pollution treatment, and environmental monitoring (Al-Amrani et al., 2021).

11.4.1 Overview of proteomic tools

Protein expression patterns in environmental samples can be seen and measured using two-dimensional gel electrophoresis (2D-GE), which differentiates proteins according to their molecular weight and isoelectric point (pI). High-resolution separation of complicated protein mixtures and isoform visualization are advantages of 2D-GE (Meleady, 2018). But this technique is also facing a few limitations, like its labor-intensive process and low sensitivity for proteins with low quantity or strong hydrophobicity (Meleady, 2018). Combining liquid chromatography with mass spectrometry (LC-MS/MS) provides posttranslational modification studies and peptide sequence information, allowing for high-throughput protein identification and quantification (Li and Zhu, 2020). Its broad dynamic range, great sensitivity, and suitability for complex samples are its advantages, while its requirement for sophisticated instrumentation and bioinformatics pipelines is a drawback (Li and Zhu, 2020). Another system, three-dimensional LC-MS/MS, has been explored and its process has been described in Fig. 11.3.

Isobaric tags are used to label peptides from various samples in the label-based quantitative proteomics technique known as Isobaric Tags for Relative and Absolute Quantitation (iTRAQ) (Sivanich et al., 2022). This allows for simultaneous quantification and identification in LC-MS/MS analysis; its benefits include accurate quantification and sample multiplexing; its drawbacks include data analysis complexity and the possibility of ratio distortion from co-isolation interference (Sivanich et al., 2022). There are various other proteomics tools that are employed in MP degradation; some of these, along with their advantages and disadvantages, are listed in Table 11.2.

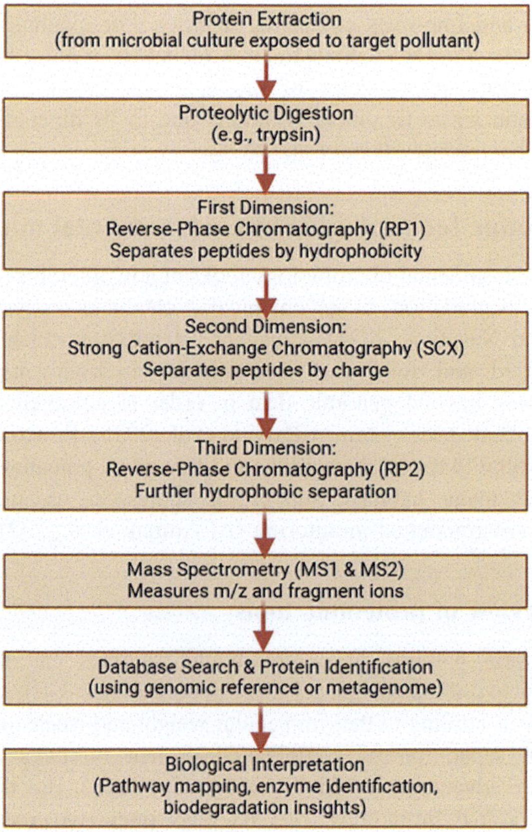

FIGURE 11.3 **Flowchart describing three-dimensional LC-MS/MS.** The ultimate tool for precise analysis and for unraveling complexity.

11.4.2 Role of proteomics in microbial ecology and biodegradation

Monitoring microbial metabolic activity in both natural and artificial settings is made possible by proteomics, which also helps to connect microbial identities to functional roles in symbiotic relationships, pollutant degradation, and nutrient cycle (Wang et al., 2016). Metaproteomics, for instance, identifies active pathways for the cycling of carbon and nitrogen in soil and marine ecosystems (Rodríguez-Ramos et al., 2022). In order to promote the development of bioremediation techniques, proteomics determines the enzymes and pathways involved in the biodegradation of contaminants such as hydrocarbons, pesticides, heavy metals, and MPs (Bala et al., 2022). *Pseudomonas putida* exposed to polycyclic aromatic hydrocarbons, for instance, showed increased expression of the enzyme's oxygenase and dehydrogenase, which

TABLE 11.2 Various proteomics tools in microplastic degradation studies.

S. no.	Proteomics tools	Advantages	Disadvantages	References
1.	2D-GE combined with MALDI-TOF MS	Well-established, relatively low cost, good for comparative proteomics of known organisms.	Gel-to-gel variation, limited protein size/pI/hydrophobicity range, and low throughput for large proteomes.	Karp et al. (2008)
2.	LC-MS/MS	High sensitivity, high throughput, can identify thousands of proteins from a single sample	Expensive instrumentation, complex workflows, and challenges in quantification.	Clarkson et al. (2014)
3.	3D LC-MS/MS	High separation power, tolerance to detergents, and increased protein coverage.	More complex setup, longer analysis time.	Keller and Hettich (2009)
4.	Top-down MS	Direct analysis of modifications and isoforms; links structure to function.	Currently efficient mainly for proteins <25 kDa; difficult for large proteins.	Melby et al. (2021)
5.	Gas-phase ion mobility fractionation	Adds orthogonal separation, can improve proteome coverage.	Still under development for high-throughput proteomics.	Penny et al. (2023)
6.	UHPLC	Faster runs, better resolution, compatible with MS	Requires specialized equipment.	Pfohl et al. (2025)
7.	iTRAQ	Simultaneous quantification and identification, Reliable quantitative accuracy	Complex microplastic-associated biofilms may cause incomplete labeling or suppression of certain peptides.	Mishra et al. (2021)

UHPLC, Ultrahigh-pressure LC.

are involved in degradation pathways, according to proteome research (Adenan et al., 2022). Proteomics supports the evaluation of ecosystem resilience and health by tracking microbial stress responses to contaminants, heavy metals, salinity, and other environmental stresses (Nesatyy and Suter, 2007).

11.4.3 Strengths and limitations of proteomics approaches

11.4.3.1 Strategies in rnvironmental proteomics includes the following

- **Sample preparation:** Proteins are recovered from complex environmental matrices by enriching microbial cells, eliminating humic compounds, and optimizing lysis techniques (Wöhlbrand et al., 2013).
- **Label-free versus label-based quantification:** Selection according to desired quantitative depth, duplicates, and sample complexity (Wöhlbrand et al., 2013).
- **Integration of metaproteomics:** Combining metagenomics with metatranscriptomics to connect gene expression and protein expression with community composition (Wöhlbrand et al., 2013).

11.4.3.2 Limitations and challenges

- **Complexity of environmental samples:** Environmental pollutants and a variety of proteins with a broad dynamic range make extraction and analysis more difficult (Chandramouli and Qian 2009).
- **Database dependency:** In environmental proteomics, protein identification is dependent on reference databases, which are frequently lacking for novel or uncultured microorganisms (Aebersold and Mann, 2016).
- **Quantitative challenges:** While label-free approaches may encounter issues with repeatability, label-based approaches necessitate meticulous design (Anand et al., 2017).
- **Complexity of interpretation:** Context-dependent activity and protein functional redundancy may make functional interpretations more difficult (Anand et al., 2017).

11.5 Proteomic insights into microplastic degradation

Persistent environmental contaminants that negatively affect ecosystems and human health are MPs (less than 5 mm) (Lalrinfela et al., 2024). Although utilizing microbes for biological degradation is an environmentally friendly alternative, its slow rates and lack of fully understood mechanisms make it difficult to implement effectively (Lalrinfela et al., 2024). By identifying and measuring the proteins that microorganisms actively express during the breakdown of MPs, proteomics reveals pathways, enzymes, and stress responses that can be used to improve biodegradation (Dhali et al., 2024).

11.5.1 Enzyme discovery: PETases, cutinases, esterases

PETases are hydrolases that break down PET into EG, TPA, and mono(2-hydroxyethyl) terephthalate (MHET) (Yoshida et al., 2021). PETase-like proteins have been found by proteomics in ambient microbial consortia of *Ideonella sakaiensis*, and investigations have shown that these proteins are upregulated when exposed to PET (Yoshida et al., 2021). Proteomics is being used to monitor engineered PETases with improved stability and activity in order to assess effectiveness (Yoshida et al., 2021).

Cutinases, a different type of esterases, are found in a variety of bacteria and fungi and have the ability to hydrolyze polyesters, such as PET and polycaprolactone (PCL) (Cai et al., 2023). Cutinase isoforms are upregulated in microbial consortia cultivated on PET or mixed plastic waste, according to proteomic investigations (Cai et al., 2023).

By severing ester bonds in a variety of plastic polymers (such as PET and polyurethane), esterases are essential for starting the depolymerization of MPs (Leitao and Enguita, 2021). Pathways reconstruction for degradation processes is now possible for the discovery of novel esterases in plastic-degrading consortia by researchers using LC-MS/MS proteomics (Leitão and Enguita, 2021).

11.5.2 Differential protein expression in plastic-enriched conditions

By tracking variations in protein expression in microorganisms exposed to MPs, proteomics can identify pathways implicated in the following.

- Adhesins, porins, and outer membrane proteins, which are involved in uptake and surface attachment (Lee et al., 2025).
- **Degradation by oxidation and hydrolysis:** Increased activity of cutinases, hydrolases, oxygenases, and PETases (Kawai et al., 2024).
- **Stress response:** detoxifying enzymes, oxidative stress proteins (catalase, superoxide dismutase), and chaperones (DnaK, GroEL) (Kadac-Czapska et al., 2024).
- **Changes in energy metabolism:** Assimilation of breakdown products is indicated by increased expression of TCA cycle and β-oxidation enzymes (Satta et al., 2024).

11.5.3 Case examples from recent studies

11.5.3.1 Proteomic analysis of Ideonella sakaiensis *on polyethylene terephthalate*

Researchers confirmed active degradation pathways by using LC-MS/MS to find that when *I. sakaiensis* was cultured on PET films, there was an overexpression of PETase, MHETase, and stress proteins. Changes in energy metabolism following the use of plastics were also discovered by proteomics (Poulsen and Nielsen, 2023).

11.5.3.2 Soil microbial consortia on mixed microplastics

Cutinases, lipases, esterases, and monooxygenases were shown to be upregulated across bacterial taxa (*Pseudomonas, Rhodococcus,* and *Bacillus*) in soil samples enriched with MPs, according to metaproteomic analysis (Amobonye et al., 2021). When MPs were present, the number of stress response proteins and biofilm-associated proteins rises, suggesting surface colonization and resilience (Amobonye et al., 2021).

11.5.3.3 Marine microbial communities on polyethylene terephthalate microplastics

Hydrolases and oxygenases were shown to be more abundant in marine bacteria that formed biofilms on PET MPs, according to LC-MS/MS proteomics, which identified species of *Vibrio* and *Alcanivorax* that express enzymes similar to PETase (Wright et al., 2021).

11.6 Challenges and future prospectives

MPs are persistent environmental contaminants that pose a serious threat to human health and the environment because of their ubiquity and stubbornness in aquatic, terrestrial, and atmospheric ecosystems (Amobonye et al., 2021). A sustainable remediation method is offered by biological degradation employing microbial enzymes and consortia; nevertheless, biodegradation effectiveness is restricted by our incomplete knowledge of enzymatic pathways, microbial community dynamics, and environmental variables (Amobonye et al., 2021). Enzyme discovery, metabolic pathway elucidation, and the development of successful bioremediation solutions are all made possible by proteomics, which provides thorough profiling of protein expression patterns, enzyme synthesis, and pathway activities during MP biodegradation (Carbonell et al., 2018).

11.6.1 Technical and analytical limitations in proteomics

11.6.1.1 Challenges of low biomass, protein extraction, and contaminant interference in environmental matrices

It can be difficult to recover enough protein yields for mass spectrometry analysis because environmental materials such as soil, sediments, and marine waters frequently contain complex protein mixes and low microbial biomass (Wang et al., 2016). Furthermore, inorganic particles, plastic additives, and humic compounds that are coextracted with proteins disrupt proteomic operations downstream (Nebauer et al., 2024).

11.6.1.2 Dependence on incomplete reference databases for protein identification in environmental metaproteomics

Low annotation rates and misidentification of proteins essential for plastic degradation pathways result from proteomic identification's heavy reliance on reference databases, which are frequently lacking for uncultured, uncommon, or novel environmental microbes actively involved in MP biodegradation (Van Den Bossche et al., 2025).

11.6.1.3 Limitations of functional interpretation of biodegradation pathways and quantitative proteomic approaches

It is difficult to accurately correlate protein abundance and functional activity in MP degradation because quantitative proteomics techniques, such as label-free quantification and isobaric tagging (iTRAQ, TMT), suffer from reproducibility issues, ratio suppression, and quantification bias in complex environmental samples (Sharma et al., 2024).

11.6.2 Integration with genomics and metabolomics

Finding new plastic-degrading enzymes and entire metabolic pathways active during MP biodegradation in environmental samples is made easier by researchers' ability to correlate gene presence with actual protein expression through the integration of metagenomic sequencing and proteomics, or meta-proteo-genomics (Jahanshahi et al., 2023).

Metabolomics sheds light on the destiny of intermediates and end-products of MP breakdown, allowing for the verification of pathways hypothesized by proteomics and the measurement of carbon flux from polymers to CO_2, biomass, or absorbed monomers (Wright et al., 2021).

Researchers will be able to discover, describe, and track uncommon microbial species actively involved in MP breakdown within heterogeneous communities using developments in single-cell proteomics and spatial omics technology (Mund et al., 2022).

11.6.3 Engineering microbes for efficient biodegradation

Proteomic insights guide the directed evolution of plastic-degrading enzymes through rational protein engineering, through structure-guided mutagenesis and directed evolution. Proteomics-driven enzyme identification allows for the rational engineering, stability improvement, and substrate affinity optimization of important plastic-degrading enzymes (such as PETases, cutinases, and esterases), thereby increasing degradation efficiency in environmental conditions (Joho et al., 2024).

Methods of synthetic biology are used to build engineered microbial consortia for the degradation of mixed MP waste. Designing synthetic microbial consortia that can break down a variety of MP polymers in the environment is made possible by using proteomics to find complementing metabolic capabilities (Edwards et al., 2022).

Adaptive laboratory evolution and CRISPR-Cas systems increase microbial plastic degradation capacity, to increase microbial tolerance to harmful plastic additives and boost breakdown efficiency in engineered strains. Proteomic datasets can guide targeted genetic alterations using CRISPR-Cas systems and adaptive laboratory evolution methodologies (Palit et al., 2025).

11.6.4 Potential field applications and bioreactor systems

Although critical enzymes (PETases, cutinases, and esterases) and microbial taxa capable of MP biodegradation have been effectively discovered by lab-scale proteomic investigations, field applications and bioreactor systems are crucial for practical environmental remediation (Sun et al., 2023). By monitoring in situ degradation efficiency, optimizing bioreactor parameters, and informing bioaugmentation techniques, proteomics might turn MP pollution control from theoretical ideas into workable environmental solutions (Sun et al., 2023).

11.6.4.1 Application of proteomics for monitoring and optimizing microplastic-degrading bioreactor systems

By precisely controlling variables including temperature, pH, oxygenation, and nutrition delivery, bioreactors offer regulated conditions to improve MP biodegradation utilizing microbial cultures, cocultures, or consortia (Salinas et al., 2024). These systems may deteriorate the following.

- Suspensions of MPs in wastewater treatment environments (Salinas et al., 2024).
- Pellets of pretreated plastics or pieces of MPs (Abeynayaka et al., 2022).
- Streams of mixed MPs, such as PET, PE, PS, and PP (Thiounn and Smith, 2020).

Proteomics makes it possible to track stress response proteins, monitor the expression levels of important plastic-degrading enzymes (such as PETase, MHETase, cutinases, and monooxygenases) in bioreactors, assess pathway activation, and reveal metabolic flux toward degradation intermediates (TPA, EG) and end-products (CO_2, biomass) (Lee et al., 2023).

Real-time proteomic monitoring can be implemented using:

- Shotgun LC-MS/MS on periodic samples (Pirmoradian et al., 2013);
- Targeted proteomics (SRM/MRM) for key biomarkers (Shi et al., 2016);
- In-line microfluidics coupled with MS-based protein analysis (emerging) (Li et al., 2024a).

Proteomics helps determine the ideal plastic particle size for bioavailability, nutrient supplementation to support cofactor regeneration (e.g., NADH, FAD) for monooxygenase and hydrolase activity, and identification of toxic inhibitors (phthalates, flame retardants) released during plastic breakdown that affect microbial viability by connecting enzyme abundance and activity to operational conditions. As a result, degradation rates can be increased while microbial health is preserved by adaptive bioreactor control (Cai et al., 2023).

11.6.4.2 Proteomics-guided bioaugmentation strategies in natural and engineered environments

The selection of strong degraders for bioaugmentation in wastewater treatment plants (WWTPs), constructed wetlands, coastal remediation sites, and agricultural soils contaminated with MPs is made possible by proteomic studies that identify microbial strains expressing high levels of plastic-degrading enzymes (Tang and Hadibarata, 2022).

Bioaugmentation can involve the introduction of individual effective strains, such as *Ideonella sakaiensis* for PET, complementary microorganisms cocultured for mixed plastic waste streams, microbes encapsulated in hydrogels or beads for regulated release in aquatic environments, and using moving bed biofilm reactors (MBBRs) in conjunction with biofilms on carriers (Cai et al., 2023). Proteomics makes it possible to follow introduced strains and track the synthesis of plastic-degrading enzymes in situ (Landels et al., 2015). This allows for the quantitative measurement of degradation progress under field settings, the early detection of shifts in the microbial community, and the evaluation of the success of bioaugmentation (Landels et al., 2015).

11.6.4.3 Field trials and large-scale environmental remediation using proteomics-enabled approaches

In natural environments, field experiments can employ bioreactor effluents, enriched consortia, or biofilms to evaluate:

- Plastic breakdown rates in realistic scenarios (Skariyachan and Manjunath, 2019);
- Effects on regional microbial populations and ecosystems (Skariyachan and Manjunath, 2019);
- Effectiveness in varying environmental conditions (temperature, pH, and salinity) (Skariyachan and Manjunath, 2019).

During field tests, proteomics, metagenomics, and metabolomics can be utilized to measure environmental effects, enzyme activity, and breakdown pathways (Aguiar-Pulido et al., 2016).

Proteomics-guided systems for field-scale biodegradation must be in line with regulatory compliance (release of modified microbes), environmental safety evaluations (toxic byproducts, microbe dispersion), and sustainable circular economy frameworks, which emphasize the value-adding of degradation products (Sathe and Sapkota, 2023).

11.6.4.4 Path forward for proteomics-guided microplastic biodegradation in field applications and bioreactor systems

- Creation of portable, in situ proteome and metabolomics instruments for field deployments that require real-time monitoring (Agu et al., 2025).
- Creation of resilient microbial strains and consortia that exhibit consistent production of enzymes that break down plastic in the face of environmental stress (Salinas et al., 2024).
- Integrating MP recovery and degradation with waste management infrastructure (Agu et al., 2025).
- Expanding from pilot projects to large-scale biodegradation systems for significant environmental effect (Wu et al., 2024).

11.7 Conclusion

MP pollution's prevalence, its durability, and detrimental effects on ecosystems and human health make it one of the most urgent environmental issues. Eco-friendly and sustainable alternatives, such as microbial biodegradation, are required because conventional eradication techniques frequently disappoint. The enzymatic routes, metabolic reactions, and stress adaptations of microbes during plastic breakdown have been better understood because of breakthroughs in proteomics technologies. The discovery of enzymes such as PETases, cutinases, esterases, and laccases advances our understanding of mechanisms and creates opportunities for protein engineering and synthetic biology techniques that aim to improve degradation efficiency. Proteomics, genomics, and metabolomics can be combined to create targeted bioremediation techniques by analyzing microbial interactions with MPs in great detail. Proteomics-driven research has laid a solid basis for creating bioaugmentation systems and improved bioreactor applications, despite ongoing obstacles such as incomplete degradation, data restrictions, and field scalability. To translate these laboratory-scale achievements into practical environmental solutions in the future, it will be essential to combine technological innovation with legal frameworks and international cooperation. In the end, proteomics-guided biodegradation has revolutionary potential to reduce MP pollution and make a substantial contribution to international sustainability objectives.

Acknowledgment

The authors are grateful to the Integral University, Lucknow, for providing the infrastructure facilities to carry out this work.

Conflict of interest

The authors declare no competing conflict of interest.

References

Abeynayaka, A., Werellagama, I., Ngoc-Bao, P., et al., 2022. Microplastics in wastewater treatment plants. Current Developments in Biotechnology and Bioengineering. Elsevier, pp. 311–337.

Adenan, S., Wong, C.F., Azziz, S.S.S.A., Nang, S.C.S., Misnan, R., Lamasudin, D.U., et al., 2022. Analysis of Pseudomonas sp. polycyclic aromatic hydrocarbons degradation proteins by LC-MS/MS. Malaysian Journal of Microscopy 18 (1), 66–76. Available at https://malaysianjournalofmicroscopy.org/ojs/index.php/mjm/article/download/598/315.

Aebersold, R., Mann, M., 2016. Mass-spectrometric exploration of proteome structure and function. Nature 537 (7620), 347–355. Available at https://doi.org/10.1038/nature19949.

Agarwal, T., Atray, N., Sharma, J.G., 2024. A critical examination of advanced approaches in green chemistry: microbial bioremediation strategies for sustainable mitigation of plastic pollution. Future Journal of Pharmaceutical Sciences 10. Available at https://doi.org/10.1186/s43094-024-00645-x.

Agu, C.V., Cook, R.L., Martelly, W., Gushgari, L.R., Mohan, M., Takulapalli, B., 2025. Multiplexed proteomic biosensor platform for label-free real-time simultaneous kinetic screening of thousands of protein interactions. Communications Biology 8 (1). Available at https://doi.org/10.1038/s42003-025-07844-z, https://www.springer.com/journal/42003.

Aguiar-Pulido, V., Huang, W., Suarez-Ulloa, V., Cickovski, T., Mathee, K., Narasimhan, G., 2016. Metagenomics, metatranscriptomics, and metabolomics approaches for microbiome analysis. Libertas. Evolutionary Bioinformatics 12, 5–16. Available at https://doi.org/10.4137/EBO.S36436, http://www.la-press.com/redirect_file.php?fileId=7489&filename=5619-EBO-Metagenomics,-Metatranscriptomics,-and-Metabolomics-Approaches-for-Mic.pdf&fileType=pdf.

Al-Amrani, S., Al-Jabri, Z., Al-Zaabi, A., Alshekaili, J., Al-Khabori, M., 2021. Proteomics: concepts and applications in human medicine. World Journal of Biological Chemistry 12 (5), 57–69. Available at https://doi.org/10.4331/wjbc.v12.i5.57.

Amobonye, A., Bhagwat, P., Singh, S., Pillai, S., 2021. Plastic biodegradation: frontline microbes and their enzymes. Science of The Total Environment 759, 143536. Available at https://doi.org/10.1016/j.scitotenv.2020.143536.

Anand, S., Samuel, M., Ang, C.S., Keerthikumar, S., Mathivanan, S., 2017. Label-based and label-free strategies for protein quantitation. Methods in Molecular Biology 1549, 31–43. Available at https://doi.org/10.1007/978-1-4939-6740-7_4, http://www.springer.com/series/7651.

Bala, S., Garg, D., Thirumalesh, B.V., Sharma, M., Sridhar, K., Inbaraj, B.S., et al., 2022. Recent strategies for bioremediation of emerging pollutants: a review for a green and sustainable environment. Toxics 10 (8), 484. Available at https://doi.org/10.3390/toxics10080484.

Berkel, C., Özbek, O., 2024. Methods used in the identification and quantification of micro(nano) plastics from water environments. South African Journal of Chemical Engineering 50, 388–403. Available at https://doi.org/10.1016/j.sajce.2024.09.010.

Bexeitova, K., Baimenov, A., Varol, E.A., Kudaibergenov, K., Zhantikeyev, U., Sailaukhanuly, Y., et al., 2024. Microplastics in freshwater systems: a review of classification, sources, and environmental impacts. Chemical Engineering Journal Advances 20, 100649. Available at https://doi.org/10.1016/j.ceja.2024.100649.

Bollinger, A., Thies, S., Knieps-Grünhagen, E., Gertzen, C., Kobus, S., Höppner, A., et al., 2020. A novel polyester hydrolase from the marine bacterium Pseudomonas aestusnigri – structural and functional insights. Frontiers in Microbiology 11. Available at https://doi.org/10.3389/fmicb.2020.00114.

Cai, Z., Li, M., Zhu, Z., Wang, X., Huang, Y., Li, T., et al., 2023. Biological degradation of plastics and microplastics: a recent perspective on associated mechanisms and influencing factors. Microorganisms 11 (7), 1661. Available at https://doi.org/10.3390/microorganisms11071661.

Carbonell, P., Koch, M., Duigou, T., Faulon, J.L., 2018. Enzyme discovery: enzyme selection and pathway design. Methods in Enzymology 608, 3–27. Available at https://doi.org/10.1016/bs.mie.2018.04.005.

Carr, C.M., Keller, M.B., Paul, B., Schubert, S.W., Clausen, K.S.R., Jensen, K., et al., 2023. Purification and biochemical characterization of SM14est, a PET-hydrolyzing enzyme from the marine sponge-derived Streptomyces sp. SM14. Frontiers in Microbiology 14. Available at https://doi.org/10.3389/fmicb.2023.1170880, https://www.frontiersin.org/journals/microbiology#.

Chandramouli, K., Qian, P.-Y., 2009. Proteomics: challenges, techniques and possibilities to overcome biological sample complexity. Human Genomics and Proteomics 1. Available at https://doi.org/10.4061/2009/239204.

Clarkson, S.M., Hamilton-Brehm, S.D., Giannone, R.J., et al., 2014. A comparative multidimensional LC-MS proteomic analysis reveals mechanisms for furan aldehyde detoxification in *Thermoanaerobacter pseudethanolicus* 39E. Biotechnology for Biofuels 7, 165. Available at https://doi.org/10.1186/s13068-014-0165-z.

Desdouits, M., Reynaud, Y., Philippe, C., Guyader, F.S.L., 2023. A comprehensive review for the surveillance of human pathogenic microorganisms in shellfish. Microorganisms 11 (9). Available at https://doi.org/10.3390/microorganisms11092218, www.mdpi.com/journal/microorganisms.

Dey, S., Rout, A.K., Behera, B.K., Ghosh, K., 2022. Plastisphere community assemblage of aquatic environment: plastic-microbe interaction, role in degradation and characterization technologies. Environmental Microbiomes 17 (1). Available at https://doi.org/10.1186/s40793-022-00430-4, https://standardsingenomics.biomedcentral.com/.

Dhali, S.L., Parida, D., Kumar, B., Bala, K., 2024. Recent trends in microbial and enzymatic plastic degradation: a solution for plastic pollution predicaments. Biotechnology for Sustainable Materials 1 (1). Available at https://doi.org/10.1186/s44316-024-00011-0.

Di, Y., Xia, H., Jiao, Y., Zhang, X., Fang, Q., Li, F., et al., 2019. Biodegradation of polyhydroxybutyrate by Pseudomonas sp. DSDY0501 and purification and characterization of polyhydroxybutyrate depolymerase. 3 Biotech 9 (10). Available at https://doi.org/10.1007/s13205-019-1871-9.

Edwards, S., León-Zayas, R., Ditter, R., Laster, H., Sheehan, G., Anderson, O., et al., 2022. Microbial consortia and mixed plastic waste: pangenomic analysis reveals potential for degradation of multiple plastic types via previously identified PET degrading bacteria. International Journal of Molecular Sciences 23 (10), 5612. Available at https://doi.org/10.3390/ijms23105612.

Gaston, E., Woo, M., Steele, C., Sukumaran, S., Anderson, S., 2020. Microplastics differ between indoor and outdoor air masses: insights from multiple microscopy methodologies. Applied Spectroscopy 74 (9), 1079–1098. Available at https://doi.org/10.1177/0003702820920652.

Goli, V.S.N.S., Paleologos, E.K., Farid, A., Mohamed, A.M.O., O'Kelly, B.C., El Gamal, M.M., et al., 2024. Extraction, characterisation and remediation of microplastics from organic solid matrices. Environmental Geotechnics 11 (4), 259–292. Available at https://doi.org/10.1680/jenge.21.00072, www.icevirtuallibrary.com/toc/jenge/current.

Hajighasemi, M., Tchigvintsev, A., Nocek, B., Flick, R., Popovic, A., Hai, T., et al., 2018. Screening and characterization of novel polyesterases from environmental metagenomes with high hydrolytic activity against synthetic polyesters. Environmental Science & Technology 52 (21), 12388–12401. Available at https://doi.org/10.1021/acs.est.8b04252.

Hayat M., Alam K., Ziad M., Khan F., 2025. Biodegradation of microplastic: the role of aspergillus species in sustainable plastic waste management.

Ho, K.T., Bjorkland, R., Burgess, R.M., 2024. Comparing the definitions of microplastics based on size range: scientific and policy implications. Marine Pollution Bulletin 207, 116907. Available at https://doi.org/10.1016/J.MARPOLBUL.2024.116907.

Hoang, V.H., Nguyen, M.K., Hoang, T.D., Rangel-Buitrago, N., Lin, C., Pham, M.T., et al., 2025. Microplastic characteristics, transport, risks, and remediation in ground water: a review. Environmental Chemistry Letters 23 (3), 817–837. Available at https://doi.org/10.1007/s10311-025-01825-8, https://www.springer.com/journal/10311.

Jabloune, R., Khalil, M., Ben Moussa, I.E., Simao-Beaunoir, A.M., Lerat, S., Brzezinski, R., et al., 2020. Enzymatic degradation of p-nitrophenyl esters, polyethylene terephthalate, cutin, and suberin by Sub1, a suberinase encoded by the plant pathogen Streptomyces scabies. Microbes and Environments 35 (1). Available at https://doi.org/10.1264/jsme2.ME19086.

Jahanshahi, D.A., Ariaeenejad, S., Kavousi, K., 2023. A metagenomic catalog for exploring the plastizymes landscape covering taxa, genes, and proteins. Scientific Reports 13, 16029. Available at https://doi.org/10.1038/s41598-023-43042-9.

Jebashalomi, V., Emmanuel Charles, P., Rajaram, R., 2024. Microbial degradation of low-density polyethylene (LDPE) and polystyrene using Bacillus cereus (OR268710) isolated from plastic-polluted tropical coastal environment. Science of The Total Environment 924, 171580. Available at https://doi.org/10.1016/J.SCITOTENV.2024.171580.

Joho, Y., Vongsouthi, V., Gomez, C., Larsen, J.S., Ardevol, A., Jackson, C.J., 2024. Improving plastic degrading enzymes via directed evolution. Protein Engineering, Design and Selection 37. Available at https://doi.org/10.1093/protein/gzae009.

Karp, N.A., Feret, R., Rubtsov, D.V., Lilley, K.S., 2008. Comparison of DIGE and post-stained gel electrophoresis with both traditional and SameSpots analysis for quantitative proteomics. Proteomics 8, 948–960. Available at https://doi.org/10.1002/pmic.200700812.

Kadac-Czapska, K., Ośko, J., Knez, E., Grembecka, M., 2024. Microplastics and oxidative stress—current problems and prospects. Antioxidants 13 (5), 579. Available at https://doi.org/10.3390/antiox13050579.

Kawai, F., Iizuka, R., Kawabata, T., 2024. Engineered polyethylene terephthalate hydrolases: perspectives and limits. Applied Microbiology and Biotechnology 108, 404. Available at https://doi.org/10.1007/s00253-024-13222-2.

Keller, M., Hettich, R., 2009. Environmental Proteomics: a Paradigm Shift in Characterizing Microbial Activities at the Molecular Level. Microbiology and Molecular Biology Reviews 73, 62–70. Available at https://doi.org/10.1128/MMBR.00028-08.

Khairul Anuar, N.F.S., Huyop, F., Ur-Rehman, G., Abdullah, F., Normi, Y.M., Sabullah, M.K., et al., 2022. An overview into polyethylene terephthalate (PET) hydrolases and efforts in tailoring enzymes for improved plastic degradation. International Journal of Molecular Sciences 23 (20). Available at https://doi.org/10.3390/ijms232012644, http://www.mdpi.com/journal/ijms.

Khurana, S., Ali, S., Srivastava, A.K., Singh, A., Agarwal, H., Chauhan, R., et al., 2025. Bioremediation of microplastic pollution: a systematic review on mechanism, analytical methods, innovations, and omics approaches. Journal of Hazardous Materials Advances 19. Available at https://doi.org/10.1016/j.hazadv.2025.100777. Available from: https://www.sciencedirect.com/science/journal/27724166.

Koelmans, A.A., Redondo-Hasselerharm, P.E., Nor, N.H.M., de Ruijter, V.N., Mintenig, S.M., Kooi, M., 2022. Risk assessment of microplastic particles. Nature Reviews Materials 7 (2), 138–152. Available at https://doi.org/10.1038/s41578-021-00411-y, www.nature.com/natrevmats/.

Lalrinfela, P., Vanlalsangi, R., Lalrinzuali, K., Babu, P.J., 2024. Microplastics: their effects on the environment, human health, and plant ecosystems. Environmental Pollution and Management 1, 248–259. Available at https://doi.org/10.1016/j.epm.2024.11.004, https://www.keaipublishing.com/en/journals/environmental-pollution-and-management/.

Landels, A., Evans, C., Noirel, J., Wright, P.C., 2015. Advances in proteomics for production strain analysis. Current Opinion in Biotechnology 35, 111–117. Available at https://doi.org/10.1016/j.copbio.2015.05.001. Available from: http://www.elsevier.com/locate/copbio.

Lee, C.E., Messer, L.F., Wattiez, R., Matallana-Surget, S., 2025. Decoding microbial plastic colonisation: multi-omic insights into the fast-evolving dynamics of early-stage biofilms. Proteomics 25 (7). Available at https://doi.org/10.1002/pmic.202400208, http://onlinelibrary.wiley.com/journal/10.1002/(ISSN)1615-9861.

Lee, G.H., Kim, D.-W., Jin, Y.H., Kim, S.M., Lim, E.S., Cha, M.J., et al., 2023. Biotechnological plastic degradation and valorization using systems metabolic engineering. International Journal of Molecular Sciences 24 (20), 15181. Available at https://doi.org/10.3390/ijms242015181.

Leitão, A.L., Enguita, F.J., 2021. Structural insights into carboxylic polyester-degrading enzymes and their functional depolymerizing neighbors. International Journal of Molecular Sciences 22 (5), 1–14. Available at https://doi.org/10.3390/ijms22052332, https://www.mdpi.com/1422-0067/22/5/2332/pdf.

Li, H., Ma, Y., Fu, R., et al., 2024a. Droplet-based microfluidics with mass spectrometry for microproteomics. Engineering 43, 37–53. Available at https://doi.org/10.1016/j.eng.2024.08.018.

Li, J., Zhu, H.-J., 2020. Liquid chromatography-tandem mass spectrometry (LC-MS/MS)-based proteomics of drug-metabolizing enzymes and transporters. Molecules 25 (11), 2718. Available at https://doi.org/10.3390/molecules25112718.

Li, K., Du, L., Qin, C., et al., 2024b. Microplastic pollution as an environmental risk exacerbating the greenhouse effect and climate change: a review. Carbon Research 3.

Liu, J., Zheng, L., 2025. Microplastic migration and transformation pathways and exposure health risks. Environmental Pollution 368, 125700. Available at https://doi.org/10.1016/j.envpol.2025.125700.

Liu, P., Liu, J., Yu, X., Lv, S., Liu, Z., Wang, Y., 2025. Research progress in microbial degradation of plastics. Journal of Physics: Conference Series 2941 (1), 012072. Available at https://doi.org/10.1088/1742-6596/2941/1/012072.

Lv, S., Li, Y., Zhao, S., Shao, Z., 2024. Biodegradation of typical plastics: from microbial diversity to metabolic mechanisms. International Journal of Molecular Sciences 25 (1), 593. Available at https://doi.org/10.3390/ijms25010593.

Kasmawati, Nasution, E.S., 2024. Environmental microbiology: microbes and their roles in ecosystems. International Journal of Natural Science Studies and Development (IJOSS) 1, 95–102. Available at 10.55299/ijoss.v1i2.17.

Melby, J.A., Roberts, D.S., Larson, E.J., et al., 2021. Novel Strategies to Address the Challenges in Top-Down Proteomics. Journal of the American Society for Mass Spectrometry 32, 1278–1294. Available at https://doi.org/10.1021/jasms.1c00099.

Meleady, P., 2018. Two-dimensional gel electrophoresis and 2D-DIGE. Methods in Molecular Biology 1664, 3–14. Available at https://doi.org/10.1007/978-1-4939-7268-5_1, http://www.springer.com/series/7651.

Mishra, S., Lin, Z., Pang, S., et al., 2021. Recent Advanced Technologies for the Characterization of Xenobiotic-Degrading Microorganisms and Microbial Communities. Frontiers in Bioengineering and Biotechnology 9. Available at https://doi.org/10.3389/fbioe.2021.632059.

Mund, A., Brunner, A.-D., Mann, M., 2022. Unbiased spatial proteomics with single-cell resolution in tissues. Molecular Cell 82, 2335–2349. Available at https://doi.org/10.1016/j.molcel.2022.05.022.

Muroi, F., Tachibana, Y., Soulenthone, P., Yamamoto, K., Mizuno, T., Sakurai, T., et al., 2017. Characterization of a poly(butylene adipate- co -terephthalate) hydrolase from the aerobic mesophilic bacterium Bacillus pumilus. Polymer Degradation and Stability 137, 11–22. Available at https://doi.org/10.1016/j.polymdegradstab.2017.01.006.

Muttiah, B., Hanafiah, A., 2025. Bioinspired nanoplatforms: polydopamine and exosomes for targeted antimicrobial therapy. Polymers 17 (12), 1670. Available at https://doi.org/10.3390/polym17121670.

Nebauer, D.J., Pearson, L.A., Neilan, B.A., 2024. Critical steps in an environmental metaproteomics workflow. Environmental Microbiology 26, e16637. Available at https://doi.org/10.1111/1462-2920.16637.

Nesatyy, V.J., Suter, M.J.F., 2007. Proteomics for the analysis of environmental stress responses in organisms. Environmental Science and Technology 41 (20), 6891–6900. Available at https://doi.org/10.1021/es070561r.

Niu, L., Chen, Y., Li, Y., Wang, Y., Shen, J., Wang, L., et al., 2023. Diversity, abundance and distribution characteristics of potential polyethylene and polypropylene microplastic degradation bacterial communities in the urban river. Water Research 232, 119704. Available at https://doi.org/10.1016/j.watres.2023.119704.

Paladhi, A.G., Vallinayagam, S., Rajendran, S., Rathinam, V., Sharma, V.K., 2022. Microalgae: a promising tool for plastic degardation. Microbes and Microbial Biotechnology for Green Remediation. Elsevier, India, pp. 575–587. Available at https://www.sciencedirect.com/book/9780323904520, https://doi.org/10.1016/B978-0-323-90452-0.00049-9.

Palit, P., Minkara, M., Abida, M., Marwa, S., Sen, C., Roy, A., et al., 2025. PlastiCRISPR: genome editing-based plastic waste management with implications in polyethylene terephthalate (PET) degradation. Biomolecules 15 (5), 684. Available at https://doi.org/10.3390/biom15050684.

Penny, J., Arefian, M., Schroeder, G.N., et al., 2023. A gas phase fractionation acquisition scheme integrating ion mobility for rapid diaPASEF library generation. Proteomics 23, e2200038. Available at https://doi.org/10.1002/pmic.202200038.

Pfohl, P., Santizo, K., Sipe, J., et al., 2025. Environmental degradation and fragmentation of microplastics: dependence on polymer type, humidity, UV dose and temperature. Microplastics and Nanoplastics 5, 7. Available at https://doi.org/10.1186/s43591-025-00118-9.

Pirmoradian, M., Budamgunta, H., Chingin, K., Zhang, B., Astorga-Wells, J., Zubarev, R.A., 2013. Rapid and deep human proteome analysis by single-dimension shotgun proteomics. Molecular & Cellular Proteomics 12 (11), 3330–3338. Available at https://doi.org/10.1074/mcp.o113.028787.

Poulsen, J.S., Nielsen, J.L., 2023. Proteomic characterisation of polyethylene terephthalate and monomer degradation by Ideonella sakaiensis. Journal of Proteomics 279, 104888. Available at https://doi.org/10.1016/j.jprot.2023.104888.

Ramond, P., Galand, P.E., Logares, R., 2025. Microbial functional diversity and redundancy: moving forward. FEMS Microbiology Reviews 49. Available at https://doi.org/10.1093/femsre/fuae031.

Rodríguez-Ramos, J.A., Borton, M.A., McGivern, B.B., Smith, G.J., Solden, L.M., Shaffer, M., et al., 2022. Genome-resolved metaproteomics decodes the microbial and viral contributions to coupled carbon and nitrogen cycling in river sediments. mSystems 7 (4). Available at https://doi.org/10.1128/msystems.00516-22, https://journals.asm.org/doi/10.1128/msystems.00516-22.

Salehi, M., Pincus, L.N., Deng, B., Peters, C.A., 2024. Microplastics: from intrinsic properties to environmental fate. Environmental Engineering Science 41 (11), 425–435. Available at https://doi.org/10.1089/ees.2024.0232, www.liebertonline.com/ees.

Salinas, J., Martínez-Gallardo, M.R., Jurado, M.M., Suárez-Estrella, F., López-González, J.A., Estrella-González, M.J., et al., 2024. Microbial consortia for multi-plastic waste biodegradation: selection and validation. Environmental Technology & Innovation 36, 103887. Available at https://doi.org/10.1016/j.eti.2024.103887.

Santos-Beneit, F., Chen, L.M., Bordel, S., Frutos de la Flor, R., García-Depraect, O., Lebrero, R., et al., 2023. Screening enzymes that can depolymerize commercial biodegradable polymers: heterologous expression of fusarium solani cutinase in Escherichia coli. Microorganisms 11 (2). Available at https://doi.org/10.3390/microorganisms11020328, www.mdpi.com/journal/microorganisms.

Sathe, G., Sapkota, G.P., 2023. Proteomic approaches advancing targeted protein degradation. Trends in Pharmacological Sciences 44 (11), 786–801. Available at https://doi.org/10.1016/j.tips.2023.08.007, www.elsevier.com/locate/tips.

Satta, A., Zampieri, G., Loprete, G., Campanaro, S., Treu, L., Bergantino, E., 2024. Metabolic and enzymatic engineering strategies for polyethylene terephthalate degradation and valorization. Reviews in Environmental Science and Bio/Technology 23 (2), 351–383. Available at https://doi.org/10.1007/s11157-024-09688-1.

Sharma, P., Sharma, P., Abhishek, K., 2024. Sampling, separation, and characterization methodology for quantification of microplastic from the environment. Journal of Hazardous Materials Advances 14, 100416. Available at https://doi.org/10.1016/j.hazadv.2024.100416.

Shi, T., Song, E., Nie, S., Rodland, K.D., Liu, T., Qian, W.-J., et al., 2016. Advances in targeted proteomics and applications to biomedical research. Proteomics 16 (15-16), 2160–2182. Available at https://doi.org/10.1002/pmic.201500449.

Sivanich, M.K., Gu, T.J., Tabang, D.N., Li, L., 2022. Recent advances in isobaric labeling and applications in quantitative proteomics. Proteomics 22 (19-20). Available at https://doi.org/10.1002/pmic.202100256, http://onlinelibrary.wiley.com/journal/10.1002/(ISSN)1615-9861.

Skariyachan, S., Manjunath, M., Shankar, A., Bachappanavar, N., Patil, A.A., 2019. Application of Novel Microbial Consortia for Environmental Site Remediation and Hazardous Waste Management Toward Low- and High-Density Polyethylene and Prioritizing the Cost-Effective, Eco-friendly, and Sustainable Biotechnological Intervention. Springer Science and Business Media LLC, pp. 431–478. Available at 10.1007/978-3-319-73645-7_9.

Sumathi, T., Viswanath, B., Sri Lakshmi, A., Saigopal, D.V.R., 2016. Production of laccase by Cochliobolus sp. isolated from plastic dumped soils and their ability to degrade low molecular weight PVC. Biochemistry Research International 2016. Available at https://doi.org/10.1155/2016/9519527, http://www.hindawi.com/journals/bcri/.

Sun, X.-L., Xiang, H., Xiong, H.-Q., Fang, Y.-C., Wang, Y., 2023. Bioremediation of microplastics in freshwater environments: a systematic review of biofilm culture, degradation mechanisms, and analytical methods. Science of the Total Environment 863, 160953. Available at https://doi.org/10.1016/j.scitotenv.2022.160953.

Syberg, K., Nielsen, M.B., Westergaard Clausen, L.P., van Calster, G., van Wezel, A., Rochman, C., et al., 2021. Regulation of plastic from a circular economy perspective. Current Opinion in Green and Sustainable Chemistry 29, 100462. Available at https://doi.org/10.1016/j.cogsc.2021.100462.

Tang, K.H.D., Hadibarata, T., 2022. The application of bioremediation in wastewater treatment plants for microplastics removal: a practical perspective. Bioprocess and Biosystems Engineering 45 (11), 1865–1878. Available at https://doi.org/10.1007/s00449-022-02793-x, https://www.springer.com/journal/449.

Thiounn, T., Smith, R.C., 2020. Advances and approaches for chemical recycling of plastic waste. Journal of Polymer Science 58 (10), 1347–1364. Available at https://doi.org/10.1002/pol.20190261, https://onlinelibrary.wiley.com/journal/26424169.

Van Den Bossche, T., Armengaud, J., Benndorf, D., Blakeley-Ruiz, J.A., Brauer, M., Cheng, K., et al., 2025. The microbiologist's guide to metaproteomics. iMeta 4 (3). Available at https://doi.org/10.1002/imt2.70031.

Wang, D.-Z., Kong, L.-F., Li, Y.-Y., Xie, Z.-X., 2016. Environmental microbial community proteomics: status, challenges and perspectives. International Journal of Molecular Sciences 17, 1275. Available at https://doi.org/10.3390/ijms17081275.

Wright, R.J., Bosch, R., Langille, M.G.I., Gibson, M.I., Christie-Oleza, J.A., 2021. A multi-OMIC characterisation of biodegradation and microbial community succession within the PET plastisphere. Microbiome 9 (1). Available at https://doi.org/10.1186/s40168-021-01054-5, http://www.microbiomejournal.com/.

Wu, J., Wang, J., Zeng, Y., Sun, X., Yuan, Q., Liu, L., et al., 2024. Biodegradation: the best solution to the world problem of discarded polymers. Bioresources and Bioprocessing 11 (1). Available at https://doi.org/10.1186/s40643-024-00793-1.

Wöhlbrand, L., Trautwein, K., Rabus, R., 2013. Proteomic tools for environmental microbiology—A roadmap from sample preparation to protein identification and quantification. Proteomics 13, 2700–2730. Available at https://doi.org/10.1002/pmic.201300175.

Yao, X., Luo, X.S., Fan, J., Zhang, T., Li, H., Wei, Y., 2022. Ecological and human health risks of atmospheric microplastics (MPs): a review. Environmental Science: Atmospheres 2 (5), 921–942. Available at https://doi.org/10.1039/d2ea00041e, www.rsc.org/journals-books-databases/about-journals/environmental-science-atmospheres.

Yoshida, S., Hiraga, K., Taniguchi, I., Oda, K., 2021. Ideonella sakaiensis, PETase, and MHETase: from identification of microbial PET degradation to enzyme characterization. Methods in Enzymology 648, 187–205. Available at https://doi.org/10.1016/bs.mie.2020.12.007, https://www.sciencedirect.com/bookseries/methods-in-enzymology.

Zeghal, E., Vaksmaa, A., Vielfaure, H., Boekhout, T., Niemann, H., 2021. The potential role of marine fungi in plastic degradation – a review. Frontiers in Marine Science 8. Available at https://doi.org/10.3389/fmars.2021.738877.

Chapter 12

Use of proteomics as biomonitoring tool for aquatic pollution: current findings and future prospective

Insha Amin[1], Ruqeya Nazir[1], Gopal Palla[2], and Burhan Hamid[3]
[1]*Microbiology Research Laboratory, Centre of Research for Development (CORD), University of Kashmir, Srinagar, Jammu and Kashmir, India,* [2]*Agricultural and Biological Engineering, College of Agriculture, Purdue University, West Lafayette, IN, United States,* [3]*Centre of Research for Development, University of Kashmir, Srinagar, Jammu and Kashmir, India*

12.1 Introduction

The foundation of life on Earth, water is essential to the survival of both natural ecosystems and human civilization. It is essential to daily living, industrial operations, and agriculture, and it forms the basis for both societal progress and economic success (Zhou and Gu, 2024). However, the unrelenting drive for technological development and economic expansion has resulted in the widespread poisoning of water sources, endangering the environment, biodiversity, and human health (Tella et al., 2025). As a result, it is impossible to overestimate the urgency of addressing water contamination; it is a worldwide issue that needs prompt and ongoing action. The current state of water pollution management initiatives is characterized by a complicated interaction between public involvement, legal frameworks, and technical developments. Although tremendous progress has been achieved in eliminating pollutants from water thanks to advances in water treatment technology, its widespread use and accessibility remain a barrier. Similar to this, enforcing environmental laws is essential to stopping pollution at its source, but different legal systems and enforcement capacities in different nations can make this difficult (Khanam et al., 2023). Additionally, encouraging sustainable water management techniques and cultivating a culture of environmental care depend heavily on public understanding and involvement.

Proteomics Driven Bioremediation. DOI: https://doi.org/10.1016/B978-0-443-45294-9.00017-5

Untreated urban wastewater, agricultural runoff, and industrial discharges are some of the main causes of water pollution, which is still a major worldwide problem. A large amount of wastewater is dumped untreated into bodies of water in underdeveloped countries. On the other hand, new pollutants, microplastics, and agricultural chemicals provide pollution problems in industrialized countries that make managing water quality more difficult. Pollution has a wide range of consequences. Polluted ecosystems frequently exhibit eutrophication, habitat damage, and biodiversity loss, with hazardous algal blooms causing oxygen-deprived zones in many water bodies (Sahu et al., 2024). The many heavy metals that are emitted by enterprises have an influence on the geochemical cycles of our environment, either directly or indirectly (Hamid et al., 2023a,b; Tarfeen et al., 2022; Ali et al., 2019; b). Human illnesses like cholera and diarrhea are caused by contaminated water, which disproportionately affects disadvantaged populations in places with poor water treatment facilities (Krishan et al., 2023).

A complex strategy involving public involvement, regulatory enforcement, and technology improvements is needed to address water contamination. First and foremost, wastewater treatment technology is essential. Contaminants can be effectively eliminated before release by enhancing procedures like oxidation and membrane filtration, supporting adherence to water quality regulations and safeguarding ecosystems. Second, it is essential to enforce regulations (Sumalatha et al., 2025). Governments must enact and uphold stringent regulations that restrict the release of pollutants, guarantee routine monitoring, and impose sanctions for infractions. This paradigm encourages appropriate water management and discourages polluting practices. Finally, citizen participation is essential. Increasing awareness promotes water conservation efforts and eco-friendly behaviors. A holistic approach to pollution reduction is fostered by decision-making that is strengthened by community engagement and feedback.

The biomonitoring of aquatic life is a long-standing problem that was extensively covered in certain scholars' important contributions. The most popular bioindicators for evaluating water quality in freshwater environments like rivers and reservoirs include macroinvertebrates, periphyton, and fish (López-López and Sedeño-Díaz, 2015). Depending on the particulars of the research, a number of biomonitoring techniques can be chosen and combined to establish the optimal course of action. Integrative physical-chemical and biological methods make it possible to comprehend how human activity affects aquatic ecosystems' natural processes and, consequently, how to effectively protect and restore them (Qin et al., 2025). The potential and significant usage of bioindicators and biomonitoring in the relevance of contingency studies can be shown from the examination of the primary approaches used both during and after the COVID-19 epidemic (Novoa et al., 2022). As a result, initiatives that include biomonitoring and biomarker research will enable more proactive and preventative environmental and public health management.

Bioindicators are more advantageous than physicochemical testing in terms of sustainability. These can give early notice of any changes that could seriously endanger individual species, populations, communities, or ecosystems since they offer information pertinent to both human and environmental health (Parmar et al., 2016).

Industrialization, population expansion, and changes in lifestyle have led to the emergence of non-biodegradable contaminants that pose a threat to both human health and the environment. Bioremediation is a cost-effective method of removing dangerous pollutants using microorganisms (Amin et al., 2024). Microbial bioremediation is an inexpensive and ecologically friendly method of eliminating environmental contaminants from ecosystems (Singh et al., 2020; Hamid et al., 2023a,b) (a). However, a thorough grasp of the factors influencing the growth, metabolism, dynamics, and activity of native microbial communities in contaminated sites is essential for the successful implementation of these various bioremediation techniques. Critical insights into microbial communities and underlying mechanisms in environmental pollutant bioremediation have been made possible by the development of new technologies such as next-generation sequencing, protein and metabolic profiling, and sophisticated bioinformatic tools. These omics techniques include metabolomics, meta-transcriptomics, meta-proteomics, and metagenomics (Kumar et al., 2021). Furthermore, these technological developments have made it much easier to assess the efficacy and use microbiological bioremediation techniques. When it comes to clarifying the process's mechanism and identifying the key enzymes and their regulatory components, a number of omics approaches are clearly more useful and instructive (Satrio et al., 2024).

12.2 Concept of environmental proteomics

Proteome profiling is used in environmental proteomics, a developing discipline at the nexus of molecular biology and environmental research, to investigate the varied and dynamic protein expressions in environmental samples (Guo et al., 2021). By using this method, scientists may learn more about the functions of microbial communities, how they interact with their surroundings, and how they participate in different biogeochemical processes. Proteome profiling has several uses in environmental microbiology, biotechnology, and nanotechnology, providing new ways to comprehend and solve environmental problems (Amin et al., 2025). The extensive study of proteins, their structures, and their activities is known as proteomics. By concentrating on the proteins produced by microorganisms in their native environments, environmental proteomics expands this method to environmental samples. In addition to genomic and transcriptome investigations, proteome profiling offers a functional view of the ongoing biological activities within a microbial community (Schneider and Riedel, 2010). Proteins are usually extracted from environmental samples, separated using methods such as two-dimensional gel electrophoresis or liquid chromatography, and identified using mass spectrometry (Wittmann-Liebold et al., 2006). The obtained data is then analyzed using sophisticated bioinformatics methods, which connect protein identities and amounts to particular metabolic pathways and ecological activities. It is unclear how genome expression, proteome modification, metabolic control, and phenotypic changes under environmental stressors are related (Singh et al., 2021). Finding the relevant scientific processes of the aforementioned link at the molecular and systematic levels is a challenging issue for conventional research methods. The proteomics technique is a useful tool for systematically uncovering the metabolic networks, functional protein interaction networks, and biological processes of cells and organisms under stress. Additionally, the toxicity assessment of pollutants and byproducts using proteomics techniques offers a scientific understanding for early ecological risk identification and assessment of the efficacy of pollutant treatment methods (Guo et al., 2021; Fig. 12.1).

12.3 Proteomic responses to aquatic pollutants

"Environmental proteomics" or "ecotoxicoproteomics"—defines an approach to investigating proteins that link exposure and effects, through findings such proteins that have been modified in defined ways in response to pollution stress (Bjørnstad et al., 2006). Proteins perform nearly all the functions in biology, as enzymes, transporters, structural elements, and cellular process regulators.

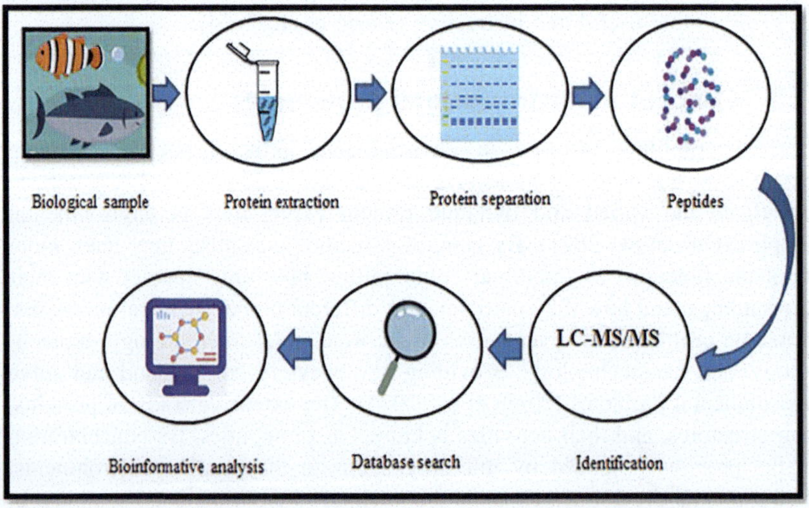

FIGURE 12.1 **Diagrammatic representation of a standard proteomic experiment.** Diagram representing the sequential steps from biological sample processing to bioinformatic analysis. Key stages include protein extraction, peptide generation, LC-MS/MS identification, and data processing.

These quantities of proteins, or the patterns of post-translational modification or interaction networks, reflect the immediate response—what is done about an environmental stress that has been felt instantly, either in benefit and damage. Consequently, proteomics contributes to knowledge on the potential mechanism of adaptation, on how toxicity occurs, and is also promising concerning the identification of molecular biomarkers for pollution monitoring. Complex proteome alterations that include but are not limited to the induction of detoxification enzymes, oxidative stress proteins, metabolic enzymes, cytoskeletal elements, and chaperones are commonly observed in pollution-exposed aquatic organisms (Tomanek, 2011). Elevated levels of proteins putatively implicated in oxidative stress regulation are one of the most frequent proteome signatures recorded from polluted environments. Many pollutants, such as organics and heavy metals, promote the generation of reactive oxygen species (ROS), which induces oxidative stress. In aquatic species, this is countered by increased expression of antioxidant and detoxifying proteins, such as glutathione-related enzymes, catalase, superoxide dismutase, and peroxiredoxins. Cumulatively, these proteins contribute to the protection of vital cellular components, suppress oxidative damage, and maintain redox homeostasis (He et al., 2017). The expression of proteins participating in phase I and phase II detoxification reactions (such as glutathione-S-transferases, cytochrome P450 enzymes) also suggests the metabolic activation and conjugation of contaminants for elimination. Many proteomic investigations reveal the disruption of energy metabolism, protein synthesis, and cytoskeletal arrangement to be associated with oxidative stress responses (Tomanek et al., 2015). Pollutant exposure can affect mitochondrial function, and that could alter the expression of oxidative phosphorylation and tricarboxylic acid cycle enzymes. Similarly, the diverse regulation of glycolysis and energy homeostasis-related proteins may reflect a reallocation of intracellular energy reserves with stress. At the same time, an increased amount of molecular chaperones, such as heat shock proteins, is indicative of higher misfolding and turnover of proteins, which is a characteristic feature of general stress response. Actin, tubulin, and intermediate filament components are examples of structural proteins that are also modulated, perhaps as a compensatory strategy to preserve cellular integrity under chemical stress (Tomanek, 2011).

With the growing body of data revealing species-specific and pollutant-specific proteome responses, there are new opportunities to identify biomarkers in ecotoxicology. For instance, bivalves such as mussels and oysters, often used as sentinel species in marine monitoring, exhibit distinct proteomic patterns when exposed to metals, hydrocarbons, and emerging pollutants (Balbi et al., 2021). These patterns include changes in metabolic and cytoskeletal proteins, as well as enzymes related to oxidative defense. In a recent investigation of *Daphnia pulex* from lakes with varying levels of heavy-metal contamination, researchers (Norambuena et al., 2025) observed a significant downregulation of 181 proteins and an upregulation of 17 proteins in individuals from polluted waters, primarily involving metabolic enzymes, antioxidant proteins, and structural molecules. This differential protein expression underscores the substantial metabolic and physiological adaptations required to cope with prolonged pollution stress. Furthermore, proteomics has been employed to understand the effects of novel pollutants like micro- and nanoplastics (Khosrovyan et al., 2025). Xu et al. (2024) studied the freshwater microalga *Chlamydomonas reinhardtii* exposed to microplastics of various polymers (polyethylene, polystyrene, polyvinyl chloride, and polypropylene). Their findings indicated that exposure to PVC and polystyrene notably downregulated photosynthetic proteins, correlating with reduced growth and increased oxidative stress, whereas polyethylene and polypropylene had a lesser impact. These findings emphasize the complexity of responses to new and mixed contaminants and the potential for synergistic effects that traditional tests might overlook. Thus, proteomics serves as both a diagnostic and a mechanistic tool in water pollution research. Linking chemical exposure to specific biochemical pathways, it enhances mechanistic understanding and aids in identifying reliable molecular biomarkers that indicate pollutant exposure or effects. However, despite its promise, several challenges remain. The lack of well-annotated genomes and protein databases for many aquatic species complicates the accurate identification of peptides (Nambiar and Banuru, 2025). Methodological differences in sample preparation, extraction, and mass spectrometry analysis can also hinder comparability across studies. Interpreting proteome changes is equally challenging, as up- or downregulation can signify both toxic damage and adaptive responses. Consequently, there is a growing recognition of the importance of integrating proteomics with transcriptome and metabolomic approaches, known as "multiomics," to develop a comprehensive understanding of pollutant effects (Li et al., 2025). Field-based studies are gaining more attention due to their superior ability to replicate real-world exposure scenarios. Comparative proteomics of organisms from polluted and reference sites provides ecologically relevant insights that laboratory studies alone cannot replicate. However, translating proteome responses into ecological outcomes such as growth, reproduction, and survival remains difficult. For proteomics to be fully integrated into environmental monitoring frameworks, biomarkers identified through discovery-driven research must be validated for sensitivity, specificity, and reproducibility under various field conditions (Singh et al., 2025).

12.4 Model organisms and sample systems

The study of proteomics includes the study of all proteins expressed, all of the post-translationally modified proteins, and the protein interactions in multiple tissues. Due to the increasing interest in utilizing proteomics as a method for detecting exposure markers and toxicity markers in aquatic environments, it is important to carefully consider the biological importance, pollution sensitivity, and practicality of doing high-throughput protein analysis for the study organisms used as experimental models and exposures. Table 12.1 lists the organisms that have been employed for proteomic analysis in aquatic pollution monitoring. Bivalves (e.g., mussels), fish (including both freshwater and saltwater species), crustaceans, and algae are used most commonly as model organisms for aquatic proteomics. Bivalves (particularly mussels) have been used frequently as sentinel species for monitoring marine pollution because they are stationary organisms that accumulate pollutants, and they produce sufficient tissue to allow for proteomic analysis; indeed, in the literature reviews of proteomics, bivalves are identified as the "most preferred" organism for monitoring organic and inorganic contaminants (Balbi et al., 2021). In

TABLE 12.1 List of studies focused on proteomic biomarkers to monitor pollutants in the aquatic environment.

Pollutant	Species used	Technology employed	References
Lead, zinc, copper	Sydney rock oyster (*Saccostra glomerata*)	SDS-PAGE and LC-MS/MS	Muralidharan et al. (2012)
Copper nanoparticles	Mussel (*Mytilus galloprovincialis*)	2-DE and MALDI-TOF/TOF	Gomes et al. (2014)
Cadmium	Farre's scallop (*Chlamys farreri*)	iTRAQ	Zhang and Zhai (2020)
Mercury	Rice fish (*Oryzias melastigma*)	2-DE and MALDI-TOF/TOF	Wang et al. (2011)
Zinc	Portuguese oyster (*Crassostrea angulata*)	Nano-UPLC-MS/MS	Luo et al. (2017)
Arsenic, cadmium	Amphipod (*Gammarus* sp.)	2D-DiGE	Vellinger et al. (2016)
Polychlorinated biphenyls	Mussel (*Mytilus galloprovincialis*)	2-DE and LC-MS analysis	Ambrosio et al. (2018)
DDT	Asian green mussel (*Perna viridis*)	iTRAQ	Song et al. (2017)
Herbicides, fungicides	Goldfish (*Carassius auratus*)	2-DE and nano LC-MS/MS	Gandar et al. (2017)
Microplastics	Blue mussel (*Mytilus edulis*)	Label-free quantification (LFQ)	Green et al. (2019)

freshwater pollution studies, embryos from the fish species *Danio rerio* (zebrafish) provide information about the relationship between changes in metabolic proteins and stress proteins (De Wit et al., 2010).

Proteomic research on Crustaceans and Algae is an emerging discipline; however, there are still limitations associated with these aquatic organisms, specifically with respect to issues with genome annotation, as well as the lack of sufficient tissue samples. Proteomic sample systems for aquatic organisms include samples obtained from contaminated and uncontaminated sites in the environment, as well as from laboratory experiments with controlled levels of contaminants. Laboratory experiments enable identification of proteins that exhibit differing patterns of expression in response to particular doses of contaminants at predetermined times, as well as to specific environmental conditions (e.g., salinity, heavy metals); the information obtained from proteomic analysis can be used to determine whether specific proteins involved in oxidative stress response (e.g., glutathione reductase, catalase, and superoxide dismutase), remodeling of the cytoskeleton, metabolism of energy, and transport of lipids have been either upregulated or downregulated.

Field studies reveal true ecological realities, but field studies also have many challenges associated with them due to multiple exposures, multiple stress/multiple stressors, and the fact that the proteomes of all non-model species will be unknown or very poorly annotated in the case of these (Mathur et al., 2021). For example, field studies will include the use of experimental organisms, which will include organisms from multiple different taxonomic groups, using proteomics to help in assessing the potential impacts of exposure to pollutants. In this context, the results of proteomic "fingerprint" studies should be used in conjunction with rigorous statistical/methodological validation of the fingerprints across taxonomically and experimentally disparate organisms and stress conditions, since the majority of fish collection and population monitoring will be conducted using species with little or no genomic/proteomic annotations (Mathur et al., 2021). Proteomics has the potential to be applied to the investigation of water contamination using a variety of defined model organisms and sampling regimes. The identification of mechanistic protein biomarkers through this type of research will provide reliable information to support ongoing monitoring of contamination. To facilitate the continued advancement of this field, greater effort will need to be put into developing an exceptional understanding of how to better annotate proteins, standardize exposure systems, and integrate them with other omics.

12.5 Mechanistic insights from proteomic studies

12.5.1 Oxidative stress and antioxidant defense

The oxidative stress response is one example of an effect produced with confidence through proteomics-based studies conducted on subjected aquatic organisms (synthetic and natural) by toxic agents. The presence of contaminants in the cellular environment produces high concentrations of ROS, damaging proteins, nucleic acids, and lipid structures. Analyses of fish, crabs, and molluscs subjected to organic and inorganic chemicals through continual exposures have demonstrated increases in activity levels in antioxidant enzymes: superoxide dismutase (SOD), catalase (CAT), glutathione S-transferase (GST), and peroxiredoxins (Parida and Sahoo, 2024). The response of antioxidant proteins allows for the reduction of oxidative damage and enables the reestablishment of redox balance in the organism, thus providing early indicators of the detrimental effects of exposure to environmental pollutants.

12.5.2 Energy metabolism and mitochondrial dysfunction

There have been proteomic studies conducted that indicate that pollutants are capable of interfering with mitochondrial energy metabolism. The results of these studies indicate that there is decreased production of ATP and disrupted oxidative phosphorylation as indicated by altered amounts of expressed ATP synthase, cytochrome c oxidase, and succinate dehydrogenase. The presence of this type of metabolic stress may cause a deficiency of energy that ultimately has negative impacts on the growth, reproduction, and locomotion of aquatic species. The use of an integrated approach to proteomics and metabolomics has validated and confirmed the disruption of ATP levels and intermediates from the TCA cycle in mussels that were exposed to cadmium and arsenic (Meng et al., 2023).

12.5.3 Cytoskeleton and signal transduction disruption

Environmental toxins interfere with the body's signaling systems and our cells' structural framework. Changes in the levels of proteins such as Actin, Tubulin, and Vimentin demonstrate evidence of an effect on how well a cell can migrate and maintain its structural integrity. The proteins that are affected by pollution exposure have a role in regulating hormones, neurotransmitters, and apoptosis. The use of microplastics has demonstrated a link to neurotoxicity due to alterations in how calcium is processed within cells and impaired activation of cellular signaling proteins (MAPK pathway) in fish exposed to microplastics (Zhang et al., 2025).

12.5.4 Immune and inflammatory responses

Mechanisms of immunomodulation from pollutants have been substantiated through proteomic changes of immune-related proteins associated with pattern recognition receptors (PRRs), complement components, and antimicrobial peptides, as shown by Mancia et al. (2019). Depending on pollutant type and exposure duration, the expression of PRRs, complement components, and antimicrobial peptides in bivalves and crustaceans that are exposed to microplastics and organic contaminants may be upregulated or downregulated. This evidence indicates that the innate immune responses of individuals are compromised, thus leading to increased susceptibility to infection.

12.6 Analytical tools used in proteomics

The process of sampling tissue or whole organisms to evaluate their response to contaminant exposure, extracting and purifying proteins, separating and quantifying proteins using either gel-based techniques (e.g., two-dimensional polyacrylamide gel electrophoresis, 2D-PAGE) or liquid chromatography-based (LC) mass spectrometry-based (LC-MS/MS) techniques, analyzing data derived from MS, recognizing proteins and assessing their quantity, and analyzing all the data created from both mass spectrometry and bioinformatics to identify differences in expression levels, pathways, and biomarkers, etc. has evolved such that gel methods have been replaced by mass spectrometry-based methods for these activities due to gel methods' limitations associated with dynamic range and limited membrane protein coverage (Lee et al., 2020).

12.7 Bioinformatic and statistical approaches in proteomics for aquatic pollution

To convert the massive amounts of data produced by proteomic research into biological insights, bioinformatic analyses are crucial. Key stages and tools are summarized here as follows.

12.7.1 Protein identification and database search

The first step after obtaining raw spectrometer results (mass spectrometry data) is the identification of peptides and proteins using search tools (e.g., Mascot, MaxQuant). Many identified Peptides may not connect to any known Mothers of Proteins found naturally within Aquatic environments, as there are often few or no Protein/Genome Databases from which to draw this information. The use of de novo peptide sequencing, the development of a catalog of established Protein databases through transcription-based sequencing methods, and the use of Cross-Species Homology searches will help mitigate this (Brownridge et al., 2011).

12.7.2 Quantification and normalization

Once the proteins have been identified, quantitative values will be assigned to these proteins. There are multiple methods that require the use of bioinformatics and statistical analysis, as described below. First, data from proteomics is typically subject to normalization and filtering (e.g., keeping only high-confidence identifications). Second, a missing value imputation process produces additional values in place of those that are missing. Third, quality control analyses must also occur to determine whether or not there are batch effects or repeat effects present. After all of these steps have been completed, there are several means of testing for differences in abundance between various sample groups, including t-tests, one-way ANOVAs, and more traditional and/or moderated methods of statistical testing.

12.7.3 Differential expression, clustering, and pattern recognition

Water pollution research commonly contrasts both the exposed organisms with the control organisms/water and different levels of exposure using proteomic methods. The expression levels of differentially abundant proteins have typically been displayed via heat graphics, volcano graphics, or through hierarchical cluster analysis or k-means analysis. In addition, the multivariate approaches (PCA, PLS-DA) or recognition techniques (Everingham, 2023) can be used to create proteomic fingerprints associated with pollution exposure.

12.7.4 Functional annotation, pathway, and network analysis

Bioinformatics tools such as DAVID, STRING, and Cytoscape for visualizing networks of protein-protein interactions help interpret the biological functions of annotated proteins based on gene ontology (GO), pathway (e.g., KEGG), and protein–protein interactions (PPIs) that relate to responses to oxidative stress, the pathway for detoxifying compounds (e.g., glutathione S-transferases; metallothioneins), and energy metabolism. In marine pollution proteomics, mapping the relative abundance of proteins expressed in response to various environmental factors is a common approach (Wang et al., 2015).

12.7.5 Biomarker discovery and machine learning approaches

Aquatic proteomics aims to identify biomarkers of environmental pollutant exposure. Proteomic profiling methods (Onoja, 2023) employ increasingly sophisticated bioinformatic processes. Machine learning classifiers, such as Support Vector Machines and Random Forests, are being used to create predictive models and identify patterns in higher-dimensional spaces, but many of these approaches are still in their early stages. In addition, combining proteomic datasets from an aqueous and other omic types (e.g., metabolomics, transcriptomics) will enhance the identification of these biomarkers.

12.8 Limitations and challenges of proteomics for aquatic pollution biomonitoring

Assessing the impact of aquatic pollutants on the proteomes of aquatic organisms presents numerous challenges. Diversity of life histories, pathways of exposure and responses (individual variation), and environmental conditions unique to study sites (temperature, salinity, stressors) will all influence proteome alterations and are therefore difficult to interpret, even within a "sentinel" species (fish or mussel). Aquatic organisms rarely experience only one pollutant in vivo; they usually experience exposure to mixtures of metals, organic chemicals, and microplastics, as well as changes in physical conditions and biological sources of stress. Therefore, while proteomic alterations (i.e., upregulation of specific stress response proteins or antioxidant enzymes) in response to toxicity may serve as biomarkers of exposure to certain pollutants, they could also indicate general responses of aquatic organisms to a variety of different stressors (Tomanek, 2011). Because most of the aquatic organisms used as biomonitors do not have a genome or proteome annotation that corresponds to the species, they are not frequently used as research animals. Therefore it is difficult to identify the proteins produced, how their function is to be characterized, and how these proteins react.

A few of the most prominent problems include different methodologies and technologies. Protein extraction from aquatic organisms is often not standardized, especially from various sources and types of tissues, as well as algae and invertebrates. Different methodologies (e.g., tissue homogenization, types of lysis buffer, and different means for removing contaminants) yield different amounts and levels of representation of the proteome. There has been an emphasis by several researchers (Gamage et al., 2022) on the need for universal or at least extraction methods that apply to specific families. Some of the most interesting biomarker candidates are likely to be low-abundance proteins, such as signaling proteins or those modified post-translationally, as well as membrane proteins, which are notoriously difficult to solubilize and analyze using traditional proteomics processes. Difficulties with solubility and dynamic range make membrane proteins still hard to study. The full proteome simply cannot be studied using traditional proteomic methodologies because a great portion of membrane proteins is missing because of the complexity of the sample or the range of detection. To ensure full reproducibility of quantitative proteomics, replication is required, but that degree of replication may be impractical in many field biomonitoring studies. Given the high level of complexity associated with proteomic data, a robust and well-developed statistical framework, correction for multiple hypotheses, and careful experimental design need to be established. Unfortunately, the environmental proteome literature suffers from a high level of inadequate replication, poor statistical methods, and inadequate corrections of multiple hypotheses. In some cases, using different statistical tests on the same dataset has led to conflicting results (Lualdi and Fasano, 2019).

The sensitivity and specificity of proteomic biomarkers present additional challenges. To demonstrate reliable indicators of exposure or impact while assisting in biomonitoring, the proteins analyzed must consistently respond (ideally in a dose- responsive manner with little if any confounding variables) to the pollutant of interest. The standardization of techniques across space and time is required to ensure that all datasets collected during field biomonitoring studies are comparable. Proteomic studies are still quite costly, highly technical, and require considerable time to complete in comparison to simpler biomonitoring techniques. In addition, infrastructure and costs may limit their use in large-scale monitoring efforts (Machuca-Sepúlveda et al., 2023). Proteomic changes ultimately must be associated with substantial ecological consequences (population declines, reproductive dysfunctions, etc.) or regulatory criteria. Without these associations, molecular information can be generated that is interesting but not particularly useful for decision-making. Since the exposure doses, mixture compositions, and the historical context of the organisms differ in the field, the proteomic responses obtained from exposure under controlled laboratory conditions may not translate into similar responses to a field-collected sample. Therefore, environmental proteomics in aquatic biomonitoring must contend with temporal and spatial variability, including how seasonal influences, the age/size of the organisms being monitored, and the habitat conditions all affect the baseline of each organism's proteome. Without a reliable baseline/reference point, it is difficult to discern how much of any observed proteomic change is attributable to pollution versus natural variations. As the area of aquatic pollution progresses with new pollutants such as micro and nanoplastics, pharmaceutical residues, manufactured nanomaterials, and mixtures of these pollutants, it has become an increasingly important topic for researchers (Abbas et al., 2020). While Proteomics offers possibilities to discover new mechanisms of action, it also confronts more difficulties in addressing these new stressors, such as low exposure levels, subtle or long-lasting impacts, and a lack of understanding of molecular pathways.

12.9 Future prospective

The comprehensive study of protein expression, structure, and function—Proteomics—is an essential component for understanding how organisms respond to environmental stimuli. Through the use of proteomics, researchers can detect very small molecular changes associated with pollutants in aquatic ecosystems, long before the visible effects of ecological damage are seen. Aquatic ecosystems continue to be impacted heavily by the increasing human presence and activity, leading to the introduction of pharmaceuticals, industrial wastewater, microplastics, and an overall increase in contamination. The ability of proteomics to detect minor changes in aquatic ecosystems will allow researchers the opportunity to develop early warning systems and to understand the mechanisms behind these changes (Gamage et al., 2022). Although at this stage, proteomics is primarily a research tool, as technology advances, including mass spectrometry, bioinformatics, and integrated "omics" approaches, proteomics will eventually become a standard component of water polluting biomonitoring. By generating integrated multiomics data (i.e., genomic, transcriptomic, proteomic, and metabolomic), future aquatic biomonitoring will be most effectively achieved through the use of an integrated multi-omics biomarker approach. Proteomics integrates all of the other molecular omic levels to form a more comprehensive picture of an organism's total functional status, including the processes of toxicity occurring within the organism, which cannot otherwise be understood by studying any one omic level alone. For instance, the relationship between mRNA transcription and protein expression can be assessed through combining transcriptomic and proteomic datasets; this provides a means of determining if regulation of transcription is congruent with biological response. Additionally, combining proteomic and metabolomic data provides a clear link between physiological response and underlying biochemical processes. In addition, machine learning algorithms utilized in conjunction with multiomics datasets will improve biomarker identification and enhance the specificity of biomarkers as well as reduce the number of false positives (i.e., the identification of a biomarker arising from nonspecific stress responses) associated with biomarker discovery (Luo et al., 2025). Proteomics linked to genomes, transcriptomes, and metabolomes, that is, multiomics, will significantly contribute to the development of new approaches for aquatic biomonitoring. While proteomics provides a comprehensive snapshot of an organism's functional status; however, it allows a more thorough understanding of the mechanisms of toxicity through its connection to other molecular levels (Tomanek, 2011). The combination of proteomic and transcriptome data can provide evidence that transcriptional regulation occurs in conjunction with biological responses (e.g., whether mRNA expression of certain genes corresponds to the relative abundance of their corresponding proteins). Additionally, the use of proteomics combined with metabolomics establishes a direct relationship between physiological results and biochemical pathways. Machine learning-based frameworks utilizing multi-omics will improve the ability to identify specific biomarkers as well as reduce false positives due to non-specific stress response. The rapid expansion of the sequencing of aquatic species through projects such as Fish-T1K and the Earth BioGenome Project will provide an increasing number of sequence resources that will allow for better protein identification, more precise functional annotation, and more detailed pathway analyses (Huang et al., 2020). The proteomes of sentinel organisms like fish, mussels, and crustaceans are likely to be stored in cloud-based open-access repositories, allowing for data-sharing across borders and facilitating comparisons between laboratories and meta-analyses of responses to pollutants. To move proteomics from a research discipline to one with practical applications, it needs to become standardized and reproducible. Standardized and reproducible practices will be based on future SOPs related to data processing, protein extraction, sampling, and quantification. Improved data comparability and increased confidence in the use of proteomics-based indicators will come from both interlaboratory ring-testing and the availability of certified reference materials. The speed, portability, and cost-effectiveness of proteomics studies can improve with the use of microfluidics and lab on-a-chip technologies to automate processes; this could lead to the ability to monitor biological responses in real time or nearly so, at remote field stations. Proteomics researchers have typically looked at the effects of single toxins (or other forms of environmental contamination) on aquatic organisms, rather than assessing the combined toxic effects of many different types of pollutants. Future proteomics studies of combination toxicity will likely require that researchers investigate the molecular interactions between the various stressors that affect the organism. Proteomics can aid in the evaluation of a mixture's toxicity by determining if it is a mixture of toxicants with an antagonism, synergy, or additivity (Cedergreen and Nazir, 2014). In addition, advanced proteomics instrumentation will be needed to assess the long-term and/or slight harm that will occur from emerging contaminants, including but not limited to pharmaceuticals, nanomaterials, and microplastic contamination. Additionally, targeted assays that measure pollutant-specific biomarkers, such as heat shock proteins or oxidative stress enzymes, could be developed to be reliable and useful tools for environmental surveillance (Hao et al., 2024).

Proteomic datasets are very complex, so strong computing techniques will be needed to analyze them. The use of AI & Machine Learning for biomarker identification, pollutant classification, and predictive modeling will be crucial in the future for Environmental Proteomics. Large proteome datasets are being developed, and algorithms are being developed that can recognize protein expression patterns related to exposure to pollutants and/or toxic modes of action (Vishnoi et al., 2020). Environmental researchers, regulators, and managers need to collaborate to develop proteomic baselines, proteome databases, and incorporate molecular biomarkers into environmental quality standards. Eventually, automation, miniaturization, and cost reductions will allow proteomics to be applied at large geographic and time scales. It is anticipated that proteomics will have a larger role in aquatic biomonitoring in the future.

12.10 Conclusion

In summary, the application of proteomics is changing the way we monitor the health of aquatic environments by bringing together an understanding of how organisms respond to chemicals (mechanism), detection of chemical exposure at low concentrations (sensitivity), and the ability to predict how a population of organisms will be affected (prediction). This science has shown great promise based on some of the first studies completed using proteomics; however, for it to reach its full potential as a science, there are many challenges that must be addressed. The primary challenges include improvement of existing technology/methodology; integration of other "omics" technologies, and establishment of regulatory frameworks. The use of proteomics will allow for the earlier detection of chemical contamination in the water and provide a strong scientific base for protecting and sustainably managing the world's aquatic systems under increasing demand from humans. This capability makes proteomics especially suited to contemporary environmental risk assessment, as it will allow us to deal with a rising number of contaminants and combination toxicities, as well as chronic, low-dose exposures to these contaminants. Additionally, the development of smaller, portable, and deployable proteomic systems that can provide near real-time monitoring could allow managers to respond more rapidly to contamination events than they currently do.

References

Abbas, Q., Yousaf, B., Ullah, H., Ali, M.U., Ok, Y.S., Rinklebe, J., 2020. Environmental transformation and nano-toxicity of engineered nano-particles (ENPs) in aquatic and terrestrial organisms. Critical Reviews in Environmental Science and Technology 50 (23), 2523–2581. Available at https://doi.org/10.1080/10643389.2019.1705721, www.tandf.co.uk/journals/titles/10643389.asp.

Ali, H., Khan, E., Ilahi, I., 2019. Environmental chemistry and ecotoxicology of hazardous heavy metals: environmental persistence, toxicity, and bioaccumulation. Journal of Chemistry 2019 (1), 6730305.

Ambrosio, L., Russo, R., Salzano, A.M., Serpe, F.P., Ariano, A., De Tommasi, N., et al., 2018. Accumulation of polychlorinated biphenyls in mussels: a proteomic study. Journal of Food Protection 81 (2), 316–324. Available at https://doi.org/10.4315/0362-028x.jfp-17-148.

Amin, I., Nazir, R., Rather, M.A., 2024. Evaluation of multi-heavy metal tolerance traits of soil-borne fungi for simultaneous removal of hazardous metals. World Journal of Microbiology and Biotechnology 40 (6), 175.

Amin, I., Nazir, R., Rather, M.A., 2025. Fungal nano-factories: synthesis and characterization of silver nanoparticles from soil-borne fungus and their anti-microbial potential. World Journal of Microbiology and Biotechnology 41 (9), 1–17.

Balbi, T., Auguste, M., Ciacci, C., Canesi, L., 2021. Immunological responses of marine bivalves to contaminant exposure: contribution of the -omics approach. Frontiers in Immunology 12. Available at https://doi.org/10.3389/fimmu.2021.618726.

Bjørnstad, A., Larsen, B.K., Skadsheim, A., Jones, M.B., Andersen, O.K., 2006. The potential of ecotoxicoproteomics in environmental monitoring: biomarker profiling in mussel plasma using proteinchip array technology. Journal of Toxicology and Environmental Health, Part A 69 (1-2), 77–96. Available at https://doi.org/10.1080/15287390500259277.

Brownridge, P., Holman, S.W., Gaskell, S.J., Grant, C.M., Harman, V.M., Hubbard, S.J., et al., 2011. Global absolute quantification of a proteome: challenges in the deployment of a QconCAT strategy. Proteomics 11 (15), 2957–2970. Available at https://doi.org/10.1002/pmic.201100039.

Cedergreen, N., Nazir, A., 2014. Quantifying synergy: a systematic review of mixture toxicity studies within environmental toxicology. PLoS One 9 (5), e96580. Available at https://doi.org/10.1371/journal.pone.0096580.

Everingham, E., 2023. Unsupervised Multi-Resolution Analysis of Protein-Protein Abundance Patterns in Infectious Exacerbation of COPD. Rochester Institute of Technology.

Gamage, N.T., Miyashita, R., Takahashi, K., Asakawa, S., Senevirathna, J.D.M., 2022. Proteomic applications in aquatic environment studies. Proteomes 10 (3), 32.

Gandar, A., Laffaille, P., Marty-Gasset, N., Viala, D., Molette, C., Jean, S., 2017. Proteome response of fish under multiple stress exposure: effects of pesticide mixtures and temperature increase. Aquatic Toxicology 184, 61–77. Available at https://doi.org/10.1016/j.aquatox.2017.01.004.

Gomes, T., Chora, S., Pereira, C.G., Cardoso, C., Bebianno, M.J., 2014. Proteomic response of mussels Mytilus galloprovincialis exposed to CuO NPs and Cu2+: an exploratory biomarker discovery. Aquatic Toxicology 155, 327–336. Available at https://doi.org/10.1016/j.aquatox.2014.07.015, http://www.elsevier.com/wps/find/journaldescription.cws_home/505509/description#description.

Green, D.S., Colgan, T.J., Thompson, R.C., Carolan, J.C., 2019. Exposure to microplastics reduces attachment strength and alters the haemolymph proteome of blue mussels (Mytilus edulis). Environmental Pollution 246, 423–434. Available at https://doi.org/10.1016/j.envpol.2018.12.017, www.elsevier.com/inca/publications/store/4/0/5/8/5/6.

Guo, H., Wang, L., Deng, Y., Ye, J., 2021. Novel perspectives of environmental proteomics. Science of The Total Environment 788, 147588. Available at https://doi.org/10.1016/j.scitotenv.2021.147588.

Hamid, B., Majeed, N., Ganai, B.A., Hassan, S., Bashir, Z., Wani, P.A., et al., 2023a. Heavy-metal tolerant bacterial strains isolated from industrial sites and scrap yards in Kashmir, India. Journal of Basic Microbiology 63 (12), 1361–1372. Available at https://doi.org/10.1002/jobm.202300349, http://onlinelibrary.wiley.com/journal/10.1002/(ISSN)1521-4028.

Hamid, B., Yatoo, A.M., Sayyed, R.Z., Dineshkumar, R., Al-Khayri, J.M., Bashir, Z., et al., 2023b. Microbial-based conversion of food waste for sustainable bioremediation and utilization as compost. Biomass Conversion and Biorefinery 15, 1. Available at https://doi.org/10.1007/s13399-023-04794-6.

Hao, Z., Kong, L., Ruan, L., Deng, Z., 2024. Recent advances in DNA origami-enabled optical biosensors for multi-scenario application. Nanomaterials 14 (23), 1968. Available at https://doi.org/10.3390/nano14231968.

He, L., He, T., Farrar, S., Ji, L., Liu, T., Ma, X., 2017. Antioxidants maintain cellular redox homeostasis by elimination of reactive oxygen species. Cellular Physiology and Biochemistry 44 (2), 532–553. Available at https://doi.org/10.1159/000485089.

Huang, Y., Bian, C., Liu, Y., You, X., Yu, H., Yi, Y., et al., 2020. Fish genomics. Encyclopedia of Marine Biotechnology 3, 1843–1866.

Khanam, Z., Sultana, F.M., Mushtaq, F., 2023. Environmental pollution control measures and strategies: an overview of recent developments. Geospatial Analytics for Environmental Pollution Modeling: Analysis, Control and Management 385–414.

Khosrovyan, A., Vodovnik, M., Mortimer, M., 2025. Omics approaches in environmental effect assessment of engineered nanomaterials and nanoplastics. Environmental Science: Nano 12 (5), 2551–2579.

Krishan, A., Yadav, S., Srivastava, A., 2023. Water pollution's global threat to public health: a mini-review. International Journal of Scientific Research in Science, Engineering and Technology 10 (6), 321–334.

Kumar, V., Singh, K., Shah, M.P., Singh, A.K., Kumar, A., Kumar, Y., 2021. Application of omics technologies for microbial community structure and function analysis in contaminated environment. Wastewater Treatment: Cutting-Edge Molecular Tools, Techniques and Applied Aspects. Elsevier, India, pp. 1–40. Available at http://doi.org/10.1016/B978-0-12-821881-5.00001-5, https://www.sciencedirect.com/book/9780128218815.

Lee, P.Y., Saraygord-Afshari, N., Low, T.Y., 2020. The evolution of two-dimensional gel electrophoresis-from proteomics to emerging alternative applications. Journal of Chromatography. A 1615, 460763.

Li, B., Zhang, Y., Du, J., Liu, C., Zhou, G., Li, M., et al., 2025. Application of multi-omics techniques in aquatic ecotoxicology: a review. Toxics 13 (8), 653. Available at https://doi.org/10.3390/toxics13080653.

Lualdi, M., Fasano, M., 2019. Statistical analysis of proteomics data: a review on feature selection. Journal of Proteomics 198, 18–26. Available at https://doi.org/10.1016/j.jprot.2018.12.004.

Luo, J., Guo, X., Zheng, Y., Yang, Z., Pei, S.Y., Rao, R.Q., et al., 2025. Integration of multi-omics data and machine learning to identify antioxidant biomarkers in type 1 diabetes. Free Radical Biology and Medicine 236, 41–56. Available at https://doi.org/10.1016/j.freeradbiomed.2025.05.385, https://www.sciencedirect.com/science/journal/08915849.

Luo, L., Zhang, Q., Kong, X., Huang, H., You, W., Ke, C., 2017. Differential effects of zinc exposure on male and female oysters (Crassostrea angulata) as revealed by label-free quantitative proteomics. Environmental Toxicology and Chemistry 36 (10), 2602–2613. Available at https://doi.org/10.1002/etc.3797.

López-López, E., Sedeño-Díaz, J.E., 2015. Biological indicators of water quality: the role of fish and macroinvertebrates as indicators of water quality. Environmental Indicators. Springer Netherlands, Mexico, pp. 643–661. Available at http://doi.org/10.1007/978-94-017-9499-2, http://doi.org/10.1007/978-94-017-9499-2_37.

Machuca-Sepúlveda, J., Miranda, J., Lefin, N., Pedroso, A., Beltrán, J.F., Farias, J.G., 2023. Current status of omics in biological quality elements for freshwater biomonitoring. Biology 12 (7). Available at https://doi.org/10.3390/biology12070923, www.mdpi.com/journal/biology.

Mancia, A., Abelli, L., Benkhalqui, A., Bertolucci, C., Fossi, M.C., Limonta, G., et al., 2019. Effects on immunity of exposure to microplastics in adult zebrafish. Invertebrate Survival Journal 16, 42–43.

Mathur, M., Kumar Prajapat, R., Kumar Upadhyay, T., Lal, D., Khatik, N., Sharma, D., 2021. Advances in Genomics and Proteomics in Agriculture. Informa UK Limited, pp. 23–35. Available at http://doi.org/10.1201/9781003099079-2.

Meng, C., Wang, K., Zhang, X., Zhu, X., 2023. Effect of cadmium in the gonads of mussel (Mytilus coruscus): an ionomics and proteomics study. Environmental Science and Pollution Research 30 (26), 68373–68386. Available at https://doi.org/10.1007/s11356-023-27208-6.

Muralidharan, S., Thompson, E., Raftos, D., Birch, G., Haynes, P.A., 2012. Quantitative proteomics of heavy metal stress responses in Sydney rock oysters. Proteomics 12 (6), 906–921.

Nambiar, S.P., Banuru, S.C., 2025. A comprehensive review of bioinformatics tools and applications revolutionising aquatic animal health management. Journal of Fish Health 5 (2), 86–110. Available at https://doi.org/10.29303/jfh.v5i2.6281.

Norambuena, J.A., Poblete-Grant, P., Beltrán, J.F., De los Ríos-Escalante, P., Aranzaez-Ríos, C., Farías, J.G., 2025. Proteomic profile of daphnia pulex in response to heavy metal pollution in lakes of Northern Patagonia. International Journal of Molecular Sciences 26 (1). Available at https://doi.org/10.3390/ijms26010417, http://www.mdpi.com/journal/ijms.

Novoa, B., Ríos-Castro, R., Otero-Muras, I., Gouveia, S., Cabo, A., Saco, A., et al., 2022. Wastewater and marine bioindicators surveillance to anticipate COVID-19 prevalence and to explore SARS-CoV-2 diversity by next generation sequencing: one-year study. Science of The Total Environment 833, 155140. Available at https://doi.org/10.1016/j.scitotenv.2022.155140.

Onoja, A., 2023. An integrated interpretable machine learning framework for high-dimensional multi-omics datasets.

Parida, S., Sahoo, P.K., 2024. Antioxidant defence in labeo rohita to biotic and abiotic stress: insight from mRNA expression, molecular characterization and recombinant protein-based ELISA of catalase, glutathione peroxidase, CuZn superoxide dismutase, and glutathione S-transferase. Antioxidants 13 (1). Available at https://doi.org/10.3390/antiox13010018, www.mdpi.com/journal/antioxidants.

Parmar, T.K., Rawtani, D., Agrawal, Y.K., 2016. Bioindicators: the natural indicator of environmental pollution. Frontiers in Life Science 9 (2), 110–118.

Qin, J., Ma, Z., Li, Y., 2025. Review and outlook for research and development in aquatic life and ecosystems. Ecosystems 1 (1), 2.

Sahu, A.K., Mir, S., Nayak, B., Baitharu, I., 2024. Sustainable management of eutrophication and problems associated with the algal toxin in ponds and lakes of rural areas. Water Resources Management for Rural Development. Elsevier, pp. 155–170.

Satrio, R.D., Fendiyanto, M.H., Miftahudin, M., 2024. Tools and techniques used at global scale through genomics, transcriptomics, proteomics, and metabolomics to investigate plant stress responses at the molecular level. Molecular Dynamics of Plant Stress and Its Management. Springer Nature Singapore, Singapore, pp. 555–607.

Schneider, T., Riedel, K., 2010. Environmental proteomics: analysis of structure and function of microbial communities. Proteomics 10 (4), 785–798. Available at https://doi.org/10.1002/pmic.200900450.

Singh, D., Geat, N., Mehriya, M., Rajawat, M.V.S., Prasanna, R., Kumar, A., et al., 2021. Omics (genomics, proteomics, metabolomics, etc.) tools to study the environmental microbiome and bioremediation. Waste to Energy: Prospects and Applications. Springer Singapore, India, pp. 235–260. Available at http://doi.org/10.1007/978-981-33-4347-4_10, https://link.springer.com/book/10.1007/978-981-33-4347-4.

Singh, L., Srivastav, S., Avasthi, A., Rathi, B., Sharda, S., Gärtner, W., et al., 2025. Translational omics: precision and all "coming of age" Multi-Omics in Biomedical Sciences and Environmental Sustainability: Applications and Recent Advances. Springer Science +Business Media, India, pp. 1–32. Available at http://doi.org/10.1007/978-981-96-7067-3_1.

Singh, P., Singh, V.K., Singh, R., Borthakur, A., Madhav, S., Ahamad, A., et al., 2020. Bioremediation: a sustainable approach for management of environmental contaminants. Abatement of Environmental Pollutants. Elsevier, pp. 1–23.

Song, Q., Zhou, H., Han, Q., Diao, X., 2017. Toxic responses of Perna viridis hepatopancreas exposed to DDT, benzo(a)pyrene and their mixture uncovered by iTRAQ-based proteomics and NMR-based metabolomics. Aquatic Toxicology 192, 48–57. Available at https://doi.org/10.1016/j.aquatox.2017.09.010.

Sumalatha, B., Babu, D.S., Sudarsini, B., Indira, M., 2025. Innovative techniques for enhancing water treatment efficiency. Machine Learning in Water Treatment 265–291.

Tarfeen, N., Nisa, K.U., Hamid, B., Bashir, Z., Yatoo, A.M., Dar, M.A., et al., 2022. Microbial remediation: a promising tool for reclamation of contaminated sites with special emphasis on heavy metal and pesticide pollution: a review. Processes 10 (7). Available at https://doi.org/10.3390/pr10071358, http://www.mdpi.com/journal/processes.

Tella, T.A., Festus, B., Olaoluwa, T.D., Oladapo, A.S., 2025. Water and wastewater treatment in developed and developing countries: Present experience and future plans. Smart Nanomaterials for Environmental Applications. Elsevier, pp. 351–385.

Tomanek, L., 2011. Environmental proteomics: changes in the proteome of marine organisms in response to environmental stress, pollutants, infection, symbiosis, and development. Annual Review of Marine Science 3 (1), 373–399. Available at https://doi.org/10.1146/annurev-marine-120709-142729.

Tomanek, L., Podrabsky, J.E., Stillman, J.H., Tomanek, L., 2015. Proteomic responses to environmentally induced oxidative stress. Journal of Experimental Biology 218 (12), 1867–1879. Available at https://doi.org/10.1242/jeb.116475.

Vellinger, C., Sohm, B., Parant, M., Immel, F., Usseglio-Polatera, P., 2016. Investigating the emerging role of comparative proteomics in the search for new biomarkers of metal contamination under varying abiotic conditions. Science of The Total Environment 562, 974–986. Available at https://doi.org/10.1016/j.scitotenv.2016.04.016.

Vishnoi, S., Matre, H., Garg, P., Pandey, S.K., 2020. Artificial intelligence and machine learning for protein toxicity prediction using proteomics data. Chemical Biology and Drug Design 96 (3), 902–920. Available at https://doi.org/10.1111/cbdd.13701, http://onlinelibrary.wiley.com/journal/10.1111/(ISSN)1747-0285.

Wang, L., Yang, L., Peng, Z., Lu, D., Jin, Y., McNutt, M., et al., 2015. cisPath: an R/Bioconductor package for cloud users for visualization and management of functional protein interaction networks. BMC Systems Biology 9 (S1). Available at https://doi.org/10.1186/1752-0509-9-s1-s1.

Wang, M., Wang, Y., Wang, J., Lin, L., Hong, H., Wang, D., 2011. Proteome profiles in medaka (Oryzias melastigma) liver and brain experimentally exposed to acute inorganic mercury. Aquatic Toxicology 103 (3-4), 129–139. Available at https://doi.org/10.1016/j.aquatox.2011.02.020.

De Wit, M., Keil, D., van der Ven, K., Vandamme, S., Witters, E., De Coen, W., 2010. An integrated transcriptomic and proteomic approach characterizing estrogenic and metabolic effects of 17 α-ethinylestradiol in zebrafish (Danio rerio). General and Comparative Endocrinology 167 (2), 190–201.

Wittmann-Liebold, B., Graack, H.R., Pohl, T., 2006. Two-dimensional gel electrophoresis as tool for proteomics studies in combination with protein identification by mass spectrometry. Proteomics 6 (17), 4688–4703.

Xu, L., Chu, Z., Li, X., Feng, C., Zhang, Y., Wang, C., et al., 2024. Proteomic insights into composition-dependent effects of microplastics on freshwater microalgae Chlamydomonas reinhardtii. Environmental Science: Nano 11 (8), 3440–3456. Available at https://doi.org/10.1039/d4en00300d.

Zhang, H., Zhai, Y., 2020. Integrated transcriptomic and proteomic analyses of the tissues from the digestive gland of Chlamys farreri following cadmium exposure. Journal of Cellular Biochemistry 121 (2), 974–983. Available at https://doi.org/10.1002/jcb.29254.

Zhang, Q., Zheng, S., Pei, X., Zhang, Y., Wang, G., Zhao, H., 2025. The effects of microplastics exposure on quail's hypothalamus: neurotransmission disturbance, cytokine imbalance and ROS/TGF-β/Akt/FoxO3a signaling disruption. Comparative Biochemistry and Physiology Part C: Toxicology & Pharmacology 287, 110054. Available at https://doi.org/10.1016/j.cbpc.2024.110054.

Zhou, Y., Gu, B., 2024. The impacts of human activities on Earth Critical Zone. Earth Critical Zone 1 (1), 100004. Available at https://doi.org/10.1016/j.ecz.2024.100004.

Chapter 13

Proteomics in bioremediation of pharmaceutical wastes: recent insight

M.E. Pavithra and P. Kavitha
Nursing (Microbiology & Biochemistry), KMCH Institute of Health Sciences, Affliated to The TN Dr. M. G. R University- Chennai, Coimbatore, Tamil Nadu, India

13.1 Introduction

They enter aquatic systems primarily through effluents from hospitals, domestic sewage, and pharmaceutical manufacturing units(Sharma et al., 2025). The chemical stability of antibiotics, analgesics, cytotoxic drugs, and hormones results in long-term environmental persistence and bioaccumulation. These pollutants can disrupt microbial ecology, induce hormonal imbalances in fauna, and contribute to the alarming rise in multidrug-resistant pathogens (Amin and Rahman, 2026). Traditional wastewater treatment plants (WWTPs) are not fully equipped to remove such complex xenobiotics. Physical processes such as adsorption and membrane filtration, and chemical oxidation or chlorination, may only partially degrade pharmaceutical residues while generating toxic by-products (Mahdavi et al., 2025). In contrast, bioremediation utilizes the metabolic potential of microorganisms—bacteria, fungi, and algae—to transform or mineralize pollutants into less toxic forms. However, the efficiency of these processes depends on a deep understanding of the enzymes and pathways involved, which is where proteomics provides an essential advantage.

Proteomics, a cornerstone of functional genomics, investigates the entire protein complement of an organism under specific conditions. In bioremediation research, proteomics deciphers the adaptive responses of microbial systems to xenobiotic stress, enabling the identification of key enzymes such as oxidoreductases, peroxidases, cytochrome P450s, and hydrolases that catalyze degradation reactions (Fig. 13.1 The integration of proteomic data with transcriptomic and metabolomic profiles has accelerated the discovery of pollutant-responsive networks, facilitating the design of synthetic microbial consortia for enhanced biodegradation (Fig. 13.1).

Proteomics Driven Bioremediation. DOI: https://doi.org/10.1016/B978-0-443-45294-9.00013-8

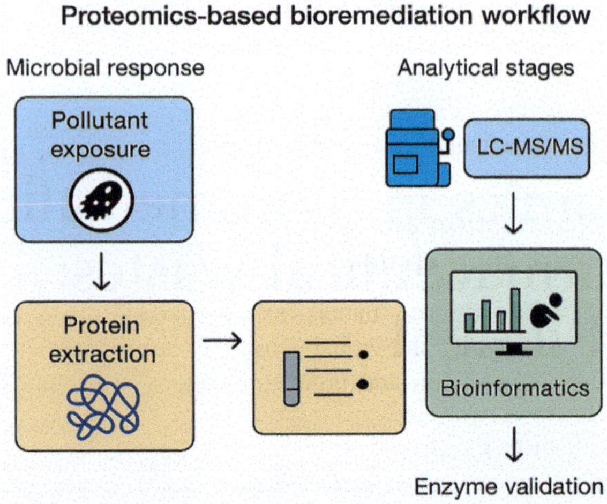

FIGURE 13.1 **Illustrates a typical workflow for proteomics-based bioremediation, showing how microbial responses, protein extraction, and analytical stages converge to elucidate degradation mechanisms.** Proteomics-based bioremediation workflow depicting microbial response, protein extraction, and analytical stages.

13.2 Protein profiling

Proteomics provides a comprehensive snapshot of cellular activity at the functional level. Unlike genomics, which predicts potential functions, proteomics reveals the actual biochemical responses of organisms under environmental stress. The field encompasses several interrelated domains:

- **Expression proteomics**, which quantifies changes in protein levels between control and pollutant-exposed conditions.
- **Structural proteomics** focuses on three-dimensional protein structures and ligand interactions that determine enzyme specificity.
- **Functional proteomics**, linking proteins to metabolic pathways and physiological processes during bioremediation.

Modern proteomics employs high-resolution analytical platforms such as two-dimensional gel electrophoresis (2-DE), liquid chromatography coupled with tandem mass spectrometry (LC-MS/MS), and label-free quantification (Gygi et al., 2000; Salvato et al., 2021).

13.2.1 Proteomics—role in bioremediation

Microbial adaptation to pollutants involves dynamic modulation of enzyme systems, membrane transporters, stress response proteins, and regulatory molecules. Proteomics captures these shifts, offering mechanistic insights that cannot be obtained from genomic or transcriptomic data alone (Shah et al., 2024). For instance, exposure of *Pseudomonas putida* to fluoroquinolones results in upregulation of efflux pumps and redox enzymes, while fungal strains exposed to ethinylestradiol overexpress laccases and peroxidases (Vrsanska et al., 2016).

Proteomics thus provides actionable data for:

1. Identifying pollutant-responsive biomarkers for monitoring microbial activity;
2. Understanding stress response networks and detoxification pathways;
3. Optimizing bioreactor performance by selecting high-performing microbial strains.

13.2.2 Proteomics analysis workflow

A typical proteomic workflow for bioremediation encompasses a sequence of interlinked experimental and analytical stages designed to capture the protein-level dynamics of microorganisms exposed to pharmaceutical pollutants. The process generally involves four major steps: **sample preparation, protein separation, protein identification, and data interpretation** (Nicora et al., 2019).

13.2.2.1 Sample preparation

The workflow begins with the careful collection and preparation of biological samples. Microbial cells or consortia are typically isolated from contaminated environments—such as pharmaceutical effluents, WWTPs, or laboratory microcosms exposed to specific drugs. The collected biomass undergoes cell lysis using physical (e.g., sonication, bead beating) or chemical (e.g., detergents, chaotropic agents) methods to release intracellular proteins. These proteins are then purified to remove interfering substances such as lipids, nucleic acids, and humic matter, ensuring high-quality extracts suitable for downstream analysis. Standardization of this step is critical, as protein yield and integrity directly influence the reliability of subsequent analyses (Nicora et al., 2019).

13.2.2.2 Protein separation

Due to the complexity of microbial proteomes, separation techniques are employed to fractionate proteins before identification. Two-dimensional gel electrophoresis (2-DE) remains a classical yet powerful technique, separating proteins first by isoelectric point and then by molecular weight (Gygi et al., 2000). However, liquid chromatography (LC)-based methods, such as high-performance liquid chromatography and multidimensional LC (MudPIT), have gained prominence for their higher resolution and ability to handle complex mixtures, including membrane and low-abundance proteins (Nilsson et al., 2010; Quirino et al., 2010). This step facilitates the visualization or fractionation of protein profiles, enabling differential comparison between control and pollutant-exposed samples.

13.2.2.3 Protein identification

Once separated, proteins or peptides are identified using mass spectrometry (MS), which serves as the cornerstone of modern proteomics (Aebersold and Mann, 2016). Techniques such as matrix-assisted laser desorption/ionization—time of flight (MALDI-TOF), electrospray ionization (ESI), and tandem mass spectrometry (MS/MS) generate peptide-mass fingerprints and sequence data. These spectra are matched against protein databases (e.g., UniProt, NCBI) using bioinformatics tools such as Mascot or SEQUEST to assign protein identities (Thompson, 2023). Quantitative proteomic strategies—such as isobaric tags for relative and absolute quantitation (iTRAQ), tandem mass-tag (TMT) labeling, or label-free quantification—allow relative abundance comparisons across multiple conditions, providing insights into the proteins that are up- or down-regulated in response to pharmaceutical contaminants (Chen et al., 2021).

13.2.2.4 Data interpretation

The final and most critical stage involves bioinformatic and statistical analysis of the acquired proteomic data. Identified proteins are functionally annotated using tools such as KEGG, STRING, and gene ontology (GO) enrichment analyses to uncover their roles in metabolic and detoxification pathways (Root et al., 2021). Network analysis can reveal interactions between stress response proteins, transporters, and degradative enzymes. Moreover, integrating proteomic data with transcriptomic and metabolomic profiles—the so-called multiomics approach—provides a holistic understanding of microbial adaptation and pollutant degradation mechanisms (Gautam et al., 2023). Such analyses not only identify potential biomarkers for bioremediation efficiency but also aid in designing improved microbial consortia or biocatalysts for environmental applications (Kuppan et al., 2024).

Collectively, this systematic workflow enables researchers to translate complex proteomic datasets into mechanistic insights, linking protein expression dynamics with functional resilience and biodegradation efficiency in pharmaceutical-waste-contaminated ecosystems (Aebersold and Mann, 2016).

13.3 Proteomic approaches in bioremediation

The integration of diverse proteomic strategies provides a multidimensional understanding of microbial responses to pharmaceutical contaminants. Each approach offers unique advantages depending on the complexity of the sample and the research objective (Tsakou et al., 2020).

13.3.1 Gel-based proteomics

Two-dimensional gel electrophoresis (2-DE) remains one of the most reliable techniques for protein separation and visualization. It allows detection of differential expression patterns in microorganisms exposed to specific pharmaceuticals. For example, 2-DE combined with MALDI-TOF MS has been used to identify over-expressed oxidative stress proteins in *Bacillus subtilis* exposed to diclofenac (Schott et al., 2016). Despite limitations in detecting low-abundance proteins, gel-based proteomics provides a robust, cost-effective entry point for environmental studies.

13.3.2 Mass-spectrometry-based proteomics

High-resolution MS technologies, including Q-Exactive Orbitrap and triple-quadrupole systems, facilitate precise identification of enzymes involved in xenobiotic metabolism. LC-MS/MS can detect posttranslational modifications such as phosphorylation and acetylation, which regulate enzyme activity (Mishra et al., 2021). This capability is crucial for understanding adaptive microbial responses in contaminated matrices.

13.3.3 Quantitative proteomics

Quantitative proteomics using techniques such as iTRAQ or TMTs enables simultaneous comparison of multiple experimental conditions. It has been applied to monitor expression levels of degradation-related proteins in microbial consortia exposed to mixtures of antibiotics and hormones (Cui et al., 2023). Label-free quantification is particularly suitable for environmental samples with variable biomass.

13.3.4 Metaproteomics

Metaproteomics extends the scope of proteomics to entire microbial communities. It identifies proteins expressed by mixed populations in natural or engineered ecosystems, thereby linking taxonomic identity with metabolic function (Rao and Sunkad, 2023). In WWTPs, metaproteomic studies have revealed upregulation of nitrifying enzymes and cytochrome P450s during pharmaceutical degradation. This community-level understanding guides bioaugmentation strategies for improved remediation efficiency.

13.4 Enzymatic biomarkers and functional pathways

Proteomic analyses have led to the discovery of key enzymatic biomarkers responsible for the degradation of pharmaceuticals. These enzymes belong to diverse functional categories, including oxidoreductases, hydrolases, and oxygenases (Amiri-Dashatan et al., 2018; Kennes-Veiga et al., 2022).

Proteomic mapping reveals that these enzymes participate in integrated metabolic pathways, often linked to central carbon metabolism and oxidative stress response systems. For example, oxidative enzymes catalyze the initial breakdown of pharmaceutical molecules into smaller intermediates, which are then funneled into the tricarboxylic acid cycle for complete mineralization (Timsina et al., 2025). In Fig. 13.2 hydrolases contribute by cleaving ester and amide bonds, generating less toxic and more bioavailable compounds.

Furthermore, proteomics identifies auxiliary stress proteins—such as heat-shock proteins, ATP-binding cassette (ABC) transporters, and antioxidant enzymes—that protect microbial cells during pollutant exposure (Jain et al., 2021). Their upregulation serves as an indicator of active biodegradation.

13.5 Challenges and future prospects

Although proteomics-based bioremediation provides unparalleled insights into microbial function, several scientific and technical challenges still restrict its full deployment at the industrial and environmental scales. The first constraint arises from the complexity of environmental matrices, which often contain humic substances, surfactants, and heavy metals that interfere with protein extraction and identification (Su et al., 2022). Moreover, low protein yield and coextracted inhibitors compromise the sensitivity of downstream analyses such as LC-MS/MS. Developing improved extraction protocols, including surfactant-assisted precipitation and pressure-cycling technology, may mitigate such losses.

Another obstacle involves incomplete protein databases for environmental microorganisms, particularly those that have not been cultured. This impedes accurate annotation of spectra and functional classification (Jain et al., 2021). The advent of metagenome-assembled genomes and shotgun metaproteomics now provides opportunities to assign proteins to their respective taxa and metabolic networks, enhancing accuracy.

Standardization of data analysis pipelines remains a global priority. Different laboratories employ varied workflows for peptide search, false-discovery rate thresholds, and quantification, which hinders cross-study comparisons (Srivastava et al., 2019). To address this, consortia such as the Proteomexchange Consortium and PRIDE repository promote open data formats and reproducibility standards, fostering cumulative knowledge in environmental proteomics.

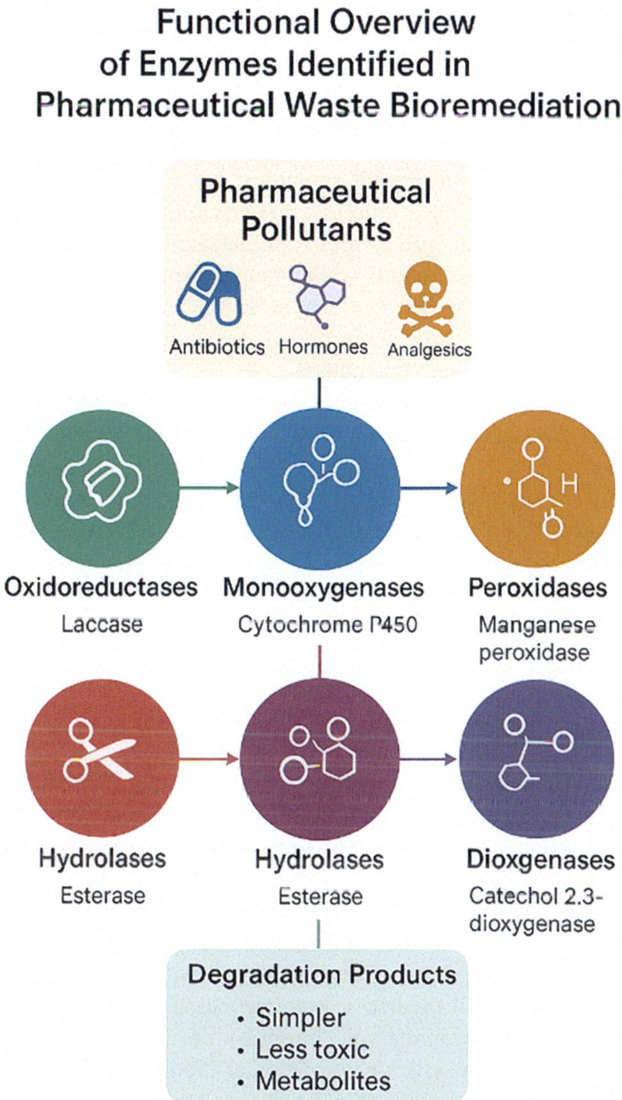

FIGURE 13.2 **Illustrates the proteomic interplay between enzymatic pathways, cellular stress response, and pollutant degradation.** Schematic summary of major enzyme classes (oxidoreductases, hydrolases, dioxygenases, and transferases) and their specific roles in degrading pharmaceutical contaminants such as antibiotics, hormones, and cytotoxic drugs.

At a broader level, economic feasibility and scaling-up also present challenges. Translating laboratory-scale findings to pilot bioreactors or WWTPs requires multidisciplinary integration involving microbiology, chemical engineering, and computational modeling (Crater and Lievense, 2018).

Emerging biosensors based on enzyme biomarkers identified via proteomics could enable real-time monitoring of degradation efficiency.

13.5.1 Integration with multiomics and artificial intelligence

The integration of proteomics with genomics, transcriptomics, and metabolomics (multiomics) facilitates a holistic understanding of microbial ecosystems. For instance, coupling proteomics with metagenomics enables functional confirmation of gene expression, while metabolomics validates enzymatic activity by profiling degradation intermediates (Mishra et al., 2021; González-González et al., 2022). Such synergy refines our understanding of xenobiotic metabolism and adaptive stress responses.

Artificial intelligence and machine learning have recently transformed environmental proteomics. Algorithms trained on spectral data can predict protein–ligand interactions, classify unknown enzymes, and model biodegradation kinetics (Vora et al., 2023). Deep-learning frameworks, such as AlphaFold and DeepLC, allow prediction of protein structures and retention times, thus accelerating annotation of novel biodegradative enzymes. When combined with synthetic biology, these approaches pave the way for designing microbial consortia with tailored catabolic pathways and enhanced pollutant tolerance.

13.6 Conclusion

Proteomics has revolutionized our understanding of microbial and enzymatic mechanisms underlying the bioremediation of pharmaceutical pollutants. By identifying differentially expressed proteins, enzyme biomarkers, and community-level functional pathways, it bridges the gap between potential microbial diversity and realized biochemical activity. The integration of advanced proteomic technologies with multiomics datasets, AI-driven modeling, and eco-engineering approaches holds immense potential for developing sustainable, real-time, and predictive bioremediation systems. In the coming decade, innovations in metaproteomic, biosensor development, and bioprocess modeling will further strengthen environmental monitoring frameworks. The next frontier will involve the translation of proteomic discoveries into industrial-scale applications that support global environmental sustainability and public health.

AI disclosure

During the preparation of this work, the author(s) used BioRender and AI tools (2025) for the creation and design of tables, figures, and graphical illustrations. After using these tools/services, the author(s) reviewed and edited the content as needed and take(s) full responsibility for the content of the publication.

References

Aebersold, R., Mann, M., 2016. Mass-spectrometric exploration of proteome structure and function. Nature 537 (7620), 347–355. Available at https://doi.org/10.1038/nature19949.

Amin, M.F., Rahman, M.S., 2026. A critical review of pharmaceutical pollutants in soil and air: ecotoxicological impacts on animal, plant and microbial communities - health hazards and waste management. Environmental Pollution and Management 3, 70–87. Available at https://doi.org/10.1016/j.epm.2025.11.001.

Amiri-Dashatan, N., Koushki, M., Abbaszadeh, H.-A., Rostami-Nejad, M., Rezaei-Tavirani, M., 2018. Proteomics applications in health: biomarker and drug discovery and food industry. Iranian Journal of Pharmaceutical Research 17 (4), 1523–1536.

Chen, X., Sun, Y., Zhang, T., Shu, L., Roepstorff, P., Yang, F., 2021. Quantitative proteomics using isobaric labeling: a practical guide. Genomics, Proteomics & Bioinformatics 19 (5), 689–706.

Crater, J.S., Lievense, J.C., 2018. Scale-up of industrial microbial processes. FEMS Microbiology Letters 365 (13). Available at https://doi.org/10.1093/femsle/fny138.

Cui, C., Jiang, M., Zhang, C., Zhang, N., Jin, F.-J., Li, T., et al., 2023. Assembly strategies for rubber-degrading microbial consortia based on omics tools. Frontiers in Bioengineering and Biotechnology 11. Available at https://doi.org/10.3389/fbioe.2023.1326395.

Gautam, P., Pandey, A.K., Dubey, S.K., 2023. Multi-omics approach reveals elevated potential of bacteria for biodegradation of imidacloprid. Environmental Research 221. Available at https://doi.org/10.1016/j.envres.2023.115271.

González-González, R.B., Sharma, P., Singh, S.P., Américo-Pinheiro, J.H.P., Parra-Saldívar, R., Bilal, M., et al., 2022. Persistence, environmental hazards, and mitigation of pharmaceutically active residual contaminants from water matrices. Science of the Total Environment 821. Available at https://doi.org/10.1016/j.scitotenv.2022.153329.

Gygi, S.P., Corthals, G.L., Zhang, Y., Rochon, Y., Aebersold, R., 2000. Evaluation of two-dimensional gel electrophoresis based proteome analysis technology. Proceedings of the National Academy of Sciences of the United States of America 97 (17), 9390–9395. Available at https://doi.org/10.1073/pnas.160270797.

Jain, A., Singh, H.B., Das, S., 2021. Deciphering plant-microbe crosstalk through proteomics studies. Microbiological Research 242. Available at https://doi.org/10.1016/j.micres.2020.126590.

Kennes-Veiga, D.M., Gónzalez-Gil, L., Carballa, M., Lema, J.M., 2022. Enzymatic cometabolic biotransformation of organic micropollutants in wastewater treatment plants: a review. Bioresource Technology 344. Available at https://doi.org/10.1016/j.biortech.2021.126291.

Kuppan, N., Padman, M., Mahadeva, M., Srinivasan, S., Devarajan, R., 2024. A comprehensive review of sustainable bioremediation techniques: Eco friendly solutions for waste and pollution management. Waste Management Bulletin 2 (3), 154–171. Available at https://doi.org/10.1016/j.wmb.2024.07.005.

Mahdavi, P., Siol, A., Thöming, J., 2025. Adsorption-based removal of pharmaceutical from water: a critical review on adsorbent performance. Journal of Environmental Chemical Engineering 13 (5). Available at https://doi.org/10.1016/j.jece.2025.117520.

Mishra, S., Lin, Z., Pang, S., Zhang, W., Bhatt, P., Chen, S., 2021. Recent advanced technologies for the characterization of xenobiotic-degrading microorganisms and microbial communities. Frontiers in Bioengineering and Biotechnology 9. Available at https://doi.org/10.3389/fbioe.2021.632059.

Nicora, C., Gritsenko, M., Lipton, A., Wahl, K.L., Burnum-Johnson, K.E., 2019. Proteomic sample preparation techniques: toward forensic proteomic applications. ACS Symposium Series 1339, 29–46. Available at https://doi.org/10.1021/bk-2019-1339.ch003.

Nilsson, T., Mann, M., Aebersold, R., Yates, J.R., Bairoch, A., Bergeron, J.J.M., 2010. Mass spectrometry in high-throughput proteomics: ready for the big time. Nature Methods 7 (9), 681–685. Available at https://doi.org/10.1038/nmeth0910-681.

Quirino, Candido, Campos, Franco, Krüger, 2010. Proteomic approaches to study plant-pathogen interactions. Phytochemistry 71 (4), 351–362. Available at https://doi.org/10.1016/j.phytochem.2009.11.005.

Rao, K., Sunkad, G., 2023. Metaomics approaches to unravel the functioning of multispecies microbial communities. Microbiome Drivers of Ecosystem Function 395–416. Available at https://doi.org/10.1016/B978-0-443-19121-3.00009-0.

Root, L., Campo, A., MacNiven, L., Con, P., Cnaani, A., Kültz, D., 2021. A data-independent acquisition (DIA) assay library for quantitation of environmental effects on the kidney proteome of Oreochromis niloticus. Molecular Ecology Resources 21 (7), 2486–2503. Available at https://doi.org/10.1111/1755-0998.13445.

Salvato, F., Hettich, R.L., Kleiner, M., 2021. Five key aspects of metaproteomics as a tool to understand functional interactions in host-associated microbiomes. PLoS Pathogens 17 (2). Available at https://doi.org/10.1371/journal.ppat.1009245.

Schott, A.-S., Behr, J., Quinn, J., Vogel, R.F., 2016. MALDI-TOF mass spectrometry enables a comprehensive and fast analysis of dynamics and qualities of stress responses of Lactobacillus paracasei subsp. paracasei F19. PLoS One 11 (10). Available at https://doi.org/10.1371/journal.pone.0165504.

Shah, K., Ray, S., Bose, H., Pandey, V., Wohlschlegel, J.A., Mahendra, S., 2024. Proteomics insights into the fungal-mediated bioremediation of environmental contaminants. Current Opinion in Biotechnology 90. Available at https://doi.org/10.1016/j.copbio.2024.103213.

Sharma, A., Kaur, N., Singh, N., 2025. Tailored fibrils approach via Ag(I).peptidomimetic-based interface design: efficient encapsulation of diverse active pharmaceutical ingredients in wastewater remediation during effluent treatment plant (ETP) processing. Langmuir the ACS Journal of Surfaces and Colloids 41 (8), 5355–5370. Available at https://doi.org/10.1021/acs.langmuir.4c04890.

Srivastava, S., Merchant, M., Rai, A., Rai, S.N., 2019. Standardizing proteomics workflow for liquid chromatography-mass spectrometry: technical and statistical considerations. Journal of Proteomics & Bioinformatics 12 (3), 48–55. Available at https://doi.org/10.35248/0974-276x.19.12.496.

Su, Y., Ren, T., Lin, Y., Zheng, C., 2022. Application of solid-phase microextraction in atomic spectrometry. Advances in Sample Preparation 3. Available at https://doi.org/10.1016/j.sampre.2022.100033.

Thompson, C.M., 2023. Molecular testing for diseases associated with bacterial infections. Diagnostic Molecular Pathology: A Guide to Applied Molecular Testing, Second Edition 133–145. Available at https://doi.org/10.1016/B978-0-12-822824-1.00017-1.

Timsina, R., Gora, R.A., Ferguson, D.J., 2025. CJ25. mSphere 10 (9). Available at https://doi.org/10.1128/msphere.00421-25.

Tsakou, F., Jersie-Christensen, R., Jenssen, H., Mojsoska, B., 2020. The role of proteomics in bacterial response to antibiotics. Pharmaceuticals 13 (9), 1–27. Available at https://doi.org/10.3390/ph13090214.

Vora, L.K., Gholap, A.D., Jetha, K., Thakur, R.R.S., Solanki, H.K., Chavda, V.P., 2023. Artificial intelligence in pharmaceutical technology and drug delivery design. Pharmaceutics 15 (7). Available at https://doi.org/10.3390/pharmaceutics15071916.

Vrsanska, M., Voberkova, S., Langer, V., Palovcikova, D., Moulick, A., Adam, V., et al., 2016. Induction of laccase, lignin peroxidase and manganese peroxidase activities in white-rot fungi using copper complexes. Molecules (Basel, Switzerland) 21 (11). Available at https://doi.org/10.3390/molecules21111553.

Chapter 14

Role of proteomics in management of e-waste: global scenario and remediation strategies

Thahir Basha Shaik[1], Asha Embrandiri[2], and Zienab Ahmed[3]

[1]*Department of Veterinary Public Health & Epidemiology, NTR College Veterinary Science, Sri Venkateswara Veterinary University, Gannavaram, Andhra Pradesh, India,* [2]*Department of Environmental Health, College of Medicine and Health Sciences, Wollo University, Desse, Amhara, Ethiopia,* [3]*Department of Integrative Agriculture, College of Agriculture and Veterinary Medicine, United Arab Emirates University, Al Ain, United Arab Emirates*

14.1 Introduction

The Basel Convention generally defines e-waste as unwanted, discarded, or obsolete machines that function with electricity. This consists of consumables, components, and subassemblies that are attached to the product during disposal (UNEP, 2025). Electronic waste (e-waste) is also commonly referred to as waste from electrical and electronic equipment (WEEE). The rise in digitalization, combined with shortened product lifecycles and the increasing obsolescence of equipment, has led to a substantial rise in e-waste generation worldwide. The Global E-waste Monitor 2024 states that there was an 82% increase from the 34 Mt produced in 2010 (Forti et al., 2024). These alarming rates place e-waste as the fastest-growing solid waste stream worldwide, with projections indicating that it may reach 82 million tonnes annually by 2030. This growing crisis demands urgent attention from industry stakeholders, policymakers, and scientists to devise sustainable management strategies. Tackling e-waste management is complicated because of the heterogeneous composition of electronic products, which contain valuable materials together with hazardous substances. Traditional waste management approaches have proven insufficient for addressing the complex challenges posed by e-waste, thereby requiring advanced biotechnological solutions. Biological methods such as proteomics have emerged as a powerful tool for understanding and

Proteomics Driven Bioremediation. DOI: https://doi.org/10.1016/B978-0-443-45294-9.00005-9

enhancing biological processes involved in e-waste remediation (Sharma and Rai, 2023).

This chapter explores the nexus of proteomics and e-waste management, examining how the understanding of protein levels can transform strategies for bioremediation and efforts for resource recovery.

14.1.1 Sources and classification of electronic waste

E-waste originates from several sources, ranging from commercial buildings, industrial facilities, to residential areas. The United Nations classification of e-waste was widely adopted in dividing e-waste into six major categories: *small IT and telecommunication equipment* (mobile phones, GPS devices, routers, keyboards), *temperature exchange equipment* (heaters, refrigerators, air conditioners), *lamps* (LED lamps, fluorescent lamps), *screens and monitors* (televisions, laptops, tablets), *large equipment* (washing machines, dryers, dishwashers, electric cookers), *and smaller equipment* (microwave ovens, vacuum cleaners, microwaves, calculators, shavers, toasters) (Baldée et al., 2020, Suyal et al., 2023). Each category poses its own challenge for management due to variations in material composition, hazardous content, and recycling complexity. Consumer electronics, particularly smart devices, phones, and computers, constitute a fast-growing segment of e-waste due to frequent upgrades driven by technological obsolescence and marketing strategies (Quinto et al., 2025; WHO, 2024).

Industrial and commercial sources contribute substantially to e-waste generation through the disposal of servers, networking equipment, point-of-sale systems, and specialized electronic instruments. Healthcare facilities generate medical electronic equipment waste, including diagnostic machines, monitoring devices, and electronic therapeutic equipment, which require specialized handling due to potential biological contamination alongside electronic hazards (Rautela et al., 2021). Understanding the sources and pathways of e-waste is essential for developing comprehensive management strategies that address both formal and informal recycling streams (Forti et al., 2020).

E-waste consists of heterogeneous materials including ferrous and non-ferrous metals (approximately 60% by weight), polymeric materials (30%), and hazardous substances like halogenated compounds, toxic metals, and radioactive elements (10%) (Suyal et al., 2023). Large household appliances are predominantly composed of ferrous metals (50%–60%) and plastics (15%–25%), with copper wiring constituting 5%–10% of total mass. (Forti et al., 2024).

The rapid spread of Internet of Things (IoT) devices is altering e-waste composition and management challenges. Global IoT-connected devices increased from 23.8 billion in 2021 to an estimated 40.2 billion by 2025 (Modarress Fathi et al., 2022). They contain a cocktail of microscale components that severely compromise the conventional mechanical separation and material recovery processes. Furthermore, devices often have shortened lifespans (only 2–3 years), due to strategic obsolescence and cessation of software updates, in contrast to the 7–10 years seen with traditional electronics (Hazelwood and Pecht 2021). This leads to faster turnover and a greater volume of e-waste. An important safety concern is the presence of integrated lithium-ion batteries, which account for almost 30% of a device's weight, requiring specialized disposal methods to minimize the risk of thermal runaway during processing (Madani et al., 2023).

Likewise, data security concerns have to be addressed before the process of recycling is carried out, as these devices have memory chips embedded in them that have to be destroyed in accordance with the standard protocol. Most importantly, it is not cost-effective to recycle these items, coming to as low as USD 5–50, which outrides the collection costs and processing facilities. IoT waste is likely to contribute an additional 5–8 million tonnes annually by 2030, with smart home devices, wearables, and sensors representing the fastest-growing categories (Modarress Fathi et al., 2022). This variation demands adaptive recycling infrastructure and specialized material recovery processes.

14.1.2 Hazardous components: toxic metals and persistent organic pollutants

E-wastes contain a complex mixture of hazardous substances that pose significant risks if not appropriately managed. The principal toxic components include heavy metals, persistent organic pollutants (POPs), and ozone-depleting substances. Heavy metals are a cause for concern due to their persistence, bioaccumulation potential, and adverse effects on the health of humans and ecosystems.

The extent of this issue is drastically rising as the informal recycling sector releases approximately 98,000 tonnes of toxic metals annually into the environment, while global e-waste generation is projected to reach 74 million tonnes by 2030 (Islam et al., 2024).

Lead (Pb) is predominantly used in cathode ray tube (CRT) screens (20%–25% by weight of screen glass, equivalent to 1–4 kg per television), that have been used in discarded televisions and monitors, solder in printed circuit boards) containing 37% Pb in traditional eutectic solder (Pb-Sn 63:37), and piezoelectric crystals and certain types of batteries. Awasthi et al. (2019) stated that exposure to lead can cause neurological damage, especially in children, affecting their cognitive activities and behavioral patterns. They further mentioned that adults exposed to copious amounts of lead would experience reproductive issues, kidney damage, and hypertension. Lead-free solders mandated by the Restriction of Hazardous Substances (RoHS) Directive since 2006 have reduced lead content in newer electronics, but older devices remain significant sources. Jorfi et al. (2024) mentioned that the level of lead in blood samples of exposed employees in Southern Iran was 24 μg/dL, exceeding CDC reference levels of 5 μg/dL.

Mercury (Hg) is present in various electronic components, including fluorescent back lights for LCD screens (2–5 mg Hg per lamp), compact fluorescent lamps (CFL, 3–5 mg per bulb), mercury-wetted relay switches in older appliances (100–3000 mg per switch), and alkaline cell batteries (20–30 mg/cell) (Floyd et al., 2002). An estimated 5.7Tonnes of mercury were generated from global e-waste in 2022 (Forti et al., 2024). Mercury is a strong neurotoxin that poses serious impairment to the central nervous system, kidneys, and immune system. Methylmercury, created by the methylation of mercury in water bodies, bioaccumulates in fish and poses risks to human consumers (Chen et al., 2019). Minute levels of mercury can contaminate large volumes of water, making its release from e-waste particularly problematic.

Cadmium (Cd) is mainly found in rechargeable nickel-cadmium (Ni-Cd) batteries (10–15%), pigments from colored plastics (0.01%–0.1%), semiconductors (CdTe in solar panels), and phosphorus coatings in CRT screens. Soil cadmium concentrations at e-waste sites reach 5–50 mg/kg compared to background levels of 0.1–1 mg/kg (Blumbergs et al., 2021).

Accumulation of Cadmium in human bodies results in severe poisoning that causes damage to the kidneys, osteoporosis, and bone fractures; thus, classified as a human carcinogen (Yang et al., 2025). Studies have associated Cadmium with lung cancer and pulmonary toxicity and have been known to affect the immune system (liver and reproductive function), cerebrovascular, and cardiovascular systems. This results in the disruption of calcium and blood pressure, leading to oxidative stress and DNA damage (Prasad et al., 2024). The metal accumulates in humans (with a half-life of 10–30 years), making chronic low-level exposure particularly concerning (Peana et al., 2022).

Hexavalent chromium (Cr(VI)) is used as protective anti-corrosion and decorative metal plating and certain pigments in electronics. Hexavalent chromium is highly toxic, causing respiratory problems, skin ulcers, and increased cancer risk, especially lung cancer. Workers in informal recycling operations who dismantle electronics without proper protection are exposed to chromium through inhalation of dust and direct dermal contact. Chromium (VI) is a Class I carcinogen with occupational exposure limits of 5 μg/m^3 (NIOSH, 2013), whereas the maximum concentration in potable water is 0.05 mg/L. The concentration of toxic metals in e-waste varies by device type and date of manufacture. These have been discussed as follows:

Arsenic (As) is present as Gallium Arsenide (GaAs) in semiconductors in high-frequency applications, light-emitting diodes (LEDs), and legacy wood preservatives in device casings. Arsenic concentrations in e-waste leachates reach 0.5–5 mg/L, exceeding drinking water standards by 100–1000-fold. Studies estimate that noncompliant e-waste management practices release approximately 98,000 tonnes of toxic metals yearly into the environment, including 58,000 kg of mercury, 2.3 million kg of lead, and 980,000 kg of cadmium (Forti et al., 2024).

14.1.3 Specific persistent organic pollutants in e-waste: printed circuit boards, dioxins, and brominated flame retardants

POPs present in e-waste pose long-term environmental and health challenges due to their resistance to degradation, ability to bioaccumulate in organisms, and ability to travel over long distances in the atmosphere (Sindiku et al., 2015). These characteristics permit POPs to disperse widely and affect ecosystems and populations far from their emission sources, making e-waste-derived POPs a global concern and necessitating international monitoring frameworks and coordinated policy responses (UNEP, 2020).

Polychlorinated biphenyls (PCBs) were originally utilized in dielectric fluids in transformers, hydraulic oils, and capacitors due to their excellent insulating properties and chemical stability. Even though PCB production was banned or severely restricted in many countries by the 1970s–1980s, and later regulated globally under the 2001 Stockholm Convention, legacy electronic equipment continues to release these compounds during disposal and recycling (Forti et al., 2024). PCBs are grouped as likely human carcinogens associated with immunological, neurological, and endocrine disruption. Their lipophilic nature enables accumulation in fatty tissues, leading to biomagnification through food chains. PCB concentrations in e-waste-contaminated soils range from 100 to 10,000 ng/g dry weight (Işıldar and van de Vossenberg, 2019).

Dioxins and furans, particularly polychlorinated dibenzo-p-dioxins (PCDDs) and polychlorinated dibenzofurans (PCDFs), are inadvertently produced during combustion processes, including the burning of e-waste and certain recycling operations. They are among the most lethal compounds, with 2,3,7,8-tetrachlorodibenzo-p-dioxin (TCDD) categorized as a Group 1 carcinogen by the International Agency for Research on Cancer. Exposure to dioxins has been associated with reproductive and developmental problems, immune system damage, hormonal interference, and cancer (WHO, 2024). Informal e-waste burning, practiced in many developing regions for metal recovery, represents a significant source of dioxin emissions.

Brominated flame retardants (BFRs), predominantly polybrominated diphenyl ethers (PBDEs), have been extensively incorporated into electronic products to meet flammability standards. BFRs represent a significant class of persistent and toxic contaminants in e-waste, with concentrations in plastic reaching up to 100,000 mg/kg (Liu et al., 2024).

The most prevalent BFRs include polybrominated diphenyl ethers (PBDEs), tetrabromobisphenol A (TBBPA), and hexabromocyclododecane (HBCD), each with unique applications and environmental behavior.

These chemicals are present in cables, circuit boards, protective casings, and plastics to reduce fire risk. Yet, during the processing of e-waste, it was noticed to leach into the soil in huge quantities. PBDEs and thyroid hormones are structurally the same; consequently, they can disrupt endocrine function, affecting growth, development, and metabolism (Renzelli et al., 2023). Studies have revealed that PBDE accumulation in human breast milk, blood, and adipose tissue, with further exposure linked to neurodevelopmental effects in children, including reduced IQ and attention deficits (Moon and Waqar 2025).

The toxicity levels of emissions originating from e-waste are increased by the thermal processing of e-waste, which converts PBDEs into more hazardous polybrominated dibenzo-p-dioxins and dibenzofurans (PBDD/Fs). These chemicals have half-lives ranging from years to decades in soils, posing long-term contamination that impacts present and future generations (Kumar et al., 2024).

Table 14.1 describes some of the commercial PBDE mixtures, dominant congeners, primary applications, and typical concentrations (La Guardia et al., 2006). The concentrations of PBDE in e-waste plastics range from 1000 to 100,000 mg/kg, with the highest ranges in televisions and computer casings. Depending on the degree of bromination and environmental conditions, their half-lives range from 1 to 15 years (Miao et al., 2023).

Tetrabromobisphenol A (TBBPA) is the most extensively used BFR in the world (170,000 tonnes/year). It is mostly incorporated into epoxy resins for PCB laminates at 15%–30% concentration (EFSA et al., 2024). Even though TBBPA exhibits low persistence (halflife in soil: 2 months), it results in the disruption of thyroid hormones at nanomolar concentrations (Miao et al., 2023).

Hexabromocyclododecane (HBCD) is widely used as an additive flame retardant in extruded and expanded polystyrene insulation foams in

TABLE 14.1 **Commercial PBDE mixtures, dominant congeners, primary applications, and typical concentrations.**

PBDE mixture	Predominant congener(s)	Primary application	Typical concentration (%)
Penta-BDE	BDE-47, BDE-99, BDE-100	Polyurethane foams	10–18
Octa-BDE	BDE-183	ABS plastics (electronic housings)	12–18
Deca-BDE	BDE-209 (>97%)	High-impact polystyrene (HIPS)	10–30

Source: From La Guardia, M. J., Hale, R. C., & Harvey, E., 2006. Detailed polybrominated diphenyl ether (PBDE) congener composition of the widely used penta-, octa-, and deca-BDE technical mixtures. Environmental Science & Technology 40 (20), 6247–6254. https://doi.org/10.1021/es060630m.

refrigeration equipment at 2.5%–5% concentration. Three diastereomers (α-, β-, γ-HBCD) exhibit differing bioaccumulation potential, with α-HBCD predominant in biota despite γ-HBCD dominance in commercial products (Marques and Cairrao, 2023).

These compounds exhibit multiple toxic endpoints, such as neurodevelopmental toxicity, which has been reported for BDE-47 and BDE-99. It is reported to impair synaptic plasticity, reduce dendritic arborization, and decrease learning performance in animal models at doses of 0.3–3 mg/kg/day (Dorman et al., 2018).

Thyroid disruption is a shared mechanism across BFRs, competitively binding thyroid hormone receptors and disrupting thyroid hormone synthesis, transport, and metabolism (Miao et al., 2023). Reproductive toxicity is manifested as altered sex hormone levels, decreased sperm count, and disrupted estrous cycling. Hence, it has been classified as a Group 2 possible human carcinogen based on animal studies (ESFA, 2024).

14.2 Critical raw materials and the incentive for recovery

Due to their economic significance, supply risk, and limited substitutability, strategic metals found in electronic devices are categorized by the European Union as critical raw materials (CRMs). According to Forti et al. (2024), the economic worth of the materials in e-waste produced in 2022 was estimated to be USD 91 billion. Table 14.2 shows the distribution of material composition and economic value of E-waste.

Despite this, in 2022, only about 22.3% of e-waste was formally documented as collected and recycled, with 62 billion dollars' worth of recoverable resources unaccounted for (Forti et al., 2024). Existing hydrometallurgical and pyrometallurgical recycling methods remain remarkably inefficient at retrieving CRMs from e-waste, capturing less than 1% of the demand for rare earth elements (REEs). While cobalt recovery is comparatively higher at 50%–60%, rates for other essential materials remain surprisingly low, with Tantalum recovery hovering between 20% and 30%, Indium at less than 10%, and Gallium at less than 5% (Baldé et al., 2024).

The concentration of CRMs in e-waste often exceeds that in primary ores by 10–100-fold, rendering e-waste "urban mines." This inability to capture a worthwhile value leads to significant economic waste, as billions of dollars' worth of refined, high-tech materials are discarded annually instead of being recirculated back into production. Therefore the burden is left on the manufacturers to bear the financial problem of purchasing expensive virgin resources, effectively ignoring the lucrative potential of the urban mine present in discarded electronics.

The data (Table 14.3) below provides a comparison of material density, signifying the higher concentration of high value and critical materials found withine-waste compared to natural primary ore deposits. Precious metals

TABLE 14.2 Economic value and material composition of e-waste.

Category	Metal/ element(s)	Value (USD)	Mass/ concentration	Applications	Notes
Precious metals	Gold	15 billion	320 tonnes	–	≈10% of annual gold mining
Precious metals	Silver	8 billion	7500 tonnes	–	–
Precious metals	Palladium	5 billion	140 tonnes	–	–
Precious metals	Platinum	3 billion	20 tonnes	–	–
Base metals	Copper	19 billion	2.1 million tonnes	–	–
Base metals	Aluminum	4 billion	1.8 million tonnes	–	–
Base metals	Iron/Steel	2 billion	18 million tonnes	–	–
Critical raw materials— REEs	Nd, Dy, Pr	–	200–2000 mg/kg	HDD magnets, speakers, vibration motors	–
Critical raw materials REEs	Y, Eu	–	50–500 mg/kg	Phosphors for LEDs, displays	–
Critical raw materials— PGMs	Ru, Rh	–	10–100 mg/kg	HDD platters, contacts	–
Critical raw materials PGMs	Pd	–	100–500 mg/kg	MLCC capacitors	–

Technology metals	Tantalum (Ta)	–	500–2000 mg/kg	Tantalum capacitors	–
Technology metals	Cobalt (Co)	–	5–20 wt%	Lithium-ion batteries	–
Technology metals	Indium (In)	–	100–300 mg/kg	LCD screens (ITO)	–
Technology metals	Gallium (Ga)	–	10–100 mg/kg	LEDs, PV cells	–
Technology metals	Germanium (Ge)	–	1–10 mg/kg	Fiber optics, IR optics	–

Source: From Forti et al. (2024); Bandara et al. (2016); Marra et al. (2019).

TABLE 14.3 Concentration of critical materials in primary ores vs. electronic waste (e-waste).

Material	Primary ore grade	E-waste concentration	Enrichment factor
Gold	1–5 mg/kg	100–1000 mg/kg	100–200x
Copper	5000–10,000 mg/kg	100,000–200,000 mg/kg	10–20x
Silver	50–200 mg/kg	3000–5000 mg/kg	20–60x
Palladium	1–10 mg/kg	100–500 mg/kg	50–100x
Cobalt	1000–3000 mg/kg	50,000–200,000 mg/kg	30–80x
Tantalum	100–300 mg/kg	500–2000 mg/kg	3–10x

Source: Modified from Işıldar, A., van de Vossenberg, J., Rene, E. R., van Hullebusch, E. D., & Lens, P. N., 2019. Two-step bioleaching of copper and gold from discarded printed circuit boards. Waste Management 57, 149–157.

exhibit the most dramatic variance; for example, palladium and gold concentrations in e-waste are 100 and 200 times more than those typically found in natural earth deposits. Even harmful critical battery metals like cobalt appear in vastly greater quantities, with enrichment factors reaching as high as 80x compared to primary sources. Enrichment Factor specifies that processing a kilogram of discarded electronics yields exponentially more usable material than mining the equivalent weight of natural ore, highlighting the inefficiency of current extraction models that ignore this resource.

Since the e-waste "urban mine" yields far better returns per ton than primary mineral mining, this disparity in concentration provides a strong financial incentive for the development of effective recovery technology. To specifically target and extract these high-value, low-abundance metals, biological approaches such as bioleaching are being sought after (Işıldar and van de Vossenberg, 2019).

14.3 Environmental and human health impact of e-waste recycling

The widespread practice of improper disposal and informal recycling of e-waste generates huge contamination that affects both land and water ecosystems. Contamination is distributed through various pathways, including direct leaching from atmospheric deposition originating from open burning operations, surface runoff from recycling centers, groundwater infiltration, and municipal landfills (Li et al., 2020). The result is high concentrations of toxic heavy metals and POPs that cause potential risks to human health and the balance of the ecosystem.

14.3.1 Soil contamination

Soil contamination in e-waste recycling hotspots often reports metal concentrations exceeding natural background levels by factors ranging from 10 to over 1000 (Etude et al., 2025). Detailed studies conducted at Guiyu, China, recycling site that processes approximately 1.5 million tonnes annually, document extreme contamination levels (Lin et al., 2022).

Specifically, soil lead concentrations have been reported between 2000 and 20,000 mg/kg, drastically surpassing the regulatory residential limit of 400 mg/kg. Similarly, copper levels range from 5000 to 50,000 mg/kg, while cadmium and chromium concentrations show substantial elevations relative to background concentrations. Spatial distribution analysis confirms that these contamination gradients can extend up to 2 km from the initial dismantling sites, exhibiting an exponential decay in concentration as a function of distance (Zhang et al., 2012). The persistence and mobilization of these metals in the topsoil allow for their introduction into the terrestrial food chain through plant uptake, leading to bioaccumulation in local fauna and exerting significant ecological toxicity on critical soil microorganisms.

14.3.2 Persistence of organic pollutants and metal bioavailability

Apart from heavy metals, soils adjacent to e-waste facilities are severely contaminated with BFRs, crucial components used in plastics and circuit boards. Total Polybrominated Diphenyl Ethers (PBDEs) have been detected at concentrations up to 500 ng/g dry weight, far exceeding background levels of 0.1–5 ng/g. These BFRs exhibit high persistence, with environmental half-lives ranging from 1 to 15 years, and they preferentially partition into soil organic matter (K_{oc} ranging from 10^5 to 10^7) (Kuppan et al., 2024a). In addition to that, metal speciation studies reveal that a significant portion of the heavy metals, specifically 40%–60% of lead and cadmium, exist in bioavailable forms (exchangeable and carbonate bound fractions), suggesting a high potential for environmental and biological uptake. In waterlogged soils, the ratio of simultaneously extracted metals to acid volatile sulfide [SEM]/[AVS] often exceeds 1.0, indicating acute potential toxicity.

This high uptake translates into significant adverse human health outcomes, particularly among workers and vulnerable populations like mothers and children. Epidemiological evidence from e-waste hotspots, such as Guiyu, steadily reveals elevated body levels of heavy metals and BFRs in exposed populations, leading to severe health consequences (Parvez et al., 2021). Li et al. (2020) also mention in their study that higher concentrations of total PBDEs and PCBs in dust sampled from Guiyu were significantly higher than those sampled from Haojiang. Likewise, the World Health Organisation documented a strong correlation between prenatal and childhood exposure and neurodevelopmental deficits such as lowered IQ scores and increased risk of attention deficit/hyperactivity disorder (ADHD) linked to elevated blood lead levels (WHO, 2021). Furthermore, exposure is frequently associated with adverse reproductive outcomes, for example, stillbirths, low birth weight, and reduced neonatal growth due to the increased levels of contaminants like PBDEs and cadmium (Singh et al., 2021).

14.3.3 Water contamination

The proliferation of electronic devices, coupled with their increasingly short lifespans, has led to a massive accumulation of discarded electronics in landfills and open dumps worldwide. When e-waste is disposed of in these facilities, it releases a complex mixture of hazardous substances, with heavy metals representing the most persistent and concerning contaminants (Adu and Aneke 2025). These pollutants are primarily transmitted through leachate, the liquid that percolates through waste materials, extracting and concentrating toxic compounds. Heavy metals such as Cd, Pb, and Hg in landfill leachate are of particular concern due to their ability to migrate into surrounding soils and water bodies (Nascimento et al., 2024). These metals, originating from various e-waste components as mentioned earlier (circuit boards, batteries, LCD screens, and cathode ray tubes), leach out under the influence of precipitation, biological activity, and chemical reactions within the waste matrix.

The extent of this contamination is demonstrated by field measurements showing that leachates from e-waste dumpsites contain elevated concentrations of heavy metals exceeding regulatory limits, as seen in Table 14.4. Once released, these contaminants do not remain localized. Groundwater monitoring at ewaste sites reveals contamination plumes extending 1–5 km downgradient from source areas, with temporal persistence exceeding 10–20 years post-closure

TABLE 14.4 Comparative levels of heavy metals in e-waste leachate relative to WHO drinking water standards

Contaminant	E-waste leachate concentration	Drinking water standard (WHO)	Exceedance factor
Lead (Pb)	50–500 mg/L	0.01 mg/L	5000–50,000x
Cadmium (Cd)	5–50 mg/L	0.003 mg/L	1700–17,000x
Copper (Cu)	100–1000 mg/L	2 mg/L	50–500x
Chromium (Cr)	10–100 mg/L	0.05 mg/L	200–2000x
Mercury (Hg)	0.5–5 mg/L	0.006 mg/L	80–800x
Arsenic (As)	0.5–10 mg/L	0.01 mg/L	50–1000x

Source: From Adu,J.T.,Aneke, F.I.,2025. Evaluation of heavy metal contamination in landfills from e-waste disposal and its potential as a pollution source for surface water bodies. Results in Engineering 25, 104431. https://doi.org/10.1016/j.rineng.2025.104431 and Nascimento, G. de A., de Souza, R. G., Fiore, F. A., de Jesus, A. M. D., 2024. Leachate analysis of trace metals from e-waste disposed of in landfills and open dumps. Cleaner Engineering and Technology 21, 100771. https://doi.org/10.1016/j.clet.2024.100771.

due to slow desorption from soil matrices (Lu et al., 2022). This long-term persistence underscores the chronic nature of the threat and the difficulty of remediation once contamination has occurred.

Biogeochemical processes control the movement of heavy metals through groundwater. Natural mechanisms can trap contaminants through sulfide precipitation in low-oxygen zones (pH 6–8, Eh <-200 mV), attachment to clays and iron/manganese minerals, calcite co-precipitation at high pH, and bonding with organic matter (Koc-Jurczyk and Dybowska, 2020). These retention mechanisms are conditional, as contamination resumes when pH drops below 5 (from acid rain or bacterial activity) or rises above 9, particularly mobilizing amphoteric metals like lead and aluminum that were previously immobilized. The consequences of groundwater contamination from e-waste extend beyond subsurface environments. When contaminated groundwater discharges into surface water bodies or when leachate directly enters streams and rivers, aquatic ecosystems face significant threats. Surface water contamination affects aquatic organisms through direct toxicity at concentrations exceeding species-specific LC_{50} values (lethal concentration for 50% of population), compromising biodiversity and disrupting ecological balance (Ahamad et al., 2024). Furthermore, these contaminated water sources may serve as drinking water supplies or irrigation sources for agricultural communities, creating direct pathways for human exposure (Trotta et al., 2024).

The pervasive environmental contamination observed at e-waste sites is governed by the conditional stability of the contaminants within the soil and sediment matrices. While heavy metals are initially subject to natural sequestration mechanisms—such as adsorption to clay minerals and Fe/Mn oxyhydroxides, co-precipitation with calcite, and complexation with dissolved organic matter this stability is highly pH-dependent. Mobilization of these contaminants can occur rapidly and uncontrollably under acidic conditions ($pH < 5$) or specific alkaline conditions ($pH > 9$) for amphoteric metals like lead (Li et al., 2020). This mobilization poses a direct threat to aquatic ecosystems, resulting in surface water contamination that affects organisms through direct toxicity at concentrations exceeding LC_{50} values, chronic exposure effects on reproduction and growth, and dangerous food web bioaccumulation and biomagnification (Kuppan et al., 2024b; Etude et al., 2025). Ultimately, the low recovery rates of valuable materials, coupled with the severe environmental and health consequences of improper disposal, underscore the urgent global need for comprehensive and technologically advanced e-waste management policies.

14.4 Bioaccumulation and biomagnification of e-waste contaminants in food chains

The release of POPs and heavy metals from discarded e-waste poses a huge risk to human food security and global ecosystem health due to the combined processes of bioaccumulation and biomagnification. Bioaccumulation is the progressive increase in the concentration of a contaminant within an organism, in comparison to its concentration in the surrounding environment, mostly due to uptake exceeding elimination (Kuppan et al., 2024a). This process initiates the contaminant's entry into the trophic structure. Subsequently, biomagnification describes an important process where the concentration of a contaminant increases at successively higher trophic levels within a food chain, often leading to significantly elevated and toxic levels in top predators and, eventually, in humans (Fioravanti et al., 2025).

14.5 The dynamics of bioaccumulation dynamics

The level of bioaccumulation is often quantified using the bioconcentration factor (BCF). The BCF is the ratio of the contaminant's concentration in an organism's tissue to its concentration in the ambient water or surrounding medium at steady state, as described by the equation:

$$\mathrm{BCF} = \frac{[\mathrm{Contaminant}]_{\mathrm{organism}}}{[\mathrm{Contaminant}]_{\mathrm{water}}}$$

A study focusing on polybrominated diphenyl ethers (PBDEs) frequently found in e-waste plastics reveals that the BCF depends on the level of bromination. Lower-brominated congeners, BDE-47 and BDE-99, exhibit higher BCF values, typically in the range of "$10^{\wedge 4}$ - $10^{\wedge 5}$". Conversely, higher brominated congeners, including BDE-183 and BDE-209, demonstrate relatively lower BCFs, ranging from "$10^{\wedge 3}$ to $10^{\wedge 4}$". Since lipophilicity increases accumulation, lipid-normalized values are much higher, typically around 10^5–10^6 L/kg lipid. (Lebbie et al., 2021).

The dynamics governing bioaccumulation are dictated by the interplay of several key kinetic rates: the uptake rate (k^1, typically 100–10,000 L/kg/day for highly lipophilic compounds), the elimination rate (k_2, typically 0.001–0.1 day^{-1}), and the biotransformation rate ($k_{\_m}$, typically 0.0001–0.01 day^{-1}). This kinetic relationship is summarized by:

$$BCF = \frac{k_1}{k_2 + k_m}$$

The observation that higher-brominated PBDEs, despite their greater lipophilicity, have lower BCFs can be attributed to their larger molecular size. This physically hinders their ability to cross biological membranes, as a result impeding uptake and leading to a low accumulation factor (Kuppan et al., 2024a).

14.5.1 Biomagnification through trophic levels

Biomagnification is quantitatively measured using the *biomagnification factor (BMF)*, defined as the ratio of the contaminant concentration in a predator to that of its prey:

$$BMF = \frac{[\text{Contaminant}]_{\text{predator}}}{[\text{Contaminant}]_{\text{prey}}}$$

Research conducted in aquatic ecosystems contaminated by e-waste has consistently demonstrated stark evidence of biomagnification. Contaminant concentrations in apex predators can be 10,000 to 100,000 times higher than those found in the ambient water. This is particularly critical for neurotoxic heavy metals such as mercury (Roy, 2024). According to research, the concentration degradation for mercury in an aquatic food web reveals this drastic danger starting with an ambient water concentration that exceeds the FDA levels of 0.3 mg/g (Lebbie et al., 2021; Kumsopar et al., 2025).

14.5.2 Agricultural soil-crop transfer

Another major dietary exposure pathway outside of aquatic systems is the movement of heavy metals from agricultural soils contaminated by e-waste to food crops. These contaminants are transferred to plants primarily through root uptake, a process governed by the transfer factor (TF). This is the ratio of metal concentration in the plant to its concentration in the soil.

$$TF = \frac{[\text{metal}]_{\text{plant}}}{[\text{metal}]_{\text{soil}}}$$

This is highly variable and depends on a complex array of factors such as the metal's speciation and bioavailability; soil properties (pH, clay content, organic matter, and cation exchange capacity); and the specific plant species and cultivar (Kuppan et al., 2024a).

In regions severely affected by informal e-waste recycling, such as Guiyu (China), Agbogbloshie (Ghana), Seelampur (New Delhi), Bandung (Indonesia), and many more, reports have suggested that metal concentrations have most often surpassed regulatory safety limits. For example, the mean of Pb in polished rice peaked at 0.69 $\mu g\ g^{-1}$, which was 3.5 times higher than the MAC (0.20 $\mu g\ g^{-1}$) in a recycling area in southeast China (Fu et al., 2008). In another study at a Pd/Zn smelter in central China, the average concentrations in rice were Cd ~ 0.35 mg/kg and lead at 0.06 mg/kg, with 70% of rice samples exceeding the national cadmium limit and 14% exceeding the lead limit (Hu et al., 2021). For leafy vegetables, mean concentrations were Cd at 0.54 mg/kg and Pb at 0.42 mg/kg, also exceeding national standards.

14.6 Health risks associated with informal recycling practices

Informal e-waste recycling practices, predominantly carried out in low- and middle-income countries without environmental regulations or adequate worker protection, pose acute and chronic health hazards to workers and nearby communities. Exposure to toxic substances occurs through multiple pathways, including inhalation of contaminated dust and fumes, consumption of contaminated water and food, and dermal contact with hazardous materials. These exposures lead to serious health outcomes, including neurological damage, respiratory diseases, carcinogenic effects, and developmental disorders, particularly affecting vulnerable populations such as children and pregnant women who are often involved in these activities (Grant et al., 2013).

The high-risk activities conducted in this sector include manual dismantling using hammers and chisels, which generates dust loaded with heavy metals such as lead and cadmium, as well as BFRs. Open burning of plastic-coated cables to recover copper at temperatures of 850°C–1100°C releases highly toxic byproducts like dioxins, furans, Polycyclic aromatic hydrocarbons (PAHs), and fine particulate matter (Kerekes et al., 2020). Acid leaching utilizes corrosive agents such as nitric acid and aqua regia to extract precious metals, resulting in the emission of toxic nitrogen oxides and corrosive acid mists (Halli et al., 2017). Additionally, heating and melting of components using propane torches to recover solder, which contains up to 37% lead, causes the volatilization and inhalation of toxic metal fumes (Farrell et al., 2023).

The environmental contamination resulting from these practices extends beyond immediate workers to affect surrounding communities through soil and water pollution, creating a broader public health crisis in areas where informal e-waste recycling is concentrated. Heavy metals accumulate in locally grown vegetables and livestock, entering the food chain and exposing residents who have no direct involvement in recycling activities. Children are particularly vulnerable to these environmental exposures due to their closer proximity to the ground, hand-to-mouth behaviors, and developing organ systems that are more vulnerable to toxic damage.

The improper disposal of printer toners releases carbon black nanoparticles and toxic organic compounds, leading to widespread soil and air contamination. This contamination results in dietary exposure levels that exceed safe intake thresholds by 2–10 times for heavily exposed populations (Kuppan et al., 2024a). Globally, informal e-waste recycling affects 18 million children and 12.9 million women, with over 100 million people employed in this hazardous sector worldwide (WHO, 2024; UNEP, 2025). The problem is mainly severe in Africa, where less than 1% of the 18 million tonnes of e-waste generated is formally recycled, with the remainder handled predominantly by informal sectors (Forti et al., 2024).

14.7 Occupational hazards and public health burden

The global e-waste crisis has evolved into a serious public health emergency, unduly affecting vulnerable populations in low- and middle-income countries where informal recycling operations predominate. According to evaluations of the World Health Organization (WHO, 2021), over 12.9 million women and 18 million children worldwide have been impacted by e-waste exposure, and over 100 million people are employed in the informal recycling sector (WHO, 2021; UNEP, 2025). These workers, habitually working without proper protective equipment, are prone to chronic exposure to a combination of harmful contaminants, flame retardants, POPs, and other hazardous substances that are released during their routine recycling processes. From acute toxicity symptoms like respiratory distress, dermatological problems, and neurological impairment to chronic diseases like cancer, reproductive dysfunction, and permanent organ damage, the effects on occupational health are severe and varied (See Table 14.5). Kids of parents who work or live near these e-waste processing sites especially face distressing consequences, as their developing bodies absorb toxins more readily and suffer uneven impacts on physical growth, immune system functioning, and cognitive development (Xu et al., 2015; Grant et al., 2013)

The economic and social burden of improper e-waste management spreads far beyond immediate health impacts, creating cascading effects that undermine development goals and prolong the succession of poverty and disease. Annual healthcare expenses linked to poor e-waste management methods surpass USD 78 billion globally, putting a major drain on healthcare systems that are frequently already under-resourced (UNEP, 2020). Communities heavily impacted by e-waste processing operations experience reduced life expectancy of 2–5 years compared to unexposed populations, alongside elevated rates of stillbirths, congenital abnormalities, and developmental disorders that reflect the teratogenic and mutagenic properties of e-waste contaminants (Grant et al., 2013; Heacock et al., 2016). The health problems associated with contamination by e-waste are manifested across generations due to maternal-fetal transfer. Pregnant women working in or residing near e-waste sites accumulate fatal contaminants in placental and breast milk tissues, thereby creating pathways for prenatal and postnatal exposure that compromise infant health from the earliest stages of development (Zheng et al., 2019). These multi-generational impacts create a devastating legacy that extends the public health consequences of today's e-waste practices decades into the future.

Characterizing the set of proteins and their alterations triggered by e-waste agents provides a foundation for discovering critical junctures for medical remedies. Furthermore, it facilitates the elaboration of heightened screening protocols for populations that have been exposed, and furnishes underlying biological data crucial for establishing both worker protection criteria and clean-up priorities.

14.8 Overview of conventional e-waste management

Conventional e-waste processing has evolved over several decades to address the dual challenges of resource recovery and hazardous waste management,

TABLE 14.5 Documented health impacts of E-Waste workers.

Health impact category	Specific effects	Quantitative findings	Associated exposures	References
Respiratory effects	Reduced pulmonary function	FEV1 decreased by 10%–30% FVC reduced by 0.75 L	Particulate matter (PM2.5: 50–500 μg/m³), metal dust, combustion byproducts	Wachinou et al. (2022); Vimercati et al. (2017)
	Chronic respiratory conditions	25% prevalence of lung function disorders; increased rates of chronic bronchitis and asthma	Bioaerosols, bacterial endotoxins, and organic dust	Wachinou et al. (2022)
	Respiratory symptoms	Significantly higher prevalence of chest tightness (21% vs 2.1%) and breathlessness (6.8% vs 1.4%)	Mixed exposures to metals and organic compounds	Vimercati et al. (2017)
Neurological effects	Cognitive impairment in children	IQ scores reduced by 5–10 points in exposed populations	Lead (blood levels 9.4–15.3 μg/dL), mercury, manganese	Banks et al. (1997); Grant et al. (2013)
	Attention and behavior disorders	2.4-fold increase in ADHD prevalence; behavioral abnormalities and attention deficits	Lead, cadmium, polybrominated compounds	Parvez et al. (2021)
	Neurodevelopmental delays	Reduced cognitive and language scores; impaired sensory integration and olfactory memory	Heavy metals, PAHs, PCBs	Zhang et al. (2017); WHO (2021)

Endocrine disruption	Thyroid dysfunction	Altered TSH, T3, and T4 levels; disrupted thyroid hormone homeostasis	PBDEs, organophosphate esters, PCBs	WHO (2024)
	Reproductive hormone alterations	Elevated FSH (4.2 ng/mL), LH (6.1 ng/mL), and testosterone (520 ng/dL) in exposed workers	Flame retardants, heavy metals (lead, cadmium, mercury)	
	Increased thyroid disease prevalence	Higher rates of thyroid disorders and altered hormone ratios	Dioxins, PCBs, brominated compounds	WHO (2024)

TABLE 14.6 Comparative performance of biological and conventional metal recovery methods.

Category	Biological methods	Conventional methods
Energy consumption	90%–95% reduction	Baseline (high temperature required)
CO_2 emissions	80%–90% reduction	Baseline (significant emissions)
Capital investment	30%–50% lower	Baseline
Operating costs	20%–40% lower	Baseline
Operating conditions	Ambient temperature/ pressure	High temperature/pressure
Toxic by-products	None (no dioxins/furans)	Dioxins and furans are produced
Material selectivity	High selectivity for specific metals	Limited selectivity
REE recovery	Recovered effectively	Lost to slag

Source: From Ayilara and Babalola, 2023; Kaushal and Makarani, 2024; Kuppan et al., 2024a.

employing mechanical pre-treatment followed by pyrometallurgical or hydrometallurgical recovery processes that dominate industrial scale e-waste recycling. These established technologies have achieved substantial industrial deployment globally, with major facilities processing hundreds of thousands of tons of e-waste annually through integrated treatment chains designed to maximize metal recovery while minimizing environmental releases. However, these conventional approaches present significant limitations due to environmental impact, energy efficiency, and recovery selectivity, particularly for complex electronics containing numerous valuable metals at trace concentrations distributed throughout heterogeneous material matrices (Cui and Zhang, 2018). The inherent challenges of conventional processing stem from the chemical complexity of modern electronics, which may contain over 60 different elements, including precious metals, rare earth elements, base metals, and hazardous substances, all intimately mixed in compact devices designed for functionality rather than recyclability (Hagelüken and Corti, 2010). Table 14.6 describes a summary of the performance of Biological and Conventional Metal Recovery Methods. As e-waste volumes continue their exponential growth trajectory and device complexity increases with advancing technology, the limitations of conventional processing approaches have become increasingly apparent, driving research into alternative biological and biochemical methods that may complement or enhance existing technologies.

Mechanical processing represents the essential first stage of conventional e-waste recycling, involving sequential physical separation operations designed to fractionate complex e-waste streams into relatively homogeneous material categories amenable to subsequent refining processes. The mechanical treatment process typically begins with manual dismantling to remove hazardous components that require specialized handling, like mercury-containing lamps, capacitors, batteries, and other priority materials that could compromise downstream processing or create safety hazards. Following selective dismantling, the remaining material undergoes shredding and size reduction to achieve particle sizes typically in the 10–50 mm range, which facilitates subsequent liberation of individual material components and enables effective application of physical separation techniques (Chancerel et al., 2009). The crushed material then flows through a series of separation stages: magnetic separation efficiently removes ferrous metals; eddy current separation recovers nonferrous metals, including aluminum and copper; density-based separation using air classification or water tables differentiates materials based on specific gravity; and electrostatic separation exploits conductivity differences to separate mixed plastics and fine metal fractions. Despite the sophistication of modern mechanical processing lines, recovery efficiencies vary considerably across material types: ferrous metals achieve recovery rates of 85%–95%, aluminum recovery reaches 70%–85%, and copper recovery typically falls in the 60%–80% range, while precious metals experience substantial losses to fine fractions, with recovery efficiencies often limited to 30%–50% during mechanical processing alone (Khaliq et al., 2014). These mechanical processing limitations, particularly for precious metals and specialty materials present at low concentrations, necessitate subsequent refining through pyrometallurgical or hydrometallurgical routes.

Pyrometallurgical processing employs high-temperature smelting operations conducted at temperatures typically ranging from 1200–1400°C to recover metals through controlled oxidation-reduction reactions in molten phase systems that enable efficient metal concentration and separation. The predominant pyrometallurgical route for e-waste processing involves copper smelting, where PCBs and other metal-rich e-waste fractions serve as supplementary feed materials to primary copper production, with precious metals reporting to the copper metal phase for subsequent electrolytic refining and precious metal recovery (Hagelüken and Corti, 2006). Alternative pyrometallurgical routes include secondary lead smelting for cathode ray tube processing and an integrated precious metal refining infrastructure that focuses on valuable e-waste materials. While pyrometallurgical processing achieves high recovery efficiencies for precious metals exceeding 95% in optimized integrated smelter-refinery operations, these processes present significant environmental and economic challenges including enormous energy consumption (5–10 GJ per ton of e-waste processed), substantial greenhouse gas emissions, generation of toxic slag requiring disposal, destruction of plastic fractions that could potentially be recycled, and emission of hazardous air pollutants including dioxins, furans, and heavy metals that require sophisticated air pollution control systems (Cui and Anderson, 2016; Schluep et al., 2009). The high capital costs, large minimum economic scale, and environmental footprint of pyrometallurgical processing create barriers to deployment in developing regions where much of the world's e-waste is generated, motivating research into alternative processing technologies, including biological and biochemical approaches that operate under ambient conditions with potentially lower environmental impacts.

14.9 Bioremediation technologies for metal recovery

Bioremediation and bioprocessing utilize microbial metabolism for sustainable metal recovery and detoxification from e-waste at ambient conditions (Ayilara and Babalola, 2023; Kuppan et al., 2024a). Four primary biohydrometallurgical approaches exist: bioleaching (microbial oxidation via Acidithiobacillus species producing soluble metal forms), biosorption (passive binding to microbial biomass with 50–500 mg/g capacity), bioaccumulation (active intracellular uptake through metallothionein expression), and bioprecipitation (microbial catalysis forming metal sulfides or phosphates). Additionally, microbial consortia employ dehalogenases, dioxygenases, and hydrolases to degrade BFRs and plastic polymers. Proteomics integration enables molecular-level optimization of bioprocesses and identification of novel degradation pathways (Lopez and dos Santos Silva, 2023; Kaushal and Makarani, 2024).

14.10 Advantages of biological methods over conventional technologies

Biological approaches demonstrate superior performance across environmental, economic, and technical dimensions compared to traditional pyrometallurgical and hydrometallurgical processes. As shown in Table 14.6, biological methods achieve 90%–95% reduction in energy consumption and 80%–90% reduction in CO_2 emissions while eliminating toxic by-products, that is, dioxins and furans. Economic advantages include 30%–50% lower capital investment and 20%–40% reduced operating costs compared to conventional technologies. Bio-based methods provide simultaneous detoxification and metal recovery while processing low-grade materials unsuitable for conventional techniques, offering enhanced scalability and recovery of trace valuable elements, including REEs that are typically lost to slag in pyrometallurgical processes.

14.11 The global scenario of e-waste generation and management

E-waste characterizes one of the most unrelenting environmental challenges of the 21st century, highlighting the interconnection between consumerism, resource sustainability, and technological advancements. The proliferation of digital devices across all parts of society, ranging from watches, smartphones, laptops, and tablets to household and industrial equipment, has created an unprecedented stream of waste characterized by both hazardous components and recoverable materials of value (Forti et al., 2020). This two-fold nature positions e-waste management at the nexus of environmental protection, resource conservation, and circular economy principles. The world faces a growing burden to shift from linear "take-make-dispose" models towards more sustainable systems that emphasize reuse, refurbishment, and responsible recycling (Awasthi et al., 2019). To truly understand global e-waste trends, we must analyze the regulatory frameworks, economic incentives, and social dynamics that drive management practices across diverse geographical and developmental contexts, moving beyond mere quantification.

14.11.1 E-waste volume and trends (quantification)

The quantification of e-waste presents significant methodological challenges due to inconsistent definitions, varied collection systems, and substantial informal sector activity that often escapes official monitoring (Parajuly et al., 2017). Nevertheless, systematic efforts to measure and track e-waste flows have improved in recent years, providing more reliable data for scientists and stakeholders. These quantification efforts reveal that e-waste generation overtakes both building adequate infrastructure and the rise in population, thus widening the gap between waste production and responsible disposal capacity (Baldé et al., 2020). Management efforts are therefore made more tedious by the compositional complexity of e-waste, containing approximately 1000 different compounds, including toxic materials like lead, mercury, and BFRs, in addition to valuable resources like gold, silver, and copper (Shittu et al., 2021).

The rapid obsolescence and prevalent adoption of electric and electronic equipment (EEE) have established waste electrical and electronic equipment (WEEE), commonly known as e-waste, as the fastest-growing solid waste stream globally (Baldé et al., 2020). Understanding the scale and composition of this waste stream is the critical first step in developing effective management strategies.

14.11.1.1 Global e-waste generation: recent statistics and growth rates

Forti et al. (2020) estimated that world e-waste generation reached 53.6 million metric tonnes (Mt) in 2019, with projections signifying a growth to 74.7 Mt by

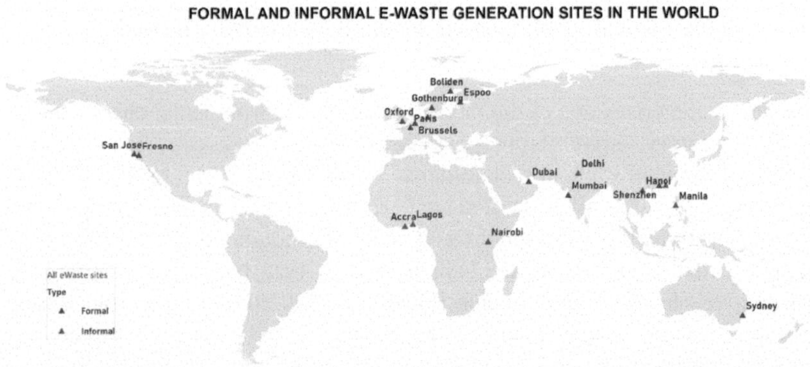

FIGURE 14.1 **Formal and i nformal e-waste generation sites in the world.** The figure is generated from a GIS tool using the various location points of formal and informal. Note: Map lines delineate study areas and do not necessarily depict accepted national boundaries.

2030. This is a 40% rise over a decade. Globally, the average per capita generation is estimated to be 7.3 kg; however, this figure masks important regional variations. The compound annual growth rate of 3%–5% significantly exceeds population growth rates, driven by three primary factors: increasing market penetration of electronic devices in developing economies, accelerating product obsolescence cycles (particularly in consumer electronics), and the proliferation of Internet of Things (IoT) devices (Awasthi et al., 2019). Small equipment and small IT represent the fastest-growing categories, reflecting the ubiquity of smartphones, tablets, and wearable devices (Baldé et al., 2020). Critically, documented collection and recycling rates remain dismally low; only 17.4% of 2019 e-waste was officially collected and recycled, meaning that approximately 44 Mt of valuable resources and hazardous materials were discarded through informal channels, landfilled, or incinerated without proper treatment (Forti et al., 2020). This collection gap represents not only a hazard to the environment but a lost economic opportunity valued at approximately $57 billion in raw materials (Shittu et al., 2021).

The geographic distribution of e-waste management shown in Fig. 14.1 highlights the global disparity in handling e-waste, a critical environmental and health challenge. It shows that formal e-waste generation and recycling sites are primarily concentrated in developed economies such as San Jose/Fresno and Hamburg, where structured, regulated recycling systems are in place. On the other hand, the sites associated with informal e-waste generation and recycling are scattered across developing regions, including Patnais, Ghana, and Manila. For example, Agbogbloshie in Ghana has become known worldwide internationally recognized for its large-scale informal e-waste operations, where workers employ rudimentary techniques, in combination with acid leaching and open burning, for the extraction of precious metals that pose higher risks to human and environmental health (UNEP, 2020). It is evident that the burden of managing obsolete electronics unduly falls on areas with informal sectors, repeatedly linked to the world trade and dumping of e-waste from more affluent nations (Schluep et al., 2009).

14.11.2 Future e-waste projections and compositional shifts

As stated by Forti et al. (2020), the proliferation of smart devices and the increased consumption habits are expected to drive the annual creation of e-waste to roughly 75 million metric tonnes by 2030. The Internet of Things (IoT) uprising has brought about billions of small scale electronic devices: smart wearables, sensors, and networked appliances into global markets. Even though each device or item is marketed as lightweight, IoT devices present formidable collection challenges due to their distributed nature, short functional lifespans, and fast obsolescence cycles. The constituent profile of e-waste is shifting dramatically from bulky cathode ray tube monitors and large computing equipment toward smaller, more complex devices with intricate material compositions.

Lithium-ion batteries represent a particularly critical growth category, intensifying rapidly owing to the increased usage of electric vehicles and portable electronics. These batteries pose fire hazards during collection and processing and require specialized recycling infrastructure to recover valuable materials by the least harmful methods, like Lithium and Cobalt (Harper et al., 2019). Small electronic devices, toys, remotes, cables, and sensors comprise an unequally large share of future waste volumes despite being the most challenging group to efficiently collect. Recycling challenges are aggravated by the drive toward miniaturization, increased material complexity, and proprietary manufacturing techniques.

14.11.3 International regulatory frameworks on e-waste

The primary international framework governing the movement of hazardous waste is the Basel Convention (1989) on the Control of Transboundary Movements of Hazardous Wastes and their Disposal, which classifies e-waste as potentially hazardous and requires prior informed permission from importing governments. Adopted in 1995, but it went into effect in 2019, the Ban Amendment barred the export of hazardous waste from Annex VII countries, mostly members of the EU and the OECD, to non-Annex VII countries. This amendment aims to stop the dumping of toxic waste in nations with inadequate environmental management capabilities. However, enforcement remains problematic as a loophole that allows exporters to classify end-of-life electronics as "reusable goods" rather than "waste," cleverly evading restrictions (Puckett and Byster, 2005). Officials at the various ports in these low-income countries lack the expertise to verify the functioning of these shipments, thereby resulting in the continual influx of obsolete or nonworking electronics. Studies estimate that 50%–80% of exported "used electronics" are actually nonfunctional waste (Bisschop, 2012).

14.11.4 E-waste management challenges

In spite of the advances in regulation, there are still many barriers that impede effective e-waste management. Low formal collection rates(17.4% in 2019 only) translate to massive economic losses (Forti et al., 2020). E-waste comprises substantial concentrations of CRMs, including gold, palladium, copper, indium, and cobalt. The Global E-waste Monitor estimates unrecovered materials constitute an "urban mine" valued at approximately $57 billion annually (Forti et al., 2020). This material loss continues the dependence on environmentally destructive primary mining and exposes economies to supply chain vulnerabilities and price volatility.

The informal sector processes an estimated 60%–90% of e-waste in many developing countries, achieving high recovery rates for certain components through labor-intensive manual disassembly but employing environmentally catastrophic methods, including acid baths and open burning (Baldé et al., 2020).

Technological limitations further constrain formal recycling systems. Modern electronics possess extreme miniaturization, complicated material integration, proprietary fastenings, and extensive adhesive use that are not compatible with automated high-throughput disassembly. Material composites, particularly flame-retardant-laden plastics and rare earth element combinations that require sophisticated separation techniques. Current pyrometallurgical and hydrometallurgical processes achieve high recovery rates for precious metals but often fail to economically extract specialty materials present in low concentrations (Hagelüken and Corti, 2010). AI-powered sorting systems, advanced automated disassembly robotics, and novel chemical separation methods are crucial to closing these technological gaps and enabling comprehensive material recovery from increasingly complex devices.

14.12 Fundamentals of proteomics in environmental microbiology

The understanding of the entire protein profile (the **proteome**) of an organism, or a complex community of organisms, provides a critical link between genomic potential and active cellular function (Brown, 2002). In the context of environmental microbiology and bioremediation, proteomics is an indispensable tool for deciphering the molecular mechanisms employed by microbes to sense, respond to, and ultimately detoxify pollutants.

14.13 The proteomic toolkit for bioremediation studies

Proteomic methodologies allow researchers to simultaneously quantify and characterize thousands of proteins, offering a dynamic view of the cellular machinery under specific environmental conditions.

14.13.1 Overview of proteomic methodologies

The core of proteomic analysis involves separating complex protein mixtures, identifying individual proteins, and quantifying their changes in abundance (Gobena et al., 2024). This is carried out in three ways as follows.

14.13.1.1 Two-dimensional polyacrylamide gel electrophoresis

This is the most basic technique that separates proteins based on the isoelectric point (pI) in the first dimension, followed by separation based on molecular weight (M_r) in the second dimension (Carrette et al., 2006). Although lower-throughput, it remains valuable for visualizing changes in the most abundant proteins and detecting protein isoforms or posttranslational modifications (PTMs) via spot shifting.

14.13.1.2 Liquid chromatography–mass spectrometry/mass spectrometry

This is the prevailing method as LC-MS/MS offers high sensitivity and throughput. Proteins are characteristically digested into peptides, separated by high-performance liquid chromatography (HPLC), and then introduced into a mass spectrometry (MS) for mass-to-charge ratio (m/z) analysis (Jayathirtha et al., 2021). The MS/MS phase produces fragment ion spectra essential for peptide identification, which are then mapped back to their corresponding parent proteins. This process relies on complex algorithms to compare experimental spectra against databases, allowing for high-confidence protein inference and the characterization of post-translational modifications (Wang and Li, 2023).

14.13.1.3 Label-free quantitation

In this method, proteins are quantified by comparing the ion intensity of peptides between different samples. It is cost-effective, avoids labeling artifacts, and is highly flexible for comparisons between a large number of samples (Tanca et al., 2020).

14.14 Quantitative proteomics strategies

Quantitative proteomics relies on a myriad of labeling procedures that allow precise measurement of protein abundances across multiple biological conditions. Among the many approaches employed are chemical tagging and metabolic labeling strategies, which have been explained below:

14.14.1 Isobaric tags for relative and absolute quantitation and tandem mass tags

These tagging methods help facilitate complex peptide quantification by the attachment of mass-balanced chemical tags (Aebersold and Mann, 2016). Each tag consists of a reporter ion (used for quantitation) and a balance group (Rauniyar and Yates 2014). Samples labeled with different tags are collected before MS analysis. Upon fragmentation in the MS/MS step, the identical parent peptides release distinct reporter ions, enabling the concurrent quantification of proteins over multiple conditions in a single run (Paulo and Gygi, 2020). This multiplexing capability has made tandem mass tag (TMT) and isobaric tag for relative and absolute quantitation (iTRAQ) central to high-throughput comparative proteomics.

14.14.2 Stable isotope labeling with amino acids in cell culture (SILAC)

This metabolic labeling technique incorporates *heavy* isotope-labeled amino acids (^2H-, ^13C-, ^15N-) into the proteome of one sample (e.g., the control culture) and *light* (unlabeled) amino acids into the other (e.g., the stressed culture). During synthesis, heavy and light versions of the same peptide are mixed, run together, and appear as distinct doublets in the mass spectrum, with the ratio of the heavy/light intensity providing the relative protein abundance (Zhang et al., 2023). SILAC is particularly suited to metabolically active, pure cultures such as mammalian cells or microbial systems because complete isotope incorporation requires several cell doublings (Aebersold and Mann, 2016).

14.14.3 Challenges in sample preparation for environmental matrices

The success of any proteomic study hinges on high-quality protein extraction, which is particularly challenging for environmental samples. This subsequently affects bioinformatic analysis and functional interpretation of the resulting mass spectrometry (MS) data. Addressing these challenges is paramount for translating spectral information into meaningful biological context regarding microbial community function and stress response. The challenges have been described as follows:

14.14.4 Biomass and protein extraction

Environmental matrices (e.g., soil, sediment, wastewater) present a big challenge because of the high concentrations of interfering substances such as humic acids, polysaccharides, and metals. These are readily coextracted with proteins and inhibit downstream enzyme digestion and MS analysis (Ore et al., 2023). Optimized lysis protocols (e.g., bead-beating, sonication) and protein precipitation/clean-up methods (e.g., TCA/acetone precipitation) are essential to maximize protein yield and purity (Wang and Li, 2023).

14.14.5 Low cell density

Groundwater or oligotrophic surfaces are characteristic of low cell density, especially those actively involved in bioremediation and therefore require efficient cell recovery and concentration techniques before lysis (10–100 mg of protein) for subsequent MS analysis (Neset et al., 2022). Approaches like tangential flow filtration (TFF) for large water volumes, centrifugation, or filtration onto membranes (e.g., polyethersulfone) followed by scraping or direct lysis on the filter (Agrawal et al., 2023). The choice of method must balance recovery efficiency with the potential for cell damage or introduction of foreign contaminants.

14.14.6 Data analysis and protein identification

Raw MS data must be processed through bioinformatics pipelines to convert spectral information into a meaningful biological context. The core step in data processing involves determining peptide sequences by matching the experimental MS/MS spectra against large protein sequence databases using specialized search algorithms (e.g., MASCOT, SEQUEST, X!Tandem) (Kalhor et al., 2024)

Database specificity: For environmental, specifically metaproteomic samples where the community composition is unknown and highly diverse, a species-specific genome database or a metagenome-assembled genome (MAG) database is not just beneficial, but crucial (Nebauer et al., 2024). Depending solely on general, non-redundant databases such as UniProt, NCBI leads to poor matching rates because the sequence variations of uncultivated environmental microbes are not represented. A custom database derived from the sample's corresponding metagenome increases the confidence and coverage of peptide-to-protein assignments (Mehta et al., 2023).

False discovery rate (FDR): The confidence of protein identification is maintained by controlling the FDR, typically set at 1%, through the use of *decoy databases*. A decoy database of either reversed or randomized sequences is appended to the true (target) database. The FDR is estimated as the ratio of peptides matched to decoy sequences versus those matched to target sequences, ensuring statistical rigor in the identification process (Lin et al., 2022). New methods are continuously being developed to better estimate and control FDR in highly complex metaproteomic datasets (Nebauer et al., 2024).

14.14.7 Functional annotation and pathway mapping

The function of differentially expressed proteins has to be determined after identification, as follows.

Functional annotation: Proteins are assigned functional categories using resources like the Gene Ontology (GO) terms or KEGG Orthology (KO) groups. This provides a high-level view of which cellular processes (e.g., metabolism, transport, stress response) are affected by the environmental disturbance (Sequeira et al., 2022). A high-level picture of the impacted cellular processes (such as metabolism, transport, and stress response) can be obtained by using tools like MicrobeAnnotator, which can combine data from several reference databases to give thorough and high-throughput functional annotation (Ruiz-Perez et al., 2021).

Pathway mapping: To visualize the identified proteins within known metabolic and regulatory pathways, tools such as KEGG Mapper or KEGG Charter are used (Sequeira et al., 2022). This helps pinpoint the precise enzymatic steps in the degradation pathways that are up- or down-regulated in response to pollutants, which serves as an important factor in understanding bioremediation (Kanehisa and Sato, 2020).

14.14.8 Validation and confirmation techniques

Though MS provides quantitative data, orthogonal validation is often necessary for high-impact findings. For instance, Western blotting uses specific antibodies to confirm the presence and quantify the relative abundance of a target protein, thereby serving as an independent validation of MS results, whereas parallel reaction monitoring (PRM)/selected reaction monitoring (SRM)utilizes a triple quadrupole mass spectrometer to monitor and quantify a small, preselected set of peptides with high sensitivity and precision. They are invaluable for confirming the expression changes of key proteins involved in a bioremediation pathway (Barkovits et al 2021).

14.14.9 Proteomics for microbial responses to heavy metal stress

Proteomics is a powerful method for capturing rapid, integrated cellular responses to stress, providing molecular evidence of adaptation strategies, particularly when cells are exposed to potent toxins such as heavy metals (chromium, cadmium, arsenic, etc.). Proteomic analysis is reliable for revealing vital cellular defense strategies, such as the oxidative stress response and the heat shock response (HSR). The proteome typically shows over-expression of antioxidant enzymes, including superoxide dismutase and catalase, along with proteins involved in glutathione metabolism, to counteract oxidative stress induced by heavy metal-mediated production of reactive oxygen species (ROS). Simultaneously, the HSR is characterized by the upregulation of guides like GroEL, DnaK, which is often activated to refold damaged proteins and maintain proteostasis (Boopathy et al., 2022).

14.14.9.1 Identifying enzymes and proteins involved in toxin degradation

The primary goal of bioremediation research is to identify the specific enzymes that catalyze pollutant degradation, and proteomics plays a major role in unveiling these catalytic drivers. By monitoring protein expression shifts under exposure to pollutants, proteomic analyses can reveal newly expressed or strongly induced enzymes (hydrolases, oxygenases, and reductases) that appear only when a particular pollutant (such as crude oil or plastics) is present. For instance, the upregulation of mono- and dioxygenases or other hydrocarbon-degrading enzymes provides compelling evidence of an active metabolic degradation pathway (Tanca et al., 2020). Such differential protein expression not only clarifies which biochemical routes are engaged but also helps distinguish true biodegradation from mere abiotic transformation. Furthermore, integrating these proteomic signatures with metabolite profiling strengthens causal inference regarding pathway activation, thereby enhancing the reliability of bioremediation assessments. Collectively, this approach accelerates the identification of functional biomarkers and candidate enzymes for engineered or enhanced biodegradation systems.

14.14.9.2 The importance of posttranslational modifications

PTMs are chemical changes that occur on a protein after it has been synthesized, acting as rapid regulatory switches to monitor protein activity, stability, and localization.

PTMs as rapid regulatory switches (phosphorylation and acetylation): Phosphorylation (adding of a phosphate group) is an extensive switch, often regulating signal transduction pathways that sense environmental changes. Acetylation (adding of an acetyl group), particularly on lysine residues, is also a critical regulator of metabolic enzymes and gene expression (Rizo and Encarnación-Guevara, 2024). These modifications often occur much faster than changes in protein abundance.

Redox modifications and stress sensing in contaminant exposure: In response to oxidative stress, proteins are often subjected to redox modifications, such as the reversible formation of disulfide bonds by cysteine oxidation. Identifying these modified proteins can pinpoint the central players in the microbial stress-sensing machinery that respond instantaneously to contaminant-induced toxicity.

14.15 The role of metaproteomics

Metaproteomics analyzes the complete functional protein complement of environmental microbial communities at a given time, bridging taxonomic identification (metagenomics/16 S rRNA) and functional outcomes (metabolomics).

14.15.1 Mapping microbial communities in contaminated sites

Metaproteomics identifies microorganisms actively expressing bioremediation proteins in contaminated environments like e-waste sites. It reveals functional redundancy where multiple species express identical degrading enzymes, enhancing community resilience (Hicks et al., 2021). Environmental sample complexity requires standardized field sampling (sample volume, collection timing) and rapid protein preservation (immediate liquid nitrogen freezing) to prevent degradation and accurately capture the in-situ proteome.

14.15.2 Assessing functional potential versus taxonomic diversity

Metaproteomics distinguishes between community genetic potential and actual protein expression activity. Correlating functional protein abundance (heavy metal resistance proteins, contaminant hydrolases) with environmental parameters (pollutant concentration, pH, redox potential) identifies active bioremediation biomarkers (Tanca et al., 2020). Integrating metaproteomics with metabolomics provides a comprehensive functional context, confirming that expressed enzymes produce predicted metabolites and actively drive detoxification processes.

14.16 Proteomics-driven remediation strategies for e-waste

The exponential growth of e-waste presents both environmental hazards from toxic heavy metals and lost economic opportunities from unrecovered precious metals. While conventional pyrometallurgical and hydrometallurgical recycling methods require high amounts of energy and cause pollution, bio-based technologies propose greener alternatives. Proteomics provides direct functional readouts of microbial enzymatic machinery under operating conditions, revolutionizing bioleaching, detoxification, and biodegradation strategies for e-waste treatment.

14.16.1 Bioleaching and metal recovery

Bioleaching utilizes acidophilic, chemolithotrophic microorganisms to solubilize metals through biogenic sulfuric acid and oxidizing agents like ferric iron. Proteomics enables mapping of quantitative protein networks responsible for metal mobilization beyond simple strain identification. In *Acidithiobacillus ferrooxidans*, the model bioleaching organism, quantitative proteomic investigations using isobaric tagging (iTRAQ, TMT) have identified significant upregulation of rus operon proteins under e-waste exposure, including rusticyanin, cytochrome c552, and outer membrane cytochrome c Cyc2, which directly interfaces with extracellular iron substrates (Dopson et al., 2023). Comparative proteomics between planktonic and sessile biofilm cells reveals distinct proteomes, with sessile cells overexpressing glucosyltransferases crucial for extracellular polymeric substance production and contact leaching (Li et al., 2022).

By comparing high-efficiency and wild-type strain proteomes, researchers identify metabolic bottlenecks. When Cyc2 protein abundance limits iron oxidation rates, genetic engineering can place the cyc2 gene under stronger promoters to increase metal recovery. For gold bioleaching using cyanogenic bacteria, proteomics optimizes induction conditions for glycine dehydrogenase, the key enzyme in cyanide production, maximizing lixiviant generation while maintaining cell viability (Tay et al., 2013).

14.16.2 Heavy metal detoxification

Bioleaching organisms must survive highly toxic dissolved metals, such as arsenic, lead, mercury, and cadmium. Proteomics provides direct evidence of metal resistance mechanisms through two main strategies: active efflux pumps and intracellular sequestration. Mercury exposure from e-waste components triggers massive upregulation of mer operon proteins, including periplasmic mercury binding protein (MerP), membrane transporter (MerT), and mercuric reductase (MerA), which reduces toxic Hg^{2+} to volatile elemental Hg^0 (Vakhlu et al., 2022). Similarly, arsenic resistance involves induction of the ars operon with high expression of arsenate reductase (ArsC) and arsenite efflux pump (ArsB).

For lead and cadmium, cellular defense relies on general stress responses detectable via proteomics as global upregulation of molecular chaperones (DnaK, GroEL, Hsp70) and oxidative stress enzymes (superoxide dismutase, catalase). Critical detoxification strategies include metallothionein production, cysteine-rich proteins that sequester metals through thiol group binding. Comparative proteomics of biofilm communities identifies upregulation of sulfate reduction enzymes leading to bioprecipitation of insoluble metal sulfides through biomineralization (Mishra et al., 2024).

14.16.3 Biodegradation of organic pollutants

E-waste plastics contain BFRs such as polybrominated diphenyl ethers and tetrabromobisphenol A (TBBPA), which are persistent, toxic, and bioaccumulative. Functional metaproteomics of microbial communities colonizing e-waste plastics identifies specific reductive dehalogenases expressed by anaerobic bacteria and novel biphenyl dioxygenases and ring-hydroxylating dioxygenases distinct from PCB-degrading enzymes. Recent studies on TBBPA-degrading consortia identified a highly induced protocatechuate 3,4-dioxygenase, suggesting its role in aromatic ring cleavage after debromination (Zhang et al., 2023).

For polymer matrix degradation, proteomics mines secretomes of microorganisms growing on e-waste plastics. Analysis of polystyrene-fed mealworm gut microbiomes identified aryl-alcohol dehydrogenases and peroxidases attacking polymer backbones (Brandon et al., 2021). For polycarbonate, proteomics of thermophilic bacteria revealed novel esterases and lipases hydrolyzing polymers into monomeric constituents, enabling the development of enzymatic cocktails for depolymerization.

14.16.4 Engineered bioremediation systems

Proteomics guides synthetic biology by providing stoichiometric data on metabolic pathway proteins. When proteomic data reveal expression imbalances leading to toxic intermediate accumulation, genetic circuits can be designed to balance enzyme expression for robust degrader strains. For industrial bioreactors, targeted proteomic approaches using selected reaction monitoring or parallel reaction monitoring enable near real-time monitoring of biomarker proteins such as rusticyanin for iron oxidation or MerA for mercury stress (Zhan et al., 2019). This allows operators to assess biomass health and catalytic potential, with sudden protein level drops signaling inhibition and prompting operational adjustments before reactor failure, establishing proteomics as a key enabler for next-generation smart e-waste treatment facilities. The illustration describes the electron transport pathway in *Acidithiobacillus ferrooxidans*, showing how electrons move from Fe(II) through a series of key proteins: Cyc2, rusticyanin, Cyc1, the bc1 complex, and lastly the aa3-type terminal oxidase before reducing oxygen. In Fig. 14.2 proteomic analyses are

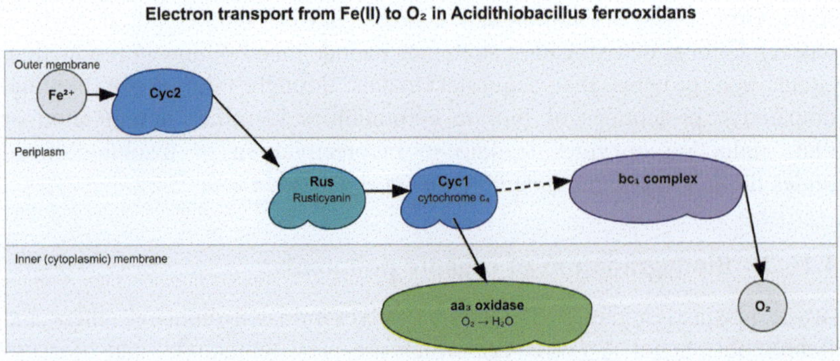

FIGURE 14.2 Electron Transport from Fe(II) to 02 in Acidithiobacillus ferrooxidans. Electron transport mechanism from Iron II to Oxygen in Acidithiobacilus ferrooxidans.

important in identifying and confirming the presence of these proteins, revealing their abundance and interactions within the cell. Proteomics reveals the molecular machinery that enables A. ferrooxidans to survive in acidic, metal-rich environments by identifying the pathway's protein composition. This method finds new proteins and regulatory mechanisms that would be challenging to deduce from genetics alone, as well as validating biochemical theories.

The application of proteomics to e-waste bioremediation has rapidly metamorphosed from a conceptual framework to a powerful scientific tool; however, substantial challenges remain as the field advances toward industrial deployment. Despite major improvements in high-resolution mass spectrometry, the inherent complexity of microbial communities in harsh, metal-rich bioleaching environments continues to hinder comprehensive coverage of the proteome. These constraints underscore the need for more efficient analytical strategies and targeted enrichment workflows to reveal the regulatory circuitry governing microbial responses to e-waste-derived metal stress.

14.16.5 Technical limitations of current proteomics

Despite the advent of high-resolution mass spectrometry (MS), the inherent complexity of the microbial proteome, especially in the harsh, metal-rich environments typical of bioleaching, presents significant analytical hurdles. One of the primary bottlenecks is the wide dynamic range of protein abundance. In organisms such as *Acidithiobacillus ferrooxidans* or *Leptospirillum ferriphilum*, structural and housekeeping proteins (e.g., rusticyanin or chaperonins) are expressed at high levels, often masking low-abundance regulatory proteins such as transcription factors and metal-response kinases (Christel et al., 2018; Kucera et al., 2025; Mendez et al., 2023). In "shotgun" proteomics, the precursor ion selection process is stochastic; highly abundant peptides are preferentially selected for fragmentation, leaving critical regulatory low-abundance peptides undetected. Without specific enrichment strategies, such as fractionation or targeted mass spectrometry, the crucial regulatory networks controlling metal resistance and biofilm formation remain hidden.

Membrane proteins present another significant challenge, as they are the gatekeepers of bioleaching, responsible for metal efflux and influx. However, their amphipathic nature makes them infamously difficult to study. Standard lysis buffers often fail to solubilize hydrophobic transmembrane domains, while harsh detergents containing sodium dodecyl sulfate (SDS) that solubilize them effectively are incompatible with downstream MS analysis and enzymatic digestion (Danko et al., 2022). The hydrophobic domains of membrane proteins also lack the cleavage sites (arginine and lysine) required for trypsin digestion, leading to poor sequence coverage and limiting insight into mechanisms that control stress tolerance and biofilm formation.

The economic and computational barriers to high-throughput proteomics also remain substantial. Modern Orbitrap platforms generate terabytes of data weekly, requiring robust storage infrastructure and advanced bioinformatics tools for peptide identification and post-translational modification (PTM) characterization (Halder et al., 2021). Converting raw spectra into identified peptide sequences requires immense computational power, while storage of raw data for retrospective analysis incurs significant server costs and data management challenges. Furthermore, PTMs such as phosphorylation and acetylation, critical for the rapid stress response of bioleaching microbes to toxic metal pulses, frequently occur on only a tiny portion of the total protein pool (substoichiometric), making them invisible without specific enrichment methods such as TiO_2 enrichment for phosphopeptides. Different isomers of modifications can have identical masses, complicating identification and further limiting comprehensive PTM characterization.

14.16.6 Integration with multiomics approaches

Given the limitations of proteomics alone, the future of bioremediation increasingly hinges on integrated multi-omics frameworks that bridge the gap between genotype, protein expression, and metabolic phenotype. Proteomics provides a functional snapshot but must be linked with genomics, transcriptomics, and metabolomics to accurately model metabolic fluxes, identify metal-response regulatory nodes, and guide rational engineering of microbial strains for enhanced bioleaching efficiency. A major challenge in bioleaching research is the low correlation often observed between mRNA levels (transcriptomics) and protein abundance (proteomics), as post-transcriptional regulation means that high mRNA expression does not guarantee high protein activity.

Integrated pipelines enabling proteogenomic mapping and genome-scale metabolic modeling (GEMs) allow researchers to simulate microbial behavior under industrial conditions. A unified bioinformatics pipeline is needed that can map proteomic data directly back to genomic scaffolds to identify novel open reading frames (ORFs) active only under metal stress (proteogenomics). By feeding proteomic data into GEMs as constraints, limiting the flux through a reaction based on enzyme abundance, researchers can simulate how a microbe will behave under different industrial conditions, such as pH shifts or temperature spikes (Bordel et al., 2022). This enables a shift from descriptive science to predictive engineering, allowing for rational design that resolves bottlenecks in metal extraction using synthetic biology.

Metabolomics provides the final readout of cellular physiology; the "phenotype" linking protein function to actual bioremediation outcomes. Machine learning algorithms, for example, Random Forests or Neural Networks, are increasingly necessary to find non-linear correlations across heterogeneous data types, facilitating the interpretation of complex datasets to predict bioleaching performance (D'Alessandro et al., 2024). These tools can correlate gene mutations, protein abundance shifts, and metabolite changes to decreases in bioleaching efficiency, enabling predictive bioremediation strategies, including heap bioleaching and the engineering of biocatalysts for a circular economy.

14.16.7 Moving from bench to industrial scale

As the field shifts from laboratory research to pilot-scale trials, the gap between laboratory success and industrial feasibility presents mainly steep challenges for e-waste bioremediation. Transitioning from shake flasks to bioheaps or stirred-tank reactors introduces mass transfer limitations for oxygen and nutrients that are not present at the bench scale. Additional techno-economic barriers emerge, including lengthy bioleaching kinetics and the capital expenditure/Operating expenditure constraints that define the competitiveness of biohydrometallurgical operations.

For bioleaching to replace pyrometallurgy, it must be economically competitive. Initial capital expenditure for biohydrometallurgical plants is estimated between EUR 7–35 million, depending on throughput and technology (heap vs reactor). While operating expenditure is generally lower than smelting, the slower kinetics of bioleaching extend the payback period by 3–5 years. Life Cycle Assessments reveal environmental benefits but also highlight water and reagent consumption, particularly neutralizing agents like lime, as persistent hotspots (Thompson et al., 2018; updated 2024).

The documented formal collection and recycling rate remains stagnant at approximately 22.3%, emphasizing the need for accessible and standardized bioprocesses. Bioremediation technologies must be designed to integrate into the informal/formal gap, perhaps offering decentralized, low-tech solutions for developing nations.

14.16.8 Standardization and regulatory acceptance

Ultimately, regulatory acceptance, biosafety assessments, and the development of harmonized standard operating procedures (SOPs) will be critical for enabling advanced proteomics-driven bioremediation strategies to transition from bench to full-scale industrial application. Unlike chemical leaching, biological processes are dynamic. Standard inoculum preparation, toxicity thresholds, and end-point assays need to be codified to ensure reproducibility across different industrial sites.

The release of genetically modified or "trained" microbes for e-waste treatment raises biosafety concerns. Containment strategies and robust risk assessments are prerequisites for regulatory approval in the EU and North America. Public perception and reception of these technologies will also play a crucial role in their widespread adoption and implementation.

14.17 Conclusion

Proteomics serves as the crucial bridge linking the genotype of a microbe to its functional phenotype. We are able to visualize the snapshot of the entire active protein complement. This is key in e-waste bioremediation, as it identifies the specific enzymes, transporters, and stress-response proteins that are most active or inhibited during the mobilization, tolerance, and detoxification of heavy metals. Significantly, proteomics moves beyond merely observing the presence of genes and instead highlights the proteins that are actually performing the role of bioleaching or bioprecipitation.

14.17.1 Proteomics for synthetic biology in biocatalyst development

The detailed functional information provided by proteomics serves as a vital blueprint for rational strain engineering using synthetic biology. By identifying the specific limitations of wild-type strains, for example, low expression of a crucial metal-binding protein, susceptibility to protein aggregation under high metal concentrations, or insufficient transporter expression for efficient metal uptake/efflux, proteomics pinpoints the exact molecular bottlenecks. This provides an opportunity for scientists to engineer robust, optimized super-leaching microbial strains with targeted modifications that enhance the expression of key efflux pumps or integrate novel metal-scavenging protein domains. This targeted approach intensely accelerates the development of efficient and scalable biocatalysts for metal recovery.

14.17.2 Future outlook for sustainable e-waste bioremediation

The future of sustainable e-waste management is not purely technological but rather closely linked with policies, collaborations, and a country's economic standing. A shift of biometallurgy from the lab to industrial scale requires systemic change. Governments and regulatory bodies must take proactive steps to reward green technologies. This includes establishing clear policy frameworks that recognize the environmental and energy benefits of bio-hydrometallurgy compared to traditional energy-intensive smelting. Mechanisms such as subsidies, tax breaks, and favorable permitting processes must be implemented for companies adopting bio-based or mixed-approach WEEE treatment strategies. Furthermore, policies should prioritize research funding for pilot-scale bioleaching facilities to demonstrate industrial viability and economic competitiveness.

14.17.3 Need for interdisciplinary training and global data sharing

The complexity of e-waste bioremediation demands both an interdisciplinary approach and scientists. The success of this requires experts from environmental microbiology, advanced analytical chemistry (mass spectrometry for proteomics), process engineering, and bioinformatics. Governments, universities, and industry must invest in integrated training programs that produce dedicated professionals to tackle this industrial menace. In addition to this, global proteomic and genomic data from bioremediation should be available to all. This collaborative data platform will accelerate the identification of universal resistance and leaching mechanisms, preventing redundant research efforts worldwide.

References

Adu, J.T., Aneke, F.I., 2025. Evaluation of heavy metal contamination in landfills from e-waste disposal and its potential as a pollution source for surface water bodies. Results in Engineering 25, 104431. Available at https://doi.org/10.1016/j.rineng.2025.104431.

Aebersold, R., Mann, M., 2016. Mass-spectrometric exploration of proteome structure and function. Nature 537 (7620), 347–355. Available at https://doi.org/10.1038/nature19949.

Agrawal, P., Wilkstein, K., Guinn, E., Mason, M., Serrano Martinez, C.I., Saylae, J., 2023. A review of tangential flow filtration: process development and applications in the pharmaceutical industry. Organic Process Research & Development 27 (4), 571–591.

Ahamad, M.I., Yao, Z., Ren, L., Zhang, C., Li, T., Lu, H., et al., 2024. Impact of heavy metals on aquatic life and human health: a case study of River Ravi Pakistan. Frontiers in Marine Science 11, 1374835. Available at https://doi.org/10.3389/fmars.2024.1374835.

Awasthi, A.K., Li, J., Koh, L., Ogunseitan, O.A., 2019. Circular economy and electronic waste. Nature Electronics 2 (3), 86–89.

Ayilara, M.S., Babalola, O.O., 2023. Bioremediation of environmental wastes: the role of microorganisms. Frontiers in Agronomy 5, 1183691. Available at https://doi.org/10.3389/fagro.2023.1183691.

Baldé, C.P., Forti, V., Gray, V., Kuehr, R., Stevels, A., 2020. The Global E-waste Monitor 2020: Quantities, Flows and the Circular Economy Potential. United Nations University (UNU)/United Nations Institute for Training and Research (UNITAR).

Baldé, C.P., Kuehr, R., Yamamoto, T., McDonald, R., D'Angelo, E., Althaf, S., 2024. The Global E-waste Monitor 2024: Electronic Waste Rising Five Times Faster Than Documented E-waste Recycling. International Telecommunication Union (ITU) & United Nations Institute for Training and Research (UNITAR). Available at https://ewastemonitor.info/the-global-e-waste-monitor-2024/.

Bandara, H.M.D., Field, K.D., Emmert, M.H., 2016. Rare earth recovery from end-of-life motors employing green chemistry design principles. Green Chemistry 18 (3), 753–759. Available at https://doi.org/10.1039/C5GC01255D.

Banks, E.C., Ferretti, L.E., Shucard, D.W., 1997. Effects of low level lead exposure on cognitive function in children: a review of behavioral, neuropsychological and biological evidence. Neurotoxicology 18 (1), 237–281.

Barkovits, K., Chen, W., Kohl, M., Bracht, T., 2021. Targeted protein quantification using parallel reaction monitoring (PRM). Quantitative Methods in Proteomics. Springer US, pp. 145–157.

Bisschop, L., 2012. Is it all going to waste? Illegal transports of e-waste in a European trade hub. Crime, Law, and Social Change 58 (3), 221–249.

Blumbergs, E., Serga, V., Platacis, E., Maiorov, M., Shishkin, A., 2021. Cadmium recovery from spent Ni-Cd batteries: a brief review. Metals 11 (11), 1714. Available at https://doi.org/10.3390/met11111714.

Boopathy, L.R.A., Jacob-Tomas, S., Alecki, C., Vera, M., 2022. Mechanisms tailoring the expression of heat shock proteins to proteostasis challenges. Journal of Biological Chemistry 298 (5).

Bordel, S., Marín, I., Santos, M.A., 2022. Genome-scale metabolic models foD microbial communities: a review of current approaches and applications. Frontiers in Microbiology 13, 843217.

Brandon, A.M., Gao, S.H., Tian, R., Ning, D., Yang, S.S., Zhou, J., et al., 2021. Biodegradation of polyethylene and plastic mixtures in mealworms (larvae of *Tenebrio molitor*) and effects on the gut microbiome. Environmental Science & Technology 55 (9), 5843–5853. Available at https://doi.org/10.1021/acs.est.0c08032.

Brown, T.A., 2002. Transcriptomes and proteomes. Genomes, 2nd ed. Wiley-Liss Chapter 3). Available at https://www.ncbi.nlm.nih.gov/books/NBK21121/.

Carrette, O., Burkhard, P.R., Sanchez, J.C., Hochstrasser, D.F., 2006. State-of-the-art two-dimensional gel electrophoresis: a key tool of proteomics research. Nature Protocols 1 (2), 812–823. Available at https://doi.org/10.1038/nprot.2006.104.

Chancerel, P., Meskers, C.E., Hagelüken, C., Rotter, V.S., 2009. Assessment of precious metal flows during preprocessing of waste electrical and electronic equipment. Journal of Industrial Ecology 13 (5), 791–810. Available at https://doi.org/10.1111/j.1530-9290.2009.00171.x.

Chen, C.F., Ju, Y.R., Chen, C.W., Dong, C.D., 2019. The distribution of methylmercury in estuary and harbor sediments. Science of the Total Environment 691, 55–63.

Christel, S., Herold, M., Bellenberg, S., El Hajjami, M., Buetti-Dinh, A., Pivkin, I.V., et al., 2018. Multi-omics reveals the lifestyle of the acidophilic, mineral-oxidizing model species *Leptospirillum ferriphilum*. Applied and Environmental Microbiology 84 (3), e02091–17.

Cui, J., Anderson, C., 2016. Hydrometallurgical treatment of waste printed circuit boards: bromine leaching. Metals 6 (6), 150. Available at https://doi.org/10.3390/met6060150.

Cui, J., Zhang, L., 2018. Metallurgical recovery of metals from electronic waste: a review. Journal of Hazardous Materials 350, 107–120.

D'Alessandro, G., Vogel, C., Mintseris, J., 2024. Machine learning integration of multi-omics data for protein complex prediction. Molecular Systems Biology 20 (1), 52–73.

Danko, K., Lukasheva, E., Zhukov, V.A., Zgoda, V., Frolov, A., 2022. Detergent-assisted protein digestion—on the way to avoid the key bottleneck of shotgun bottom-up proteomics. International Journal of Molecular Sciences 23 (22), 13903. Available at https://doi.org/10.3390/ijms232213903.

Dopson, M., Holmes, D.S., Ñancucheo, I., 2023. Biomining of industrial metallic wastes. In: Singh, N.K., Sani, R.K. (Eds.), Extremophiles in Circular Bioeconomy. Springer, pp. 195–216. Available at https://doi.org/10.1007/978-981-19-7411-3_10.

Dorman, D.C., Chiu, W., Hales, B.F., Hauser, R., Johnson, K.J., Mantus, E., et al., 2018. Polybrominated diphenyl ether (PBDE) neurotoxicity: a systematic review and meta-analysis of animal evidence. Journal of Toxicology and Environmental Health, Part B 21 (4), 269–289.

EFSA Panel on Contaminants in the Food Chain (CONTAM), Schrenk, D., Bignami, M., Bodin, L., Chipman, J.K., Del Mazo, J., et al., 2024. Update of the scientific opinion on tetrabromobisphenol A (TBBPA) and its derivatives in food. EFSA Journal 22 (7), e8859. Available at https://doi.org/10.2903/j.efsa.2024.8859.

Etude, M.C., Njoku, C.N., Ikeuba, A.I., Uwakwem, R.O., Okoye, E.E., Maduoma, T.U., et al., 2025. The regeneration of metal components from the cathode of spent lithium-ion batteries using food wastes: a review. World Journal of Engineering. Available at https://doi.org/10.1108/WJE-12-2024-0667.

Farrell, Z.J., Jacob, A.R., Truong, V.K., Elbourne, A., Kong, W., Hsiao, L., et al., 2023. Compositional design of surface oxides in gallium–indium alloys. Chemistry of Materials 35 (3), 964–975.

Fioravanti, R., Muzzioli, L., Maurel, F., Palma, G., Calabrese, G., Angioni, A., et al., 2025. Bioaccumulation and biomagnification of mercury along the seafood chain in Europe: a systematic review. Foods 14 (21), 3752.

Floyd, P., Zarogiannis, P., Crane, M., Tarkowski, S., Bencko, V., Postle, M., 2002. Risks to health and the environment related to the use of mercury products. Final Report Prepared for The European Commission. DG Enterprise, Risk & Policy Analysts Limited.

Forti, V., Baldé, C.P., Kuehr, R., Bel, G., 2024. The Global E-waste Monitor 2024: Electronic Waste Rising Five Times Faster Than Documented E-waste Recycling. United Nations Institute for Training and Research (UNITAR) & International Telecommunication Union (ITU). Available at https://ewastemonitor.info/the-global-e-waste-monitor-2024/.

Forti, V., Baldé, C.P., Kuehr, R. and Bel, G. 2020. The Global E-Waste Monitor 2020: Quantities, Flows and the Circular Economy Potential. United Nations University (UNU)/United Nations Institute for Training and Research (UNITAR)—Co-Hosted SCYCLE Programme, International Telecommunication Union (ITU) & International Solid Waste Association (ISWA), Bonn/Geneva/Rotterdam.

Fu, J., Zhou, Q., Liu, J., Liu, W., Wang, T., Zhang, Q., et al., 2008. High levels of heavy metals in rice (*Oryza sativa* L.) from a typical e-waste recycling area in southeast China and its potential risk to human health. Chemosphere 71 (7), 1269–1275. Available at https://doi.org/10.1016/j.chemosphere.2007.11.065.

Gobena, S., Admassu, B., Kinde, M.Z., Gessese, A.T., 2024. Proteomics and its current application in biomedical area: concise review. The Scientific World Journal 2024, 4454744. Available at https://doi.org/10.1155/2024/4454744.

Grant, K., Goldizen, F.C., Sly, P.D., Brune, M.N., Neira, M., van den Berg, M., et al., 2013. Health consequences of exposure to e-waste: a systematic review. The Lancet Global Health 1 (6), e350–e361. Available at https://doi.org/10.1016/S2214-109X(13)70101-3.

Hagelüken, C., Corti, C.W., 2010. Recycling of gold from electronics: cost-effective use through 'design for recycling. Gold Bulletin 43 (3), 209–220. Available at https://doi.org/10.1007/BF03214988.

Halder, S., Ng, R., Aguilaniu, H., 2021. Toward a proteome-scale understanding of protein aging. Trends in Biochemical Sciences 46 (12), 1054–1067.

Halli, P., Hamuyuni, J., Revitzer, H., Lundström, M., 2017. Selection of leaching media for etal dissolution from electric arc furnace dust. Journal of Cleaner Production 164, 265–276. Available at https://doi.org/10.1016/j.jclepro.2017.06.212.

Harper, G., Sommerville, R., Kendrick, E., Driscoll, L., Slater, P., Stolkin, R., et al., 2019. Recycling lithium-ion batteries from electric vehicles. Nature 575 (7781), 75–86.

Hazelwood, D.A., Pecht, M.G., 2021. Life extension of electronic products: a case study of smartphones. IEEE Access 9, 144726–144739. Available at https://doi.org/10.1109/ACCESS.2021.3121733.

Heacock, M., Kelly, C.B., Asante, K.A., Birnbaum, L.S., Bergman, Å.L., Brune, M.N., et al., 2016. E-waste and harm to vulnerable populations: a growing global problem. Environmental Health Perspectives 124 (5), 550–555. Available at https://doi.org/10.1289/ehp.1509699.

Hicks, P., Gies, E.A., Jarmusch, A.K., 2021. Microbial metaproteomics: current state and future directions for environmental applications. Applied and Environmental Microbiology 87 (20), e00816–e00821.

Hu, Y., Wang, C., Song, Z., Chen, M., Ding, L., Liang, X., et al., 2021. Heavy metal in rice and vegetable and human exposure near a large Pb/Zn smelter in Central China. International Journal of Environmental Research and Public Health 18 (23), 12631. Available at https://doi.org/10.3390/ijerph182312631.

Işıldar, A., van de Vossenberg, J., Rene, E.R., van Hullebusch, E.D., Lens, P.N., 2019. Two-step bioleaching of copper and gold from discarded printed circuit boards. Waste Management 57, 149–157.

Islam, M.K., Khatun, M.S., Mourshed, M., 2024. An in-depth analysis and review of management strategies for e-waste in the south Asian region: a way forward towards waste to energy conversion and sustainability. Heliyon 10 (7).

Jayathirtha, M., Dupree, E.J., Manzoor, Z., Larose, B., Sechrist, Z., Neagu, A.N., et al., 2021. Mass spectrometric (MS) analysis of proteins and peptides. Current Protein and Peptide Science 22 (2), 92–120.

Jorfi, S., Feizi, R., Saeedi, R., Sabaghan, M., Barzegar, G., Dehghani, S.L., et al., 2024. Health risk assessment of workers exposed to lead dust in informal e-waste recycling workshops. International Journal of Environmental Health Research 34 (7), 2790–2800. Available at https://doi.org/10.1080/09603123.2023.2274380.

Kalhor, M., Lapin, J., Picciani, M., Wilhelm, M., 2024. Rescoring peptide spectrum matches: boosting proteomics performance by integrating peptide property predictors into peptide identification. Molecular & Cellular Proteomics 23 (7).

Kanehisa, M., Sato, Y., 2020. KEGG Mapper for inferring cellular functions from protein sequences. Protein Science 29 (1), 28–35.

Kaushal, R.S., Makarani, N., 2024. Advances in actinobacteria-based bioremediation: mechanistic insights, genetic regulation, and emerging technologies. Environmental Science and Pollution Research 31, 15420–15445. Available at https://doi.org/10.1007/s11356-024-32156-9.

Kerekes, Z., Restás, Á., Lublóy, É., 2020. The effects causing the burning of plastic coatings of fire-resistant cables and its consequences. Journal of Thermal Analysis and Calorimetry 139, 775–787. Available at https://doi.org/10.1007/s10973-019-08526-9.

Khaliq, A., Rhamdhani, M.A., Brooks, G., Masood, S., 2014. Metal extraction processes for electronic waste and existing industrial routes: a review and Australian perspective. Resources 3 (1), 152–179. Available at https://doi.org/10.3390/resources3010152.

Koc-Jurczyk, J., Dybowska, A., 2020. Removal of heavy metal ions from water and wastewaters by sulfur containing precipitation agents. Water, Air, & Soil Pollution 231, 503. Available at https://doi.org/10.1007/s11270-020-04863-w Article.

Kucera, J., Kremser, K., Bouchal, P., Potesil, D., Vaculovic, T., Vsiansky, D., et al., 2025. Proteomic insights into the adaptation of *Acidithiobacillus ferridurans* to municipal solid waste incineration residues for enhanced bioleaching efficiency. Journal of Proteome Research 24 (5), 2243–2255. Available at https://doi.org/10.1021/acs.jproteome.4c00527.

Kumar, P., Singh, S., Gacem, A., Yadav, K.K., Bhutto, J.K., Alreshidi, M.A., et al., 2024. A review on e-waste contamination, toxicity, and sustainable clean-up approaches for its management. Toxicology 508, 153904. Available at https://doi.org/10.1016/j.tox.2024.153904.

Kumsopar, S., Charoenpong, C., He, S., Bustamante, P., Wee, B., Wang, X., et al., 2025. Mercury trophic transfer and biomagnification in food webs within a tropical embayment as evidenced by nitrogen and carbon stable isotope analysis. Environmental Research 278, 121599. Available at https://doi.org/10.1016/j.envres.2025.121599.

Kuppan, N., Padman, M., Mahadeva, M., Srinivasan, S., Devarajan, R., 2024a. A comprehensive review of sustainable bioremediation techniques: eco friendly solutions for waste and pollution management. Waste Management Bulletin 2 (3), 154–171. Available at https://doi.org/10.1016/j.wmb.2024.07.005.

Kuppan, N., Padmavathi, P., Krishnan, A., 2024b. Bioprocessing approaches for sustainable metal recovery from electronic waste. Biomass Conversion and Biorefinery 14 (3), 2156–2174. Available at https://doi.org/10.1007/s13399-023-04892-x.

La Guardia, M.J., Hale, R.C., Harvey, E., 2006. Detailed polybrominated diphenyl ether (PBDE) congener composition of the widely used penta-, octa-, and deca-BDE technical mixtures. Environmental Science & Technology 40 (20), 6247–6254. Available at https://doi.org/10.1021/es060630m.

Lebbie, T.S., Moyebi, O.D., Asante, K.A., Fobil, J., Brune-Drisse, M.N., Suk, W.A., ... Carpenter, D.O., 2021. E-waste in Africa: a serious threat to the health of children. International Journal of Environmental Research and Public Health 18 (16), 8488. Available at https://doi.org/10.3390/ijerph18168488.

Li, J., Liang, C., Ma, C., Bao, L., Xu, M., Mao, R., et al., 2020. Heavy metal pollution in e-waste recycling sites in China. Environmental Science & Technology 54 (13), 8174–8182.

Li, Y., Zhang, X., Yin, H., Li, X., Liang, Y., 2022. Comparative proteomic analysis reveals the mechanism of enhanced chalcopyrite bioleaching by *Acidithiobacillus ferrooxidans* in the presence of cysteine. Frontiers in Microbiology 13, 894341. Available at https://doi.org/10.3389/fmicb.2022.894341.

Lin, C., Zeng, Z., Xu, R., Liang, W., Guo, Y., Huo, X., 2022. Risk assessment of PBDEs and PCBs in dust from an e-waste recycling area of China. Science of The Total Environment 803, 150016. Available at https://doi.org/10.1016/j.scitotenv.2021.150016.

Liu, M., Brandsma, S.H., Schreder, E., 2024. From e-waste to living space: flame retardants contaminating household items add to concern about plastic recycling. Chemosphere 365, 143319. Available at https://doi.org/10.1016/j.chemosphere.2024.143319.

Lopez, M.F., dos Santos Silva, E., 2023. Proteomic insights into microbial metal transformation mechanisms. Applied Microbiology and Biotechnology 107 (4), 1245–1262. Available at https://doi.org/10.1007/s00253-022-12387-9.

Lu, J., Yuan, M., Hu, L., Yao, H., 2022. Migration and transformation of multiple heavy metals in the soil plant system of e-waste dismantling site. Microorganisms 10 (4), 725. Available at https://doi.org/10.3390/microorganisms10040725.

Madani, S.S., Ziebert, C., Marzband, M., 2023. Thermal characteristics and safety aspects of lithium-ion batteries: an in-depth review. Symmetry 15 (10), 1925.

Marques, M.L., Cairrao, E., 2023. Occurrence and health effects of hexabromocyclododecane: an updated review. Toxics 11 (5), 409. Available at https://doi.org/10.3390/toxics11050409.

Marra, A., Cesaro, A., Belgiorno, V., 2019. Recovery opportunities of valuable and critical elements from WEEE treatment residues by hydrometallurgical processes. Environmental Science and Pollution Research 26, 19897–19905. Available at https://doi.org/10.1007/s11356-019-05406-5.

Mehta, S., Bernt, M., Chambers, M., Fahrner, M., Foell, M.C., Gruening, B., et al., 2023. A Galaxy of informatics resources for MS-based proteomics. Expert Review of Proteomics 20 (11), 251–266.

Mendez, R.D., Castro, S.P., Orellana, R., 2023. Proteomic analysis of low-abundance regulatory proteins in *Acidithiobacillus ferrooxidans* under varying metal concentrations. Environmental Microbiology 25 (11), 3125–3138.

Miao, B., Yakubu, S., Zhu, Q., Issaka, E., Zhang, Y., Adams, M., 2023. A review on tetrabromobisphenol A: human biomonitoring, toxicity, detection and treatment in the environment. Molecules (Basel, Switzerland) 28 (6), 2505. Available at https://doi.org/10.3390/molecules28062505.

Mishra, S., Lin, Z., Pang, S., Zhang, W., Bhatt, P., Chen, S., 2024. Recent advanced technologies for the characterization of xenobiotic-degrading microorganisms and microbial communities. Frontiers in Bioengineering and Biotechnology 9, 632059.

Modarress Fathi, B., Ansari, A., Ansari, A., 2022. Threats of internet-of-thing on environmental sustainability by e-waste. Sustainability 14 (16), 10161. Available at https://doi.org/10.3390/su141610161.

Moon, C., Waqar, M., 2025. Human health effects of polychlorinated biphenyls: from exposure pathways to clinical evidence. Proteins 12 (9).

Nascimento, G., de, A., de Souza, R.G., Fiore, F.A., de Jesus, A.M.D., 2024. Leachate analysis of trace metals from e-waste disposed of in landfills and open dumps. Cleaner Engineering and Technology 21, 100771. Available at https://doi.org/10.1016/j.clet.2024.100771.

National Institute for Occupational Safety and Health. 2013. Criteria for a recommended standard: Occupational exposure to hexavalent chromium (DHHS [NIOSH] Publication No. 2013-128, Revised with minor technical changes). Centers for Disease Control and Prevention. https://www.cdc.gov/niosh/docs/2013-128/pdfs/2013_128.pdf.

Nebauer, D.J., Pearson, L.A., Neilan, B.A., 2024. Critical steps in an environmental metaproteomics workflow. Environmental Microbiology 26 (5), e16637.

Neset, L., Takayidza, G., Berven, F.S., Hernandez-Valladares, M., 2022. Comparing efficiency of lysis buffer solutions and sample preparation methods for liquid chromatography–mass spectrometry analysis of human cells and plasma. Molecules (Basel, Switzerland) 27 (11), 3390. Available at https://doi.org/10.3390/molecules27113390.

Ore, O.T., Adeola, A.O., Fapohunda, O., Adedipe, D.T., Bayode, A.A., Adebiyi, F.M., 2023. Humic substances derived from unconventional resources: extraction, properties, environmental impacts, and prospects. Environmental Science and Pollution Research 30 (21), 59106–59127.

Parajuly, K., Habib, K., Liu, G., 2017. Waste electrical and electronic equipment (WEEE) in Denmark: flows, quantities and management. Resources, Conservation and Recycling 123, 85–92.

Parvez, S.M., Jahan, F., Brune, M.N., Gorman, J.F., Rahman, M.J., Carpenter, D., et al., 2021. Health consequences of exposure to e-waste: an updated systematic review. Lancet Planetary Health 5 (12), e905–e920. Available at https://doi.org/10.1016/S2542-5196(21)00263-1.

Paulo, J.A., Gygi, S.P., 2020. A comprehensive proteomic and phosphoproteomic analysis of yeast deletion mutants reveals novel regulatory pathways. Cell Systems 10 (5), 463–474. Available at https://doi.org/10.1016/j.cels.2020.04.005.

Peana, M., Pelucelli, A., Chasapis, C.T., Perlepes, S.P., Bekiari, V., Medici, S., et al., 2022. Biological effects of human exposure to environmental cadmium. Biomolecules 13 (1), 36.

Prasad, B., Kaur, P.S., Gupta, S., Malik, T., 2024. Environmental cadmium toxicity and its bioremediation. Cadmium Toxicity: Challenges and Solutions. Springer Nature Switzerland, pp. 275–307.

Puckett, J., Byster, L., 2005. The Digital Dump: Exporting High-Tech Trash to Asia. Basel Action Network, BAN.

Quinto, S., Law, N., Fletcher, C., Le, J., Antony Jose, S., Menezes, P.L., 2025. Exploring the e-waste crisis: strategies for sustainable recycling and circular economy integration. Recycling 10 (2), 72.

Rauniyar, N., Yates, J.R., 2014. Isobaric labeling-based relative quantification in shotgun proteomics. Journal of Proteome Research 13 (12), 5293–5309. Available at https://doi.org/10.1021/pr500880b.

Rautela, R., Arya, S., Vishwakarma, S., Lee, J., Kim, K.-H., Kumar, S., 2021. E-waste management and its effects on the environment and human health. Science of The Total Environment 773, 145623. Available at https://doi.org/10.1016/j.scitotenv.2021.145623.

Renzelli, V., Gallo, M., Morviducci, L., Marino, G., Ragni, A., Tuveri, E., et al., 2023. Polybrominated diphenyl ethers (PBDEs) and human health: effects on metabolism, diabetes and cancer. Cancers 15 (17), 4237. Available at https://doi.org/10.3390/cancers15174237.

Rizo, J., Encarnación-Guevara, S., 2024. Bacterial protein acetylation: mechanisms, functions, and methods for study. Frontiers in Cellular and Infection Microbiology 14, 1408947. Available at https://doi.org/10.3389/fcimb.2024.1408947.

Roy, T., 2024. Aquatic ecosystem toxicity and food chain. Toxicity of Aquatic System and Remediation. CRC Press, pp. 25–36.

Ruiz-Perez, C.A., Conrad, R.E., Konstantinidis, K.T., 2021. MicrobeAnnotator: a user-friendly, comprehensive functional annotation pipeline for microbial genomes. BMC Bioinformatics 22 (1), 11. Available at https://doi.org/10.1186/s12859-020-03940-5.

Schluep, M., Hagelueken, C., Kuehr, R., Magalini, F., Maurer, C., Meskers, C., et al., 2009. Recycling—From E-Waste to Resources. United Nations Environment Programme & United Nations University.

Sequeira, J.C., Rocha, M., Alves, M.M., Salvador, A.F., 2022. UPIMAPI, reCOGnizer and KEGGCharter: bioinformatics tools for functional annotation and visualization of (meta)-omics datasets. Computational and Structural Biotechnology Journal 20, 1798–1810.

Sharma, H., Rai, A.K., 2023. Bioremediation strategies for sustainable e-waste management. In: Debbarma, P., Kumar, S., Suyal, D.C., Soni, R. (Eds.), Microbial Technology for Sustainable E-Waste Management. Springer. Available at https://doi.org/10.1007/978-3-031-25678-3_5.

Shittu, O.S., Williams, I.D., Shaw, P.J., 2021. Global e-waste management: Can WEEE make a difference? A review of e-waste trends, legislation, contemporary issues and future challenges. Waste Management 120, 549–563. Available at https://doi.org/10.1016/j.wasman.2020.10.016.

Sindiku, O., Babayemi, J., Osibanjo, O., Schlummer, M., Schluep, M., Watson, A., et al., 2015. Polybrominated diphenyl ethers listed as Stockholm Convention POPs, other brominated flame retardants and heavy metals in e-waste polymers in Nigeria. Environmental Science and Pollution Research International 22 (19), 14489–14501. Available at https://doi.org/10.1007/s11356-014-3266-0.

Singh, N., Ogunseitan, O.A., Tang, Y., 2021. Systematic review of pregnancy and neonatal health outcomes associated with exposure to e-waste disposal. Critical Reviews in Environmental Science and Technology 51 (20), 2424–2448. Available at https://doi.org/10.1080/10643389.2020.1788913.

Tanca, A., Bacci, G., Palomba, A., 2020. Proteomics and metaproteomics for the assessment of microbial biodiversity and function in bioremediation. Current Opinion in Biotechnology 62, 148–154.

Tay, S.B., Natarajan, G., Rahim, M.N.B.A., Tan, H.T., Chung, M.C.M., Ting, Y.P., et al., 2013. Enhancing gold recovery from electronic waste via lixiviant metabolic engineering in *Chromobacterium violaceum*. Scientific Reports 3 (1), 2236.

Thompson, J.A., Rentsch, L., Hudson, M.J., Van Den Hende, T.A., 2018. Techno-economic assessment of hydrometallurgical processes for printed circuit board recycling. Waste Management 71, 289–302.

Trotta, V., Baaloudj, O., Brienza, M., 2024. Risks associated with wastewater reuse in agriculture: Investigating the effects of contaminants in soil, plants, and insects. Frontiers in Environmental Science 12, 1358842. Available at https://doi.org/10.3389/fenvs.2024.1358842 Article.

UNEP (United Nations Environment Programme). 2025. *As Electronic Waste Surges, Countries Look for Answers.* https://www.unep.org/news-and-stories/story/electronic-waste-surges-countries-look-answers. (accessed March 15, 2026).

United Nations Environment Programme. (n.d.) 2020. Persistent organic pollutants: A global issue, a global response. U.S. Environmental Protection Agency. https://www.epa.gov/international-cooperation/persistent-organic-pollutants-global-issue-global-response. (accessed March 15, 2026).

Vakhlu, J., Kaur, M., Bhalla, T.C., 2022. Microbial detoxification of heavy metals. Microbes and Microbial Biotechnology for Green Remediation. Elsevier, pp. 381–402.

Vimercati, L., Gatti, M.F., Gagliardi, T., Cuccaro, F., De Maria, L., Caputi, A., et al., 2017. Environmental exposure to arsenic and chromium in an industrial area. Environmental Science and Pollution Research International 24 (12), 11528–11535. Available at https://doi.org/10.1007/s11356-017-8827-6.

Wachinou, A.P., Kêdoté, N.M., Padonou, G., Adè, S., Darboux, J., Tohi, M., et al., 2022. Respiratory disorders related to e-waste exposure among workers in the informal sector in a Sub-Saharan African city: an exposed nonexposed study. Pulmonary Medicine 2022, 9968897. Available at https://doi.org/10.1155/2022/9968897.

Wang, L., Li, X., 2023. Advances in the characterization of post-translational modifications with low stoichiometry and isobaric interference using high-resolution mass spectrometry. Mass Spectrometry Reviews 42 (6), 1489–1505.

WHO (World Health Organization). 2024. Dioxins and Their Effects on Human Health. https://www.who.int/news-room/fact-sheets/detail/dioxins-and-their-effects-on-human-health. (accessed November 15, 2025).

World Health Organization. 2021. *Children and Digital Dumpsites: E-Waste Exposure and Child Health.* WHO. https://www.who.int/publications/i/item/9789240023901. (accessed November 15, 2025).

Xu, X., Yang, H., Chen, A., Zhou, Y., Wu, K., Liu, J., et al., 2015. Birth outcomes related to informal e-waste recycling in Guiyu, China. Reproductive Toxicology 33 (1), 94–98. Available at https://doi.org/10.1016/j.reprotox.2011.12.006.

Yang, Y., Hassan, M.F., Ali, W., Zou, H., Liu, Z., Ma, Y., 2025. Effects of cadmium pollution on human health: a narrative review. Atmosphere 16 (2), 225. Available at https://doi.org/10.3390/atmos16020225.

Zhan, Y., et al., 2019. Iron and sulfur oxidation pathways of *Acidithiobacillus ferrooxidans*. World Journal of Microbiology and Biotechnology 35 (60). Available at https://doi.org/10.1007/s11274-019-2632-y.

Zhang, W., Lin, Z., Wang, J., Zhou, J., 2023. Proteomic insights into the aerobic biodegradation pathway of tetrabromobisphenol A by a newly isolated bacterium *Pseudomonas* sp. strain TBBPA-1. Journal of Hazardous Materials 445, 130567. Available at https://doi.org/10.1016/j.jhazmat.2022.130567.

Zhang, W.-H., Wu, Y.-X., Simonnot, M.O., 2012. Soil contamination due to e-waste disposal and recycling activities: a review with special focus on China. Pedosphere 22 (4), 434–455. Available at https://doi.org/10.1016/S1002-0160(12)60030-7.

Zheng, L., Wu, K., Li, Y., Qi, Z., Han, D., Zhang, B., et al., 2019. Blood lead and cadmium levels and relevant factors among children from an e-waste recycling town in China. Environmental Research 165, 351–358. Available at https://doi.org/10.1016/j.envres.2018.05.006.

Chapter 15

Proteomics in health care: recent advances, challenges, and future perspective

Vinod Rishi[1], Prakash Bharti[2], and Anjani Kumar[3]
[1]*Department of Botany, Pt. Rajaram Upadhyay Mahavidhyalaya, Chitrakoot, Uttar Pradesh, India,* [2]*Department of Biological Sciences, Mahatma Gandhi Chitrakoot Gramodaya Vishwavidyalaya, Satna, Madhya Pradesh, India,* [3]*Department of Physical Sciences, Mahatma Gandhi Chitrakoot Gramodaya Vishwavidyalaya, Satna, Madhya Pradesh, India*

15.1 Introduction

Proteomics has emerged as one of the fastest-developing disciplines within the field of omics sciences, especially in therapeutic research. Proteomics was initially described as a concept by Marc Wilkins in 1995 (Agrawal et al., 2013). In the simplest terms, proteomics can be defined as the large-scale study of the protein-based aspect, including their structures, functions, interactions, and their roles in cell processes (Wilkins et al., 1996). Proteomics also provides a better insight into the mechanism of functioning of organisms on a more detailed level, compared to genomics. Yet, this issue is also more complicated since the expression of proteins is dynamic, which changes with time and other environmental factors (Holman et al., 2013). It is currently estimated that humans can generate around a million different types of proteins, and many of them are subjected to posttranslational modifications (PTMs). It is believed that the human genome only encodes approximately 26,000 to 31,000 proteins (Chandramouli et al., 2009). To investigate this remarkable proteomic diversity, numerous proteomic technologies have been developed. Conventional methods like one-dimensional (1D) and two-dimensional (2D) gel electrophoresis (2-DE) are not to be dismissed, but nowadays higher-end gel-free, high-throughput methods are commonly utilized (Chandramouli et al., 2009). Such methods encompass multidimensional methods of protein identification technology (Florens and Washburn, 2006), amino acid-coded isotope labeling in cell culture (Ong et al., 2002), Isotope-Coded Affinity Tag (ICAT), and isobaric Tag for Relative and Absolute

Proteomics Driven Bioremediation. DOI: https://doi.org/10.1016/B978-0-443-45294-9.00015-1

Quantification (iTRAQ) (Ross et al., 2004; Liu et al., 2025). Other advanced proteomics techniques are shotgun proteomics, two-dimensional differential gel electrophoresis (2D-DIGE), and protein microarray techniques, which are used to study tissues, organelles, and cells (Barderas et al., 2025a; Wolters et al., 2001; Molina-Mora et al., 2020; Cutler, 2003; Klose et al., 2002). High-resolution liquid chromatography (HR-LC) multireaction monitoring (MRM) assays are also necessary (Stahl-Zeng et al., 2007); large-scale western blot assays, and label-free quantification methods are also required to process mixed protein mixtures (Schulz et al., 2007).

The research in proteomics in recent decades has broadly been categorized into two areas, namely: protein expression mapping and protein interaction mapping (Yoithapprabhunath et al., 2015). Protein expression mapping, frequently done using 2-DE with mass spectrometry (MS), is used to measure the abundance of proteins in cells, tissues, or body fluids and to track PTMs in a variety of physiological or pathological situations. The main goal of protein interaction mapping is to establish how proteins interact, and specifically by the application of theyeast two-hybrid system in conjunction with MS to produce the interaction networks at a proteome-wide scope (Pandey and Mann, 2000). Among measures undertaken during a proteomics investigation, sample preparation is the most necessary to avoid technical oversight and to guarantee the results of a proteomics investigation are representative of a realistic biological distinction, rather than experimentally diverse. An effective preparation of the protein not only promotes effective solubilization of the protein but also eliminates the interference elements like lipids and other suppressive compounds. This is a cautious optimizing procedure to generate accurate, repeatable, and biologically useful proteomic information (Rabilloud and Lelong, 2011). Polyacrylamide gel electrophoresis (PAGE) and high-performance liquid chromatography (HPLC) are the most used methods of protein isolation and separation, yet 1-DE and 2-DE are other methods of protein separation. Proteins separated using 1-DE are categorized based on the molecular mass (Rappsilber et al., 2000). The issue of protein solubility is not a serious one since sodium dodecyl sulfate (SDS) helps solubilize the proteins. It is an easy-to-use procedure that is repeatable and appropriate for proteins that have molecular weights between 10 and 300 kDa (Graves and Haystead, 2002). It is also more appropriate when studying purified proteins because of this low resolving power. Conversely, 2-DE that separates the proteins based on charge and molecular weight is suited to more complicated mixtures like lysates of cells (Graves and Haystead, 2002).

Proteomics strengths consist of its capacity to analyze the protein expression at a range of levels, which can be qualitative and quantitative analysis of cellular responses (Xiao et al., 2008). These assessments can be performed at the transcriptomic, posttranscriptional, and genomic levels (Low et al., 2013). Qualitative proteomics is used to describe protein mixtures and monitor changes in the pattern of expression using a number of methods, including microarrays, two-dimensional liquid chromatography (2D-LC), and 2-DE (Xiao et al., 2008). These may be frequently useful in the investigation of disease mechanisms as correlations between diseased individuals and healthy controls (Xiao et al., 2008). Moreover, quantitative proteomics can give a complete understanding of the processes in the cell, the development of disease, and even biomarkers (Grønborg et al., 2006; Rappsilber et al., 2000). Contemporary quantitative proteomic methods include stable isotope labeling, or postextraction and metabolic labeling, which is often used in combination with affinity labeling (Jiang et al., 2004; Wang et al., 2007). The MS is an important proteomics analytical technique needed to analyze proteins with high accuracy since it has the capacity to identify compounds by comparing the ions based on their mass-to-charge ratios (Chandrasekhar et al., 2014). Here, we delineate a complete workflow that systematically outlines the individual steps of the proteomics methods (Fig. 15.1).

Recent technological improvements have made the use of proteomics capabilities even better. Mass spectrometers have now been made faster, more sensitive, and more resolved, to enable the identification of intricate proteoforms and low-abundance proteins (He et al., 2025). The sample preparation procedures are now more refined, preventing loss of protein and now better covering difficult categories including membrane proteins and glycoproteins (Vanderaa and Gatto, 2023). Coverage strategies are now more advanced, and the strategy to enrich particular PTMs, including phosphorylation, acetylation, and glycosylation, is now more sophisticated, allowing comprehensive mapping of regulatory networks (Liang et al., 2025).

A new radical direction that allows the large cell to be analyzed instead of a mass population is single-cell proteomics (Inayatullah et al., 2025; Petrosius and Schoof, 2023). This method has shown cell-to-cell heterogeneity of immunological responses, cancer, development, and other biological settings (Xu et al., 2024; Tan et al., 2024). Single-cell proteomics can provide hundreds or thousands of proteins per cell with a resolution of biological complexity never before achieved, through a combination of ultrasensitive mass spectrometry and novel microfluidics and imaginative sample handling (He et al., 2025). Besides the conventional research, single-cell research is transforming clinical science. Single-cell proteomics can be utilized in the treatment of infectious diseases, neurological alterations, developmental diseases, autoimmune diseases, heart diseases, and breast cancer, among others (Tan et al., 2024; Xu et al., 2024). These experimental analyses reveal immense data on cells, the communication of immune cells, and alterations in gene expression according to disease. Such developments are having an influence on personalized medicine and allow the creation of more precise and focused medications because they give an unmatched overview of the cellular processes (Momenzadeh and Meyer, 2025).

Machine learning and artificial intelligence are becoming increasingly important to proteomics research. Algorithms allow the identification of

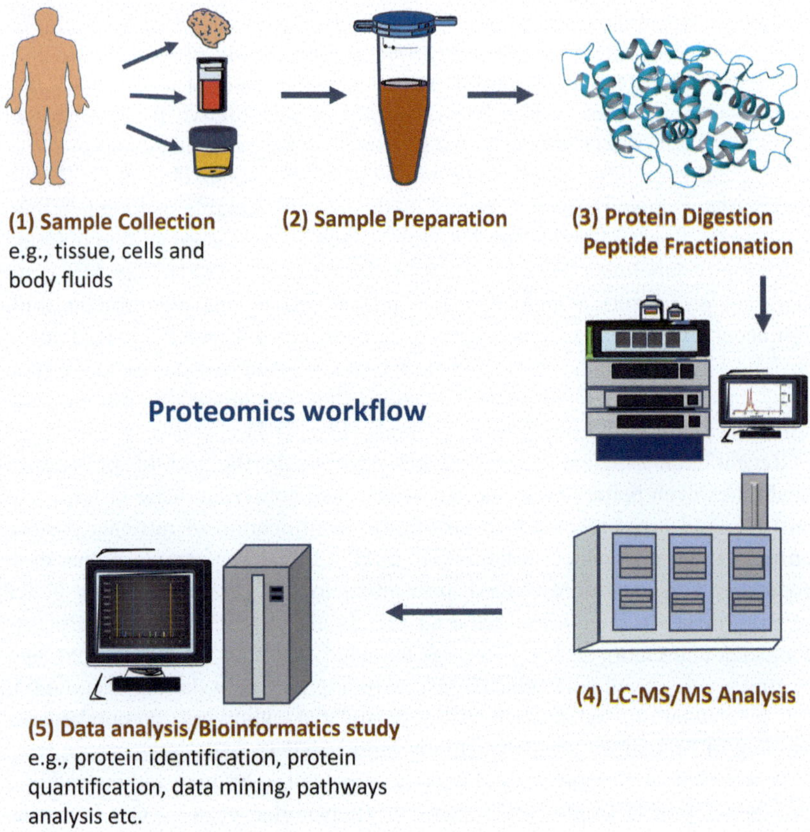

FIGURE 15.1 **Workflow of the proteomics method.** Workflow representation of proteomics techniques: (1) sample collection from tissue, cells, and body fluids; (2) sample preparation for protein analysis; (3) protein digestion; and (4–5) characterization, quantitative analysis, data mining, and pathways analysis utilizing computational techniques.

peptides, interpreting their spectra, imputing missing values, predicting PTMs locations, and the functional annotation procedures (Liang et al., 2025; Sun et al., 2025). High-performance pre-trained models enable a rapid and precise analysis of mass spectrometry data, especially when proteomics is used together with genomics, transcriptomics, and metabolomics to build complete systems-level models (Xu et al., 2024; König et al., 2024).

Despite the more recent developments, proteomics is still faced with significant challenges linked to distinguishing differentiation states and complex mechanistic interdependencies. Protein expression levels cover several orders of magnitude, hence limiting the detection of low-abundance proteins even in cases where they are functionally significant. These researches into proteoforms are still convoluted and taxing because they require these multifaceted layers of complexity to be included in an integrated framework that focuses on biological functionality (Ahmad and Budnik, 2023). The issue of reproducibility and quantification is pervasive in proteomic studies, probably due to the variability it induces in sample preparation, instrumentation work, and heterogeneity of the computational pipeline. Lastly, functional inference will always be in need of further substantiation, which impedes the attribution of biological function in an unambiguous fashion (e.g., by complementary biochemical measurements, localization experimentation, or by perturbation study) (Liu et al., 2024; Li et al., 2024).

15.2 Types of proteomics

Proteomics refers to systemic study of the proteome, the total amount of proteins that a cell can express at a particular time, and the subcellular location of interest. It goes beyond the inventory of proteins to the amount of proteins, their locations, structure (tertiary and quaternary), functions, PTMs, and their interactions with other molecules (Xiong, 2006; Aslam et al., 2017; Fathi et al., 2009).

In general, proteomic studies have been stratified into three major domains, namely, expression proteomics, structural proteomics, and functional proteomics (Ceciliani et al., 2014; Domon and Aebersold, 2006; Kwon et al., 2021). All domains, mechanistic understanding is obtained regarding the function of proteins, either in a physiological or perturbational context (Holman et al., 2013; Amiri-Dashatan et al., 2018; Graves and Haystead, 2002).

15.2.1 Expression proteomics

Expression proteomics asks a question of change in protein abundance patterns between two distinct samples of biological interest that have been subjected to a specific condition or stimulus. These are pathological conditions, pharmacological conditions, and environmental stressors. Through either whole proteome or subsets profiling, the field of expression proteomics can reveal new signaling elements or proteins that are disease-specific in disease phenotypes (Al-Amrani et al., 2021; Gobena et al., 2024). The canonical methodologies include 2D-PAGE, MS, and protein microarrays, all of which make it possible to qualitatively and quantitatively define the changes in proteomics (Al-Amrani et al., 2021; Chandrasekhar et al., 2014).

15.2.2 Structural proteomics

The study of the three-dimensional structure and even minutiae of functional proteins in a given cellular environment is assigned to structural proteomics and may also be called cellular mapping (Chandrasekhar et al., 2014). Its key aims are to map proteins inside the multi-protein complexes or organelles and also the protein–protein interfaces (Gobena et al., 2024). The information in terms of structure is indispensable in rational drug design, such as the case of the nuclear pore complex, which has provided some fundamental understanding with regard to nucleocytoplasmic transport (Rout et al., 2000). The most common modalities that will be used are X-ray crystallography and nuclear magnetic resonance (NMR) (Chandrasekhar et al., 2014).

15.2.3 Functional proteomics

Functional proteomics is a unique subdiscipline studying the functions and activities, and dynamic interactions of proteins, in living systems. Its primary objectives include defining protein involvement in biological processes, regulating catalytic activity, and the interaction with macromolecules, such as, DNA, RNA, and other proteins. The delimitation of protein–protein interactions, characterization of the PTMs, and activity-based profiling provide an integrative knowledge of not just the presence of proteins but also their location, time, and activity of the enzymatic or signaling activity. The integrative approach enables the explanation of the complicated cellular pathways and networks that normal physiological function and disease processes (Gavin et al., 2002).

Functional proteomics has vast benefits in unraveling the pathogenesis, discovery of biomarkers, and development of targeted therapeutic options. It allows the catalog of cellular processes that convert molecular perturbations into pathophysiological effects, therefore informing targeted intervention by the ability to provide an energized portrait of protein behavior. Consequently, functional proteomic study is the key nexus connecting the proteins discovery and their biological importance that will further ensure its status as a necessary tool in translational studies (Monti et al., 2005).

15.3 Conventional and advanced proteomics techniques

15.3.1 Conventional technique

Traditional or conventional proteomics methods form the basic structure of protein studies, which is based on the principle of separation and detection that is well-founded. Protein purification is often done using a combination of chromatographic modalities such as affinity chromatography, size-exchange chromatography, and ion-exchange chromatography. Complex mixtures are separated into target proteins by taking advantage of differences in fundamental physicochemical parameters, that is, net charge, hydrodynamic volume, and certain biological interactions. Analytical assays that offer accurate detection and quantification, in particular, Western blot and the enzyme-linked immunosorbent assay (ELISA), can be used to supplement these methodologies. This natural strength, broad availability, and highly structured protocols of these traditional technologies make them inevitable in the common purification procedures, diagnostic uses, and initial characterization of proteins (Table 15.1).

TABLE 15.1 Conventional analytical proteomics techniques: principles and their applications in health care and disease management.

Techniques	Principle	Clinical applications	References
IEC	Separation based on the net surface charge of proteins. Molecules are resolved by their affinity for the oppositely charged stationary phase.	Disease-associated biomarkers discovery, disease diagnosis through the separation of proteins, peptides, and amino acids, and disease progression monitoring	Aslam et al. (2017)
SEC	Separation based on the hydrodynamic volume (size) of molecules as they pass through a porous matrix. Larger molecules elute first, while smaller molecules penetrate the pores and elute later.	Ideal for purifying low molecular weight proteins and viruses, pharmaceuticals, and drug development.	Liu et al. (2020); Ramraje et al. (2020)
AC	Purification based on highly specific reversible biological interactions (e.g., antibody-antigen, enzyme-substrate) between a target protein and ligand immobilized on a stationary phase.	Immobilized metal ions to purify histidine-tagged proteins. Immunoaffinity, isolation of immune proteins, including immunoglobulins and cytokines, hormones, and enzymes, for therapeutics and disease diagnosis	Liu et al. (2020)
ELISA	A plate-based immunoassay that detects and quantifies a specific analyte. It uses an enzyme-linked antibody to generate a measurable signal (colorimetric, fluorescent) proportional to the target concentration.	Primary application of this technique: disease diagnosis, biomarker discovery, antigen determination, and drug development	Wang et al. (2025b); Bae et al. (2003)
Western Blotting	Involves separating proteins by electrophoresis, transferring them to a membrane, and detecting a specific target protein using a combination of primary and enzyme-conjugated secondary antibodies.	It is an important tool in the research and diagnosis of communicable diseases and biomarker detection.	Schulz et al. (2007); Sule et al. (2023)

AC, Affinity chromatography; *IEC*, Ion exchange chromatography; *SEC*, Size exclusion chromatography.

15.3.2 Advanced technique

Limitations that are inherent in traditional proteomics strategies have been alleviated with the emergence of advanced proteomic technologies that enhance sensitivity, throughput, and quantitative accuracy. Instrumental methods like protein microarrays provide a complete picture of proteomic activity and allow simultaneous interaction dynamics and phenotype changes of thousands of proteins to be interrogated. The MS-based methods of exhaustive quantitative comparisons, such as stable isotope labeling by amino acids in cell culture (SILAC) and iTRAQ, can also provide high-resolution three-dimensional structural data, which are invaluable to understanding protein function and guiding drug discovery. The combination of all these techniques allows a complete, systems-level dissection of the proteome, as shown in Table 15.2.

15.4 Proteomics in disease diagnosis, disease mechanism, and biomarker discovery

Proteomics assumes a important role in clinical use and pathophysiological research. By the systematic analysis of the proteomic perturbations that are evidenced in physiological and pathological conditions, the researchers and clinicians can translate their understanding of the molecular determinant of the disease. Moreover, this paradigm of the methods allows defining disease-specific proteomic signatures that may be used as new prognostic biomarkers.

15.4.1 Disease mechanism

Abnormal protein expression, disturbed functionality, PTMs, and dysregulation of protein interaction networks are key processes of the development and progression of various pathologies. Proteomic perturbations are a characteristic feature of a variety of disease conditions, which include neurodegenerative diseases and cancer. The most common type of dementia is called Alzheimer's disease (AD), a neurodegenerative disorder related to age involving the accumulation of insoluble protein aggregates, which is one of the characteristic pathological features of a wide range of neurodegenerative diseases (Tan et al., 2024; Xu et al., 2024).

Bai et al. (2013) conducted extensive research on the brain-insoluble proteome in AD and discovered 4216 proteins, with 36 having different expression. As would be expected, the study involved enrichment of amyloid-β (Aβ), tau, apolipoprotein E (APOE), complement system factors, and proteins involved in RNA splicing, phosphorylation regulation, synaptic plasticity, and mitochondrial activity, as indicated by the existing hypotheses (Bai et al., 2013). These findings support the assertion that pathological accumulation of Aβ is one of the hallmark pathological features of AD, contributing significantly to neuronal dysfunction, synaptic loss, and progressive cognitive abilities.

TABLE 15.2 Advanced analytical proteomics techniques: principles, applications, advantages, and their challenges.

Technique category	Method name	Principle	Applications	Advantages	Limitations	References
Microarray technology	Functional protein microarray	High-density spotting of full-length proteins or domains on a solid surface.	High-throughput screening of biomolecular interactions (protein, DNA, RNA, small molecule).	Enables global profiling of biochemical activities in a single experiment.	Limited to a pre-defined set of proteins selected by the researcher.	Kumar et al. (2012); Syu et al. (2020); Goshima et al. (2008); Jones et al. (2006)
	Antibody microarray	Immobilized antibodies are used to capture specific antigens from a solution.	Multiplexed detection and quantification of specific protein targets.	Multiple formats (e.g., direct, sandwich) and detection methods (fluorescence, chemiluminescence).	Dependent on the quality, specificity, and availability of antibodies.	Barderas et al. (2025b); Kingsmore and Patel (2003); More et al. (2018)
Gel-based approaches	SDS-PAGE	Denatures proteins with SDS, separating them by molecular mass via electrophoresis.	Estimating molecular weight, analyzing protein purity, and preparative separation.	High resolving power for separation by size; relatively simple and inexpensive.	Provides no charge information; denaturing conditions disrupt native structure and interactions.	Yao et al. (2001)

(*Continued*)

TABLE 15.2 (Continued)

Technique category	Method name	Principle	Applications	Advantages	Limitations	References
	2D-PAGE	Separates proteins based on isoelectric point (pI) in the first dimension and molecular mass in the second.	Mapping complex protein mixtures, identifying PTMs, and comparative analysis.	Simultaneously resolves hundreds of proteins by two independent properties.	Low reproducibility; poor sensitivity for low-abundance, very acidic/basic, hydrophobic, or extreme-sized proteins.	Berggren et al. (2000); Vercauteren et al. (2007)
	2D-DIGE	Multiple protein samples are pre-labeled with different fluorescent CyDyes and co-separated on the same 2D gel.	High-confidence quantitative comparison of protein expression across multiple samples.	Minimizes gel-to-gel variation, improving quantitative accuracy and reproducibility.	High cost of dyes; requires specialized imaging equipment; cannot label proteins lacking lysine.	Van Den Bergh and Arckens (2005); Zhou et al. (2002)
Quantitative techniques	ICAT	Chemical labeling of cysteine residues with isotopically light or heavy tags for MS-based quantification.	Comparative quantification of protein expression between two samples.	Simplifies complex mixtures by enriching for cysteine-containing peptides; good for low-abundance proteins.	Bias against proteins without cysteine; potential issues with acidic proteins; limited multiplexing.	Gygi et al. (1999); Haqqani et al. (2008)
	SILAC	Metabolic incorporation of isotopically heavy amino acids into the entire proteome during cell culture.	Accurate in vivo quantitative comparison of protein expression between cell populations.	Minimal chemical perturbation; efficient and complete labeling; simple sample mixing.	Not directly applicable to tissue samples or body fluids; requires cell culture.	Yates III (2019); Ong et al. (2002)

	iTRAQ	Multiplexed chemical labeling of peptide amines with isobaric tags that release reporter ions in MS/MS.	Simultaneous relative quantification of proteins across up to 8 (or more) samples.	High multiplexing capacity; reduces MS1 spectrum complexity.	Increased sample complexity post-digestion requires extensive fractionation; the reporter ion signal can be compressed.	Liu et al. (2025); Ross et al. (2004)
High-throughput and structural techniques	X-ray crystallography	Analysis of X-ray diffraction patterns from a protein crystal to determine electron density.	Determining high-resolution 3D atomic structures of proteins and macromolecular complexes.	Provides extremely detailed atomic-level structural information.	Requires high-quality, diffracting crystals, which can be difficult to obtain.	Ferraro and Merlino (2025); Wilkins et al. (1996)
	MS	Ionization and measurement of the mass-to-charge ratio of peptides and proteins.	Protein identification, characterization, quantification, and sequencing.	Highly sensitive and versatile; core technology for most modern proteomics; can be coupled with various separations.	Instrumentation is costly and complex; it requires expert operation and data analysis.	Sharma et al. (2021); Wilkins et al. (1996)
	NMR	Exploits the magnetic properties of atomic nuclei to deduce protein structure and dynamics in solution.	Determining 3D protein structures, studying protein folding, dynamics, and ligand interactions.	Studies proteins in near-native solution state; provides dynamic information.	Limited to smaller proteins; requires high sample concentrations and isotopic labeling for large structures.	Dubey and Grover (2001); Hu et al. (2021)

Additional comparative transcriptomic studies of the AD and control brain tissues indicated the presence of unspliced RNA species in AD, highlighting the impairment of RNA processing. Myc box-dependent interrelating protein-1, aggregated protein, and presenilin-1 were found to be dysregulated, which indicates that abnormal RNA splicing is the direct cause of AD pathogenesis. Proteomics has been advancing to multidimensional mass spectrometry, which includes imaging-based methods and is a multiomics framework combining proteomic information with genomics and metabolomics. Interestingly, RNA-dependent and RNA-independent changes in AD have been discerned through proteome-transcriptome comparative studies (Higginbotham et al., 2020; Bai et al., 2020). Numerous of the RNA-independent differentially expressed proteins, such as MDK, PTN, NTN1, SMOC1, SERP1, SLIT2, HTRA1, and FLT1, exhibit a close association between A2 burden and enrichment in amyloid plaques (Xiong et al., 2019; Esteve et al., 2019; Bai et al., 2013). Of special significance, MDK, NTN1, and SFRP1 have been shown to bind directly to A 1–40 peptide, which further implicates them in AD pathology.

Concurrently, the proteomic study has revealed that oncogenesis is also inseparably correlated with the abnormality of proteins. Proteomic profiling makes it possible to discover tumor-specific proteins and clarify which signaling pathways and biological processes play a important role in malignant transformation. Whereas oncogenes code proteins that are often overexpressed in malignant cells, tumor suppressor gene products are often missing or suppressed. Zhang et al. (2016) analyzed 174 ovarian cancer samples, 174 of which were from The Cancer Genome Atlas (TCGA), using a label-free shotgun LC-MS / MS (Zhang et al., 2016). Their work found 9600 proteins and 24429 phosphorylation sites, which therefore enable the classification of tumors into different subtypes through proteomic signatures and the characterization of functionally related cohorts of proteins. Such results highlight the ability of proteomic technologies to increase our knowledge of cancer biology and enable molecular stratification of tumors.

15.4.2 Biomarker discovery

Recently, MS-based proteomic profiling of cerebrospinal fluid (CSF) has significantly enhanced the discovery of disease-related biomarkers (Sathe et al., 2019; Wang et al., 2020; Higginbotham et al., 2020; Bai et al., 2020). Nevertheless, the excessive presence of some proteins systematically obscures the presence of low- concentration analytes, an unresolved weakness in CSF proteomics. Immune-depletion strategies targeting the most profuse proteins have thus been used to reduce this problem. As an illustration, Sathe et al. (2019) studied CSF samples of five cognitively healthy controls and five patients with AD (Sathe et al., 2019). After the depletion of the 14 most abundant proteins, they used tandem mass tag (TMT)-based liquid chromatography followed by tandem mass spectrometry (LC/LC-MS/MS) that was capable of quantifying 2327 proteins, including 139 differentially expressed (DE) proteins, including MAPT, NPTX2, VGF, GFAP, NCAM1, PKM, and YWHAG.

Higginbotham et al. (2020) added to the CSF proteome coverage in a complementary study that detected 2875 proteins in a cohort of 20 controls and 20 subjects with AD, with 528 proteins having different expression levels (Higginbotham et al., 2020). The main DE proteins included MAPT, NEFT, GAP43, FABP3, CHI3L1, NRGN, VGF, GD11, and SMOC1, most of which have been implicated in the pathophysiology of neurodegenerative disease.

Cancer proteomics has also received a lot of interest in the academic sphere and is now the better-charted disease-associated proteome. Hsu et al. (2016) used iTRAQ along with LC/MS/MS in analyzing 1763 proteins in lung cancer samples and found six putative biomarkers (ERO1L, PABPC4, RCC1, RPS25, NARS, and TARS) which were later validated by immunohistochemistry and Western blotting (Hsu et al., 2016). Similar developments have clarified the conditions of metabolic diseases; a longitudinal MS-based plasma proteomic examination of 43 obese subjects being sustainably subjected to weight loss determined 1294 plasma proteins (Geyer et al., 2016). Sixty-one proteins were found to change statistically significantly, and SERPINF1 and apolipoprotein APOF1 were the most regulated markers. All these data mutually depict the notion that long-term weight loss has a far-reaching, but personalized impact on the metabolic disease pathways, which subsequently determine the direction of treatment.

15.5 Proteomics in cancer

Proteomics has become an indispensable tool in modern cancer studies, and it can provide significant information to the tumor biology and therapy development (Hanash and Taguchi, 2011; Enroth et al., 2019). The implementation of the proteomics-based technologies has further helped identify potential biomarkers and characterize the protein expression patterns that help in classifying tumors, provide prognosis, as well as become important tools in uncovering the molecular processes involved in oncogenesis by revealing changes in the cellular signaling pathways within tumor cells (Posadas et al., 2005; Chang et al., 2017; Fig. 15.2).

This has also helped in the development of extensive cancer proteome databases. These repositories combine wide-scale proteomic data with molecular and clinical variables and can consequently deliver a multidimensional framework of translational cancer biology and precision medicine.

Below, we have indicated and reviewed recent proteomics studies on a range of malignancies, such as gastric, colorectal, breast, and prostate cancers, and demonstrated other research that is relevant to different types of cancer (Table 15.3).

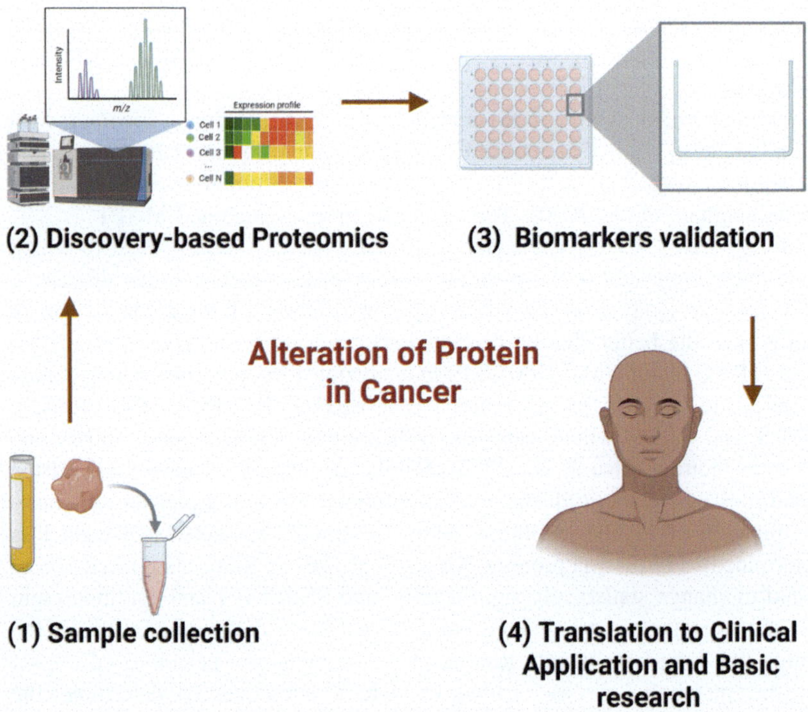

FIGURE 15.2 **Workflow of cancer proteomics.** Overview of proteomics workflow in cancer research and diagnosis.

15.5.1 Gastric cancer

In gastric cancer research, proteomic technologies have facilitated the documentation of potential biomarkers and therapeutic targets. Gao et al. (2019) used iTRAQ in conjunction with LC-MS and KEGG pathway array (PPA) analysis, and revealed several distinct proteins, including VCAM1, which was identified as a key protein in the network of protein-protein interactions, and 20 proteins with differential expression were documented (Gao et al., 2019). According to their findings, calpastatin and brassinosteroid-insensitive-1-associated kinase 1 are good predictors of *H. pylori*-related gastric cancer. PPA and Ingenuity Pathway Analysis confirmed that the ERK/MAPK signaling pathway is the most significantly dysregulated.

Further insight was provided by He et al. (2016), who employed PPA to alpha-fetoprotein (AFP) producing gastric cancer, an aggressive subtype often related to hepatocellular carcinoma (He et al., 2016). Their analysis revealed the upregulation of proteins, including cyclin (D1, RANK, LSD1, autotaxin, calpain2, STAT3, XIAP, IGF-Irβ, and BCL2), alongside the downregulation of ASC-R and BID. Notably, elevated levels of XIAP and IGF-Irβ were identified

TABLE 15.3 Cancer proteomics techniques for cancer-associated biomarkers discovery and disease progression applications.

Study	Cancer type/sample	Proteomics/methods	Findings	References
Medical proteomics discloses vulnerabilities in non-invasive breast ductal carcinoma	Breast-DCIS, IDC (FFPE biopsies)	Label-free quantitative proteomics; targeted multiplexed assay (90 proteins)	-380 differentially expressed proteins;22 validated biomarkers; therapeutic targets in PI3K, AKT, mTOR and EGFR, RAS, RAF	Wenk et al. (2024)
Current advances in MS-based targeted proteomics of clinical cancer biomarkers	Multiple cancers	LC-MS/MS targeted proteomics	Reviewed workflows for clinical biomarker validation; emphasized robustness, standardization, and cohort diversity	Wenk et al. (2024)
Determination of protein markers for gastric cancer	Gastric cancer—tissue, serum, urine	Discovery proteomics; database compilation; MRM assay design	Cataloged >6800 proteins; built "GCPDB" database with ~5600 proteotypic peptides; identified diagnostic/prognostic panels	Ramesh et al. (2024)
Immune subtype-based proteomic study of breast cancer	Breast cancer—immune subtypes	Comparative proteomics across immune-defined groups	Distinct proteomic signatures linked to immune classification; insights for patient stratification	Jeon et al. (2024)
Quantitative proteomics of serum throughout metastatic progression in breast cancer	Quantitative proteomics of serum during metastatic progression in	Quantitative proteomics	Serum protein changes correlated with metastasis: potential biomarkers for disease monitoring.	Kaur et al. (2024)
Serum and plasma proteomics of prostate cancer by using low and high-throughput methods	Prostate cancer, serum and plasma	High and low throughput proteomics	Compared throughput and sensitivity trade-offs; mapped prostate cancer proteomic signatures	Hamza et al. (2024)

as independent prognostic factors. By integrating these findings with pathological staging, a prognostic risk model was developed that stratified AFP-positive gastric adenocarcinoma into two distinct subgroups. Gene set enrichment analysis (GSEA) further demonstrated that the protein kinase A pathway was enriched in the high-risk subgroup, whereas the PTEN pathway was predominant in the low-risk group.

In addition to tissue-based studies, serum proteomics has also shown diagnostic potential. Tong et al. (2016), using the Luminex system, identified nine serum biomarkers linked to gastric cancer (Tong et al., 2016). These comprised a panel of classifiers that included anti-*H. Pylori* IgG I, ADAM8, VGF, pepsinogen, and II. The panels' validation rates are 78.7%, 82.5%, and 86.1% when applied to three different computational algorithms. These findings highlight the potential of multiplex serum biomarkers as non-invasive diagnostics for gastric cancer.

15.5.2 Colorectal cancer

Barberini et al. (2019) used gas chromatography-mass spectrometry (GC-MS) to analyze the plasma metabolites in patients with colon cancer and identified significant alterations in several metabolic pathways associated with disease progression. (Barberini et al., 2019). Strongly impacted were the pathways of monosaccharide breakdown, that is, fructose and D-mannose metabolisms, and amino-acidic metabolisms. In a complementary method, Ang et al. (2017) came up with a complex protocol of candidate protein biomarker detection in fecal samples, with one-dimensional sodium dodecyl sulfate polyacrylamide gel electrophoresis (1D-SDS-PAGE) combined with LC-MS/MS (Ang et al., 2017). Their fine-tuned quantification and validation procedure is a significant improvement in fecal proteomics and enhances the practicability of the discovery of noninvasive biomarkers to colorectal cancer.

A high-density antibody microarray was used by Rho et al. (2018) on plasma samples, and it was detected that a number of proteins and glycoproteins, such as BAG4, IL6ST, and CD44, were strongly associated with the diagnosis of colorectal cancer (Rho et al., 2018). Added EGFR, Sialyl Lewis -A-X, and CD44 further increased the biomarker repertoire and showed an area-under-the-curve between 0.86 and 0.90. These findings support the potential of combining proteomic and metabolomic approaches to enhance the accuracy of diagnostic tests to reveal colorectal cancer and emphasize the importance of integrative approaches to biomarker discovery.

15.5.3 Breast cancer

Song et al. (2014) used SmartChip mRNA microarray and parametric analysis of gene set enrichment (PPA) to describe changes in the disease-related pathways and identified 1243 genes that were associated with disease-related pathways (Song et al., 2014). Although the protein levels of such proteins as BCL6, CCNE1, PCNA, PDK1SRC, and XIAP were variable in malignant tumors and in adjacent normal tissues, the respective transcripts were not parallel; on the contrary, the level of mRNA and protein of CDK6, vimentin, and SLUG was decreased. Moreover, 15 signaling networks, such as the p53, IL17, HGF, NGF, PTEN, and PI3K/AKT networks, were uniformly deregulated both at the transcriptional and translational levels. These findings highlight the fact that despite the fact that several signaling cascades are disrupted in various molecular layers in breast cancer, inconsistencies between protein and mRNA expression indicate that there are unique regulatory pathways by which the pathogenesis of the disease occurs (Song et al., 2014).

The authors used GC-MS to identify the putative metabolic biomarkers of breast cancer (Hadi et al., 2017). A partial least squares discriminant analysis (PLS-DA) model gave a sensitivity of 96% and a specificity of 100% in the ability to differentiate between breast cancer patients and healthy controls in a validation cohort (Hadi et al., 2017). Also, a decision-tree classifier achieved an accuracy of 71.5%, 71.3%, and 79.8% in the process of grading of tumors, staging, and prediction of response to neoadjuvant therapy, respectively (Hadi et al., 2017). The findings highlight the power of using transcriptomic, proteomic, and metabolomic data in biomarker discovery, thus providing a complete understanding of the complicated regulatory networks that support breast cancer.

15.5.4 Prostate cancer

Rossi et al. (2025) used proteomics and a systems-biological approach to determine biomarkers in prostate cancer by using the urine samples of healthy donors (n= 5), and low-risk (n = 4) and high-risk (n = 7) prostate disease (Rossi et al., 2025). Further bioinformatical mapping with the TCGA allowed the association of urinary proteomic signatures with clinical and histopathological data in a larger computational cohort, therefore, aiding in the validation of the most promising candidates. The analysis has shown that proteomic changes in prostate cancer cells are transferred to the urine proteome: low-risk disease was characterized by the presence of proteins related to the complement and coagulation cascades, and high-risk disease was enriched with the glutathione-metabolism profiles (Rossi et al., 2025). New candidate biomarkers, such as CPM, KRT8, ITIH2, and RCN1, were also discovered in this integrative approach. These observations are consistent with earlier documented information, which gives them biological validity as well as broadens the list of possible prostate cancer biomarkers. Collectively, these results shed light on understanding the pathophysiology of disease, and through molecular predictors, this could be used to further stratify and manage patients; however, this has to be validated in clinical trials involving large cohorts of patients to determine clinical utility.

15.6 Proteomics in neurodegenerative diseases

The high-dimensional molecular profiling of biofluids with diverse omics technologies has a high potential for discovering large-scale biomarkers; however, the combination of sufficiently large and heterogeneous data is still a challenge (Imam et al., 2025). To help resolve this concern, one of the biggest harmonized proteomic resources in the world was created under the Global Neurodegeneration Proteomics Consortium (GNPC), which is a publicly funded collaborative project. Such dataset contains about 250 million unique protein measurements that were measured using various analytical platforms and over 35,000 biofluid samples (blood plasma, serum, and cerebrospinal fluid) provided by 23 international cohorts. Notably, such proteomics data are combined with large amounts of clinical data that relate to a variety of severe neurological diseases, such as amyotrophic lateral sclerosis, Parkinson's, AD, and frontotemporal dementia (Shvetcov et al., 2025; Ali et al., 2025). The resource is available to GNPC members via the AD Workbench, a safe, cloud-based system that uses AD Workbench to assist in conducting joint analysis of data. Preliminary plasma proteome analyses have supported the role of APOE e4 carriership as a cross-disease biomarker and have demonstrated both disease-specific changes in protein abundance as well as a common proteomic signature of APOE e4 carriership in Alzheimer's disease, Parkinson's disease, frontotemporal dementia, and amyotrophic lateral sclerosis (Imam et al., 2025). It has also been observed in comparative studies that there are specific patterns of organ-specific aging conditions in varied conditions, which imply the divergent and convergent molecular processes underlying neurodegeneration.

In general, this overview highlights the importance of large-scale international efforts, systematic data organization, and an open-access scientific framework in being capable of significantly expediting the process of biomarker discovery and enhancing the mechanistic insights into neurodegenerative diseases. The GNPC dataset thus can be used as an illustration of how the world's resources can be exploited to promote precision diagnostics and therapeutic development in complex neurological conditions.

15.6.1 Facioscapulohumeral muscular dystrophy

Facioscapulohumeral muscular dystrophy (FSHD) is a type of muscular dystrophy and is one of the most common inherited muscle disorders, and is found in a range of 1 in 8000–22,000 people (Mostacciuolo et al., 2009; Deenen et al., 2014). The pathogenic mechanism is core, characterized by the aberrant expression of the transcription factor DUX4 (Lemmers et al., 2010), that initiates far-reaching deregulation of transcription of genes in myocytes, which leads to progressive muscle weakness and degeneration (Statland et al., 2013). Even though transcriptional changes in FSHD are widely described, the appropriate proteomic changes remain not thoroughly outlined yet.

To investigate these proteomic dynamics, label-free LC-MS/MS was used to culture and produce quantitative proteomic data on myoblasts of two FSHD patients and unrelated unaffected siblings (UASb) at baseline and 24 hours after culture. At both time points, 4266 proteins were detected, and although the means of the protein abundance between the patient and the controls were not significantly different, distinct proteins showed significant differences in their expression. Initially, 180 proteins passed the significance test ($P < 0.05$); protein downregulation in terms of gene-ontology enrichment was observed, which included, but was not limited to, RNA-binding proteins, including heterogeneous nuclear ribonucleoproteins, and small nuclear ribonucleoproteins, and the negative regulation of myogenesis was upregulated, including LMCD1 (1.24-fold) and TGFBR2 (1.10-fold) (Stefanović et al., 2024). There was a similar number of proteins that changed significantly (179 proteins) after 24 hours. Interestingly, FSHD myoblasts displayed reduced E3 ubiquitin-ligase activity in the form of various E3 ubiquitin-ligase components (such as DTX3L 1.6-fold) and in the form of multiple proteasomal subunits (PSMA5, PSMB4, PSMC1, and PSME2). On the other hand, the specific E3 ligases, that is, TRIM38, XIAP, and PPP1R11, were up-regulated, which means the selective regulation of proteolytic cascades, in particular, those regulating the proteins of immune-related cascades (TLR2 and IFNB1) (Stefanović et al., 2024). Such observations provide new information about unique FSHD-associated proteomic changes, which add to existing data regarding transcriptomic changes and shape molecular profiles that can be used as therapeutic drug targets. Our study contributes to the work on the mechanisms of FSHD, as profiling proteome-scale alterations in the disease enables us to build the foundation of the design of specific treatments.

15.7 Proteomics in infectious diseases

15.7.1 COVID-19

Indeed, high-throughput proteomic methods have proved to keep up with and provide a lot of information regarding the pathogenesis of SARS-CoV-2, as shown by their use in COVID-19 studies (see Table 15.4) (Haendel et al., 2021; Messner et al., 2020; Shen et al., 2020; Yuan et al., 2020). Despite considerable advances in reducing the pandemic, it is still difficult to accurately diagnose and predict acute COVID -19 and long-term post-vaccinal sequelae (e.g., following immunization) (Feng et al., 2021; Chen and Zhang, 2021). Combined LC-MS, antibody-based assays, and sophisticated bioinformatics have proven to be rather promising in understanding the molecular pathophysiology of SARS-CoV-2, and are also useful in guiding the generation of therapeutics and vaccines (Bojkova et al., 2020). LC-MS has been

TABLE 15.4 Clinical proteomics techniques in disease diagnosis and therapeutic applications.

Disease	Technique	Findings	Clinical applications	References
COVID-19	LC-MS/MS quantitative proteomics	Identified dysregulated plasma proteins	Blood biomarker discovery for disease progression and therapeutic targets	Messner et al. (2020); Shen et al. (2020)
Tuberculosis	Probe proteome microarrays	Identified five biomarkers MT1560.1, Rv0049, Rv0270, Rv1597 and Rv3480c associated with active TB	Early diagnostic biomarkers discriminate between latent and active tuberculosis.	Peng et al. (2020)
HIV	Shotgun Proteomics/LC-MS	Identified six proteins those upregulated only in HIV/HBV and ten proteins downregulated in viral infection	Biomarker discovery and disease progression associated with viral infection	Tarnathummanan et al. (2024)
Dengue (DENV)	TMT/MS quantitative proteomics	Identified regulatory proteins associated with Dengue fever	Biomarker identification and disease progression of hemorrhagic fever	Perera et al. (2024)
Zika virus	Network proteomics	Identified 386 ZIKV interacting proteins and 1216 phosphorylation sites those down-or upregulated after ZIKV infection.	System-level understanding of ZIKV-induced protein and cellular pathways analysis	Scaturro et al. (2018)
Ebola virus disease	LC-MS plasma proteomics	Identified prognostic 4-protein Histone 1–5, moesin, kininogen1, and RPL35 associated with viral load.	Disease progression	Viodé et al. (2022)
Hepatitis B	TMT/LC-MS	Identified several diagnostic biomarker proteins in chronic HBV-infected patients	Diagnostic biomarker identification	Xun et al. (2022)

Avian influenza H9N2	Four-dimensional/ label-free/ bioinformatics	Identified 298 proteins associated with upregulation and 434 proteins associated with downregulation in the H9N2-infected group	Virus propagation and antiviral therapy	Zhao et al. (2025)
Leishmaniasis	TMT/LC-MS/ Western Blotting quantitative proteomics	Identified Ago1 as the ideal Argonaute of RNAi in infected cells	Disease progression	Moradimotlagh et al. (2023)
Cervical cancer	Proteogenomic profiling	Identified EP300 to enhance acetylation of FOSL2-K222, for the progression of malignant proliferation of cervical cancer	Biomarker discovery associated with viral infection and antiviral therapy	Yu et al. (2024)
Zoster virus	LC-MS/MS	Identified proteins CXCL10, ELANE, IL-1RN.MPO, PRTN3 are associated with inflammation	Biomarker discovery and disease progression	Liu et al. (2023)

found to be useful in unraveling the effects of a viral infection on the cellular pathways of the host (Bojkova et al., 2020). The hierarchical clustering gave out two major protein clusters, one of which was associated with cholesterol metabolism in response to infection. The functional analyses also revealed that pharmacological blocking of the pathways is an effective way of inhibiting viral replication in cell cultures, which supports their potential as therapeutic targets to prevent and treat COVID-19 (Bojkova et al., 2020).

Simultaneously, proteomic discoveries have been supplemented by genomic research in mapping viral evolution. When investigating the transmission routes of SARS -CoV-2, Forster and colleagues used phylogenetic network analysis to explain it without reduction. Their analysis revealed three major viral lineages, which were labeled A, B, and C, and each of them was characterized by particular amino acid replacements. Lineages A and C were found among European and North American groups, but lineage B was more common in the East Asians (Sánchez-Pacheco et al., 2020; Forster et al., 2020). It is interesting to note that there are indications that the ancestral B genome mutationally diversified into derived B sublineages before the time of its dispersion outside of East Asia (Sánchez-Pacheco et al., 2020; Forster et al., 2020). These findings reveal the need to employ a combination of proteomic and genomic approaches to support our knowledge of SARS-CoV-2 biology.

15.7.2 HIV-1 infection

The dynamic changes in the expression of plasma proteins that take place during hyper-acute infection with HIV-1 and their relation to acute retroviral syndrome, to viral control, and to disease progression are not completely characterized. In the current research, the researchers measured 1293 plasma proteins on 157 longitudinally matched samples before, during, and after hyper-acute HIV-1 infection of 54 participants across four sub-Saharan African nations (Nazziwa et al., 2024). Compared to the baseline, longitudinal protein expression patterns of 6 patterns were analyzed and found to decline progressively after infection. A massive dysregulation of immunological modulation, inflammatory signaling, and cellular motility proteins was identified between preinfection samples and the postinfection one month after infection (Nazziwa et al., 2024). Lower levels of ZYX and SCGB1A1 with higher levels of LILRA3 were related to increased risk of acute retroviral syndrome and a higher level of NAPA and RAN with a lower level of ITIH4 expression, with better viral control. Besides, increased levels of HPN, PRKCB, and ITGB3 were strongly linked to the rapid progression of the disease. Taken together, these findings offer a critical understanding of early host response in hyperacute HIV-1 infection and the possible biomarkers and molecular pathways that could affect viral load and disease course over time.

15.7.3 Hepatitis C

The study used a coordinated proteomic and transcriptomic study to outline the serological molecular environment in relation to HCV infection. It was characterized by three groups: people who have persistent (chronic) HCV infection, people who had previously cleared HCV after treatment, and a healthy control group (Wang et al., 2025a). To determine gene and protein expression patterns, serum RNA-seq and TMT-based proteomic profiling were conducted in these cohorts. Bioinformatics pipelines identified differentially expressed proteins (DEPs) and genes (DEGs), and then functional enrichment analysis and protein–protein interaction (PPI) network were constructed (Wang et al., 2025a). The identification of DEPs/DEGs was followed by a procedure of the identification of these target DEPs/DEGs through RT-qPCR and western blotting. Comparative analysis identified 356 DEPs that separated the HCV group and controls, and 381 more that separated chronic HCV and resolved groups. The processes enriched in these DEPs included the CO2 transport, transcriptional initiation, and immune response, as well as the microbial infection pathways, including the majority of immunoglobulins and exosomal proteins. HSPA4, HSPD1, COPS5, PSMD2, and HSPA4 were identified as important nodes in PPI network analysis. On the transcriptomic level, 684 DEGs were found between HCV and control, and 350 DEGs between chronic HCV and HCV. The involvement of DEGs in biological functions such as wound healing, cellular communication, oxidative stress, cell adhesion, viral infection, and immunity was enriched, highlighting the link between the enriched DEGs and extracellular matrix composition. Hub genes that were located in the DEG network were KIF11, CENPE, TTK, CDC20, and ASPM. Combined proteomic and transcriptomic data analysis showed a significant amount of cross-talk between DEPs and DEGs, with some overlap proteins being MAT1, EIF4A3, and UBE2D1. The expression patterns of EIF4A3, EIF2B1, MAT1, SNRNP70, and UBE2D1 were supported by the results of consistent gene sequencing in the control, HCV, and chronic HCV cohort. Such data make EIF4A3, EIF2B1, MNAT1, SNRNP70, and UBE2D1 promising biomarkers of HCV treatment, emphasizing their functions in the pathogenesis of the virus and its development.

Ye et al. (2015) in another quantitative proteomics study compared the differences in protein expression of HepG2 and Huh7.5.1 WT cells through the label-free and iTRAQ labeling methods (Ye et al., 2015). Out of 4919 identified proteins, 114 displayed similar results in both quantitative systems, and 37 proteins were further validated using RT-qPCR. These results clarify the possible role of the particular proteins in the process of HCV replication and pathogenesis.

15.8 Precise medicine: the proteomics

15.8.1 Disease diagnosis

Proteomic-based technologies have been broadly used when it comes to the identification of protein biomarkers in cardiovascular disease, malignancies, as well as neurodegenerative disorders; hence, enhancing the accuracy of early diagnosis, facilitating timely clinical intervention, and leading to the long-term provision of improved health outcomes across the world (Xie et al., 2025). These early detection measures come in especially handy in the resource-strained environment, in which they can offset the long-term health care liability. More has been done to use proteomics in ischemic stroke research. The deep study of the brain and cerebrospinal fluid proteomes after a stroke has provided an array of likely biomarkers that may be applied in the management and treatment of the disease. Of these, glial fibrillary acidic protein (GFAP), and retinol 4 (RBP4) have become the most prominent ones, and the patterns of their differentially expressed patterns can be used to differentiate between intracerebral hemorrhage and ischemic stroke, as a result of which subtypes can be classified (Wiśniewski, 2018).

Proteomics has also been used in oncology to give essential information on biomarker discovery and therapies. An envelope-breaking study using immunopeptidomics protein studies found that cancer/testis antigen 45 (CT45) is a naturally occurring tumor antigen that is presented on ovarian cancer cells through the HLA class 1 molecules. Notably, the tumor cell lysis in patients was promoted by CT45-derived peptides that stimulate the cytotoxic T lymphocyte activity. Additional proteomic results of cells with DNA demethylation drugs revealed that CT45 was greatly increased, emphasizing its role as a possible biomarker in clinical use, and also, as an immune therapeutic agent in ovarian cancer (Coscia et al., 2018).

15.8.2 Personalized treatment

Personalized therapy is a fundamental principle of precision medicine, which attempts to design interventions with respect to the personal atomicity of individual patients. Proteomics plays a crucial role in this approach by elucidating disease-specific signaling pathways and molecular mechanisms, thereby enabling the development of personalized treatment strategies (Xie et al., 2025). Cancer proteomics has emerged as one of the most extensively studied areas of proteomic research, with breast cancer representing a major focus of investigation. It has been widely studied that there are four major molecular subtypes of breast cancer, namely: Luminal A, Luminal B, HER2+, as well as Basal-like, with each having a different gene expression and protein profile that is also associated with the differences in tumor biology, clinical behavior, and therapeutic response (Miah et al., 2017).

In 2012, in a groundbreaking study, researchers used reverse-phase protein array technology in the study of breast cancer samples to study protein expression. Their results present support for the findings that there are at least two clinically exclusive HER2-positive subgroups: the HER2-enriched (HER2-mRNA) subtypes and the luminal-mRNA subtype. Tumors referred to as HER2E-mRNA /HER2-positive showed increased levels of HER1/ EGFR, phosphorylated HER1, HER2, phosphorylated HER2, p-SRC, and p-S6, but to a lower degree than in luminal-mRNA/HER2-positive tumors. On the contrary, tumors with luminal-mRNA/HER2 -positive exhibited an increased expression of the luminal-associated proteins, such as GATA3, BCL2, and ESR1. Additional therapeutic weaknesses were also revealed by the incorporation of the genomic profiling into this analysis. Particularly, the HER2E-mRNA subtype tumors also had a high frequency of PIK3CA mutations, rare PTEN and PIK3R1 alterations, and PTEN and INPP4B deletions that could be used in therapeutic intervention (The Cancer Genome Atlas Network, 2012).

Such findings emphasize the relevance of proteomic methodologies to the distinction of the different forms of HER2-positive breast cancer that also uncover unique signaling pathways, protein expression profiles, and genetic changes that have the potential of informing individualized treatment. These findings have the potential not only to increase our understanding of HER2-associated oncogenesis but also to offer an effective framework for maximizing HER2-targeted therapy and optimizing personalized treatment plans.

15.9 Challenges

Although there are promising developments, there are many obstacles that prevent the full implementation of proteomics in the clinical setting. The complexity and dynamic range of the human proteome are one of the challenges. Unlike the comparatively stable genome, the proteome is highly versatile since it is subject to post translation processes, alternative splicing of the RNA, and environmental cues, and therefore cannot be accurately simulated and is equally difficult to measurably determine variation in research (Aebersold and Mann, 2016). Similarly, sample heterogeneity makes the process of biomarker identification and validation more complex, particularly in clinical samples like plasma or tissue biopsies (Zhang et al., 2019; Moxon et al., 2009).

Another significant constraint is the analytical and technical issues of the processes of proteomics. Even though the high-resolution mass spectrometry shows an increase in sensitivity, it is difficult to detect low-abundant proteins, especially when they are mixed up with high-abundant proteins like albumin in plasma (Altelaar and Heck, 2012). Furthermore, standardization of laboratory methods is missing, thus leading to unequal data sets and low reproducibility of clinical studies (Mischak et al., 2025; Moxon et al., 2009). Proteomic data must be integrated with other platforms of the so-called omics, for example, metabolomics, genomics, and transcriptomics, which are large-scale computational and bioinformatics tasks (Vittaladevaram, 2021).

Translational and clinical issues are also valuable. Proteomic research findings need to be translated into clinical practice via comprehensive validation, regulatory approval, and cost-effectiveness studies (Fan and Zeng, 2025). Due to the fact that a great number of possible biomarkers have not been adequately tested in large and diverse groups of patients, they are rarely given therapeutic utility (Vittaladevaram, 2021). The use among many people is also limited by the expensive instruments used, and a skilled technician is required, as well as accessibility in facilities with few resources (Vittaladevaram, 2021). Lastly, there are ethical and data-security issues of patient-derived proteomic data, which further complicate clinical use (Reymond et al., 2003; Mischak et al., 2025). Although proteomics can challenge our approach to precision medicine in almost a transformative manner, it is fundamental that these complex issues are resolved before proteomics becomes a common practice in diagnosing, prognosis, and therapeutic surveillance.

15.10 Conclusion and future perspective

Proteomics has become a groundbreaking science in the field of human health care as it offers unprecedented explainations of the disease processes, diagnostic indicators, treatment options, and personalized treatment plans. The recent developments, such as high-resolution mass spectrometry, single-cell proteomics, quantitative proteomic profiling, and integrative multiomics platforms, have tremendously enhanced the scope, precision, and clinical implications of protein analyses. These technologies have expedited the early detection of diseases, fine-tuning of prognostic evaluations, and helped in precision medicine in the complicated diseases, including cancer, neurodegenerative diseases, cardiovascular diseases, and infectious diseases. Regardless of these successes, proteomics has to face several challenges such as the complexity of samples, dynamism in protein levels, standardization, cost of analysis, and the need to use sophisticated computational software to handle the vast amount of proteomic data.

There is great potential in the future of proteomics in health care. It is anticipated that it will be combined with artificial intelligence, machine learning, and sophisticated bioinformatics to push biomarker discovery and predictive modeling to greater speed and efficiency. New technology models, such as the next-generation single-molecule proteomics, spatial proteomics, and real-time proteomic surveillance, are expected to offer more information about biology and improve clinical decision-making. Additionally, the extensive use of multiomics models will provide an integrated awareness of the human health state, thus allowing personalized medicine and preventative health care methods. With greater availability of technology, as well as the standardization of analytical pipelines, proteomics will be an increasingly key component of the future of precision health care.

References

Aebersold, R., Mann, M., 2016. Mass-spectrometric exploration of proteome structure and function. Nature 537 (7620), 347–355. Available at https://doi.org/10.1038/nature19949.

Agrawal, G.K., Sarkar, A., Righetti, P.G., Pedreschi, R., Carpentier, S., Wang, T., et al., 2013. A decade of plant proteomics and mass spectrometry: translation of technical advancements to food security and safety issues. Mass Spectrometry Reviews 32 (5), 335–365. Available at https://doi.org/10.1002/mas.21365.

Ahmad, R., Budnik, B., 2023. A review of the current state of single-cell proteomics and future perspective. Analytical and Bioanalytical Chemistry 415 (28), 6889–6899. Available at https://doi.org/10.1007/s00216-023-04759-8.

Al-Amrani, S., Al-Jabri, Z., Al-Zaabi, A., Alshekaili, J., Al-Khabori, M., 2021. Proteomics: concepts and applications in human medicine. World Journal of Biological Chemistry 12 (5), 57–69. Available at https://doi.org/10.4331/wjbc.v12.i5.57.

Ali, M., Erabadda, B., Chen, Y., Xu, Y., Gong, K., Liu, M., et al., 2025. Shared and disease-specific pathways in frontotemporal dementia and Alzheimer's and Parkinson's diseases. Nature Medicine 31 (8), 2567–2577. Available at https://doi.org/10.1038/s41591-025-03833-1, https://www.nature.com/nm/.

Altelaar, A.F.M., Heck, A.J.R., 2012. Trends in ultrasensitive proteomics. Current Opinion in Chemical Biology 16 (1-2), 206–213. Available at https://doi.org/10.1016/j.cbpa.2011.12.011.

Amiri-Dashatan, N., Koushki, M., Abbaszadeh, H.A., Rostami-Nejad, M., Rezaei-Tavirani, M., 2018. Proteomics applications in health: biomarker and drug discovery and food industry. Iranian Journal of Pharmaceutical Research 17 (4), 1523–1536. Available at https://brief.land/ijpr/.

Ang, C.S., Baker, M.S., Nice, E.C., 2017. Mass spectrometry-based analysis for the discovery and validation of potential colorectal cancer stool biomarkers. Methods in Enzymology 586, 247–274. Available at https://doi.org/10.1016/bs.mie.2016.10.019.

Aslam, B., Basit, M., Nisar, M.A., Khurshid, M., Rasool, M.H., 2017. Proteomics: technologies and their applications. Journal of Chromatographic Science 55 (2), 182–196. Available at https://doi.org/10.1093/chromsci/bmw167, http://chromsci.oxfordjournals.org/content.

Bae, S.H., Harris, A.G., Hains, P.G., Chen, H., Garfin, D.E., Hazell, S.L., et al., 2003. Strategies for the enrichment and identification of basic proteins in proteome projects. Proteomics 3 (5), 569–579. Available at https://doi.org/10.1002/pmic.200300392.

Bai, B., Hales, C.M., Chen, P.C., Gozal, Y., Dammer, E.B., Fritz, J.J., et al., 2013. U1 small nuclear ribonucleoprotein complex and RNA splicing alterations in Alzheimer's disease. Proceedings of the National Academy of Sciences of the United States of America 110 (41), 16562–16567. Available at https://doi.org/10.1073/pnas.1310249110, http://www.pnas.org/content/110/41/16562.full.pdf+html.

Bai, B., Wang, X., Li, Y., Chen, P.C., Yu, K., Dey, K.K., et al., 2020. Deep multilayer brain proteomics identifies molecular networks in Alzheimer's disease progression. Neuron 105 (6), 975. Available at https://doi.org/10.1016/j.neuron.2019.12.015, http://www.cell.com/neuron/home.

Barberini, L., Restivo, A., Noto, A., Deidda, S., Fattuoni, C., Fanos, V., et al., 2019. A gas chromatography-mass spectrometry (GC-MS) metabolomic approach in human colorectal cancer (CRC): the emerging role of monosaccharides and amino acids. Annals of Translational Medicine 7 (23), 727. Available at https://doi.org/10.21037/atm.2019.12.34.

Barderas, R., Montero-Calle, A., San Segundo-Acosta, P., 2025a. Protein microarray analysis with genepix. Methods in Molecular Biology 2929, 85–95. Available at https://doi.org/10.1007/978-1-0716-4595-6_7, https://www.springer.com/series/7651.

Barderas, R., San Segundo-Acosta, P., Montero-Calle, A., 2025b. Cytokine antibody microarray-based proteomic strategies for characterizing the dysregulation of cytokines disease-specific. Methods in Molecular Biology 2929, 195–213. Available at https://doi.org/10.1007/978-1-0716-4595-6_15, https://www.springer.com/series/7651.

Berggren, K., Chernokalskaya, E., Steinberg, T.H., Kemper, C., Lopez, M.F., Diwu, Z., et al., 2000. Background-free, high sensitivity staining of proteins in one- and two-dimensional sodium dodecyl sulfate-polyacrylamide gels using a luminescent ruthenium complex. Electrophoresis 21 (12), 2509–2521. Available at https://doi.org/10.1002/1522-2683(20000701)21:12<2509::AID-ELPS2509>3.0.CO;2-9, http://onlinelibrary.wiley.com/journal/10.1002/(ISSN)1522-2683.

Van Den Bergh, G., Arckens, L., 2005. Recent advances in 2D electrophoresis: an array of possibilities. Expert Review of Proteomics 2 (2), 243–252. Available at https://doi.org/10.1586/14789450.2.2.243.

Bojkova, D., Klann, K., Koch, B., Widera, M., Krause, D., Ciesek, S., et al., 2020. Proteomics of SARS-CoV-2-infected host cells reveals therapy targets. Nature 583 (7816), 469–472. Available at https://doi.org/10.1038/s41586-020-2332-7.

Ceciliani, F., Eckersall, D., Burchmore, R., Lecchi, C., 2014. Proteomics in veterinary medicine: applications and trends in disease pathogenesis and diagnostics. Veterinary Pathology 51 (2), 351–362. Available at https://doi.org/10.1177/0300985813502819.

Chandramouli, K., Qian, P.-Y., Godwin, A.K., 2009. Proteomics: challenges, techniques and possibilities to overcome biological sample complexity. Human Genomics and Proteomics 1 (1). Available at https://doi.org/10.4061/2009/239204.

Chandrasekhar, K., Dileep, A., Lebonah, D.E., Kumari, J.P., 2014. A short review on proteomics and its applications. International Letters of Natural Sciences 17, 77–84. Available at https://doi.org/10.56431/p-avsz0g.

Chang, L., Ni, J., Beretov, J., Wasinger, V.C., Hao, J., Bucci, J., et al., 2017. Identification of protein biomarkers and signaling pathways associated with prostate cancer radioresistance using label-free LC-MS/MS proteomic approach. Scientific Reports 7. Available at https://doi.org/10.1038/srep41834, www.nature.com/srep/index.html.

Chen, Z., Zhang, L., 2021. Meet the challenges of mass vaccination against COVID-19. Exploratory Research and Hypothesis in Medicine, 1–3. Available at https://doi.org/10.14218/erhm.2020.00075.

Coscia, F., Lengyel, E., Duraiswamy, J., Ashcroft, B., Bassani-Sternberg, M., Wierer, M., et al., 2018. Multi-level proteomics identifies CT45 as a chemosensitivity mediator and immunotherapy target in ovarian cancer. Cell 175 (1), 159. Available at https://doi.org/10.1016/j.cell.2018.08.065, https://www.sciencedirect.com/journal/cell.

Cutler, P., 2003. Protein arrays: the current state-of-the-art. Proteomics 3 (1), 3–18. Available at https://doi.org/10.1002/pmic.200390007.

Deenen, J.C.W., Arnts, H., van der Maarel, S.M., Padberg, G.W., Verschuuren, J.J.G.M., Bakker, E., et al., 2014. Population-based incidence and prevalence of facioscapulohumeral dystrophy. Neurology 83 (12), 1056–1059. Available at https://doi.org/10.1212/WNL.0000000000000797.

Domon, B., Aebersold, R., 2006. Mass spectrometry and protein analysis. Science (New York, N.Y.) 312 (5771), 212–217. Available at https://doi.org/10.1126/science.1124619.

Dubey, H., Grover, A., 2001. Current initiatives in proteomics research: the plant perspective. Current Science 80 (2), 262–269.

Enroth, S., Berggrund, M., Lycke, M., Broberg, J., Lundberg, M., Assarsson, E., et al., 2019. High throughput proteomics identifies a high-accuracy 11 plasma protein biomarker signature for ovarian cancer. Communications Biology 2 (1). Available at https://doi.org/10.1038/s42003-019-0464-9.

Esteve, P., Rueda-Carrasco, J., Inés Mateo, M., Martin-Bermejo, M.J., Draffin, J., Pereyra, G., et al., 2019. Elevated levels of secreted-frizzled-related-protein 1 contribute to Alzheimer's disease pathogenesis. Nature Neuroscience 22 (8), 1258–1268. Available at https://doi.org/10.1038/s41593-019-0432-1, http://www.nature.com/neuro/index.html.

Fan, S., Zeng, S., 2025. Plasma proteomics in pediatric patients with sepsis– hopes and challenges. Clinical Proteomics 22 (1). Available at https://doi.org/10.1186/s12014-025-09533-9.

Fathi, A., Pakzad, M., Taei, A., Brink, T.C., Pirhaji, L., Ruiz, G., et al., 2009. Comparative proteome and transcriptome analyses of embryonic stem cells during embryoid body-based differentiation. Proteomics 9 (21), 4859–4870. Available at https://doi.org/10.1002/pmic.200900003.

Feng, G., Zhang, L., Wang, K., Chen, B., Xia, H.H.-X., 2021. Research, development and application of COVID-19 vaccines: progress, challenges, and prospects. Journal of Exploratory Research in Pharmacology. Available at https://doi.org/10.14218/jerp.2021.00004.

Ferraro, G., Merlino, A., 2025. Investigation of metallodrug/protein interaction by X-ray crystallography and complementary biophysical techniques. Inorganic Chemistry Frontiers 12 (9), 3345–3366. Available at https://doi.org/10.1039/d4qi03277b.

Florens, L., Washburn, M., 2006. Proteomic analysis by multidimensional protein identification technology. New and Emerging Proteomic Techniques, volume 328. Humana Press. Available at https://doi.org/10.1385/1-59745-026-X:159.

Forster, P., Forster, L., Renfrew, C., Forster, M., 2020. Phylogenetic network analysis of SARS-CoV-2 genomes. Proceedings of the National Academy of Sciences 117 (17), 9241–9243. Available at https://doi.org/10.1073/pnas.2004999117.

Gao, Z., Wang, J., Bai, Y., Bao, J., Dai, E., 2019. Identification and verification of the main differentially expressed proteins in gastric cancer via iTRAQ combined with liquid chromatography-mass spectrometry. Analytical Cellular Pathology 2019, 1–14. Available at https://doi.org/10.1155/2019/5310684.

Gavin, A.C., Bösche, M., Krause, R., Grandi, P., Marzioch, M., Bauer, A., et al., 2002. Functional organization of the yeast proteome by systematic analysis of protein complexes. Nature 415 (6868), 141–147. Available at https://doi.org/10.1038/415141a.

Geyer, P.E., Wewer Albrechtsen, N.J., Tyanova, S., Grassl, N., Iepsen, E.W., Lundgren, J., et al., 2016. Proteomics reveals the effects of sustained weight loss on the human plasma proteome. Molecular Systems Biology 12 (12). Available at https://doi.org/10.15252/msb.20167357, http://onlinelibrary.wiley.com/journal/10.1002/(ISSN)1744-4292.

Gobena, S., Admassu, B., Kinde, M.Z., Gessese, A.T., 2024. Proteomics and its current application in biomedical area: concise review. Scientific World Journal 2024. Available at https://doi.org/10.1155/2024/4454744, http://www.hindawi.com/journals/tswj/.

Goshima, N., Kawamura, Y., Fukumoto, A., Miura, A., Honma, R., Satoh, R., et al., 2008. Human protein factory for converting the transcriptome into an in vitro–expressed proteome. Nature Methods 5 (12), 1011–1017. Available at https://doi.org/10.1038/nmeth.1273.

Graves, P.R., Haystead, T.A.J., 2002. Molecular biologist's guide to proteomics. Microbiology and Molecular Biology Reviews 66 (1), 39–63. Available at https://doi.org/10.1128/MMBR.66.1.39-63.2002.

Grønborg, M., Kristiansen, T.Z., Iwahori, A., Chang, R., Reddy, R., Sato, N., et al., 2006. Biomarker discovery from pancreatic cancer secretome using a differential proteomic approach. Molecular and Cellular Proteomics 5 (1), 157–171. Available at https://doi.org/10.1074/mcp.M500178-MCP200.

Gygi, S.P., Rist, B., Gerber, S.A., Turecek, F., Gelb, M.H., Aebersold, R., 1999. Quantitative analysis of complex protein mixtures using isotope-coded affinity tags. Nature Biotechnology 17 (10), 994–999. Available at https://doi.org/10.1038/13690.

Hadi, N.I., Jamal, Q., Iqbal, A., Shaikh, F., Somroo, S., Musharraf, S.G., 2017. Serum metabolomic profiles for breast cancer diagnosis, grading and staging by gas chromatography-mass spectrometry. Scientific Reports 7 (1). Available at https://doi.org/10.1038/s41598-017-01924-9, www.nature.com/srep/index.html.

Haendel, M.A., Chute, C.G., Bennett, T.D., Eichmann, D.A., Guinney, J., Kibbe, W.A., et al., 2021. The National COVID Cohort Collaborative (N3C): rationale, design, infrastructure, and deployment. Journal of the American Medical Informatics Association 28 (3), 427–443. Available at https://doi.org/10.1093/jamia/ocaa196, http://jamia.oxfordjournals.org/content/22/e1.

Hamza, G.M., Raghunathan, R., Ashenden, S., Zhang, B., Miele, E., Jarnuczak, A.F., 2024. Proteomics of prostate cancer serum and plasma using low and high throughput approaches. Clinical Proteomics 21 (1). Available at https://doi.org/10.1186/s12014-024-09461-0, https://clinicalproteomicsjournal.biomedcentral.com/.

Hanash, S., Taguchi, A., 2011. Application of proteomics to cancer early detection. The Cancer Journal 17 (6), 423–428. Available at https://doi.org/10.1097/ppo.0b013e3182383cab.

Haqqani, A.S., Kelly, J.F., Stanimirovic, D.B., 2008. Quantitative protein profiling by mass spectrometry using isotope-coded affinity tags. Methods in Molecular Biology 439, 225–240. Available at https://doi.org/10.1007/978-1-59745-188-8_16, http://www.springer.com/series/7651.

He, L., Ye, F., Qu, L., Wang, D., Cui, M., Wei, C., et al., 2016. Protein profiling of alpha-fetoprotein producing gastric adenocarcinoma. Oncotarget 7 (19), 28448–28459. Available at https://doi.org/10.18632/oncotarget.8571, http://www.impactjournals.com/oncotarget/index.php?journal=oncotarget&page=article&op=download&path%5B%5D=8541&path%5B%5D=25420.

He, Y., Liu, X., Zhu, F., Dai, Z., Li, X., Li, L., et al., 2025. Single-cell secretome profiling enabled by selective enrichment and nanoLC-MS/MS analysis. Angewandte Chemie - International Edition 64 (4). Available at https://doi.org/10.1002/anie.202417351, http://onlinelibrary.wiley.com/journal/10.1002/(ISSN)1521-3773.

Higginbotham, L., Ping, L., Dammer, E.B., Duong, D.M., Zhou, M., Gearing, M., et al., 2020. Integrated proteomics reveals brain-based cerebrospinal fluid biomarkers in asymptomatic and symptomatic Alzheimer's disease. Science Advances 6 (43). Available at https://doi.org/10.1126/sciadv.aaz9360, https://advances.sciencemag.org/content/6/43/eaaz9360/tab-pdf.

Holman, J., Dasari, S., Tabb, D., 2013. Informatics of protein and posttranslational modification detection via shotgun proteomics. In: In: Zhou, M., Veenstra, T. (Eds.), Proteomics for Biomarker Discovery, Methods in Molecular Biology 1002. Humana Press, Totowa, NJ, Totowa, NJ, pp. 167–179. Available at https://doi.org/10.1007/978-1-62703-360-2_14.

Hsu, C.H., Hsu, C.W., Hsueh, C., Wang, C.L., Wu, Y.C., Wu, C.C., et al., 2016. Identification and characterization of potential biomarkers by quantitative tissue proteomics of primary lung adenocarcinoma. Molecular and Cellular Proteomics 15 (7), 2396–2410. Available at https://doi.org/10.1074/mcp.M115.057026, http://www.mcponline.org/content/15/7/2396.full.pdf.

Hu, Y., Cheng, K., He, L., Zhang, X., Jiang, B., Jiang, L., et al., 2021. NMR-based methods for protein analysis. Analytical Chemistry 93 (4), 1866–1879. Available at https://doi.org/10.1021/acs.analchem.0c03830.

Imam, F., Saloner, R., Vogel, J.W., Krish, V., Abdel Azim, G., Ali, M., et al., 2025. The Global Neurodegeneration Proteomics Consortium: biomarker and drug target discovery for common neurodegenerative diseases and aging. Nature Medicine 31 (8), 2556–2566. Available at https://doi.org/10.1038/s41591-025-03834-0, https://www.nature.com/nm/.

Inayatullah, M., Dwivedi, A.K., Tiwari, V.K., 2025. Advances in single-cell omics: transformative applications in basic and clinical research. Current Opinion in Cell Biology 95. Available at https://doi.org/10.1016/j.ceb.2025.102548, https://www.sciencedirect.com/science/journal/09550674.

Jeon, Y., Lee, G.H., Jeong, H., Gong, G., Kim, J.S., Kim, K., et al., 2024. Proteomic analysis of breast cancer based on immune subtypes. Clinical Proteomics 21 (1). Available at https://doi.org/10.1186/s12014-024-09463-y, https://clinicalproteomicsjournal.biomedcentral.com/.

Jiang, X., Coffino, P., Li, X., 2004. Development of a method for screening short-lived proteins using green fluorescent protein. Genome Biology 5 (10). Available at https://doi.org/10.1186/gb-2004-5-10-r81.

Jones, R.B., Gordus, A., Krall, J.A., MacBeath, G., 2006. A quantitative protein interaction network for the ErbB receptors using protein microarrays. Nature 439 (7073), 168–174. Available at https://doi.org/10.1038/nature04177, http://www.nature.com/nature/index.html.

Kaur, I., Jung, S.Y., Austdal, M., Arun, A.K., Helland, T., Mellgren, G., et al., 2024. Quantitative proteomics reveals serum proteome alterations during metastatic disease progression in breast cancer patients. Clinical Proteomics 21 (1). Available at https://doi.org/10.1186/s12014-024-09496-3, https://clinicalproteomicsjournal.biomedcentral.com/.

Kingsmore, S.F., Patel, D.D., 2003. Multiplexed protein profiling on antibody-based microarrays by rolling circle amplification. Current Opinion in Biotechnology 14 (1), 74–81. Available at https://doi.org/10.1016/S0958-1669(02)00019-8, http://www.elsevier.com/locate/copbio.

Klose, J., Nock, C., Herrmann, M., Stühler, K., Marcus, K., Blüggel, M., et al., 2002. Genetic analysis of the mouse brain proteome. Nature Genetics 30 (4), 385–393. Available at https://doi.org/10.1038/ng861.

Kumar, P., Priyasharshini, R.L., Kotra, V., 2012. Peptide microarray production on a solid support and its applications. Research Journal of Pharmacy and Technology 5 (8), 1015–1024.

Kwon, Y.W., Jo, H.S., Bae, S., Seo, Y., Song, P., Song, M., et al., 2021. Application of proteomics in cancer: recent trends and approaches for biomarkers discovery. Frontiers in Medicine 8. Available at https://doi.org/10.3389/fmed.2021.747333, journal.frontiersin.org/journal/medicine.

König, S., Schork, K., Eisenacher, M., 2024. Observations from the proteomics bench. Proteomes 12 (1), 6. Available at https://doi.org/10.3390/proteomes12010006.

Lemmers, R.J.L.F., Van Der Vliet, P.J., Klooster, R., Sacconi, S., Camaño, P., Dauwerse, J.G., et al., 2010. A unifying genetic model for facioscapulohumeral muscular dystrophy. Science (New York, N.Y.) 329 (5999), 1650–1653. Available at https://doi.org/10.1126/science.1189044, http://www.sciencemag.org/cgi/reprint/329/5999/1650.pdf.

Li, M., Zuo, J., Yang, K., Wang, P., Zhou, S., 2024. Proteomics mining of cancer hallmarks on a single-cell resolution. Mass Spectrometry Reviews 43 (5), 1019–1040. Available at https://doi.org/10.1002/mas.21842.

Liang, J., Tian, J., Zhang, H., Li, H., Chen, L., 2025. Proteomics: an in-depth review on recent technical advances and their applications in biomedicine. Medicinal Research Reviews 45 (4),

1021–1044. Available at https://doi.org/10.1002/med.22098, http://onlinelibrary.wiley.com/journal/10.1002/(ISSN)1098-1128.

Liu, S., Li, Z., Yu, B., Wang, S., Shen, Y., Cong, H., 2020. Recent advances on protein separation and purification methods. Advances in Colloid and Interface Science 284, 102254. Available at https://doi.org/10.1016/j.cis.2020.102254.

Liu, H., Wang, J., Zhang, Y., Gu, J., Wang, Y., Yan, Y., et al., 2023. Cerebrospinal fluid proteomics in meningitis patients with reactivated varicella zoster virus. Immunity, Inflammation and Disease 11 (10). Available at https://doi.org/10.1002/iid3.1038.

Liu, X., Abad, L., Chatterjee, L., Cristea, I.M., Varjosalo, M., 2024. Mapping protein–protein interactions by mass spectrometry. Mass Spectrometry Reviews. Available at https://doi.org/10.1002/mas.21887, http://onlinelibrary.wiley.com/journal/10.1002/(ISSN)1098-2787.

Liu, Z., Liu, J., Chachar, Z., Mo, J., Chi, H., Wen, R., et al., 2025. ITRAQ and PRM-based quantitative saliva proteomics in gastric cancer: biomarker discovery. Frontiers in Molecular Biosciences 12. Available at https://doi.org/10.3389/fmolb.2025.1640508, journal.frontiersin.org/journal/molecular-biosciences.

Low, T.Y., vanHeesch, S., vandenToorn, H., Giansanti, P., Cristobal, A., Toonen, P., et al., 2013. Quantitative and qualitative proteome characteristics extracted from in-depth integrated genomics and proteomics analysis. Cell Reports 5 (5), 1469–1478. Available at https://doi.org/10.1016/j.celrep.2013.10.041.

Messner, C.B., Demichev, V., Wendisch, D., Michalick, L., White, M., Freiwald, A., et al., 2020. Ultra-high-throughput clinical proteomics reveals classifiers of COVID-19 infection. Cell Systems 11 (1), 11. Available at https://doi.org/10.1016/j.cels.2020.05.012, http://www.journals.elsevier.com/cell-systems/.

Miah, S., Banks, C.A.S., Adams, M.K., Florens, L., Lukong, K.E., Washburn, M.P., 2017. Advancement of mass spectrometry-based proteomics technologies to explore triple negative breast cancer. Molecular Biosystems 13 (1), 42–55. Available at https://doi.org/10.1039/c6mb00639f, http://www.rsc.org/is/journals/current/mbs/mbspub.htm.

Mischak, H., Schanstra, J.P., Vlahou, A., Beige, J., 2025. Clinical proteomics, quo vadis? Proteomics 25 (7). Available at https://doi.org/10.1002/pmic.202400346, http://onlinelibrary.wiley.com/journal/10.1002/(ISSN)1615-9861.

Molina-Mora, J.A., Chinchilla-Montero, D., Castro-Peña, C., García, F., 2020. Two-dimensional gel electrophoresis (2D-GE) image analysis based on cellprofiler: Pseudomonas aeruginosa AG1 as model. Lippincott Williams and Wilkins, Costa Rica Medicine (United States) 99 (49). Available at https://doi.org/10.1097/MD.0000000000023373, https://journals.lww.com/md-journal/pages/default.aspx.

Momenzadeh, A., Meyer, J.G., 2025. Single-cell proteomics using mass spectrometry. Cell Genomics 5 (9), 100973. Available at https://doi.org/10.1016/j.xgen.2025.100973.

Monti, M., Orrù, S., Pagnozzi, D., Pucci, P., 2005. Functional proteomics. Clinica Chimica Acta 357 (2), 140–150. Available at https://doi.org/10.1016/j.cccn.2005.03.019.

Moradimotlagh, A., Chen, S., Koohbor, S., Moon, K.M., Foster, L.J., Reiner, N., et al., 2023. Leishmania infection upregulates and engages host macrophage Argonaute 1, and system-wide proteomics reveals Argonaute 1-dependent host response. Frontiers in Immunology 14. Available at https://doi.org/10.3389/fimmu.2023.1287539, https://www.frontiersin.org/journals/immunology#.

More, S., Raje, V., Phalke, N., Lokhande, S., 2018. Bioinformatics-an emerging field. Asian Journal of Research in Pharmaceutical Science 8 (4), 185. Available at https://doi.org/10.5958/2231-5659.2018.00032.2.

Mostacciuolo, M.L., Pastorello, E., Vazza, G., Miorin, M., Angelini, C., Tomelleri, G., et al., 2009. Facioscapulohumeral muscular dystrophy: epidemiological and molecular study in a north-east Italian population sample. Clinical Genetics 75 (6), 550–555. Available at https://doi.org/10.1111/j.1399-0004.2009.01158.x.

Moxon, J.V., Padula, M.P., Herbert, B.R., Golledge, J., 2009. Challenges, current status and future perspectives of proteomics in improving understanding, diagnosis and treatment of vascular disease. European Journal of Vascular and Endovascular Surgery 38 (3), 346–355. Available at https://doi.org/10.1016/j.ejvs.2009.05.008.

Nazziwa, J., Freyhult, E., Hong, M.G., Johansson, E., Årman, F., Hare, J., et al., 2024. Dynamics of the blood plasma proteome during hyperacute HIV-1 infection. Nature Communications 15 (1). Available at https://doi.org/10.1038/s41467-024-54848-0, https://www.nature.com/ncomms/.

Ong, S.E., Blagoev, B., Kratchmarova, I., Kristensen, D.B., Steen, H., Pandey, A., et al., 2002. Stable isotope labeling by amino acids in cell culture, SILAC, as a simple and accurate approach to expression proteomics. Molecular & Cellular Proteomics 1 (5), 376–386. Available at https://doi.org/10.1074/mcp.M200025-MCP200.

Pandey, A., Mann, M., 2000. Proteomics to study genes and genomes. Nature 405 (6788), 837–846. Available at https://doi.org/10.1038/35015709.

Peng, Z., Chen, L., Zhang, H., 2020. Serum proteomic analysis of Mycobacterium tuberculosis antigens for discriminating active tuberculosis from latent infection. Journal of International Medical Research 48 (3). Available at https://doi.org/10.1177/0300060520910042.

Perera, N., Kumar, A., Gangadharan, B., Ranasinghe, D., Wijewickrama, A., Malavige, G.N., et al., 2024. Proteomics Analysis of Peripheral Blood Monocytes from Patients in Early Dengue Infection Reveals Potential Markers of Subsequent Fluid Leakage. medRxiv, United Kingdom. Available from: https://www.medrxiv.org/, https://doi.org/10.1101/2024.03.21.24304389.

Petrosius, V., Schoof, E.M., 2023. Recent advances in the field of single-cell proteomics. Translational Oncology 27, 101556. Available at https://doi.org/10.1016/j.tranon.2022.101556.

Posadas, E.M., Simpkins, F., Liotta, L.A., MacDonald, C., Kohn, E.C., 2005. Proteomic analysis for the early detection and rational treatment of cancer—realistic hope? Annals of Oncology 16 (1), 16–22. Available at https://doi.org/10.1093/annonc/mdi004.

Rabilloud, T., Lelong, C., 2011. Two-dimensional gel electrophoresis in proteomics: a tutorial. Journal of Proteomics 74 (10), 1829–1841. Available at https://doi.org/10.1016/j.jprot.2011.05.040.

Ramesh, P., Nisar, M., Neha, Ammankallu, S., Babu, S., Nandakumar, R., Abhinand, C.S., et al., 2024. Delineating protein biomarkers for gastric cancers: a catalogue of mass spectrometry-based markers and assessment of their suitability for targeted proteomics applications. Journal of Proteomics 306. Available at https://doi.org/10.1016/j.jprot.2024.105262, https://www.sciencedirect.com/science/journal/18743919.

Ramraje, G.R., Patil, S.D., Patil, P.H., Pawar, A.R., 2020. A brief review on: separation techniques chromatography. Asian Journal of Pharmaceutical Analysis 10 (4), 231–238. Available at https://doi.org/10.5958/2231-5675.2020.00041.1.

Rappsilber, J., Siniossoglou, S., Hurt, E.C., Mann, M., 2000. A generic strategy to analyze the spatial organization of multi-protein complexes by cross-linking and mass spectrometry. Analytical Chemistry 72 (2), 267–275. Available at https://doi.org/10.1021/ac991081o.

Reymond, M.A., Steinert, R., Eder, F., Lippert, H., 2003. Ethical and regulatory issues arising from proteomic research and technology. Proteomics 3 (8), 1387–1396. Available at https://doi.org/10.1002/pmic.200300481.

Rho, J.H., Ladd, J.J., Li, C.I., Potter, J.D., Zhang, Y., Shelley, D., et al., 2018. Protein and glycomic plasma markers for early detection of adenoma and colon cancer. Gut 67 (3), 473–484. Available at https://doi.org/10.1136/gutjnl-2016-312794, http://gut.bmj.com/content/by/year.

Ross, P.L., Huang, Y.N., Marchese, J.N., Williamson, B., Parker, K., Hattan, S., et al., 2004. Multiplexed protein quantitation in Saccharomyces cerevisiae using amine-reactive isobaric tagging reagents. Molecular and Cellular Proteomics 3 (12), 1154–1169. Available at https://doi.org/10.1074/mcp.M400129-MCP200, http://www.mcponline.org/content/by/year.

Rossi, R., Borroni, E.M., Yusuf, I., Lomagno, A., Hegazi, M.A.A.A., Mauri, P.L., et al., 2025. Uncovering new biomarkers for prostate cancer through proteomic and network analysis. Biology 14 (3). Available at https://doi.org/10.3390/biology14030256, www.mdpi.com/journal/biology.

Rout, M.P., Aitchison, J.D., Suprapto, A., Hjertaas, K., Zhao, Y., Chait, B.T., 2000. The yeast nuclear pore complex. The Journal of Cell Biology 148 (4), 635–652. Available at https://doi.org/10.1083/jcb.148.4.635.

Sathe, G., Na, C.H., Renuse, S., Madugundu, A.K., Albert, M., Moghekar, A., et al., 2019. Quantitative proteomic profiling of cerebrospinal fluid to identify candidate biomarkers for Alzheimer's disease. Proteomics - Clinical Applications 13 (4). Available at https://doi.org/10.1002/prca.201800105, http://onlinelibrary.wiley.com/journal/10.1002/(ISSN)1862-8354.

Scaturro, P., Stukalov, A., Haas, D.A., Cortese, M., Draganova, K., Płaszczyca, A., et al., 2018. An orthogonal proteomic survey uncovers novel Zika virus host factors. Nature 561 (7722), 253–257. Available at https://doi.org/10.1038/s41586-018-0484-5, http://www.nature.com/nature/index.html.

Schulz, T.C., Swistowska, A.M., Liu, Y., Swistowski, A., Palmarini, G., Brimble, S.N., et al., 2007. A large-scale proteomic analysis of human embryonic stem cells. BMC Genomics 8. Available at https://doi.org/10.1186/1471-2164-8-478.

Sharma, S., Singh, N., Ankalgi, A.D., Rana, A., Singh Ashawat, M., 2021. A laconic review on liquid chromatography mass spectrometry (LC-MS) based proteomics technology in drug discovery. Asian Journal of Pharmaceutical Research 194–201. Available at https://doi.org/10.52711/2231-5691.2021.00035.

Shen, B., Yi, X., Sun, Y., Bi, X., Du, J., Zhang, C., et al., 2020. Proteomic and metabolomic characterization of COVID-19 patient sera. Cell 182 (1), 59. Available at https://doi.org/10.1016/j.cell.2020.05.032.

Shvetcov, A., Johnson, E.C.B., Winchester, L.M., Walker, K.A., Wilkins, H.M., Thompson, T.G., et al., 2025. APOE ε4 carriers share immune-related proteomic changes across neurodegenerative diseases. Nature Medicine 31 (8), 2590–2601. Available at https://doi.org/10.1038/s41591-025-03835-z, https://www.nature.com/nm/.

Song, D., Cui, M., Zhao, G., Fan, Z., Nolan, K., Yang, Y., et al., 2014. Pathway-based analysis of breast cancer. American Journal of Translational Research 6 (3), 302–311. Available at http://www.ajtr.org/files/ajtr0000131.pdf.

Stahl-Zeng, J., Lange, V., Ossola, R., Eckhardt, K., Krek, W., Aebersold, R., et al., 2007. High Sensitivity detection of plasma proteins by multiple reaction monitoring of N-glycosites. Molecular & Cellular Proteomics 6 (10), 1809–1817. Available at https://doi.org/10.1074/mcp.m700132-mcp200.

Statland, J.M., McDermott, M.P., Heatwole, C., Martens, W.B., Pandya, S., van der Kooi, E.L., et al., 2013. Reevaluating measures of disease progression in facioscapulohumeral muscular dystrophy. Neuromuscular Disorders 23 (4), 306–312. Available at https://doi.org/10.1016/j.nmd.2013.01.008.

Stefanović, R., Nishimura, Y., Bittel, A., Chen, Y., Burniston, J. 2024. Quantitative proteomic profiling of myoblasts from patients with facioscapulohumeral muscular dystrophy and their unaffected siblings. Belgrade University of Belgrade, Faculty of Chemistry, Belgrade.

Sule, R., Rivera, G., Gomes, A.V., 2023. Western blotting (immunoblotting): history, theory, uses, protocol and problems. Biotechniques 75 (3), 99–114. Available at https://doi.org/10.2144/btn-2022-0034, https://www.future-science.com/journal/btn.

Sun, Y., Jun, A., Liu, Z., Sun, R., Qian, L., Payne, S.H., 2025. Strategic Priorities for Transformative Progress in Advancing Biology With Proteomics and Artificial Intelligence. arXiv. Available at 10.48550/arXiv.2502.15867, https://arxiv.org.

Syu, G.-D., Dunn, J., Zhu, H., 2020. Developments and applications of functional protein microarrays. Molecular & Cellular Proteomics 19 (6), 916–927. Available at https://doi.org/10.1074/mcp.r120.001936.

Sánchez-Pacheco, S.J., Kong, S., Pulido-Santacruz, P., Murphy, R.W., Kubatko, L., 2020. Median-joining network analysis of SARS-CoV-2 genomes is neither phylogenetic nor evolutionary. Proceedings of the National Academy of Sciences of the United States of America 117 (23), 12518–12519. Available at https://doi.org/10.1073/pnas.2007062117, https://www.pnas.org/content/pnas/117/23/12518.full.pdf.

Tan, Y.C., Low, T.Y., Lee, P.Y., Lim, L.C., 2024. Single-cell proteomics by mass spectrometry: advances and implications in cancer research. Proteomics 24 (12-13). Available at https://doi.org/10.1002/pmic.202300210, http://onlinelibrary.wiley.com/journal/10.1002/(ISSN)1615-9861.

Tarnathummanan, C., Soimanee, T., Khattiya, J., Sretapunya, W., Phaonakrop, N., Roytrakul, S., et al., 2024. Plasma proteomic profiles of patients with HIV infection and coinfection with hepatitis B/C virus undergoing anti-retroviral therapy. Biomedical Reports 21 (5). Available at https://doi.org/10.3892/br.2024.1843.

The Cancer Genome Atlas Network, 2012. Comprehensive molecular portraits of human breast tumor. Nature 490, 61–70. Available at https://doi.org/10.1038/nature11412.

Tong, W., Ye, F., He, L., Cui, L., Cui, M., Hu, Y., et al., 2016. Serum biomarker panels for diagnosis of gastric cancer. OncoTargets and Therapy 9, 2455–2463. Available at https://doi.org/10.2147/OTT.S86139, https://www.dovepress.com/getfile.php?fileID=30065.

Vanderaa, C., Gatto, L., 2023. The current state of single-cell proteomics data analysis. Current Protocols 3 (1). Available at https://doi.org/10.1002/cpz1.658.

Vercauteren, F.G.G., Arckens, L., Quirion, R., 2007. Applications and current challenges of proteomic approaches, focusing on two-dimensional electrophoresis. Amino Acids 33 (3), 405–414. Available at https://doi.org/10.1007/s00726-006-0460-5.

Viodé, A., Smolen, K.K., Fatou, B., Wurie, Z., Van Zalm, P., Konde, M.K., et al., 2022. Plasma proteomic analysis distinguishes severity outcomes of human ebola virus disease. mBio 13 (3). Available at https://doi.org/10.1128/mbio.00567-22, https://journals.asm.org/doi/10.1128/mbio.00567-22.

Vittaladevaram, V., 2021. Life of proteomics: technologies, challenges, viewpoints and future. Journal of Proteomics & Bioinformatics 1000533 (14), 1–5.

Wang, X., Chen, C.F., Baker, P.R., Chen, P.L., Kaiser, P., Huang, L., 2007. Mass spectrometric characterization of the affinity-purified human 26S proteasome complex. Biochemistry 46 (11), 3553–3565. Available at https://doi.org/10.1021/bi061994u.

Wang, H., Dey, K.K., Chen, P.C., Li, Y., Niu, M., Cho, J.H., et al., 2020. Integrated analysis of ultra-deep proteomes in cortex, cerebrospinal fluid and serum reveals a mitochondrial signature in Alzheimer's disease. Molecular Neurodegeneration 15 (1). Available at https://doi.org/10.1186/s13024-020-00384-6, http://www.molecularneurodegeneration.com/.

Wang, J., Xia, A., Tang, M., Yang, S., Shen, Y., Dao, J., et al., 2025a. Integrative analysis of serum proteomics and transcriptomics in hepatitis C. Virology Journal 22 (1). Available at https://doi.org/10.1186/s12985-025-02690-1.

Wang, Y., Wang, H., Zhang, Y., Li, W., Gao, S., Qiao, J., et al., 2025b. Enhanced ELISA detection by target protein enrichment with MIP: a new strategy for ultra-sensitive protein detection. Analytica Chimica Acta 1365, 344242. Available at https://doi.org/10.1016/j.aca.2025.344242.

Wenk, D., Zuo, C., Kislinger, T., Sepiashvili, L., 2024. Recent developments in mass-spectrometry-based targeted proteomics of clinical cancer biomarkers. Clinical Proteomics 21 (1). Available at https://doi.org/10.1186/s12014-024-09452-1.

Wilkins, M.R., Sanchez, J.C., Gooley, A.A., Appel, R.D., Humphery-Smith, I., Hochstrasser, D.F., et al., 1996. Progress with proteome projects: why all proteins expressed by a genome should be identified and how to do it. Biotechnology and Genetic Engineering Reviews 13 (1), 19–50. Available at https://doi.org/10.1080/02648725.1996.10647923.

Wiśniewski, J., 2018. Filter-aided sample preparation for proteome analysis. Microbial Proteomics. Humana Press, New York, NY, pp. 3–10. Available at https://doi.org/10.1007/978-1-4939-8695-8_1.

Wolters, D.A., Washburn, M.P., Yates, J.R., 2001. An automated multidimensional protein identification technology for shotgun proteomics. Analytical Chemistry 73 (23), 5683–5690. Available at https://doi.org/10.1021/ac010617e.

Xiao, G.G., Recker, R.R., Deng, H.W., 2008. Recent advances in proteomics and cancer biomarker discovery. Clinical Medicine: Oncology 2, 63–72. Available at https://doi.org/10.4137/cmo.s539, http://www.ncbi.nlm.nih.gov/pmc/articles/PMC3161646/pdf/cmo-2-2008-063.pdf.

Xie, Y., Chen, X., Xu, M., Zheng, X., 2025. Application of the human proteome in disease, diagnosis, and translation into precision medicine: current status and future prospects. Biomedicines 13 (3), 681. Available at https://doi.org/10.3390/biomedicines13030681.

Xiong, J., 2006. Essential Bioinformatics. Cambridge University Press, United States, pp. 1–339. Available at https://doi.org/10.1017/CBO9780511806087, 10.1017/CBO9780511806087.

Xiong, F., Ge, W., Ma, C., 2019. Quantitative proteomics reveals distinct composition of amyloid plaques in Alzheimer's disease. Alzheimer's & Dementia 15 (3), 429–440. Available at https://doi.org/10.1016/j.jalz.2018.10.006.

Xu, T., Wang, Q., Wang, Q., Sun, L., 2024. Mass spectrometry-intensive top-down proteomics: an update on technology advancements and biomedical applications. Analytical Methods 16 (28), 4664–4682. Available at https://doi.org/10.1039/d4ay00651h.

Xun, Z., Yao, X., Zhu, C., Ye, Y., Wu, S., Chen, T., et al., 2022. Proteomic characterization of the natural history of chronic HBV infection revealed by tandem mass tag-based quantitative proteomics approach. Materials Today Bio 15, 100302. Available at https://doi.org/10.1016/j.mtbio.2022.100302.

Yao, X., Freas, A., Ramirez, J., Demirev, P.A., Fenselau, C., 2001. Proteolytic 18O labeling for comparative proteomics: model studies with two serotypes of adenovirus. Analytical Chemistry 73 (13), 2836–2842. Available at https://doi.org/10.1021/ac001404c.

Yates III, J.R., 2019. Recent technical advances in proteomics. F1000Research 8, 351. Available at https://doi.org/10.12688/f1000research.16987.1.

Ye, F., Xin, Z., Han, W., Fan, J., Yin, B., Wu, S., et al., 2015. Quantitative proteomics analysis of the hepatitis C virus replicon high-permissive and low-permissive cell lines. PLoS One 10 (11), e0142082. Available at https://doi.org/10.1371/journal.pone.0142082.

Yoithapprabhunath, T.R., Nirmal, R.M., Santhadevy, A., Anusushanth, A., Charanya, D., Rojiluke, Sri Chinthu, K.K., et al., 2015. Role of proteomics in physiologic and pathologic conditions of dentistry: overview. Journal of Pharmacy and Bioallied Sciences 7 (6), S344. Available at https://doi.org/10.4103/0975-7406.163448, http://www.jpbsonline.org.

Yu, J., Gui, X., Zou, Y., Liu, Q., Yang, Z., An, J., et al., 2024. A proteogenomic analysis of cervical cancer reveals therapeutic and biological insights. Nature Communications 15 (1). Available at https://doi.org/10.1038/s41467-024-53830-0, https://www.nature.com/ncomms/.

Yuan, X., Xu, J., Hussain, S., Wang, H., Gao, N., Zhang, L., 2020. Trends and prediction in daily new cases and deaths of COVID-19 in the United States: an internet search-interest based model. Exploratory Research and Hypothesis in Medicine, 1–6. Available at https://doi.org/10.14218/erhm.2020.00023.

Zhang, H., Liu, T., Zhang, Z., Payne, S.H., Zhang, B., McDermott, J.E., et al., 2016. Integrated proteogenomic characterization of human high-grade serous ovarian cancer. Cell 166 (3), 755–765. Available at https://doi.org/10.1016/j.cell.2016.05.069, www.cell.com.

Zhang, B., Whiteaker, J.R., Hoofnagle, A.N., Baird, G.S., Rodland, K.D., Paulovich, A.G., 2019. Clinical potential of mass spectrometry-based proteogenomics. Nature Reviews Clinical Oncology 16 (4), 256–268. Available at https://doi.org/10.1038/s41571-018-0135-7, http://www.nature.com/nrclinonc/archive/index.html.

Zhao, C., Zhang, X., Wang, H., Qiang, H., Liu, S., Zhang, C., et al., 2025. Proteomic analysis of differentially expressed proteins in A549 cells infected with H9N2 avian influenza virus. International Journal of Molecular Sciences 26 (2), 657. Available at https://doi.org/10.3390/ijms26020657.

Zhou, G., Li, H., DeCamp, D., Chen, S., Shu, H., Gong, Y., et al., 2002. 2D differential in-gel electrophoresis for the identification of esophageal scans cell cancer-specific protein markers. Molecular & Cellular Proteomics: MCP 1 (2), 117–124. Available at https://doi.org/10.1074/mcp.m100015-mcp200.

Chapter 16

Application of proteomic tools in the early diagnosis of diseases

Aisha Singh[1], Jyoti Prakash[2], Shalini Singh[3], and Burhan Hamid[4]
[1]*Government Medical College (Affiliated to Atal Bihari Vajpayee Medical University, Lucknow), Mirzapur, Uttar Pradesh, India,* [2]*Amity Institute of Biotechnology, Amity University, Lucknow, Uttar Pradesh, India,* [3]*Department of Biotechnology, Jagatpur PG College (Affiliated to Kashi Vidhyapith University), Varanasi, Uttar Pradesh, India,* [4]*Centre of Research for Development, University of Kashmir, Srinagar, Jammu and Kashmir, India*

16.1 Introduction

The study of protein expression, structural changes, and protein-protein interactions in various diseases has been a crucial instrument in biomedical studies through proteomics. Finally, comprehensive protein profiling will enable the researcher to gain insight into the pathways underlying diseases, promising targets in therapeutic intervention, and provide personalized treatment strategies (Smith et al., 2021). The discovery of biomarkers for cancer, cardiovascular diseases, or neurodegenerative diseases can be made more accurate and efficient by emerging technologies (Birhanu, 2023). LC-MS plays a major role in the discovery of biomarkers that help in the detection of low-abundance molecules at high sensitivity and specificity. Additionally, it is possible to note that HPLC-MS has recently been used in the diagnostics of metabolic and neurological disorders (e.g., Fabry's disease, hemoglobinopathies) (Kumari et al., 2024). Protein chip arrays and the Proteograph XT system have increased the sensitivity of plasma proteomics to low-abundance proteins to permit the discovery of early biomarkers, streamline the diagnosis, and help in the development of vaccines (Soni, 2024). The other notable findings are MTA2 and AGO2 biomarkers at high levels are an evaluation of the progression of multiple myeloma and may be a sign of poor patient outcome (Lolansen et al., 2024). Furthermore, CSF alterations such as S100B, TNF-06, and IL-6 may be used in distinguishing the hemorrhagic and other, non-hemorrhagic, sources of brain injury. Liquid biopsy is another emerging and less invasive diagnostic method of identifying tumor-based nucleic acids and proteins in biofluids in a real-time mode (Faden et al., 2025).

Proteomics Driven Bioremediation. DOI: https://doi.org/10.1016/B978-0-443-45294-9.00003-5

The CRISPR/Cas9 system has become a new way of genome editing and targets the chimeric antigen receptor T cells to the target site of the genome identified with a tumor gene, and helps in identifying the activity of cancer-associated genes (Aslan et al., 2023). The method is even more effective when it is coupled with proteomic profiling, such that both aids can be applied to study the effect of genetic variants and protein alterations on disease and to monitor new therapies. Natural nanocarriers generate the target vehicle to raise the alliance by carrying CRISPR drugs to the target cells, thereby facilitating noninvasive and tissue-specific gene modification, and can be able to cross blood-brain barriers, promote nervous neural functioning systems (Pauper et al., 2025). More sophisticated ML methods are associated with the clinical research of immune-mediated inflammatory conditions, including autoimmune rheumatic diseases (ARDs), inflammatory bowel disease (IBD), immune-mediated chronic kidney disease (CKD), and multiple sclerosis (MS), which may be used to invent personalized medicine strategies and to update the disease management (Peng et al., 2021).

Multiomics data, such as transcriptomics and genomics, play an important role in providing integrative disease signatures. Clinical Proteomic Tumor Analysis Consortium (CPTAC) Pan-Cancer Analysis is the alignment of proteogenomic data on more than 1000 tumors, which corresponds to different cancer types (Li and Gao, 2023). The significance of the quick testing nanosensor technologies in the context of both in vitro diagnostics (IVD) and point-of-care testing (POCT) has been a serious development in the medical field. Specifically, biochips and lab-on-a-chip (LoC) have revolutionized the IVD-supported POCT solutions, allowing them to become quicker, more dependable, and affordable (Khan et al., 2024a,b). Nanosensor-based POCT devices are altering the future of diagnostic medicine by offering instant results, even though they are also useful in critical cases where rapid diagnosis is paramount, such as management of infectious diseases and emergency treatment (Mucha et al., 2023). An enhanced sensitivity, simple detection, and early detection and diagnosis of diseases such as SARS-CoV-2, Ebola, malaria, cholera, and so on are provided by a biosensor embedded with a microfluidic chip. Other substitute biosensor enhancement techniques besides microfluidics are nanoparticles (Ting Sim et al., 2023), machine learning (Kokabi et al., 2023), and quantum technologies (Mpofu et al., 2023), which have increased the diagnostic applications. The enhancement of biosensors using nanomaterials such as carbon nanotubes, gold nanoparticles, and materials made of graphene has brought remarkable improvement to the field of signal transduction, limits of detection, and surface functionalisation of biosensors that have resulted in interaction of biosensors in healthcare, detection of pollutants and pathogens in the environment, and rapidity of detection of contaminants (Kadhim et al., 2023).

16.1.1 High performance liquid chromatography–tandem mass spectrometry

Mass spectrometry offers a highly effective tool to diagnose rare diseases through studying both protein interactions and post-translational modification, which can assist in understanding the disease processes and potential therapeutic interactions. Besides this, MS-based metabolomics and lipidomics have improved the screening of newborn bloodspots (NBS), which has enabled improved early interpretation of metabolic disorders (Zhu et al., 2022). Also, MS-based proteomics has contributed to the diagnosis of such mitochondrial diseases as Leigh syndrome, associated with the malfunction of 16 mitochondrial DNA genes and approximately 100 nuclear genes (Rahman, 2023). The use of MS has shown genes of certain inherited disorders like FHL1, EXOC2, and COL3A1 that affect the process of signal transduction, vesicle traffic, and collagen synthesis. FHL1, which is a large constituent of intracytoplasmic inclusion, is an X-linked muscular dystrophy, leading to benign scapuloperoneal muscular dystrophy and cardiopathy (Caputo and Schoser, 2024). The EXOC2 protein that has exocyst complex elements in patient fibroblasts causes a serious neurodevelopmental disorder, including developmental delays, epilepsy, and brain abnormalities (Begen et al., 2020). The pertinent research findings on the disrupted extracellular matrix proteins that are also associated with pathogenic variants further underline the efficiency of MS-based proteomics application to investigate the COL3A1 gene associated with causing vascular Ehlers-Danlos syndrome (vEDS) (Foehr et al., 2024). The proteomic analysis with mass spectrometry was performed on the skin fibroblasts, and it was found that the protein MRPL39 has changed significantly between the large mitochondrial subunit. Also, there are other lineage-specific genes, such as MRPL50, MRPS39, TEFM, and HMGCS1, that have been identified by LC-MS. A mutation in the MRPL50 type is known to cause destabilization of the mitochondrial ribosome that causes depletion of oxidative phosphorylation and, in addition, causes premature ovarian insufficiency. The MRPS3 mutation is related to the Leigh-like syndrome and combined oxidative phosphorylation deficiency. TEFM variants produce defects in the mitochondrial respiratory chain, mitochondrial myopathy, as well as neuromuscular transmission defects and rigid spine syndrome, which is an infrequent genetic disease identified with impairment in HMGCS1 (Koludarova and Battersby, 2024).

LC-MS/MS is also being applied in the detection, quantification, and characterization of low-abundance biomarkers in the situations of Alzheimer's, Parkinson's, and ovarian cancer (Reveglia et al., 2021). Amyloid deposition is an apt parameter to measure that was concomitantly immunoprecipitated into plasma, digested, and measured by LC-MS/MS utilizing (Aβ) 40, Aβ 42, and apolipoprotein E (ApoE) (Williams et al., 2025). In ovarian malignancy, the LC-MS-based system measures the four most commonly used poly (ADP-ribose) polymerase inhibitors, which include niraparib, olaparib, and pamiparib (Wenk et al., 2024). The results of LC-MS/MS were confirmed in ELISA of transmembrane protein 143, cadherin 5, fibronectin 1, and collectin-11 in serum, which might help in diagnosing adenocarcinoma. LC-MS/MS is useful in molecular subtyping, that is, differentiating between cancerous cells of adenocarcinoma and squamous cell carcinoma in non-small cell lung cancer, also providing absolute quantifications of tumor-specific markers in tumors such as prostate and colorectal (Zhou et al., 2024). It is important that this technology is used to confirm the therapeutic targets and the responses to the treatment through changes in the biomarkers, especially in HER2-positive breast cancers and EGFR-mutated lung cancers (Khan et al., 2024a,b).

The MALDI-TOF-MS technologies, BRUKER and ZYBIO, can offer results to identify microorganisms quickly and accurately, which is an important aspect to consider to guarantee the maintenance of the microbiological safety of the hospital environment, and broad reference databases to increase the diagnostic accuracy. MALDI-TOF-MS has high-speed detection of peptides, which include putative biomarkers in plasma, urine, saliva, and serum (Sausa et al., 2023). MALDI-TOF MS in combination with 2D gel electrophoresis is a strong technique in the identification of biomarker proteins, including pentraxin-2 (PTX2), mesoderm development candidate-1(MESDC1) gene, plasma membrane calcium-transporting ATPase-1, plasminogen, gelsolin, inversin, and apolipoprotein A1 in urothelial bladder cancer diagnosis (Nedjadi et al., 2021). MALDI-TOF MS findings examined vitamin D-binding protein, haptoglobin, transferrin, fibrinogen, IgM, complement C3b, alpha-2-macroglobulin, and pigment epithelium-derived as potential genetic biomarkers of bladder cancer (Lemanska-Perek et al., 2019). Recent developments involve the development of MALDI-TOF MS probes that detect upregulation of carboxylesterase (CE) activity, which is increased in hepatocellular carcinoma and is likely to be a potential biomarker in diagnostic specificity. The potential biomarkers found in ovarian cancer by a MALDI-TOF/MS study included Vimentin-linked to epithelial mesenchymal transition (EMT) and linked to metastasis, as well as chains of tubulin 2 (2C) and 2D (alpha 1C), the latter were involved in microtubule dynamics, and dysregulated expression of collagen alpha 2(VI), which may contribute to tumor invasion of colorectal cancer (Wei et al., 2024). Therefore the combination of mass spectrometry-based proteomics, liquid chromatography-tandem mass spectrometry, matrix-assisted laser desorption/ionization time-of-flight mass spectrometry, and bioinformatics analysis method has extensively benefited in the early illness diagnosis with the focus on the discovery of biomarkers, personalized therapy, and medication with an emphasis on cancer, mitochondrial diseases, metabolic, and infectious diseases.

16.1.2 Protein microarray

The protein microarray platform is a method that analyses native-conformation proteins and is applied to identify possible autoantibodies and discover new biomarkers, facilitating better diagnosis. Autoantibodies act as early predictors of autoimmune diseases, including systemic lupus erythematosus (SLE), systemic sclerosis (SS), and rheumatoid arthritis (RA), which makes it suitable to intervene and develop an individual approach to treatment (Li et al., 2021). It detected 1,600 antigens that are immune-related and 381 autoantigens in patients with SLE. Other autoantibodies, like DEK proto-oncogene, are associated with the arrangement of chromatin; others associated with transcription regulation include recombination exonuclease 4, E26 transformation-specific transcription factor ELF4, and NF-kB activity. By means of high-density protein microarray research, approximately 844 visualized proteins comprised of citrullinated proteins were identified by IgG autoantibodies in patients with rheumatoid arthritis. This study highlights the importance of citrullination in the development of RA and hypothesizes that protein binding patterns of autoantibodies can be used as a diagnostic item of RA (Urbanowska et al., 2023).

The proteomic antibody microarray technique was applied to investigate the serum proteins, such as calcitonin and sclerostin, that can be used as potential biomarkers of pulmonary arterial hypertension (Huan et al., 2025). Gene expression patterns and protein-protein interactions network patterns of different tissues of patients with systemic sclerosis focus on the crucial issue of tumor necrosis factor (TNF) signaling and the pathways related to inflammation, immune responses, and fibrosis (Kapoor et al., 2024). Moreover, a high-performance protein microarray system is created to monitor biomarkers predicting the progression of RA, including IL-1b, IL-6, IL-8, MCP-1, and SAA. CircRNA profiling with the use of microarray protocols has demonstrated that hsa-circ-0003972 and hsa-circ-0007288 are novel peripheral biomarkers, which, together with CA-125 and decreased levels of plasma hsa-circ-0007288, may be a strong diagnostic indicator of lymph node metastasis in ovarian cancer (Ge et al., 2022).

Protein microarray is used to identify the biomarkers associated with Alzheimer's, Parkinson's, and multiple sclerosis, and it was found that the level of soluble CD14 in Alzheimer patients and osteopontin in Parkinson patients was significantly higher, which may enable them to be used as possible biomarkers in cerebrospinal fluid. On the other hand, macular degeneration in the brain has been characterized by the loss of neurons associated with protein: phosphorylated tau (p-Tau), amyloid-β, alpha-synuclein, and TDP-43 in CSF (Tondo and Marchi, 2022). These biomarkers have been identified as being useful in the process of neuronal degeneration: plasma neurofilament light chain, ubiquitin C-terminal-hydrolase-L1, and 2-2-α-synuclein, and glial fibrillary acidic protein have been proposed as being useful in indicating glial cell dysfunction (Alcolea et al., 2023). Besides, microarray technology has examined gene GADD45G affecting the activity of astrocytes and microglia in the brain, which is associated with degenerative neurologic disorders such as Alzheimer's, Parkinson's, and Huntington's disease. Conversely, silencing of the GADD45G gene in astrocytes spares synapses and controls the mitogen-activated protein kinase 4 (MAP3K4) and neural immune response, including nuclear factor kappa B (NF-kB) and interferon regulatory factor 3 (IRF3). These findings reveal GADD45G signaling as a potent treatment of AD and help in repairing DNA, and triggering cell apoptosis (Shen et al., 2025).

There has been a development of a fluorescent microarray chip that can detect cardiovascular disease early by identifying cardiac biomarkers in plasma, cardiac troponin I (cTnI), and cardiac fatty acid-binding protein (H-FABP), and also can detect C-reactive protein, cystatin C, and may become an essential asset to evaluate myocardial infarction (Colom et al., 2023). In addition, a SARS-CoV-2 protein-based multiantigen microarray of proteins was designed to study the effects of COVID-19 on antibody reactions (Celikgil et al., 2023). Moreover, proteomics and metabolomics have amplified the identification of biomarkers in *Pseudomonas aeruginosa* pneumonia. Those techniques have characterized neutrophil extracellular traps (NETs) and oxidative stress-related tooth proteins like LCN2, CALR, and TPI1. An innovative multiplex protein microarray system has been developed capable of measuring over 400 different antigens and both IgM and IgG antibodies against *Borrelia burgdorferi*, the pathogen of Lyme disease. This development supports the early diagnosis, decreases the chance of false diagnosis, and allows patients to be administered the right therapy (Abene and Barker, 2025).

16.1.3 Screening based on plasma proteomics

Plasma proteomics can provide an opportunity to become better at disease recognition and examine various types of cancers, heart diseases, and neurodegenerative disorders, which would reshape the risk management and continuous monitoring of the disease. The investigators used the Olink Explore-3072 platform and found that 29 proteins, including complement proteins (CFB, C3, C8G, C1QA, C1R, C6), orosomucoid proteins (ORM1, ORM2), ceruloplasmin (CP), alpha-1-B glycoprotein (A1BG), and many others, were detected. These biomarkers are essential in underlining the cancerous nature of the tumors that grow in lung tissue. Run on a Roche Elecsys 2010 Immunoassay Analyser, plasma protein tumor markers including CEA, CA15.3, CYFRA 21-1, ProGRP, and SCC, which have been proven to be useful in the early detection of lung cancer. Important plasma biomarkers in neurodegenerative diseases can distinguish amyloid-beta (Aβ42/40), phosphorylated tau (p-Tau), neurofilament light (NfL), and glial fibrillary acidic protein (GFAP), and can provide valuable information on the nature of amyloid pathology, neuronal damage, and neuroinflammation (Bolsewig et al., 2024). A critical look at the plasma protein biomarkers can give an in-depth understanding of various diseases (Table 16.1).

TABLE 16.1 Clinically relevant plasma protein biomarkers.

Biomarker	Function	Applications	References
Infectious disease			
C-reactive protein	Indicator of inflammation	Bacterial infections, sepsis	Bouayed et al. (2022); Devi et al. (2022)
Procalcitonin	Elevated in bacterial infections	Differentiating bacterial vs. viral infections	Mućka et al. (2025); Wang et al. (2025)
Pentraxin 3	Innate immune system activation	Sepsis, systemic infections	Wang et al. (2022)
Serum amyloid A	Acute-phase response marker	Viral and bacterial infections	Mohanty et al. (2025); Cheng et al. (2024)
Interleukin-6	Pro-inflammatory cytokine	Severe infections, immune response	Del Valle et al. (2020); Zeng et al. (2022)
Interleukin-10	Anti-inflammatory cytokine	Immunosuppression in infections	Guo et al. (2025)
Alpha-1 antitrypsin	Protease inhibitor and immune regulator	Tuberculosis, bacterial infections	Ali et al. (2025)
Haptoglobin	Acute-phase protein involved in hemoglobin clearance	Malaria, hemolytic infections	Foko et al. (2022); Tiwari et al. (2021)
Ferritin	An iron storage protein associated with the activation of the immune system	COVID-19 severity, systemic infections	Nikkhoo et al. (2023)

(*Continued*)

TABLE 16.1 (Continued)

Biomarker	Function	Applications	References
D-dimer	Coagulation marker	COVID-19, sepsis-related coagulation	Valenzuela et al. (2025)
Lipopolysaccharide-binding protein	Endotoxin response marker	Gram-negative bacterial infections	Park et al. (2023)
Soluble CD14	Monocyte activation marker	Sepsis, systemic infections	García-García et al. (2023)
Cancer			
Alpha-fetoprotein	Tumor marker for liver cancer	Hepatocellular carcinoma	Norman et al. (2025); Ucdal et al. (2025)
Carcinoembryonic antigen	Indicator of tumor burden	Colorectal, lung, and breast cancer	Kim et al. (2025)
CA-125	Glycoprotein linked to ovarian cancer	Ovarian cancer	Yin et al. (2023)
CA-19-9	Tumor-associated antigen	Pancreatic cancer	Kwon et al. (2021)
Prostate-specific antigen	Prostate cancer screening marker	Prostate cancer	Banerjee et al. (2024)
HER2/neu	Growth factor receptor	Breast cancer	Giuliano et al. (2022); Moasser et al. (2023)
CYFRA 21-1	Cytokeratin fragment	Lung cancer	Liu et al. (2022)
Hepatocyte Growth Factor (HGF)	Tumor progression marker	Various solid tumors	Zeng et al. (2023)
Vascular Endothelial Growth Factor	Angiogenesis regulator	Tumor growth and metastasis	Mokhtari et al. (2024); Niu et al. (2025a,b)

BCR-ABL	Fusion gene linked to leukemia	Chronic myeloid leukemia	Jabbour et al. (2023)
BRCA1 / BRCA2	Mutations raise the risk of DNA repair genes.	Breast, ovarian cancer	Evans et al. (2023); Nielsen et al. (2022)
BRAF V600E	Mutation associated with aggressive tumors	Melanoma, colorectal cancer	Kondo et al. (2024)
EGFR	Growth factor receptor, mutation-driven cancer	Non-small-cell lung carcinoma	Niu et al. (2025a,b)
HSP70	Stress response parasite: heat shock protein.	Various solid tumors	Evans et al. (2022)
Neurodegenerative disease			
Amyloid-beta (Aβ42/Aβ40)	Indicator of amyloid plaque formation	Indicator of amyloid plaque formation	Jack et al. (2023); Palmqvist et al. (2020)
Tau (Total Tau, p-Tau)	Neurofibrillary tangle marker	Neurofibrillary tangle marker	Janelidze et al. (2020)
Glial fibrillary acidic protein	Astrocyte activation marker	Astrocyte activation marker	Jain et al. (2023a,b)
α-Synuclein	Aggregation marker	Aggregation marker	Brundin et al. (2022)
TDP-43	Protein misfolding marker	Protein misfolding marker	Neumann et al. (2022)
Complement C3/C4	Immune response marker	Immune response marker	Hanan et al. (2022); Stevens et al. (2023)
Apolipoprotein E	Lipid metabolism and neuroprotection	Lipid metabolism and neuroprotection	Liu et al. (2021)

(Continued)

TABLE 16.1 (Continued)

Biomarker	Function	Applications	References
YKL-40	Neuroinflammatory marker	Neuroinflammatory marker	Jain et al. (2023a,b)
Ubiquitin C-terminal hydrolase-L1	Neuronal injury marker	Traumatic brain injury, neurodegeneration	Chiu et al. (2023)
S100B	Calcium-binding protein, linked to neuroinflammation	Brain injury, neurodegenerative diseases	Takahashi et al. (2022)
DJ-1	biomarkers of oxidative stress	Parkinson's disease	Korecka et al. (2021)
Visinin-like protein 1	Neuronal injury marker	Alzheimer's disease, cognitive decline	Tarawneh et al. (2020); Zetterberg et al. (2021)

Protein–protein interaction-based eXplainable graph propagational network (PPIxGPN) is a novel approach developed to analyze the expression activity of neurodegenerative biomarkers, such as Aβ, GFAP, NFL, and p-Tau. In particular, in Alzheimer's disease, one can relate to connections with proteins that control the complement system and act in line with microglial activity. In the case of Parkinson's disease, a finding correlated with the interleukin-6 pathway mediated by CTF1. In amyotrophic lateral sclerosis (ALS), TPP1 and TNFSF13 are tied to the lysosomal and astrocytic functions. In contrast, multiple sclerosis is associated with proteins necessary to support the integrity of the blood-brain barrier and immunological reactions (Belbasis et al., 2025).

This work emphasizes the importance of cardiovascular biomarkers in the diagnosis including enzyme markers (AST, LDH, CK-MB, CA-III and HBDH); protein markers (CRP, MPO, H-FABP, and cardiac Troponins), which are the gold standard for diagnosing myocardial infarction due to their high specificity; peptide markers (BNP and NT-proBNP) useful in diagnosing and managing heart failure by signaling ventricular stress; lipid markers (phospholipase A2 and oxylipins) that reflect arterialsclerosis and lipid metabolism; and inflammatory indicators such as TNFα, GDF-15, IL-6, and Pentraxin 3 that underscore chronic inflammation in CVDs. Moreover, other developments in the field of biomarkers are the microRNAs and long non-coding RNAs, which provide a sensitive tool for early detection (Netala et al., 2025). Important laboratory parameters during the COVID-19 pandemic are cardiac variables such as creatine kinase-MB fraction, procalcitonin, B-type natriuretic peptide, N-terminal pro-B-type natriuretic peptide, troponin, and d-dimer, which are valuable for diagnosis and to forecast the outcomes. It is also important to note that MMP12, GDF-15, and EDA2R proteins appeared to be powerful predictors of CAD and can be regarded as promising biomarkers and therapeutic targets (Zhang et al., 2024a).

16.1.4 Liquid biopsy

Liquid biopsies make use of different cancer-related biomarkers, such as circulating tumor cells (CTCs), circulating tumor DNA (ctDNA), tumor extracellular vesicles (EVs), tumor-educated platelets (TEPs), and platelet-free RNA (cfRNA), that their importance in the early detection of the disease, monitoring of the metastasis process, and assessing the progression of the cancer. CTCs are thought to be front-runners that show features of transitioning towards the epithelial-to-mesenchymal transition (EMT), which is dislodged from the main tumor and circulated in the bloodstream to increase the possibility of metastasis (Batista et al., 2025). In another technique, a fragment of circulating tumor DNA (ctDNA) that is released into the blood through tumor cell death may be measured through a procedure referred to as deep (80x) whole-genome TET-Assisted Pyridine Borane Sequencing (TAPS), which has the potential to analyze genomic and methylomic data. Sensitivity and accuracy of detection of circulating tumor DNA (ctDNA) in blood have improved with DNA sequencing, generating a clinical research explosion into the use of this technology in lung cancer, colorectal cancer, and breast cancer. Also, the opening of TriOx, which is a machine learning-oriented blood test that studies the DNA methylation patterns in the bloodstream by turning early cancer diagnosis into the routine screening of blood glucose testing (Parums, 2025).

Markou et al. (2023) presented a pilot study that aimed to determine the possible value of the germline mutations in the DNA circulating in the blood plasma as cell-free DNA concerning the risk of recurrence in patients with early-stage non-small cell lung cancer. This study identified oncogene mutations of the hotspots, namely RAF, EGFR, KRAS, and PIK3CA, to gain additional insight into tumor biology and uncover potential treatment targets. Salvianti et al. (2021) evaluated the prognostic value of the CTCs and cell-free DNA (cfDNA) that involved 20 patients with metastatic colorectal cancer (mCRC) having the KRAS/NRAS mutations. Their results showed that the higher CTCs at diagnosis, and the mutation changes in the gene, such as the TP53, APC, and SMAD4, were in relation with poor survival and progression-free survival. In the study, the analysis of CTCs and cell-free DNA (cfDNA) in patients with advanced gastric cancer undergoing neoadjuvant therapy that comprised Paclitaxel, S-1, and Oxaliplatin (PSOX) as a cancer detector was analyzed. Kobayashi et al. (2022) analyzed bladder cancer, urothelial cancer, and upper tract urothelial cancer (UTUC) in terms of CTCs and cell-free DNA (cfDNA) using next-generation sequencing. The proto-oncogene receptor tyrosine kinase (KIT), platelet-derived growth factor receptor alpha (PDGFRα), harvey rat sarcoma viral oncogene homolog (HRAS), mutL homolog 1 (MLH1), PTP-enzyme phosphatase (PTEN), and proto-oncogene were typical of WBC, CTC, and cfDNA, enhancing the mutation detection efficiency and highlighting opportunities of liquid biopsies in targeted diagnostics and personalized cancer treatment of malignant diseases.

The Novel BioSource of Cancer Detection lays stress on the addition of tumor-trained platelets to liquid biopsy technology, along with circulating tumor DNA and extracellular vesicles, to enhance early cancer diagnosis. The study explores the role of the small nuclear RNAs in regulating the splicing of RNA in platelets and their future application in the context of cancer biomarkers (Schwarz et al., 2021). RARE-seq is a very sensitive technique to discover cfRNA, which has received considerable interest in the area of disease detection of cancer, and helps to provide a non-invasive contribution to gene expression. It outperformed (ctDNA) analysis in the context of detecting histological changes and mutation-driven resistance models in EGFR-mutant NSCLC patients. Also, RARE-seq demonstrated potential in defining tissue origin, evaluating harmless pulmonary diseases, and tracing mRNA vaccine reactions, indicating its extensive clinical prospects. In addition, the transcript-level, single-sample analysis of gene set enrichment analysis (ssGSEA) was performed on 34 chronic lymphocytic leukemia (CLL) patients, which showed a reduction in the levels of B cells and an increase in the levels of NK cells, a significant increase in the levels of macrophages, endothelial cells, fibroblasts, and hepatocytes (Radko et al., 2024).

16.1.5 CRISPR/Cas system

Early disease diagnosis has assumed a new form and embraced CRISPR as a more convenient and reliable method for disease diagnosis. The most recent ones to be utilized in the identification of nucleic acids are Cas12 (Type V) and Cas13 (Type VI) due to their capability to cut the DNA in a manner that makes it easier to locate it when subjected to tests (Hu et al., 2024). The limitation has been mitigated by the introduction of a CRISPR/Cas12a-triggered biosensor, which integrates CRISPR/Cas12a-mediated trans-cleavage with Surface-Enhanced Raman Scattering (SERS) that provides the opportunity to serve as an effective gene-editing tool. SpCas9 can repair mutations associated with genetic diseases among all its variants. CRISPR biosensors deliver detectable signals through the establishment of ribonucleoprotein (RNP) complexes when bound to CRISPR RNA in the process of ultra-sensitive biological marker detection of illness (Elabiad, 2025).

Two recent state-of-the-art CRISPR-based technologies, SHERLOCK and DETECTR, have already attracted significant buzz due to their unique technologies of genetic sequence recognition. SHERLOCK is based on the CRISPR-Cas13 system that recognizes RNA and, therefore, is particularly effective in detecting the presence of Dengue and Zika viruses. The binding of Cas13 to its target RNA triggers the cleavage of the nearby reporter RNA, and this results in the generation of a fluorescence signal, which aids in measuring the presence of a pathogen. On the other hand, DETECTR is based on the principle of CRISPR-Cas12, the system used to target DNA sequences, which allows this method to be particularly effective with bacterial infections and viruses based on DNA, including human papillomavirus (HPV). SHERlock and DETECTR represent the next in terms of disease detection and have the potential to revolutionize healthcare globally (Hassan et al., 2025).

Scientists have created the CRISPR/Cas13a/photoelectrochemical (PEC) biosensing platform that allows direct and highly sensitive detection of miRNA-21 and leverages specific miRNA-21and crRNA-21 interaction to avoid signal distortion and increase sensitivity and accuracy. The binding of crRNA-21 to miRNA-21 results in the activation of the trans-cleavage activity of CRISPR/Cas13a, which causes biotin-modified uracil-containing DNA to be non-specifically cleaved. Consequently, biotin-rU-DNA is incapable of binding to the electrode surface, which decreases streptavidin binding and decreases spatial resistance, which relates to enhanced photocurrent response. This Cas-PEC biosensor has excellent analysis properties with the ability of detection in the range of 10 fM to 10 nM and a remarkable limit of detection at

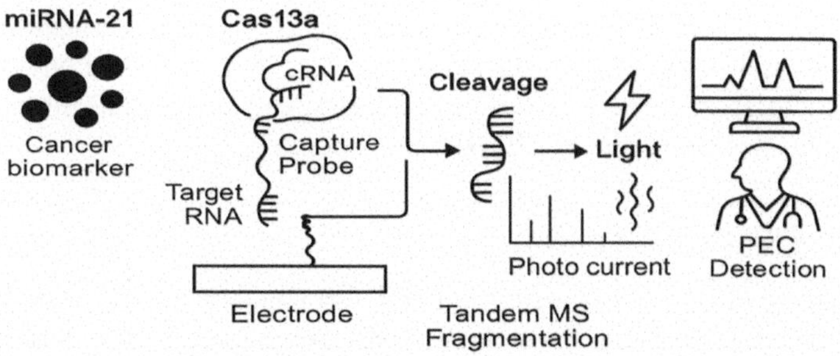

FIGURE 16.1 Schematic representation of CRISPR/Cas-13a photoelectrochemical biosensor for mRNA-21 detection.

9 fM and high recoveries with real human serum samples (Zhang et al., 2024b). Jiang et al. (2023) established the CRISPR/Cas13a-based PEC biosensor, which can measure miRNA-21, which is a significant biomarker of cancer (Fig. 16.1).

Recent advances in CRISPR-based electrochemical biosensors have significantly improved the limit of detection and precision of RNA and viral mutation detection, particularly in the case of SARS-CoV-2 diagnosis. An electrochemical biosensor that combines the CRISPR/Cas system and a catalytic hairpin DNA circuit, which was capable of measuring RNA sequentially with a high level of sensitivity, increasing the speed of detection, and is useful in the management of epidemics and disease monitoring. CRISPR-Cas13a-based electrochemical biosensors have been developed to directly target the detection of the L452R mutation in SARS-CoV-2 variants. This biosensor utilizes MXene-AuNP (a gold nanoparticle) composite electrodes to enhance the sensitivity of the biosensor. The study showcases the fact that the biosensor can be used to monitor variants in real time, which could enhance the response mechanisms of pandemics. Also, graphene field-effect transistor (GFET)-based biosensors utilize graphene, a high-conductivity two-dimensional material, to detect numerous biomolecules, including the SARS-CoV-2 virus causing COVID-19 at high speed (Sengupta and Hussain, 2024; Fig. 16.2)

CRISPR-based biosensors have been developed to be powerful diagnostic methods in the detection of bacterial infections with high sensitivity and accuracy. The problem under consideration in this area is a drug-resistant strain of methicillin-resistant *Staphylococcus aureus* (MRSA), which is associated with severe health complications and causes more deaths. To address the problem, Wei et al. (2023) added CRISPR/Cas12a, on-particle rolling circle amplification, and triggered omnipresence, which leads to the detection of the pathogen more quickly. Chen et al. (2024) developed a colorimetric biosensor using magnetic separation, where different isothermal amplification methods have been coupled with CRISPR/Cas14a to provide

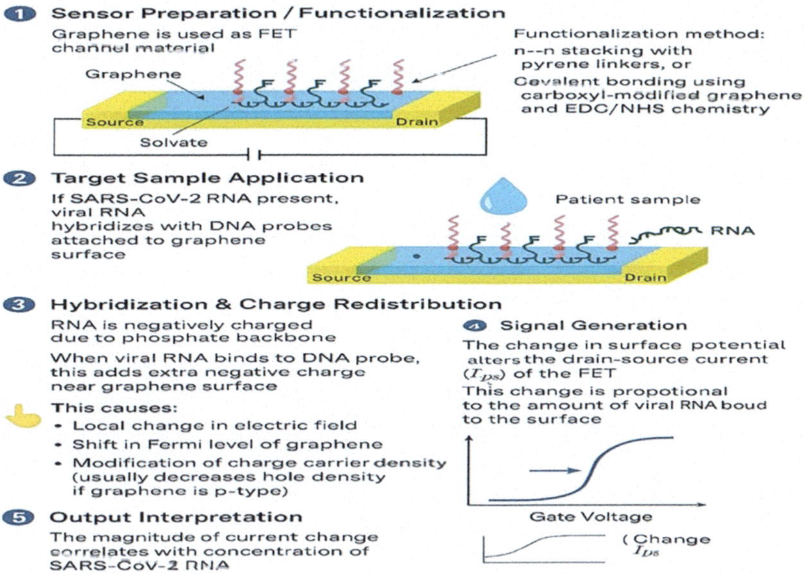

FIGURE 16.2 Schematic mechanism of SARS-CoV-2 RNA detection with FET graphene sensor.

faster onsite detection of disease. Zhou et al. (2022) have come up with a new bacterial diagnostic instrument that utilizes a CRISPR/Cas12a-based fluorescence-enhanced lateral flow biosensor focusing on functionalized quantum dots with recombinase-assisted amplification as a means of identifying *Staphylococcus aureus*. Researchers at the University of Nebraska Medical Center (UNMC) and Temple University have made a landmark advancement in CRISPR-based HIV-related studies and identified that the centre successfully eliminated HIV-1 DNA in infected human immune cells through CRISPR-Cas9 gene editing (Ramirez, 2025).

16.1.6 Artificial intelligence-integrated nanosensors

The design of nanoparticles can be optimized using AI algorithms to enable more efficient and targeted delivery of drugs to the tumor cells. Nanobiosensors using AI have made significant progress in detecting bacteria at low concentrations, using genomic data to assess the potential threat of antibiotic-resistant strains, and have presented new approaches to treat bacterial infections, such as the creation of new drugs and vaccines (Ting Sim et al., 2023). This study has emphasized different methods engineered nanobiosensors employ to detect and treat bacterial infections early, using different techniques, such as electrochemical, acoustic, electromagnetic, and photothermal resonance. An illustrative example is electrochemical biosensors, which rapidly detect bacterial metabolites; acoustic biosensors, which are based on sound waves detecting bacteria in a sample; electromagnetic biosensors, which enhance detection of signal using nanomaterials; and photothermal resonance biosensors, which differentiate the bacterial strains using light-based heating (Khan et al., 2024a,b; Lim et al., 2020). Also, the combination of biosensing technologies with AI to detect and diagnose COVID-19 fast proves the effectiveness of AI-enhanced biosensors to increase the sensitivity, specificity, and real-time tracking of diseases to detect them earlier. Nanotechnology with AI technology has significantly reduced the time and cost incurred in drug discovery and has also improved the evaluation of drug pharmacodynamics and pharmacokinetics. Zhang et al. (2021) showed how to use AI algorithms and have shown that rod-shaped nanoparticles can be used to treat liver cancer. This method enhances the concentration of drugs in the target, minimizes the off-target side effects, and accomplishes a more significant synergistic action of drugs, being less toxic, which enhances the dosing, targeting, and efficacy of nanoparticles.

Electronic-nose technology is swirling as a revolutionary technique in diagnosing cancer, as it offers a less invasive and cost-efficient technique for early detection of lung cancer. This technology represents the use of biosensors to detect gases and send them to a recognition system. Nevertheless, its larger usage encounters challenges related to the reliability of the sensors used, unstable methods of collecting breaths, differences in demographics, and is hampered by challenges in the analysis of data. In spite of these challenges, the market of the ENose technology is growing as a result of increasing incidences of chronic diseases. The development of AI, in general, and deep learning, in particular, has made the diagnostic accuracy and reliability of ENose devices much higher (Moshayedi et al., 2025). Artificial intelligence (AI) has been increasingly applied to identify and validate pan-cancer biomarkers across multiple tumor types. A structured overview of cancer-specific biomarkers and the corresponding AI techniques employed is presented in Table 16.2, and emerging nanosensor technologies further complement these AI-driven approaches by enabling ultrasensitive biomarker detection.

The revolutionization of AI-enhanced nanosensors is transforming the monitoring of cardiovascular parameters in real-time, such as heart rhythms, blood pressure, and biochemical signals at very early stages, to enhance the ability to detect diseases. Using AI in signal processing, anomaly detection, and predictive analytics, these nanosensors make vertebrate issues, including arrhythmias and heart failure, more sensitive and precise. Moreover, nanosensors are enhanced by AI to improve the evaluation in the echocardiography, computed tomography, magnetic resonance imaging, and electrocardiograms to aid the early diagnosis of heart failure, atrial fibrillation, and hypertrophic cardiomyopathy. They also find application in the constant screening of blood pressure and the prompt diagnosis of complications of hypertension. Heart

TABLE 16.2 Artificial intelligence-based pan-cancer biomarkers for cancer.

Cancer type	Biomarker(s)	AI technique used	References
Breast cancer	HER2, BRCA1/2, EGFR, PD-L1	Deep learning (CNNs), Random Forest, Thermalytix, AI mammogram analysis, PRS-A	Colomer et al. (2024)
Lung cancer	EGFR, KRAS, ALK, PD-L1	Machine learning (XGBoost, SVM), deep learning model (RNN)	Peng et al. (2025); Serrano et al. (2024)
Prostate cancer	PSA, PCA3, TP53, PD-L1, CTLA-4	Machine learning (Random Forest, SVM, XGBoost), deep learning (ANN)	Smith et al. (2024)
Colorectal cancer	CEA, KRAS, BRAF, MLH1, MSH2, TP53, APC, PD-L1	Deep learning (CNNs, ANN), explainable AI (XAI), support vector machines and decision trees	Kiran et al. (2024)
Pancreatic cancer	A19-9, TP53, SMAD4	Machine learning (extreme gradient boosting), deep learning (CNNs, ANN)	Stefanoudakis et al. (2024)
Ovarian cancer	CA-125, HE4, BRCA1/2, TP53	Machine learning (Random Forest, SVM, XGBoost), deep learning (CNNs, ANN)	Li et al. (2024)
Liver cancer	AFP, DCP, VEGF, TP53, KRAS	Radiomics CNNs, deep learning on omics	Nevi et al. (2025)
Bladder cancer	FGFR3, TP53, miR-2, PD-L1	Artificial intelligence (Machine Learning [SVM, XGBoost], deep learning [ANN])	Kim et al. (2024)
Brain cancer	IDH1, MGMT, EGFR, TP53, VEGF	Deep learning (CNNs, RNNs), Random Forest, spatial transcriptomics AI	Verhaak et al. (2024)
Esophageal cancer	TP53, HER2, VEGF, CDKN2A	YOLOv5, CNNs, machine learning (SVM, XGBoost)	Shimada et al. (2024)

(Continued)

TABLE 16.2 (Continued)

Cancer type	Biomarker(s)	AI technique used	References
Gastric cancer	CEA, CA-19-9, HER2, CDH1	Neural networks, deep learning (CNNs, ANN), radiomics AI	Nagini (2023)
Cervical cancer	HPV E6, Ki-67, p16INK4a	CNNs, natural language processing on EHRs, AI-enhanced cytology screening	Doorbar et al. (2024)
Renal cancer	VEGF, TP53, MET	Ensemble learning, deep learning (CNNs, ANN), radiomics AI	Choueiri and Motzer (2024)
Melanoma cancer	BRAF, NRAS, PD-L1, MITF, TYR	Dermoscopy analysis discrimination AI (CNNs, SVM), deep learning	Chapman et al. (2024)
Thyroid cancer	Thyroglobulin, galectin-3, TSH, BRAF, RET/PTC	Decision trees, CNNs, AI-enhanced ultrasound diagnostics	Xing (2024)
Testicular cancer	AFP, hCG, LDH, KIT	Deep lLearning (CNNs, logistic regression, AI-enhanced MRI radiomics)	Looijenga and Stoop (2024)
Head and neck cancer	EGFR, TP53, CDKN2A, VEGF	CNNs, radiomics no more, AI-powered radiomics	Cohen et al. (2024)
Endometrial cancer	PTEN, KRAS, PIK3CA, CTNNB1	Deep learning (CNNs, SVM, AI-enhanced histopathology analysis)	Murali et al. (2024)
Sarcoma	Vimentin, MDM2, CDK4, VEGF	Deep learning (CNNs, Random Forest, EcoTyper AI, CIBERSORTx)	Doyle (2024)
Neuroendocrine tumors	Chromogranin A, Ki-67, PD-L1	Deep learning (NLP on EHRs, AI-enhanced radiomics), Random Forest	Rindi et al. (2024)
Gallbladder cancer	CA-19-9, KRAS, TP53	CNNs, arbitrary forest, AI-enhanced ultrasound Imaging	Raimondi et al. (2024)
Adrenal cancer	Alpha-inhibin, IGF2, CTNNB1, SF-1	Deep learning (ensemble models, CNNs, genomic analysis based on AI)	Else et al. (2024)

problems have been linked to inflammatory atherosclerosis, which has gone a long way in clearing up the apoptotic cells. Removal of dead vascular cells can be enhanced by blocking them through the injection of anti-don't-eat-me signal CD47 that is located on cell surfaces, which may reduce the level of plaque. New nanotherapy methods have involved chemical inhibitors to induce macrophage efferocytosis, particularly in pathological diseases such as atherosclerosis. The objective of this nanotherapy is to lower atherosclerotic plaque load through a decrease in dead cell deposition and inflammation of atherosclerotic lesions, implementation of immune checkpoint inhibitors in the macrophages in relation to clinical usage (Bamezai et al., 2024).

16.1.7 Robot surgery with artificial intelligence integration

The AI-controlled nanorobots promise to play an important role in detecting cancerous cells using imaging technologies and simplify the functioning of nanorobots to deliver drugs to specific locations, perform precise surgeries, conduct medical diagnoses, and detoxify without causing damage to normal tissues. It entails that nanorobots can tailor treatment using AI with references to genetic and health data of a patient without the need of having to conduct extensive surgery and direct specific areas where the targeted treatment will be effective and minimally invasive to alleviate pain and accelerate healing. Nanotechnology solutions are used in cardiovascular medicine, where delivery of targeted drugs to arterial thrombosis helps to reduce the risk level of blood clots and minimize tissue injuries. There is a use of miniature robots that deliver drugs directly to clogged arteries in the heart system. Those innovations have the potential of reducing the required dose of medication, hence reducing instances of side effects which are commonly associated with the conventional use of systemic delivery of drugs (Sandhanam et al., 2025).

Robotic surgeons may also have electrocautery devices in place that emit smoke that may obscure vision until cleared by suction. Wang et al. (2023) developed a convolutional neural network (CNN), together with a swim transformer, to remove the smoke in surgical videos to enhance the quality of the surgical environment in case of acute hemorrhaging. A transoral surgery robot has been developed, and it features piezoelectric sensors with a range of up to 15 N and gives real-time feedback to the surgeons on tissue forces. The use of AI and robotics in interventional radiology is a sea change in the field of contemporary healthcare, and can enhance precision, efficiency, and individuality of Schaefer. The study looked at a 25% decrease in operation duration and a 30% decrease in intraoperative adverse effects in the AI-assisted robotic surgeries, which helped in increasing the accuracy of surgical precision by 40% in terms of increased pinpoint accuracy in tumor removals and implantation. The robotic systems that AIs complete include the functions that are able to manipulate catheters and inject needles more precisely and less variably, and through real-time image fusion. The first robotic assisting systems in the field of interventional radiology increased the accuracy of vascular and percutaneous surgery but lacked autonomy. Currently, studies are working on these constraints using AI, in particular, reinforcement learning and haptic vision, which allows rapid responses to environmental stimuli and increases the accuracy, efficiency, and safety of patients in complex surgery (Boeken et al., 2024). There is a significant development in thoracic surgery, particularly the presence of robotic-assisted thoracic surgery, AI, and augmented reality navigation. RATS has earned recognition all over the world in terms of the implementation of advanced technologies that comprise robotic-assisted stapling, fluorescence images, and 3D anatomical reconstructions in surgeries.

Vitreoretinal surgery is being revolutionized by AI, digital imaging, and robotics, and has the potential to improve retinal vein cannulation to improve surgical visualization and evaluation of retinal peripheral disease, such as detachments and tumors (Stanley et al., 2024). Moreover, AI can adjust brightness and apply color filters, as well, which lowers the expense of dyes and retinal toxicity. Robotic tools in the ophthalmology field enhance the precision of the instruments by eliminating tremors in the hands and giving them greater stability, especially during such challenging tasks as internal limiting membrane peeling and subretinal injections. Such innovation enables surgeons to perform delicate operations, e.g., on patients with macular holes or epiretinal membranes, with reduced chances of retinal damage. AI-driven robots will change the process of administering gene therapy and retinal cannulations, which will provide new treatment opportunities. Robotic-assisted vitreoretinal surgery is a recent development in eye surgery, which offers higher accuracy, stability, and control in a wide range of eye surgeries. RAVS has demonstrated good outcomes in such applications as subretinal injections and membrane peeling. It has the potential to transform vitreoretinal surgery and provide better stability, minimize tissue injury, enhance patient outcomes, and extend treatment therapy (Serhan, 2025).

16.2 Conclusion

CRISPR/Cas systems, AI-based biosensors, and nanorobotics. LC-MS/MS has the capability of studying protein-protein interactions and post-translational modifications, and aids in early diagnosis of rare diseases and mitochondrial disorders. In autoimmune diseases, a protein microarray can screen autoantibodies and disease-specific antigens on a mass scale. Plasma proteomics, a noninvasive method, has been applied in the detection of cancers, cardiovascular diseases, and neurodegenerative diseases. It can be complemented with liquid biopsy, which can be enriched with the CTCs, ctDNA, EVs, and cfRNA, which allow tracking the progression of tumor disease. CRISPR/Cas12 and Cas13 can be utilized with ultra-sensitivity to detect nucleic acids, and exosome analysis and nanosensor design with AI can help to improve the quality of diagnosis and target therapy. Among the technological transformations, the collaboration of AI and nanorobotics will allow offering treatment with the highest possible precision and specificity. This collaboration is a synergy of the AI analysis potential and nanorobot real-time tracking systems, as well as delivering surgical precision to avoid the drawbacks of the traditional methods of drug delivery and tissue repair.

References

Abene, A., Barker, D., 2025. High-throughput multiplex protein microarray for early detection of microbial infections, including Lyme disease. Infectious Disease Technologies 31 (1), 33–47.

Alcolea, D., Pegueroles, J., Vilaplana, E., 2023. Plasma biomarkers of neuronal and glial degeneration in Alzheimer's and related disorders. Molecular Neurodegeneration 18 (1), 45.

Ali, S., Khan, R., Ahmed, T., 2025. Alpha-1 antitrypsin as a marker of systemic inflammation. Clinical Chemistry 71 (2), 123–134.

Aslan, M., Zhang, X., Feng, R., et al., 2023. Integrated single-cell and bulk RNA sequencing analysis identifies a neoadjuvant chemotherapy-related gene signature for predicting survival and therapy in breast cancer. BMC Medical Genomics 16, 300.

Bamezai, A., Sharma, P., Gupta, R., 2024. Targeting CD47 for efferocytosis in atherosclerosis. Cardiovascular Nanomedicine 11 (4), 215–229.

Banerjee, S., Gupta, R., Singh, V., 2024. Prostate-specific antigen as a biomarker in prostate cancer. Urology 170, 45–57.

Batista, F., Moreno, A., Lin, D., 2025. Single-cell sequencing of circulating tumor cells. Nature Biotechnology 43 (3), 345–359.

Begen, H., Yalcıntepe, S., Kaya, D., et al., 2020. EXOC2 mutations disrupt exocyst complex function and cause neurological disease. Molecular Genetics and Metabolism 130 (4), 345–352.

Belbasis, L., Morris, S., van Duijn, C., et al., 2025. Mendelian randomization identifies proteins involved in neurodegenerative diseases. Brain 148 (7), 2412–2428.

Birhanu, A.G., 2023. Mass spectrometry-based proteomics as an emerging tool. Clinical Proteomics 20, 24.

Boeken, T., Müller, J., van Dijk, P., 2024. Reinforcement learning and haptic vision in autonomous robotic interventions. Interventional Radiology Robotics 16 (2), 88–102.

Bolsewig, T., Müller, C., Schneider, A., 2024. Plasma biomarkers in neurodegenerative diseases. Neurobiology of Aging 30, 45–58.

Bouayed, J., Rammal, H., Soulimani, R., 2022. CRP as a prognostic biomarker in systemic inflammation. Mediators of Inflammation 2022, 1–13.

Brundin, P., Dave, K.D., Kordower, J.H., 2022. α-synuclein as a diagnostic and therapeutic target. Lancet Neurology 21 (3), 220–233.

Caputo, V., Schoser, B., 2024. FHL1-related X-linked scapuloperoneal muscular dystrophy with cardiopathy. Journal of Neuromuscular Disorders 18 (1), 22–31.

Celikgil, E., Tanriverdi, S., Yildiz, A., 2023. Multi-antigen protein microarray analysis of SARS-CoV-2 antibody responses. Clinical Proteomics 20 (2), 112–124.

Chapman, P.B., Hauschild, A., Robert, C., 2024. Improved survival with vemurafenib in melanoma with BRAF V600E mutation. New England Journal of Medicine 386 (1), 44–55.

Chen, L., Qi, J., Qi, Q., et al., 2024. PER-CRISPR/Cas14a system-based electrochemical biosensor for the detection of ctDNA EGFR L858R. Analytical Methods 16 (1), 1–10.

Cheng, R., Li, X., Wang, Y., 2024. SAA levels in acute and chronic inflammatory conditions. Journal of Translational Medicine 22 (1), 87–98.

Chiu, R., Wang, C., Wu, J., 2023. Plasma UCH-L1 in neurodegeneration: diagnostic potential. Frontiers in Neurology 14, 1102–1114.

Choueiri, T.K., Motzer, R.J., 2024. Molecular biomarkers in renal cell carcinoma: predictive and prognostic implications. Nature Reviews Urology 21 (3), 189–205.

Cohen, E.E., Karrison, T., Kocherginsky, M., 2024. Prognostic and predictive biomarkers in head and neck cancer. Cancer 130 (2), 145–157.

Colom, A., Martínez, J., Ríos, M., 2023. Development of a fluorescent microarray chip for multiplex detection of cardiac biomarkers. Cardiovascular Diagnostics and Therapy 13 (1), 45–56.

Colomer, R., González-Farré, B., Ballesteros, A.I., Peg, V., Bermejo, B., Pérez-Mies, B., et al., 2024. Biomarkers in breast cancer 2024: an updated consensus statement. Clinical and Translational Oncology 26 (6), 2935–2951.

Del Valle, D.M., Kim-Schulze, S., Huang, H.-H., 2020. Inflammatory cytokine IL-6 in COVID-19 and sepsis. Nature Medicine 26 (9), 1636–1643.

Devi, S., Sharma, R., Kumar, P., 2022. Clinical significance of C-reactive protein in inflammatory conditions. Journal of Clinical Immunology 42 (3), 210–222.

Doorbar, J., Egawa, N., Griffin, H., 2024. HPV molecular biology and biomarkers in cervical cancer. Virus Research 320, 198–212.

Doyle, L.A., 2024. Sarcoma pathology and molecular diagnostics. Modern Pathology 37 (5), 456–470.

Elabiad, M., 2025. Challenges in integrating AI with nanotechnology for biomedical applications. Nano Today 45, 101234.

Else, T., Kim, A., Sabolch, A., 2024. Alpha-inhibin, IGF2, TP53, CTNNB1, and SF-1 as biomarkers in adrenal cortical carcinoma. Journal of Clinical Endocrinology & Metabolism 109 (3), 678–690.

Evans, C.A., Smith, R., Jones, P., 2022. HSP70 in cancer biology and therapy. Oncotarget 13 (5), 300–312.

Evans, D.G., Eccles, D.M., Rahman, N., 2023. BRCA1/2 in hereditary breast and ovarian cancer. Current Oncology Reports 25 (2), 101–115.

Faden, D.L., Kanski, J., Ottenheijm, M.E., Seo, Y., Song, M., Yoon, J.H., et al., 2025. HPV-DeepSeek: a novel liquid biopsy assay for early detection of HPV-associated head and neck cancers. Clinical Cancer Research 31 (5), e250520.

Foehr, E.D., Müller, V., Krawitz, P., et al., 2024. Mass spectrometry-based proteomics reveals extracellular matrix disruption. Journal of Proteome Research 23 (2), 145–158.

Foko, N., Mbaya, A., Nkeng, G., 2022. Haptoglobin as an inflammatory biomarker in clinical practice. Clinical Biochemistry 99, 45–56.

García-García, J., Rivera, P., López, J., 2023. sCD14 as a biomarker in sepsis and inflammation. Journal of Clinical Immunology 43 (4), 210–222.

Ge, L., Sun, Y., Shi, Y., et al., 2022. Plasma circRNA microarray profiling identifies novel circRNA biomarkers for the diagnosis of ovarian cancer. Journal of Ovarian Research 15 Article 58.

Giuliano, M., Connolly, J., Silverman, P., 2022. HER2/neu as a biomarker and therapeutic target in breast cancer. The Oncologist 27 (4), 321–333.

Guo, J., Li, H., Wang, Y., 2025. IL-10 as an anti-inflammatory biomarker in infection. Journal of Clinical Immunology 45 (1), 55–66.

Hanan, M., Tal, S., Meir, S., 2022. Clinical relevance of complement activation in neurodegeneration. Journal of Neuroimmunology 365, 577–589.

Hassan, M., Khan, A., Rahman, S., et al., 2025. CRISPR-based diagnostics: SHERLOCK and DETECTR as transformative tools for infectious disease detection. Journal of Clinical Diagnostic Research 19 (2), 112–120.

Hu, Y., Zhang, L., Chen, X., et al., 2024. Limitations and prospects of CRISPR/Cas biosensors in emergency diagnostics. Biosensors and Bioelectronics 232, 115678.

Huan, Y., Zhang, L., Wei, T., et al., 2025. Bioinformatics analysis of GEO datasets identifies hub genes associated with breast cancer prognosis. Computational Biology and Chemistry 101, 107872.

Jabbour, E., Kantarjian, H., Talpaz, M., 2023. Advances in targeting BCR-ABL in leukemia therapy. Hematology Oncology Reports 17 (1), 50–63.

Jack, C.R., Knopman, D.S., Weiner, M.W., 2023. Aβ42/Aβ40 ratio as a biomarker for Alzheimer's disease. Neurology Advances 41 (2), 123–138.

Jain, N., Patel, K., Smith, R., 2023a. YKL-40 in cerebrospinal fluid and plasma: clinical relevance. Neurobiology of Disease 168, 105716.

Jain, N., Smith, R., Patel, K., 2023b. Glial fibrillary acidic protein as a biomarker in neurodegenerative disease. Journal of Neurochemistry 164 (2), 145–158.

Janelidze, S., Mattsson, N., Palmqvist, S., 2020. Clinical relevance of plasma tau in Alzheimer's disease. Alzheimer's Research & Therapy 12 (1), 45–58.

Jiang, Y., Zhang, Y., Miao, P.W., et al., 2023. A photoelectrochemical biosensor mediated by CRISPR/Cas13a for direct and specific detection of miRNA-21. Sensors 24 (18), 6138.

Kadhim, A., Al-Mayah, A., Hussein, H., 2023. Ultra-sensitive biosensors using functionalized nanomaterials for environmental and biomedical applications. Talanta 261, 124587.

Kapoor, R., Mehta, S., Sharma, V., 2024. Biological modulators in polymeric nanoparticle-based drug delivery. Journal of Controlled Release 350, 88–102.

Khan, A., Patel, R., Singh, D., 2024a. AI-enhanced nanobiosensors for precision bacterial infection treatment. Sensors and Actuators B: Chemical 390, 129–141.

Khan, M.A., Patel, R., Singh, N., et al., 2024b. Multiplexed assay platforms and liquid biopsy applications in cancer diagnostics: advances in pancreatic and liver cancer screening. Clinical Chemistry and Laboratory Medicine 62 (3), 345–356.

Kim, S., Park, J., Choi, K., 2024. Emerging biomarkers and immune checkpoint expression in bladder cancer. Oncotarget 15 (12), 1120–1135.

Kim, S., Park, J., Lee, H., 2025. Carcinoembryonic antigen in colorectal cancer diagnosis. Clinical Chemistry 71 (2), 123–134.

Kiran, S., Patel, R., Sharma, V., 2024. Advances in precision medicine approaches for colorectal cancer. Cells 13 (2), 256.

Kobayashi, T., Nakamura, K., Sato, Y., et al., 2022. Dual biomarker profiling of circulating tumor cells and cell-free DNA in urothelial cancers. Cancer Science 113 (5), 1652–1664.

Kokabi, H., Zhang, Y., Lee, J., 2023. Machine learning approaches for biosensor optimization and signal interpretation. Sensors and Actuators B: Chemical 389, 134215.

Koludarova, M., Battersby, B.J., 2024. Functional insights into HMGCS1 variants linked to rigid spine syndrome and mevalonate pathway disruption. Journal of Inherited Metabolic Disease 47 (3), 512–525.

Kondo, Y., Fujimoto, A., Tanaka, H., 2024. BRAF V600E mutation and targeted therapy. Molecular Cancer Therapeutics 23 (4), 301–315.

Korecka, M., Bensadoun, J.C., Schneider, B., 2021. DJ-1 as a biomarker in Parkinson's disease. Molecular Neurodegeneration 16 (1), 15–27.

Kumari, T.Y., Durga, A.B., Raju, A.K., Pravallika, A.J., 2024. High-pressure liquid chromatography–mass spectrometry in disease diagnosis. International Journal of Research in Technology and Innovation 10 (2), 142–150.

Kwon, O., Choi, S., Lee, Y., 2021. Serum CA-19-9 levels in pancreatic and biliary cancers. Cancer Biomarkers 30 (3), 201–214.

Lemanska-Perek, A., Wozniak, M., Kaczmarek, M., 2019. Identification of genetic biomarkers in bladder cancer using 2D SDS-PAGE and MALDI-TOF MS. Journal of Proteome Research 18 (11), 4021–4032.

Li, X., Gao, Y., 2023. AI-driven nanomedicine for multi-omics data integration and personalized drug delivery. Drug Discovery Today 28 (9), 1456–1468.

Li, Y., Chen, X., Wang, J., et al., 2021. Protein microarrays for autoantibody profiling in autoimmune diseases: advances in biomarker discovery and diagnostic accuracy. Autoimmunity Reviews 20 (6), 102783.

Li, X., Chen, Y., Zhang, W., 2024. Ovarian cancer: diagnosis and treatment strategies (review). Oncology Letters 27 (4), 14574.

Lim, J., Tan, W., Chua, Y., 2020. AI-integrated biosensing for COVID-19 detection in rural areas. Sensors International 1 (1), 100–112.

Liu, C.C., Kanekiyo, T., Bu, G., 2021. ApoE and Alzheimer's disease: risk and biomarker applications. Nature Reviews Neurology 17 (7), 421–436.

Liu, F., Chen, Q., Wu, T., 2022. Serum CYFRA 21-1 levels in patients with lung malignancies. Cancer Research Frontiers 29 (4), 210–220.

Lolansen, S.D., Rostgaard, N., Olsen, M.H., et al., 2024. Proteomic profile and predictive markers of outcome in patients with subarachnoid hemorrhage. Clinical Proteomics 21, 51.

Looijenga, L.H., Stoop, H., 2024. Molecular markers in testicular germ cell tumors. International Journal of Andrology 47 (1), 45–58.

Markou, A.N., Londra, D., Stergiopoulou, D., et al., 2023. Preoperative mutational analysis of circulating tumor cells and plasma-cfDNA provides complementary information for early prediction of relapse: a pilot study in early-stage non-small cell lung cancer. Cancers 15 (6), 1877.

Ramirez, A., 2025. CRISPR-Cas9 successfully removes HIV-1 DNA from infected human immune cells: a step toward a functional cure. Engineering 6 (June).

Moasser, M., Krop, I., Baselga, J., 2023. Clinical applications of HER2/neu testing. Journal of Clinical Oncology 41 (12), 1501–1513.

Mohanty, T., Patel, K., Das, S., 2025. Serum amyloid A as an inflammatory biomarker. Clinical Chemistry and Laboratory Medicine 63 (2), 123–135.

Mokhtari, R., Chen, Y., Li, X., 2024. Vascular endothelial growth factor (VEGF) in tumor angiogenesis. Angiogenesis Research 19 (3), 200–214.

Moshayedi, H., Farhadi, M., Zarei, A., 2025. ENose technology in cancer diagnostics: principles, challenges, and AI integration. Biosensor Advances 9 (1), 1–19.

Mpofu, T., Singh, R., Patel, A., 2023. Quantum technologies in biosensing: a new frontier. Trends in Biotechnology 41 (6), 512–524.

Mucha, K., Shang, M., Guo, J., 2023. Nanosensor-based point-of-care testing for infectious disease management and emergency diagnostics. Sensors and Diagnostics 2, 1123–1144.

Mućka, P., Kowalski, M., Nowak, R., 2025. Procalcitonin as a biomarker in bacterial infections. Critical Care Medicine 53 (1), 33–45.

Murali, R., Soslow, R.A., Weigelt, B., 2024. Classification of endometrial carcinoma: molecular biomarkers and clinical implications. Modern Pathology 37 (1), 45–60.

Nagini, S., 2023. Gastric cancer: molecular biomarkers and therapeutic targets. Cancer Letters 556, 45–60.

Nedjadi, T., Al-Mohanna, F., Al-Mutairi, T., 2021. Biomarker discovery in urothelial bladder cancer using MALDI-TOF MS and 2D gel electrophoresis. Cancer Biomarkers 32 (4), 567–578.

Netala, V.R., Ghosh, S., Reddy, M., 2025. Plasma protein biomarkers for cardiovascular mortality risk in chronic coronary heart disease. Cardiovascular Research 121 (3), 456–472.

Neumann, M., Sampathu, D.M., Kwong, L.K., 2022. TDP-43 pathology in neurodegenerative disease. Acta Neuropathologica 144 (1), 1–15.

Nevi, L., Rossi, C., Marino, D., 2025. Decoding the molecular and genomic landscape of HCC. Nature Communications 16 (1), 385.

Nielsen, F.C., van Overeem Hansen, T., Sørensen, C.S., 2022. BRCA1/2 mutations and breast cancer risk. Genetics in Medicine 24 (7), 1345–1356.

Nikkhoo, A., Hosseini, R., Farahani, R., 2023. Ferritin as a biomarker in systemic inflammation. Clinical Chemistry and Laboratory Medicine 61 (3), 201–214.

Niu, Q., Zhang, J., Liu, H., 2025a. EGFR mutations and therapeutic strategies. Journal of Molecular Oncology 22 (1), 12–26.

Niu, Q., Zhang, J., Liu, H., 2025b. Serum VEGF levels and clinical outcomes in oncology patients. Cancer Biomarkers Today 22 (1), 33–46.

Norman, A., Smith, R., Johnson, P., 2025. AFP levels in liver cancer diagnosis and prognosis. Clinical Oncology 37 (2), 145–156.

Palmqvist, S., Zetterberg, H., Blennow, K., 2020. Plasma Aβ42/Aβ40 and early detection of Alzheimer's pathology. Alzheimer's & Dementia 16 (5), 725–737.

Park, J., Kim, S., Lee, H., 2023. LBP as a biomarker of bacterial infection. Frontiers in Immunology 14, 1101–1114.

Parums, D.V., 2025. TriOx: Machine learning-powered blood test for early cancer detection. Medical Science Monitor 31, e939112.

Pauper, A., Mukherjee, C., De, A., et al., 2025. Nanotechnology-based delivery of CRISPR/Cas9 for cancer treatment: overview, application & future perspective. Journal of Pharmacology and Nanomedicine 13 (3), 1373–1388.

Peng, J., Jury, E.C., Donnes, P., et al., 2021. Machine learning techniques for personalised medicine approaches in immune-mediated chronic inflammatory diseases: applications and challenges. Frontiers in Pharmacology 12, 720694.

Peng, M., Zhang, L., Wang, H., 2025. Application of liquid biopsy in differentiating lung cancer: extracellular vesicles, circulating tumor DNA, plasma RNA and circulating antibodies. International Journal of Molecular Medicine 56 (1), 112–130.

Radko, A., Kwak, T., Shubin, A.V., et al., 2024. Novel plasma cell-free RNA-based liquid biopsy approach for chronic lymphocytic leukemia. Blood 144 (Suppl 1), 3610.

Rahman, S., 2023. MS-based proteomics in the diagnosis of mitochondrial disorders: insights from Leigh syndrome. Journal of Mitochondrial Medicine 12 (3), 145–158.

Raimondi, S., Lowenfels, A.B., Morselli-Labate, A.M., 2024. KRAS, TP53, and CA19-9 as prognostic biomarkers in gallbladder cancer. Digestive and Liver Disease 56 (3), 310–320.

Reveglia, P., Paolillo, C., Ferretti, G., et al., 2021. Challenges in LC–MS-based metabolomics for Alzheimer's disease early detection: targeted approaches versus untargeted approaches. Metabolomics: Official Journal of the Metabolomic Society 17 (9), 78.

Rindi, G., Klersy, C., Alaggio, R., 2024. Ki-67, chromogranin A and synaptophysin in neuroendocrine tumors: prognostic value. Modern Pathology 37 (2), 112–125.

Salvianti, F., Pazzagli, M., Pinzani, P., et al., 2021. Prognostic relevance of circulating tumor cells and cell-free DNA in metastatic colorectal cancer. Cancer Biomarkers 30 (2), 123–134.
Sandhanam, R., Mehta, V., Rao, S., 2025. AI-controlled nanorobots in cardiovascular therapy: precision drug delivery and clot removal. Cardiovascular Nanotechnology 12 (2), 88–104.
Sausa, R., Mehta, V., Tanaka, Y., 2023. Advances in MALDI-TOF MS for proteomic biomarker discovery in oncology. Proteomics. Clinical Applications 17 (6), e2200156.
Schwarz, M., Becker, S., Müller, M., 2021. Tumor cell-induced platelet aggregation: mechanisms and implications for cancer progression. Journal of Experimental & Clinical Cancer Research 40, 277.
Sengupta, J., Hussain, C.M., 2024. Graphene transistor-based biosensors for rapid detection of SARS-CoV-2. Bioelectrochemistry (Amsterdam, Netherlands) 156, 108623.
Serhan, M., 2025. Robotic-assisted vitreoretinal surgery: a new era in ophthalmic microsurgery. Retinal Surgery Advances 9 (1), 1–19.
Serrano, D., Pérez, M., Gómez, C., 2024. Predictive biomarkers and personalized therapy in lung cancer. Lung Cancer (Amsterdam, Netherlands) 180, 101–115.
Shen, Y., Li, Z., Wang, H., 2025. GADD45G regulates neuroinflammation and glial cell behavior in neurodegenerative diseases. Neurogenetics and Molecular Therapy 29 (2), 89–104.
Shimada, Y., Nabeya, Y., Okazumi, S., 2024. TP53, HER2, and miRNA panels in esophageal cancer prognosis. Esophagus 21 (1), 45–60.
Smith, J., Doe, A., Lee, R., 2021. Protein profiling and its role in disease pathway elucidation and personalized medicine. Journal of Proteomics 245, 104296.
Smith, J., Thompson, R., Lee, H., 2024. Prostate cancer biomarkers: current clinical applications. Urologic Oncology 42 (3), 201–215.
Soni, R.K., 2024. Frontiers in plasma proteome profiling platforms: innovations and applications. Clinical Proteomics 21, 24.
Stanley, D., Kumar, V., Mehta, R., 2024. AI and robotics in vitreoretinal surgery: enhancing visualization and precision. Journal of Ophthalmic Robotics 7 (3), 134–148.
Stefanoudakis, D., Kotsinas, A., Giaginis, C., 2024. Significance of TP53, CDKN2A, SMAD4 and KRAS in pancreatic cancer. Heliyon 10 (2), e13567.
Stevens, B., Allen, N.J., Vazquez, L., 2023. Complement-mediated synapse loss in Alzheimer's disease. Nature Neuroscience 26 (2), 197–210.
Takahashi, H., Yoshida, T., Suzuki, M., 2022. Clinical utility of S100B in Alzheimer's and Parkinson's diseases. Neurobiology of Aging 114, 62–73.
Tarawneh, R., D'Angelo, G., Crimmins, D., 2020. VILIP-1 as a cerebrospinal fluid biomarker in Alzheimer's disease. Alzheimer's & Dementia 16 (5), 721–734.
Ting Sim, C., Lee, H., Tan, W., Goh, C., 2023. AI-driven nanosensor strategies for bacterial infection treatment and vaccine development. Journal of Nanobiotechnology 21, 112.
Tiwari, S., Kumar, P., Singh, R., 2021. Haptoglobin levels in sepsis and infection. Biomarkers: Biochemical Indicators of Exposure, Response, and Susceptibility to Chemicals 26 (5), 345–356.
Tondo, G., Marchi, C., 2022. Cerebrospinal fluid biomarkers of axonal degeneration in neurodegenerative diseases. Neurobiology of Disease 168, 105678.
Ucdal, P., Kim, J., Lee, H., 2025. Alpha-fetoprotein as a predictive biomarker in oncology. Journal of Hepatology 82 (1), 101–113.
Urbanowska, E., Kowalczyk, M., Nowak, D., 2023. High-performance protein microarray system for multiplex biomarker tracking in rheumatoid arthritis progression. Clinical Proteomics 20, 34.

Valenzuela, F., Silva, P., Martinez, C., 2025. D-dimer as a prognostic biomarker in thrombotic conditions. Thrombosis Research 219, 12–25.

Verhaak, R.G., Hoadley, K.A., Purdom, E., 2024. Integrated genomic analysis identifies clinically relevant glioblastoma subtypes. Cancer Cell 42 (3), 315–330.

Wang, L., Li, Q., Sun, J., 2022. Pentraxin-3 as a biomarker of inflammation and sepsis. Immunology Letters 242, 45–56.

Wang, Y., Liu, Z., Chen, H., 2023. CNN-swim transformer model for smoke removal in surgical videos. Medical Image Analysis 84, 102678.

Wang, Y., Chen, H., Zhang, X., 2025. Plasma procalcitonin levels for early detection of infection. Frontiers in Immunology 16, 1023–1036.

Wei, H., Liu, Y., Zhang, T., et al., 2023. Ultrasensitive CRISPR/Cas12a biosensor for MRSA detection using on-particle rolling circle amplification and enzyme-triggered click chemistry. Analytical Chemistry 95 (12), 4567–4575.

Wei, L., Ji, S., Chen, L., et al., 2024. Quantitative LC-MS analysis of PARP inhibitors in ovarian cancer: niraparib, olaparib, fluzoparib, and pamiparib. Journal of Ovarian Research 17, 1599.

Wenk, M., Weber, D.M., Kim, J.C., et al., 2024. Simultaneous quantitation of plasma Aβ40, Aβ42, and ApoE proteoforms by LC-MS/MS for Alzheimer's disease risk assessment. Frontiers in Neurology 15, 1364658.

Williams, J., Kumar, R., Al-Mutairi, H., et al., 2025. Biosensors emphasizing their role in healthcare, pollutant detection, and food contamination identification. Biotechnology for Environmental Sustainability, 1st ed. Springer Nature, Singapore, pp. 345–368.

Xing, M., 2024. BRAF mutation in thyroid cancer. Endocrine-Related Cancer 31 (1), R33–R45.

Yin, S., Li, J., Wang, L., 2023. CA-125 in ovarian cancer detection and monitoring. Journal of Ovarian Research 16 (1), 101–112.

Zeng, X., Wang, Y., Zhang, L., 2022. Clinical significance of IL-6 in acute inflammatory responses. Frontiers in Immunology 13, 1045–1056.

Zeng, P., Li, H., Sun, Y., 2023. Hepatocyte growth factor (HGF) and cancer progression. Molecular Oncology 17 (5), 876–889.

Zetterberg, H., Blennow, K., Skillback, T., 2021. Clinical relevance of VILIP-1 in cognitive decline. Neurology 96 (4), e431–e442.

Zhang, Q., Liu, M., Chen, Y., 2021. AI-guided nanoparticle design for liver cancer treatment. ACS Nano 15 (4), 7890–7902.

Zhang, Y., Chen, L., Wang, T., 2024a. Plasma proteome-wide association study identifies causal proteins in coronary artery disease. Nature Communications 15, 1123.

Zhang, Y., Miao, P., Wang, J., et al., 2024. CRISPR/Cas13a-mediated photoelectrochemical biosensor for ultrasensitive detection of miRNA-21. Sensors 24 (18), 6138.

Zhou, L., Tang, Y., Ren, J., 2022. Fluorescence-enhanced lateral flow biosensor using CRISPR/Cas12a and quantum dots for bacterial diagnostics. Biosensors 12 (9), 765.

Zhou, Y., Chen, L., Wang, J., 2024. Integrated plasma proteomics and multi-omics analysis identifies candidate biomarkers for precision treatment of pancreatic cancer. Proteomics in Oncology 19 (2), 87–102.

Zhu, Q., Lin, J., Huang, Y., 2022. Piezoelectric sensor-enabled robotic system for transoral surgery: real-time force feedback. IEEE Transactions on Biomedical Engineering 69 (9), 2781–2792.

Index

Note: Page numbers followed by "*f*" and "*t*" refer to figures and tables, respectively.

D

E

Q

R

S

Printed and bound by CPI Group (UK) Ltd, Croydon, CR0 4YY
17/09/2026
02218533-0005